AF360727

Molecular Research in Rice

Molecular Research in Rice

Agronomically Important Traits

Editors

Kiyosumi Hori
Matthew Shenton

MDPI • Basel • Beijing • Wuhan • Barcelona • Belgrade • Manchester • Tokyo • Cluj • Tianjin

Editors
Kiyosumi Hori
National Agriculture and Food
Research Organization (NARO)
Japan

Matthew Shenton
National Agriculture and Food
Research Organization (NARO)
Japan

Editorial Office
MDPI
St. Alban-Anlage 66
4052 Basel, Switzerland

This is a reprint of articles from the Special Issue published online in the open access journal *International Journal of Molecular Sciences* (ISSN 1422-0067) (available at: https://www.mdpi.com/journal/ijms/special_issues/Rice_Traits).

For citation purposes, cite each article independently as indicated on the article page online and as indicated below:

LastName, A.A.; LastName, B.B.; LastName, C.C. Article Title. *Journal Name* **Year**, *Article Number, Page Range.*

ISBN 978-3-03943-238-7 (Hbk)
ISBN 978-3-03943-239-4 (PDF)

Cover image courtesy of Kiyosumi Hori.

Contents

About the Editors

Kiyosumi Hori (Ph. D.) Rice Applied Genomics Research Unit, Institute of Crop Science, National Agriculture and Food Research Organization (NARO) from 2016-present. Department of Integrated Biosciences, Graduate School of Frontier Science, The University of Tokyo from 2019-present. Rice Applied Genomics Research Unit, National Institute of Agrobiological Sciences from 2006–2016.

Matthew Shenton (Ph. D.) Breeding Materials Development Unit, Institute of Crop Science, National Agriculture and Food Research Organization (NARO) from 2018-present. Meiji University School of Agriculture, Bioinformatics Laboratory, 2017–2018. Plant Genetics Laboratory, National Institute of Genetics, Japan, 2009–2017. Iwate Biotechnology Research Center, 2006–2009.

International Journal of
Molecular Sciences

Editorial

Recent Advances in Molecular Research in Rice: Agronomically Important Traits

Kiyosumi Hori * and Matthew Shenton *

National Agriculture and Food Research Organization, Institute of Crop Science, Tsukuba, Ibaraki 305-8518, Japan
* Correspondence: horikiyo@affrc.go.jp (K.H.); matt.shenton@affrc.go.jp (M.S.)

Received: 14 August 2020; Accepted: 17 August 2020; Published: 19 August 2020

Rice (*Oryza sativa* L.) is the most important food crop in the world, being a staple food for more than half of the world's population. Recent improvements in living standards have increased the worldwide demand for high-yielding and high-quality rice cultivars [1]. To achieve improved agricultural performance in rice, while overcoming the challenges presented by climate change, it is essential to understand the molecular basis of agronomically important traits related to increasing grain yield, grain quality, disease resistance, and stress tolerance. Recently developed techniques in molecular biology can reveal the complex fundamental mechanisms involved in the control of these agronomic traits [2–4]. As rice was the first crop genome to be sequenced in 2004 [5], molecular research tools are well established in rice, and wide-ranging molecular studies using natural variation, mutants, transgenic lines, genome editing lines, and whole-genome analyses, such as genomics, transcriptomics, proteomics, epigenetics, and metabolomics, are increasing our knowledge about factors affecting agronomic traits. These significant advances will enable the development of novel rice cultivars, combining superior agronomic performance with resilience to climate stresses.

As for many other crops, release of the reference genome sequence has facilitated the identification and cloning of genetic factors, such as genes and quantitative trait loci (QTLs) involved in the control of various agronomic traits in rice [5]. The rice genome contains more than 37,000 annotated genes. However, notwithstanding the many research achievements to date, the individual molecular functions of most genes are still unknown. Continuing research efforts in gene functional analysis are necessary to elucidate the genetic networks and molecular mechanisms controlling agronomically important traits. In the Special Issue "Molecular Research in Rice: Agronomically Important Traits", we collected several recent studies that identified genetic factors and revealed their molecular contributions to rice agronomic traits under various cultivation conditions.

Increasing grain yield is a major objective in breeding programs, because of the need to meet the increased demand for rice fueled by population growth. Changes in plant architecture, such as in grain shape, grain number per panicle, plant height, number of tillers, leaf size, and leaf angle, have been associated with improved grain yield. Seo et al. showed that the *OsbHLH079* gene is associated with the control of leaf angle and grain shape via brassinosteroid biosynthesis and signaling pathways [6]. Gull et al. focused on grain shape and grain weight, and they estimated the genetic effects of each allelic combination among seven genes—*GS2*, *GS3*, *GS5*, *GW5*, *GS7*, *SLG7*, and *GW8*—in improving grain length, grain width, grain thickness, and thousand grain weight [7]. Jiang et al. identified a semi-dwarfism gene, *OsCYP96B4*, which affected the content of γ-aminobutyrate (GABA), amino acids, saccharides, and other secondary metabolites in rice plants [8]. Yu et al. detected eight QTLs underlying heterosis in traits, including days to heading, grain yield, and plant height between the two rice subspecies *O. sativa* ssp. *indica* and *O. sativa* ssp. *japonica* [9]. Zhang et al. found a large set of genes that are differentially expressed at the shoot apical meristem in response to nitrogen application, which regulated the number of tillers and panicles in rice plants [10]. Gao et al. identified that the *OsEWL4* gene was a regulator of tiller number and plant biomass (yield of above-ground parts in rice plants) by using targeted mutations of eight rice *FLW* genes using the CRISPR/Cas9 system [11]. Fan et al. produced overexpression and knockdown transformants of the *OsNAR2.1* gene and revealed that changing expression levels of the *OsNAR2.1* gene altered global methylation at the whole genome level and the phenotypes of plant height and grain yield at the plant level [12]. These novel technologies are powerful tools to understand the molecular basis of increasing rice yields.

It is also important for the understanding of the genetic basis of yield components to characterize genes that function in chloroplast development and chlorophyll degradation during leaf greening and senescence. Shim et al. reported that the grain shape gene *GW2* was responsible for the leaf senescence QTL *qGC2*. They revealed that *GW2*, encoding a RING-type E3 ubiquitin ligase, controlled both grain shape and chlorophyll content by the transcriptional regulation of the cytokinin, brassinosteroid, auxin, and abscisic acid (ABA) phytohormone signaling pathways [13]. Zhang et al. showed that the *OsCAF1* gene encodes a chloroplast RNA splicing and ribosome maturation domain (CRM) protein and plays a key role in chloroplast development in rice [14]. Zhao et al. investigated the effects of low temperature treatment on the greening of rice seedlings. Decreased chlorophyll content was caused by the inhibition of δ-aminolevulinic acid and the suppression of conversion from protochlorophyllide into chlorophyll [15].

Rice consumers pay particular attention to high grain quality, and grain quality largely determines the market price of rice grains. Therefore, grain quality is an important trait alongside grain yield in rice. Wu et al. detected a total of 14 QTLs for protein content in rice grains. A stably inherited QTL, *qGPC1-1*, was delimited within an 862 kbp genome region on the high-density linkage map [16]. These research efforts are helpful in dissecting the genetic basis of grain components and improving rice grain quality in future breeding programs.

Besides grain yield and quality, traits in resistance to biotic and abiotic stresses are of paramount importance as climate change tests the resilience of rice production systems. Recent studies have provided important knowledge of biotic interactions between rice and pathogens. Resistance genes against blast disease caused by *Magnaporthe oryzae* and bacterial blight disease caused by *Xanthomonas oryzae* pv. *oryzae* have been genetically identified. Furthermore, there are many studies investigating the molecular mechanisms of rice immune responses, such as receptor-mediated pathogen recognition, host immune signaling, and pathogen effector-mediated susceptibility. In this Special Issue, Kanda et al. focuses on the *BSR1* gene, which encodes a receptor-like cytoplasmic kinase subfamily VII protein and concludes that the hyperactivation of microbe-associated molecular pattern (MAMP)-triggered immune responses conferred robust *BSR1*-mediated broad spectrum resistance to fungal and bacterial pathogens [17]. Liu et al. reports that the *OsAAA-ATPase1* gene enhanced the expression of pathogenesis-related genes, salicylic acid (SA)-mediated defense responses, and resistance to rice blast fungus [18]. Liang et al. found a novel allele of the durable blast resistance gene *pi21*, which was associated with basal resistance to rice blast disease. They detected four QTLs for basal resistance, one of which was identified as a novel haplotype of *pi21* [19]. Hsu et al. developed a rice line, pyramiding five bacterial blight resistance genes: *Xa4*, *xa5*, *Xa7*, *xa13*, and *Xa21*. The pyramiding line showed not only high levels of resistance to bacterial blight disease, but also increased grain yield and grain quality [20].

Abiotic stresses, including high temperature, low temperature, and saline conditions, are an important issue for rice cultivation because they cause significant decreases in yield and grain quality. Various genes have been identified, cloned, and functionally characterized with the aim of overcoming these stresses and protecting rice plants. Jiang et al. detected 11 QTLs for low temperature germination ability and identified the *DEP1*, *qLTG3-1*, and *OsSAP16* genes as being responsible for three QTLs, proposing a novel candidate gene for another QTL, *qLTG6* [21]. Yuan et al. detected a total of 36 QTLs for seed dormancy and germination behavior and identified a candidate gene for one major effect QTL, *qDOM3.1*, which was involved in an ABA signaling pathway [22]. Sheteiwy et al. evaluated the priming effects of GABA on seed gemination and seedlings under saline and osmotic stress conditions. The salinity and osmotic stress treatments resulted in obvious increases in the expression of *OsCIPK* genes that were associated with changes in cell ultra-morphology and cell cycle progression [23]. Liu et al. identified differentially expressed genes in anthers in a thermotolerant rice cultivar. The expression and polymorphism analysis of the *OsACT* gene suggested that it may be involved in high temperature tolerance in rice cultivars [24]. Finally, Schaarschmidt et al. investigated alterations in more than 140 metabolites and in agronomic traits, including grain yield and plant height under high night temperature stress conditions [25].

The significant studies mentioned here demonstrate the importance of the research community in understanding and explaining the molecular genetic basis of agronomically important traits in rice. To develop novel rice cultivars, showing climate resilience and strong agronomic performance in the future, we have to identify further important genes, elucidate their molecular functions, and design desirable genotypes based on the individual and interaction effects of important genes. In addition to the traits described in this Special Issue, there are many other agronomically important traits that should be explored using molecular genetic and biological research. The sharing of experimental results among researchers facilitates the development of new rice cultivars in future rice breeding programs. Thus, these research efforts are necessary to address the increasing food supply and security problems around the world.

Funding: This research received no external funding.

Conflicts of Interest: The authors declare no conflict of interest.

References

1. Hori, K. Genetic dissection and breeding for grain appearance quality in rice. In *Rice Genomics, Genetics and Breeding*; Sasaki, T., Ashilari, M., Eds.; Springer: Singapore, 2018; pp. 435–451.
2. Yang, Y.; Li, Y.; Wu, C. Genomic resources for functional analyses of the rice genome. *Curr. Opin. Plant Biol.* **2013**, *16*, 157–163. [CrossRef] [PubMed]
3. Song, S.; Tian, D.; Zhang, Z.; Hu, S.; Yu, J. Rice genomics: Over the past two decades and into the future. *Genom. Proteom. Bioinf.* **2018**, *16*, 397–404. [CrossRef] [PubMed]
4. Biswal, A.K.; Mangrauthia, S.K.; Reddy, M.R.; Yugandhar, P. CRISPR mediated genome engineering to develop climate smart rice: Challenges and opportunities. *Semin. Cell Dev. Biol.* **2019**, *96*, 100–106. [CrossRef] [PubMed]
5. International Rice Genome Sequencing Project. The map-based sequence of the rice genome. *Nature* **2005**, *436*, 793–800. [CrossRef] [PubMed]
6. Seo, H.; Kim, S.H.; Lee, B.D.; Lim, J.H.; Lee, S.J.; An, G.; Paek, N.C. The *rice basic helix-loop-helix 79 (OsbHLH079)* determines leaf angle and grain shape. *Int. J. Mol. Sci.* **2020**, *21*, 2090. [CrossRef] [PubMed]
7. Gull, S.; Haider, Z.; Gu, H.; Raza Khan, R.A.; Miao, J.; Wenchen, T.; Uddin, S.; Ahmad, I.; Liang, G. InDel marker based estimation of multi-gene allele contribution and genetic variations for grain size and weight in rice (*Oryza sativa* L.). *Int. J. Mol. Sci.* **2019**, *20*, 4824. [CrossRef]
8. Jiang, L.; Ramamoorthy, R.; Ramachandran, S.; Kumar, P.P. Systems metabolic alteration in a semi-dwarf rice mutant induced by *OsCYP96B4* gene mutation. *Int. J. Mol. Sci.* **2020**, *21*, 1924. [CrossRef]
9. Yu, Y.; Zhu, M.; Cui, Y.; Liu, Y.; Li, Z.; Jiang, N.; Xu, Z.; Xu, Q.; Sui, G. Genome sequence and QTL analyses using backcross recombinant inbred lines (BILs) and BILF$_1$ lines uncover multiple heterosis-related loci. *Int. J. Mol. Sci.* **2020**, *21*, 780. [CrossRef]
10. Zhang, X.; Zhou, J.; Huang, N.; Mo, L.; Lv, M.; Gao, Y.; Chen, C.; Yin, S.; Ju, J.; Dong, G.; et al. Transcriptomic and co-expression network profiling of shoot apical meristem reveal contrasting response to nitrogen rate between *indica* and *japonica* rice subspecies. *Int. J. Mol. Sci.* **2019**, *20*, 5922. [CrossRef]
11. Gao, Q.; Li, G.; Sun, H.; Xu, M.; Wang, H.; Ji, J.; Wang, D.; Yuan, C.; Zhao, X. Targeted mutagenesis of the rice *FW 2.2-like* gene family using the CRISPR/Cas9 system reveals OsFWL4 as a regulator of tiller number and plant yield in rice. *Int. J. Mol. Sci.* **2020**, *21*, 809. [CrossRef]
12. Fan, X.; Chen, J.; Wu, Y.; Teo, C.; Xu, G.; Fan, X. Genetic and global epigenetic modification, which determines the phenotype of transgenic rice? *Int. J. Mol. Sci.* **2020**, *21*, 1819. [CrossRef] [PubMed]

13. Shim, K.C.; Kim, S.H.; Jeon, Y.A.; Lee, H.S.; Adeva, C.; Kang, J.W.; Kim, H.J.; Tai, T.H.; Ahn, S.N. A RING-type E3 ubiquitin ligase, *OsGW2*, controls chlorophyll content and dark-induced senescence in rice. *Int. J. Mol. Sci.* **2020**, *21*, 1704. [CrossRef] [PubMed]

14. Zhang, Q.; Shen, L.; Wang, Z.; Hu, G.; Ren, D.; Hu, J.; Zhu, L.; Gao, Z.; Zhang, G.; Guo, L.; et al. OsCAF1, a CRM domain containing protein, influences chloroplast development. *Int. J. Mol. Sci.* **2019**, *20*, 4386. [CrossRef] [PubMed]

15. Zhao, Y.; Han, Q.; Ding, C.; Huang, Y.; Liao, J.; Chen, T.; Feng, S.; Zhou, L.; Zhang, Z.; Chen, Y.; et al. Effect of low temperature on chlorophyll biosynthesis and chloroplast biogenesis of rice seedlings during greening. *Int. J. Mol. Sci.* **2020**, *21*, 1390. [CrossRef] [PubMed]

16. Wu, Y.B.; Li, G.; Zhu, Y.J.; Cheng, Y.C.; Yang, J.Y.; Chen, H.Z.; Song, X.J.; Ying, J.Z. Genome-wide identification of QTLs for grain protein content based on genotyping-by-resequencing and verification of *qGPC1-1* in rice. *Int. J. Mol. Sci.* **2020**, *21*, 408. [CrossRef] [PubMed]

17. Kanda, Y.; Nakagawa, H.; Nishizawa, Y.; Kamakura, T.; Mori, M. Broad-spectrum disease resistance conferred by the overexpression of rice RLCK BSR1 results from an enhanced immune response to multiple MAMPs. *Int. J. Mol. Sci.* **2019**, *20*, 5523. [CrossRef] [PubMed]

18. Liu, X.; Inoue, H.; Tang, X.; Tan, Y.; Xu, X.; Wang, C.; Jiang, C.J. Rice *OsAAA-ATPase1* is induced during blast infection in a salicylic acid-dependent manner, and promotes blast fungus resistance. *Int. J. Mol. Sci.* **2020**, *21*, 1443. [CrossRef]

19. Liang, T.; Chi, W.; Huang, L.; Qu, M.; Zhang, S.; Chen, Z.Q.; Chen, Z.J.; Tian, D.; Gui, Y.; Chen, X.; et al. Bulked segregant analysis coupled with whole-genome sequencing (BSA-Seq) mapping identifies a novel *pi21* haplotype conferring basal resistance to rice blast disease. *Int. J. Mol. Sci.* **2020**, *21*, 2162. [CrossRef] [PubMed]

20. Hsu, Y.C.; Chiu, C.H.; Yap, R.; Tseng, Y.C.; Wu, Y.P. Pyramiding bacterial blight resistance genes in Tainung82 for broad-spectrum resistance using marker-assisted selection. *Int. J. Mol. Sci.* **2020**, *21*, 1281. [CrossRef]

21. Jiang, S.; Yang, C.; Xu, Q.; Wang, L.; Yang, X.; Song, X.; Wang, J.; Zhang, X.; Li, B.; Li, H.; et al. Genetic dissection of germinability under low temperature by building a resequencing linkage map in *japonica* Rice. *Int. J. Mol. Sci.* **2020**, *21*, 1284. [CrossRef]

22. Yuan, S.; Wang, Y.; Zhang, C.; He, H.; Yu, S. Genetic dissection of seed dormancy using chromosome segment substitution lines in rice (*Oryza sativa* L.). *Int. J. Mol. Sci.* **2020**, *21*, 1344. [CrossRef] [PubMed]

23. Sheteiwy, M.S.; Shao, H.; Qi, W.; Hamoud, Y.A.; Shaghaleh, H.; Khan, N.U.; Yang, R.; Tang, B. GABA-alleviated oxidative injury induced by salinity, osmotic stress and their combination by regulating cellular and molecular signals in rice. *Int. J. Mol. Sci.* **2019**, *20*, 5709. [CrossRef] [PubMed]

24. Liu, G.; Zha, Z.; Cai, H.; Qin, D.; Jia, H.; Liu, C.; Qiu, D.; Zhang, Z.; Wan, Z.; Yang, Y.; et al. Dynamic transcriptome analysis of anther response to heat stress during anthesis in thermotolerant rice (*Oryza sativa* L.). *Int. J. Mol. Sci.* **2020**, *21*, 1155. [CrossRef] [PubMed]

25. Schaarschmidt, S.; Lawas, L.M.F.; Glaubitz, U.; Li, X.; Erban, A.; Kopka, J.; Jagadish, S.V.K.; Hincha, D.K.; Zuther, E. Season affects yield and metabolic profiles of rice (*Oryza sativa*) under high night temperature stress in the field. *Int. J. Mol. Sci.* **2020**, *21*, 3187. [CrossRef] [PubMed]

International Journal of
Molecular Sciences

Article

The Rice *Basic Helix–Loop–Helix 79* (*OsbHLH079*) Determines Leaf Angle and Grain Shape

Hyoseob Seo [1,†], Suk-Hwan Kim [1,†], Byoung-Doo Lee [1,†], Jung-Hyun Lim [1], Sang-Ji Lee [1], Gynheung An [2] and Nam-Chon Paek [1,*]

[1] Department of Plant Science, Plant Genomics and Breeding Institute, Research Institute of Agriculture and Life Sciences, Seoul National University, Seoul 08826, Korea; flameseob@snu.ac.kr (H.S.); sukhwan0819@snu.ac.kr (S.-H.K.); bdlee94@snu.ac.kr (B.-D.L.); jh.lim19@cj.net (J.-H.L.); sangjee715@snu.ac.kr (S.-J.L.)
[2] Department of Plant Molecular Systems Biotechnology, Crop Biotech Institute, Kyung Hee University, Yongin 17104, Korea; genean@khu.ac.kr
* Correspondence: ncpaek@snu.ac.kr; Tel.: +82-2-880-4543; Fax: +82-2-877-4550
† These authors contributed equally to this work.

Received: 26 February 2020; Accepted: 17 March 2020; Published: 18 March 2020

Abstract: Changes in plant architecture, such as leaf size, leaf shape, leaf angle, plant height, and floral organs, have been major factors in improving the yield of cereal crops. Moreover, changes in grain size and weight can also increase yield. Therefore, screens for additional factors affecting plant architecture and grain morphology may enable additional improvements in yield. Among the basic Helix-Loop-Helix (bHLH) transcription factors in rice (*Oryza sativa*), we found an enhancer-trap T-DNA insertion mutant of *OsbHLH079* (termed *osbhlh079-D*). The *osbhlh079-D* mutant showed a wide leaf angle phenotype and produced long grains, similar to the phenotypes of mutants with increased brassinosteroid (BR) levels or enhanced BR signaling. Reverse transcription-quantitative PCR analysis showed that BR signaling-associated genes are largely upregulated in *osbhlh079-D*, but BR biosynthesis-associated genes are not upregulated, compared with its parental *japonica* cultivar 'Dongjin'. Consistent with this, *osbhlh079-D* was hypersensitive to BR treatment. Scanning electron microscopy revealed that the expansion of cell size in the adaxial side of the lamina joint was responsible for the increase in leaf angle in *osbhlh079-D*. The expression of cell-elongation-associated genes encoding expansins and xyloglucan endotransglycosylases/hydrolases increased in the lamina joints of leaves in *osbhlh079-D*. The regulatory function of OsbHLH079 was further confirmed by analyzing *35S::OsbHLH079* overexpression and *35S::RNAi-OsbHLH079* gene silencing lines. The *35S::OsbHLH079* plants showed similar phenotypes to *osbhlh079-D*, and the *35S::RNAi-OsbHLH079* plants displayed opposite phenotypes to *osbhlh079-D*. Taking these observations together, we propose that *OsbHLH079* functions as a positive regulator of BR signaling in rice.

Keywords: bHLH transcription factor; lamina joint; leaf angle; long grain; brassinosteroid signaling

---

1. Introduction

In cereal crops, leaf angle (defined as the angle between the leaf blade and the leaf sheath) is a key factor determining plant architecture, which also includes plant height, tiller number, and panicle morphology [1,2]. In cereal crops including rice (*Oryza sativa*), plant architecture has been an important agronomic trait for increasing crop yield. In particular, leaf angle is closely associated with photosynthetic capacity [3]. Plants with erect leaves capture more sunlight for photosynthesis and are amenable to much denser planting in populations with a high leaf area index for increasing total grain yield. The lamina joint, which connects the leaf blade and leaf sheath, is central in controlling leaf angle [4], as the degree of leaf inclination largely depends on cell proliferation or cell expansion as well as the cell wall composition at the lamina joint.

Brassinosteroid (BR) phytohormones affect lamina joint morphology and increase leaf angle in rice [5]. BRs are a group of steroid phytohormones that are widely distributed in plants; more than 69 types of BRs have been isolated from diverse plants [6]. BRs play pivotal roles in cell expansion, cell division, vascular bundle differentiation, male fertility, senescence, seed germination, grain filling, photomorphogenesis, flowering time, root growth, and abiotic/biotic stress responses [7–13]. In rice, BR functions in the regulation of grain size, leaf angle, and yield potential. For instance, several mutants with low BR contents or weak BR signaling, such as *dwarf2* (*d2*), *d11*, and *d61*, exhibit dwarfism and produce short grains and erect leaves [14–16]. Additionally, many genes have a role in controlling leaf angle, such as *TILLER ANGLE1* (*Ta1*), *EBISU DWARF* (*D2*), *INCREASED LAMINA INCLINATION1* (*ILI1*), *LEAF INCLINATION2* (*LC2*), *INCREASED LEAF ANGLE1* (*ILA1*), and *SLENDER GRAIN* (*SLG*) [1,17–21]. Moreover, loss-of-function mutants of BR-related genes, including *OsDWARF4* and *OsBRI1*, show improved grain yield due to their ability to be planted at a higher density and their enhanced photosynthetic rate [1,22]. Therefore, understanding the effects of BR on rice architecture has important implications for improving yield.

BR signal transduction has been intensively studied in *Arabidopsis thaliana* [23]. Under normal BR levels, BR interacts with BRASSINOSTEROID INSENSITIVE1 (BRI1) and BRASSINOSTEROID ASSOCIATED RECEPTOR KINASE1 (BAK1), forming a BRI–BR–BAK1 complex [24,25]. This complex inhibits the activity of BRASSINOSTEROID INSENSITIVE2 (BIN2) and activates PHOSPHATASE 2A (PP2A) for the activation of BRASSINAZOLE RESISTANT1 (BZR1). The activated BZR1 is translocated into the nucleus and regulates its downstream genes at the transcriptional level [23,26–28].

In rice, the BR signaling pathway remains largely unknown, since only a few components have been reported [23]. BR interacts with OsBRI1 and is involved in the formation of the OsBRI1–OsBAK1 complex [29,30], which inactivates OsBIN2 by an unknown pathway [23]. OsBIN2 phosphorylates OsBZR1, LEAF AND TILLER ANGLE INCREASED CONTROLLER (LIC), and DWARF AND LOW TILLERING (DLT) and inhibits their activities. OsBZR1 upregulates *ILI1* and downregulates *LIC* and *DLT*, thus transmitting the BR signal to their downstream genes, which affect plant growth and development [23].

BR mainly affects cell elongation and cell division; moreover, cell number and cell size largely determine organ size during organogenesis [31,32]. Grain size (GS), another key trait determining yield, is mainly determined by grain length (GL), grain width (GW), and grain thickness, all of which are closely related to cell elongation or cell division. Various genes and quantitative trait loci (QTLs) in rice, such as *GS3*, *GS5*, *GW2*, *GW5*, *GW8*, *GW6a*, *qGL3*, *THOUSAND-GRAIN WEIGHT6* (*TGW6*), and *BIG GRAIN1* (*BG1*), affect grain size by regulating cell number [33–41]. In addition, *GS2/GL2*, *GL7*, and *POSITIVE REGULATOR OF GRAIN LENGTH1* (*PGL1*) regulate grain size by influencing cell size in rice [40,42,43].

The basic helix–loop–helix (bHLH) domain transcription factors act in various biological processes in animals and plants [44]. In flowering plants, 162 bHLH proteins have been identified in *Arabidopsis thaliana* and 167 in rice [45]. These proteins are divided into two groups: typical bHLH proteins harboring both motifs (basic and HLH motif) bind to DNA through the basic region, whereas atypical, non-DNA-binding bHLH proteins lacking the basic region require other bHLH proteins to bind to DNA as protein dimers [46]. For example, rice ILI1 is an atypical bHLH protein that interacts with the typical bHLH protein OsIBH1 and represses OsIBH1 function [47]. This antagonistic regulation controls cell length in the lamina joint. Several bHLH transcription factors, such as BRASSINOSTEROID UPREGULATED1 (BU1), *O. sativa* BU1-LIKE1 (OsBUL1), and OsbHLH107, are involved in controlling leaf angle or grain size in rice [47–50].

In this study, we show that OsbHLH079 acts as a key regulator in determining leaf angle and grain length. *OsbHLH079*-overexpressing lines exhibited exaggerated leaf inclination, with longer cells on the adaxial surface of lamina joint. In addition, *OsbHLH079* is involved in modulating grain shape because the *OsbHLH079*-overexpressing mutant produced long grains. Several molecular genetic approaches showed that the function of OsbHLH079 is closely associated with the BR signaling pathway. This study provides new insight into the roles of OsbHLH079 in determining leaf angle and grain shape.

2. Results and Discussion

2.1. OsbHLH079 Increases Leaf Angle in Rice

To identify new components that regulate plant architecture, we screened a collection of T-DNA insertion lines in rice in the Rice Functional Genomic Express Database [51]. We isolated a new mutant with increased leaf angle phenotype (Figure 1a), and found that an enhancer-trap line, PFG_3A-01275, which is derived from the Korean *japonica* rice cultivar 'Dongjin (hereafter wild type, WT)' harbors a T-DNA containing four tandem repeats of the CaMV 35S promoter in the promoter of *OsbHLH079* (LOC_Os02g47660) (Figure 1b). To check whether the T-DNA insertion alters the expression of *OsbHLH079*, we compared *OsbHLH079* transcript levels in various organs between WT and the enhancer-trap T-DNA insertion line. RT-qPCR analysis revealed that the transcript levels of *OsbHLH079* in the T-DNA line were much higher in the leaf blade, leaf sheath, and root, compared with WT, although the degrees of overexpression varied among tissues (Figure 1c). Thus, the gain-of-function mutant was termed *osbhlh079-D*.

Next, to characterize the leaf angle phenotype of *osbhlh079-D* in more detail, we compared the leaf angles of the top four leaves between WT and *osbhlh079-D* in field-grown plants at heading stage. The leaf angles of all four top leaves in *osbhlh079-D* were significantly enlarged compared to those in WT, especially those of flag leaves (Figure 1d,e). These results indicate that the overexpression of *OsbHLH079* is closely associated with increase in leaf angle in rice.

To verify if the overexpression of *OsbHLH079* leads to an increase in leaf angle, we generated two independent transgenic rice lines overexpressing the full-length coding sequence of *OsbHLH079* (*35S::OsbHLH079* #2 and #12) as well as two individual RNAi-mediated knockdown lines of *OsbHLH079* (*35S::RNAi-OsbHLH079* #4 and #5). First, we checked whether the expression of *OsbHLH079* is altered in the *35S::OsbHLH079* and *35S::RNAi-OsbHLH079* lines. RT-qPCR analysis revealed that the transcript levels of *OsbHLH079* were upregulated in two *35S::OsbHLH079* lines (Figure 2a) and downregulated in two *35S::RNAi-OsbHLH079* lines (Figure 2b). Next, we compared the leaf angles of top four leaves among WT, *35S::OsbHLH079*, and *35S::RNAi-OsbHLH079* at heading stage grown under NLD conditions in the paddy field. Indeed, all the leaf angles of *35S::OsbHLH079* were much larger than WT, especially for the flag leaf, as is the case for *osbhlh079-D*. By contrast, all the leaf angles of *35S::RNAi-OsbHLH079* were significantly smaller, except for the flag leaf angle (Figure 2c,d). Collectively, these results suggested that OsbHLH079 increases leaf angle during leaf blade growth.

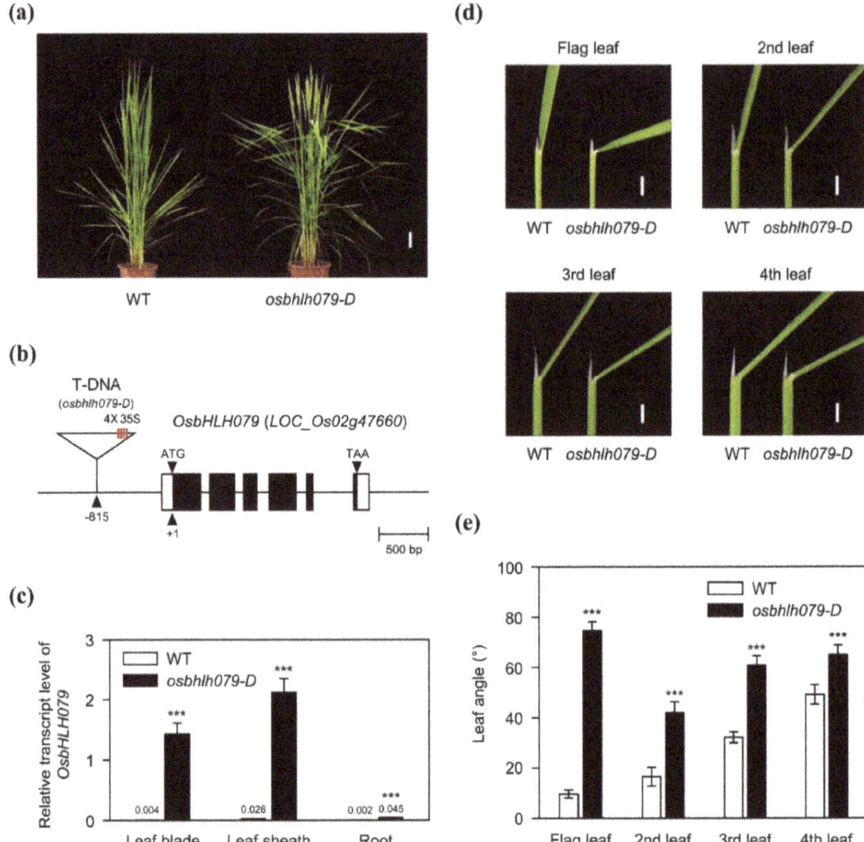

Figure 1. Phenotypic characterization of the *osbhlh079-D* mutant in rice. (**a**) Phenotypes of wild-type (WT) and *osbhlh079-D* at heading stage in plants grown under natural long day (NLD) conditions in the paddy field. Scale bar = 10 cm. (**b**) Schematic diagram illustrating the position of the T-DNA insertion in *OsbHLH079* (LOC_Os02g47660). Open boxes and filled boxes represent the untranslated region and coding sequence of *OsbHLH079*, respectively. (**c**) Comparison of the *OsbHLH079* transcript levels between 3-week-old plants of WT and *osbhlh079-D* grown under natural sunlight in the greenhouse. The transcript level of *OsbHLH079* was measured by RT-qPCR and normalized to *UBQ5*. Means and standard deviations were obtained from five biological replicates. (**d**) The leaf angle phenotypes of WT and *osbhlh079-D* at heading stage grown under NLD conditions in the paddy field. Scale bar = 1 cm. (**e**) Statistical analysis of leaf angles between WT and *osbhlh079-D* at heading stage grown under NLD conditions in the paddy field. Means and standard deviations were obtained from ten biological replicates. Significant differences between means were analyzed using Student's *t*-test (*** $p < 0.001$). These experiments were repeated twice with similar results.

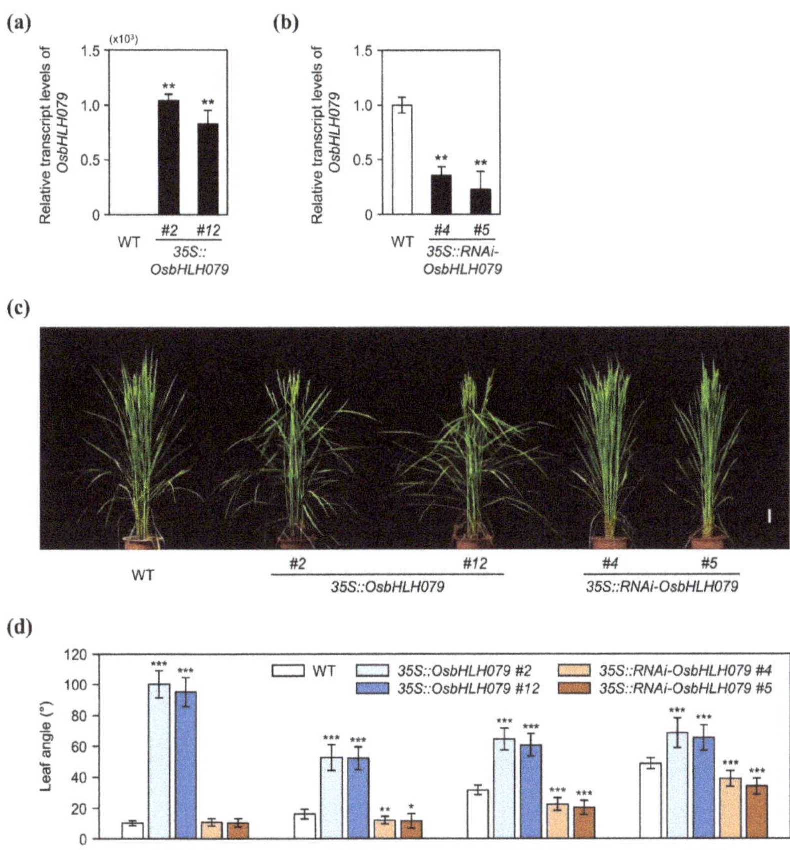

Figure 2. The leaf angles of WT, *35S::OsbHLH079*, and *35S::RNAi-OsbHLH079*. (**a**) Relative transcript levels of *OsbHLH079* in WT, *35S::OsbHLH079* #2, and *35S::OsbHLH079* #12. (**b**) Relative transcript levels of *OsbHLH079* in WT, *35S::RNAi-OsbHLH079* #4, and *35S::RNAi-OsbHLH079* #5. (**a,b**) Total RNA was extracted from the 2-cm lamina joint tissues between the leaf blade and leaf sheath of WT, *35S::OsbHLH079* #2, *35S::OsbHLH079* #12, *35S::RNAi-OsbHLH079* #4, and *35S::RNAi-OsbHLH079* #5 at heading stage in plants grown under NLD conditions in the paddy field. Relative expression levels of *OsbHLH079* were determined by RT-qPCR analysis and normalized to *UBQ5*. Means and standard deviations were obtained from five biological replicates. Differences between means were compared using Student's *t*-test (** $p < 0.01$). (**c**) Plant phenotypes of WT, *35S::OsbHLH079* #2, *35S::OsbHLH079* #12, *35S::RNAi-OsbHLH079* #4, and *35S::RNAi-OsbHLH079* #5 at heading stage in plants grown under NLD conditions in the paddy field. Scale bar = 10 cm. (**d**) Statistical analysis of leaf angles among WT, *35S::OsbHLH079* #2, *35S::OsbHLH079* #12, *35S::RNAi-OsbHLH079* #4, and *35S::RNAi-OsbHLH079* #5 at heading stage in plants grown under NLD conditions in the paddy field. Means and standard deviations were obtained from ten biological replicates. Significant differences between means were analyzed using Student's *t*-test (* $p < 0.05$, ** $p < 0.01$, *** $p < 0.001$). These experiments were repeated twice with similar results.

2.2. OsbHLH079 Increases Grain Length in Rice

In addition to their increased leaf angle, *osbhlh079-D* plants produced long grains (Figure 3a). The grain length of *osbhlh079-D* was longer than WT, while the grain width and grain thickness of *osbhlh079-D* were smaller, resulting in no significant difference in 500-grain weight between WT and *osbhlh079-D* (Figure 3b). To confirm if the long grain phenotype of *osbhlh079-D* is caused by the overexpression of *OsbHLH079*, we compared the grain length among WT, *35S::OsbHLH079*, and *35S::RNAi-OsbHLH079*. The grain lengths of two independent *35S::OsbHLH079* lines were much longer than WT, whereas the grain lengths of two independent *35S::RNAi-OsbHLH079* lines were significantly shorter than WT (Figure 3c,d). Collectively, these results suggested that *OsbHLH079* is also involved in the regulation of grain length in rice.

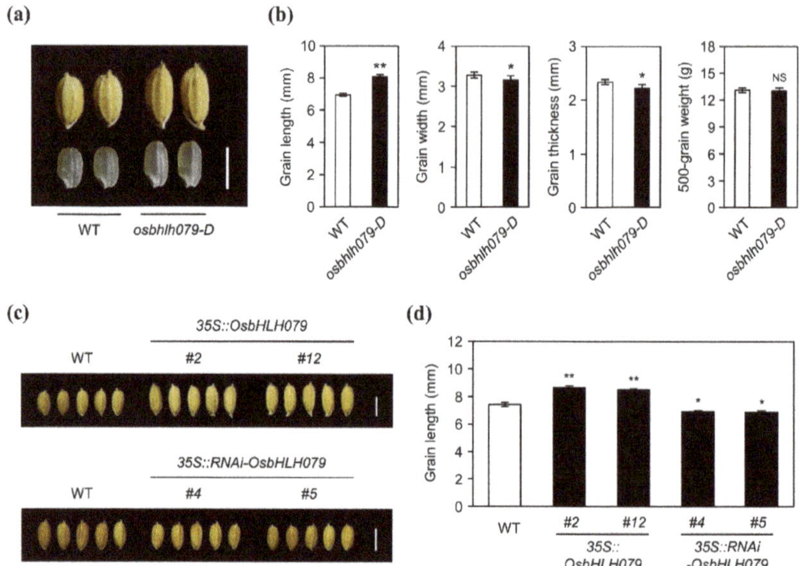

Figure 3. The grain phenotypes of WT, *osbhlh079-D*, *35S::OsbHLH079*, and *35S::RNAi-OsbHLH079*. (**a**) Unhulled and hulled grain phenotypes of the *osbhlh079-D* mutant compared to those of WT. Scale bar = 0.5 cm. (**b**) Comparison of grain length, grain width, grain thickness, and 500-grain weight between WT and the *osbhlh079-D* mutant. Means and standard deviations were obtained from twenty biological replicates. Asterisks indicate statistically significant differences (* $p < 0.05$, ** $p < 0.01$, Student's *t*-test) compared to WT. NS, not significant. (**c**) The unhulled grain phenotype of *35S::OsbHLH079* (upper panel), and *35S::RNAi-OsbHLH079* (lower panel) compared to that of WT. Scale bar = 0.5 cm. (**d**) Statistical analysis of grain lengths among the WT, *35S::OsbHLH079* #2, *35S::OsbHLH079* #12, *35S::RNAi-OsbHLH079* #4, and *35S::RNAi-OsbHLH079* #5. Means and standard deviations were obtained from twenty biological replicates. Significant differences between means were analyzed using Student's *t*-test (* $p < 0.05$, ** $p < 0.01$). These experiments were repeated twice with similar results.

2.3. OsbHLH079 is a Transcription Factor of the Basic Helix-Loop-Helix (bHLH) Family in Rice

The domains of OsbHLH079 were analyzed using the NCBI-BLASTP program [52]. OsbHLH079 has a conserved basic helix–loop–helix (bHLH) domain from the 174th to 221th amino acids (Figure 4a). Moreover, the bHLH domain was found to be a putative G-box binding type, which directly binds to the G-box motif in the rice genome, in a previous genome-wide analysis [45]. These data suggested that OsbHLH079 is a bHLH-type G-box binding transcription factor. To determine if OsbHLH079 acts as a transcription factor, we first examined its subcellular localization in onion epidermal cells. The *35S::YFP* (control) and *35S::YFP-OsbHLH079* constructs were introduced into the onion epidermal cells by particle bombardment, and, at 18 h after particle bombardment, onion nuclei were stained with DAPI to detect the nucleus. Confocal laser scanning microscopy showed that YFP-OsbHLH079 fusion proteins exclusively localized in the DAPI-stained nuclei, while YFP proteins were detected throughout the cells (Figure 4b). Next, we performed a transactivation activity assay for OsbHLH079 in yeast. The full-length cDNA of *OsbHLH079* was fused with the yeast GAL4 activation domain in the pGADT7 vector, or with the yeast GAL4 DNA-binding domain in the pGBKT7 vector. Then, the yeast strain AH109, harboring the *HIS3*, *ADE2*, and *LacZ* reporter genes, was co-transformed with a pair of plasmids and plated on each selective medium, as shown in Figure 4c. Only the yeast expressing GAL4BD-OsbHLH079 grew on the selective medium lacking histidine and adenine (Figure 4c). Furthermore, in the β-galactosidase liquid assay, LacZ activity was highly upregulated in the yeast expressing GAL4BD-OsbHLH079 compared to that in the negative control (Figure 4d), indicating that OsbHLH079 has transactivation activity. Taking these observations together, it can be concluded that OsbHLH079 functions as a transcription factor of the basic helix–loop–helix (bHLH) family in rice.

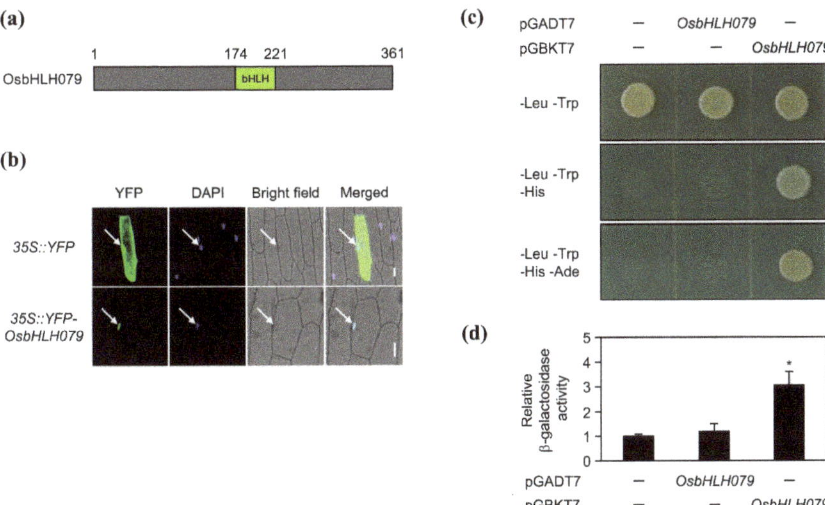

Figure 4. OsbHLH079 as a putative transcription factor. (**a**) Domain analysis of the 361-amino-acid-long OsbHLH079 protein. The green box indicates a basic helix–loop–helix domain. (**b**) Subcellular localization of OsbHLH079 in onion epidermal cells. The *35S::YFP* and *35S::YFP-OsbHLH079* constructs were introduced into onion epidermal cells and the cells were analyzed by confocal laser scanning microscopy at 18 h after particle bombardment. Onion nuclei were stained with DAPI. Arrows indicate the nucleus. Scale bar = 50 μm. DAPI, 4′,6-diamidino-2-phenylindole. (**c**,**d**) Transactivation activity assay of OsbHLH079. The full-length cDNA of *OsbHLH079* was fused with the yeast GAL4 activation domain in the pGADT7 vector, or with the yeast GAL4 DNA-binding domain in the pGBKT7 vector, and the fusion proteins were expressed in the yeast strain AH109. (**c**) Transformed yeasts were grown on the Leu⁻ Trp⁻, Leu⁻ Trp⁻ His⁻, and Leu⁻ Trp⁻ His⁻ Ade⁻ agar media for yeast cell survival assay. (**d**) LacZ activity was obtained using the β-galactosidase liquid assay. The relative β-galactosidase activity was obtained by normalizing to the activity level of the negative control. Means and standard deviations were obtained from three biological samples. Significant differences between means were analyzed using Student's *t*-test (* $p < 0.05$). These experiments were repeated twice with similar results. -, empty vector.

2.4. OsbHLH079 Enlarges Cell Size in the Adaxial Side of Leaf Lamina Joints by Upregulating Cell Expansion-Related Genes

In general, the tissue-specific expression of genes is closely associated with their biological functions. Therefore, we first checked the spatial expression patterns of *OsbHLH079* in field-grown (NLD conditions) WT at heading stage. This revealed that *OsbHLH079* is mainly expressed in the stem, node, internode, and lamina joint (Figure 5a). Previous studies showed that several genes controlling leaf angle are highly expressed in the lamina joint, as is the case of *OsbHLH079*, and the degree of leaf inclination is mainly regulated by cell proliferation and/or cell expansion in the lamina joint, especially in the adaxial side of the lamina joint [5,20,21,41,53–57]. Therefore, we speculated that the expression levels of cell proliferation- or expansion-related genes in lamina joint would be altered in *osbhlh079-D*, and thus compared the transcript levels of those genes in the lamina joint between WT and *osbhlh079-D* by RT-qPCR analysis. The expression levels of cell proliferation-related genes, including *OsCDC6*, *OsMCM3*, *OsE2F1*, and *OsCYCA3;1* [58], in the lamina joint were not significantly different between WT and *osbhlh079-D* (Figure 5b). However, the transcript levels of cell expansion-related genes, such as *OsEXPAs* and *OsXTHs* [59,60], were highly upregulated in the lamina joint of *osbhlh079-D* compared to WT (Figure 5c). Therefore, we hypothesized that the increased leaf angle of *osbhlh079-D* might be caused by expansion of cell size, mainly in the adaxial side of lamina joints.

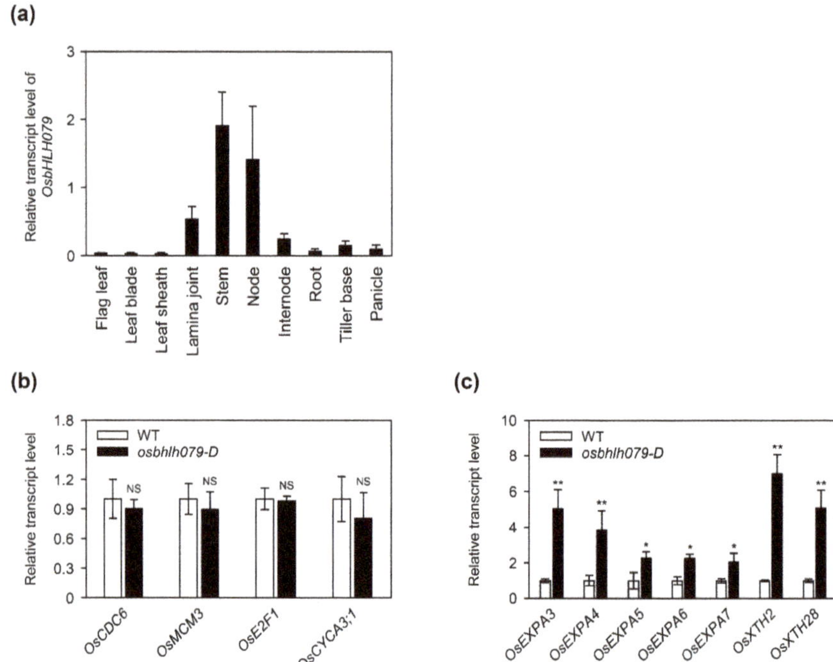

Figure 5. Expression of cell cycle- and cell elongation-related genes in *osbhlh079-D*. (**a**) Spatial expression patterns of *OsbHLH079* in WT at the heading stage grown under NLD conditions in the paddy field. The transcript level of *OsbHLH079* was determined by RT-qPCR analysis and normalized to *UBQ5*. Means and standard deviations were obtained from three biological replicates. (**b**) Expression patterns of cell cycle-related genes in the *osbhlh079-D* mutant compared to those in WT. (**b**) Altered expressions of cell elongation-related genes in the *osbhlh079-D* mutant compared to those in WT. (**b,c**) Total RNA was extracted from the 2-cm lamina joint segments between leaf blade and leaf sheath of 2-week-old WT and *osbhlh079-D* grown under long day (LD) conditions (14.5 h light, 30 °C/9.5 h dark, 24 °C) with 60% relative humidity in a growth chamber. The transcript level of each gene was determined by RT-qPCR analysis and normalized to that of *UBQ5*. Means and standard deviations were obtained from three biological replicates Significant differences between means were analyzed using Student's *t*-test (* $p < 0.05$, ** $p < 0.01$). These experiments were repeated twice with similar results. NS, not significant.

Table 079. *D*, we observed longitudinal sections of flag–leaf lamina joints in WT and *osbhlh079-D* by scanning electron microscopy. The cell length on the adaxial side in *osbhlh079-D* was much larger than WT along the adaxial–abaxial and proximal–distal axes; the abaxial cell size in *osbhlh079-D* was also slightly increased in both axes (Figure 6a–h). These results suggested that OsbHLH079 increases leaf angle by expanding the cell size on the adaxial side of the lamina joint through the upregulation of *OsEXPA* and *OsXTH* genes (Figure 5c).

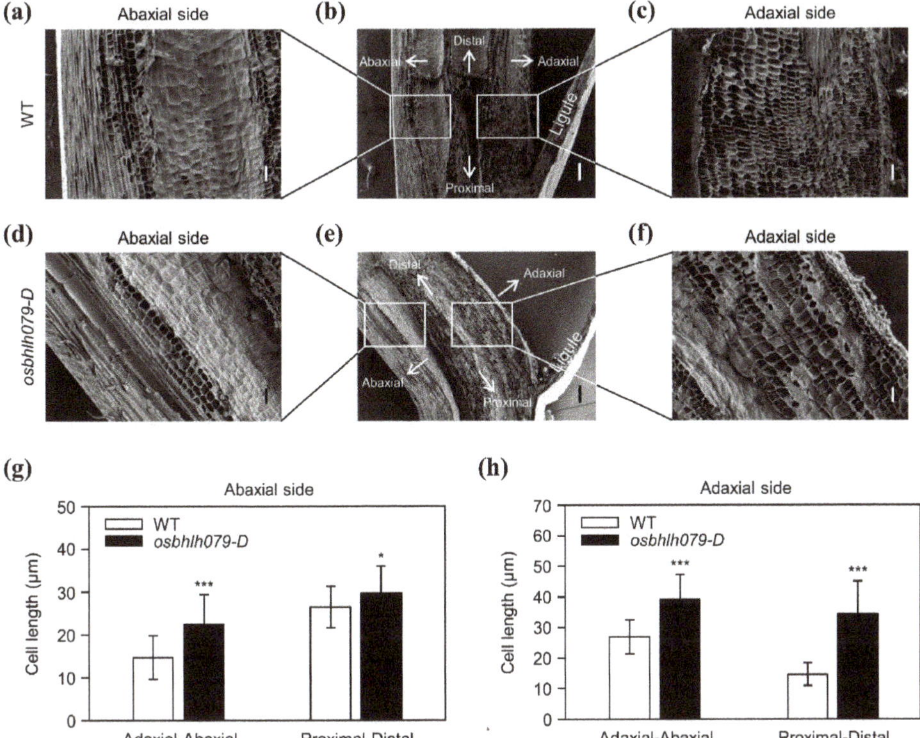

Figure 6. Scanning electron microscopy of the lamina joints of leaves in WT and *osbhlh079-D*. (**a**–**f**) Longitudinal sections of the lamina joint of flag leaf in WT (**a**–**c**) or *osbhlh079-D* (**d**–**f**) at heading stage grown under NLD conditions in the paddy field. (**a**,**d**) Close-up of abaxial regions denoted by rectangles (left side) in (**b**,**e**), respectively. (**c**,**f**) Close-up of adaxial regions denoted by rectangles (right side) in (**b**,**e**), respectively. Scale bar = 200 μm in (**b**,**e**). Scale bar = 50 μm in (**a**,**c**,**d**,**f**). (**g**) Statistical analysis of cell lengths in (**a**,**d**). (**h**) Statistical analysis of cell lengths in (**c**,**f**). (**g**,**h**) Cell lengths along the adaxial–abaxial axis and proximal–distal axis were measured on the abaxial and adaxial sides of lamina joints. Means and standard deviations were obtained from thirty cells. Significant differences between means were analyzed using Student's *t*-test (* $p < 0.05$, *** $p < 0.001$). These experiments were repeated twice with similar results.

2.5. OsbHLH079 Regulates the Expression of BR Signaling-Related Genes

The wide leaf angle phenotype of *osbhlh079-D* resembles that of mutants with elevated BR accumulation or enhanced BR signaling [11,14,15,61–63]. Moreover, the transcript levels of several XTHs and expansin genes, which are upregulated in *osbhlh079-D* (Figure 5c), are significantly increased by BR treatment in *Arabidopsis thaliana*, rice, soybean (*Glycine max*), maize (*Zea mays*), and wheat (*Triticum aestivum*) [59,64–70]. Therefore, we speculated that the increased leaf angle of *osbhlh079-D* is caused by either elevated endogenous BR accumulation or enhanced BR signaling.

To investigate whether the expression of BR biosynthesis- or BR signaling-related genes is altered in *osbhlh079-D*, we compared their transcript levels in the lamina joints of leaf blades between WT and *osbhlh079-D*. In the lamina joints, the expression of BR biosynthesis-related genes, such as *D2*, *D11*, and *BRD1* [15,16,61], was significantly downregulated in *osbhlh079-D* compared to that of WT (Figure 7a). In addition, the transcript level of *OsBRI1*, the BR receptor, was also significantly downregulated compared to that of WT (Figure 7b), indicating a negative feedback regulation by enhanced BR signaling [14,63,71]. Among the BR signaling-related genes, including *OsBAK1*, *OsBSK3*, *GSK2*, *BU1*, *OsBZR1*, *ILI1*, and *DLT* [14,30,47,48,72–75], the expression of *OsBZR1*, and its downstream genes, such as *ILI1*, and *DLT*, was significantly altered in the lamina joint of *osbhlh079-D* compared to WT (Figure 7b). For example, the expression of genes encoding positive regulators of the BR signaling pathway, such as *OsBZR1*, and *ILI1*, was highly upregulated, but the transcript level of *DLT*, which also encodes a positive regulator of BR signaling pathway but is repressed directly by OsBZR1, was significantly downregulated in *osbhlh079-D* (Figure 7b). To confirm whether the expression levels of *OsBZR1*, *ILI1*, and *DLT* are altered by the ectopic or knockdown expression of *OsbHLH079*, we compared the expression levels of *OsBZR1*, *ILI1*, and *DLT* in the lamina joint among WT, *35S::OsbHLH079*, and *35S::RNAi-OsbHLH079*. The transcript levels of *OsBZR1*, and *ILI1* were highly upregulated, while *DLT* expression was significantly downregulated in the lamina joint of *35S::OsbHLH079* lines, as in *osbhlh079-D* (Figure 7c–e). By contrast, the expression levels of *OsBZR1*, and *ILI1* were significantly decreased, while the transcript level of *DLT* was highly increased in the lamina joint of *35S::RNAi-OsbHLH079* lines (Figure 7c–e). These results indicated that the increased activity of OsbHLH079 enhances the BR signaling pathway by altering the expression of *OsBZR1* and its downstream genes, such as *ILI1*, and *DLT*.

To verify whether the response to BR treatment is enhanced by the overexpression of *OsbHLH079*, we carried out a BR-induced lamina joint inclination assay. For this assay, 2-cm lamina joint segments were detached from 10-day-old seedlings of WT and *osbhlh079-D* grown in darkness and treated with 1 µM BL for 48 h in darkness. Then, we compared the extent of lamina inclination of *osbhlh079-D* with WT. As shown in Figure 7f, *osbhlh079-D* was more sensitive to BR (24-epibrassinolide) treatment. Moreover, the difference in the extents of lamina inclination between WT and *osbhlh079-D* increased as the BR concentration increased (Figure 7b). These data indicated that BR signaling is enhanced in *osbhlh079-D*. Therefore, we concluded that OsbHLH079 enhances the BR signaling pathway, which leads to the expansion of cell size in the adaxial side of lamina joints via upregulation of *OsEXPAs* and *OsXTHs*, resulting in an increase in leaf angle in rice (Figure 8).

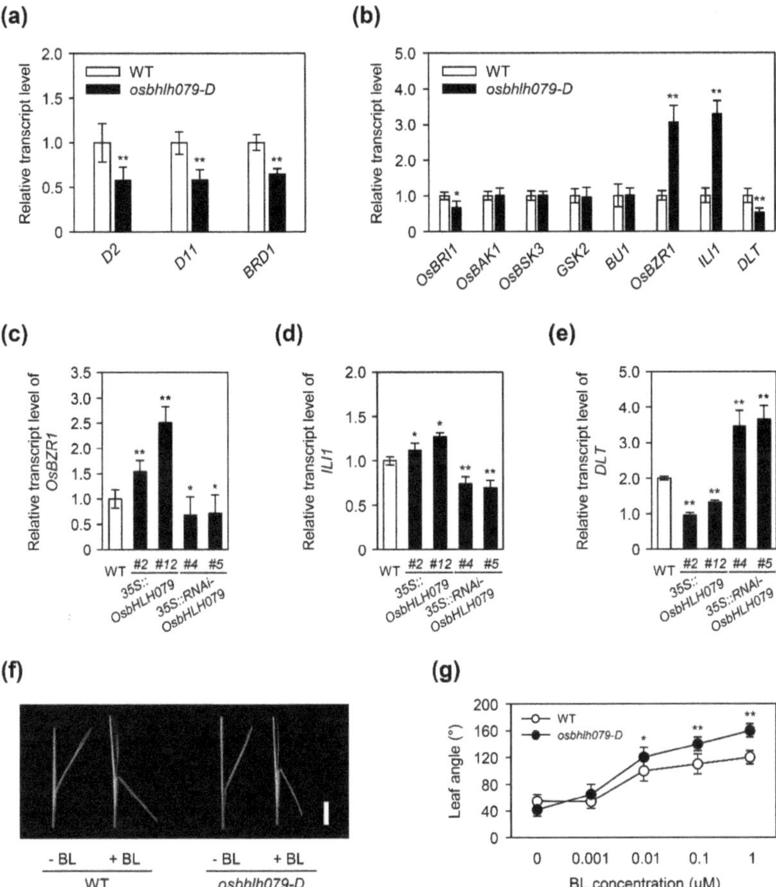

Figure 7. OsbHLH079 acts as a positive regulator of the brassinosteroid signaling pathway. (**a**) Expression patterns of brassinosteroid (BR) biosynthesis-related genes in the *osbhlh079-D* mutant compared to those in WT. (**b**) Altered expressions of BR signaling-related genes in the *osbhlh079-D* mutant compared to those in WT. (**c–e**) Altered expressions of *OsBZR1* (**c**), *ILI1* (D), and *DLT* (E) in *35S::OsbHLH079* and *35S::RNAi-OsbHLH079* compared to those in WT. (**a–e**) Total RNA was extracted from the 2-cm lamina joints between leaf blade and leaf sheath of 4-week-old plants of WT, *osbhlh079-D*, *35S::OsbHLH079*, and *35S::RNAi-OsbHLH079* grown under LD conditions (14.5 h light, 30 °C/9.5 h dark, 24 °C) with 60% relative humidity in a growth chamber. The transcript level of each gene was determined by RT-qPCR analysis and normalized to *UBQ5*. Means and standard deviations were obtained from three biological replicates. Asterisks indicate statistically significant differences (* $p < 0.05$, ** $p < 0.01$, Student's *t*-test) compared to WT. (**f**) BR-induced lamina joint inclination in WT and the *osbhlh079-D* mutant. The 2-cm lamina joint segments of 10-day-old seedlings of WT and *osbhlh079-D* grown at 30 °C in darkness were treated with 1 µM BL for 48 h in darkness. Scale bar = 0.5 cm. BL, 24-epibrassinolide. (**g**) Dose-dependent responses of the lamina joint of WT and *osbhlh079-D* to various concentrations of BL. Means and standard deviations were obtained from more than ten biological replicates. Significant differences between means were analyzed using Student's *t*-test (* $p < 0.05$, ** $p < 0.01$). These experiments were repeated twice with similar results. BL, 24-epibrassinolide.

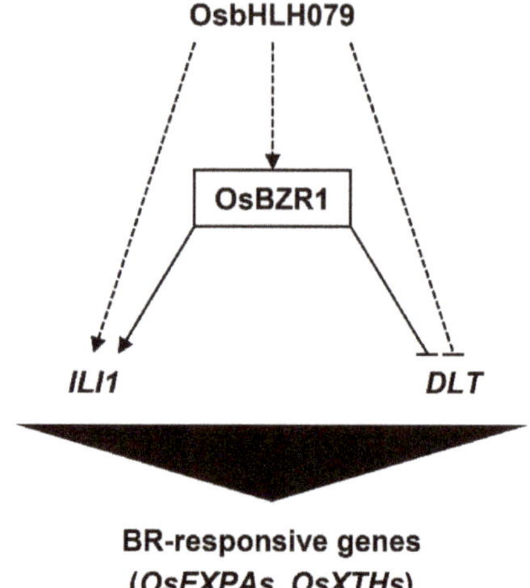

Figure 8. A proposed model of the OsbHLH079-mediated regulatory network in the BR signaling pathway. OsbHLH079 enhances brassinosteroid signaling by upregulating genes encoding positive regulators of the BR signaling pathway, such as *OsBZR1*, and *ILI1*, and downregulating *DLT*, which also encodes a positive regulator of BR signaling and downregulated directly by OsBZR1. Then, altered expressions of BR-responsive genes such as *OsEXPA*, and *OsXTH* produce changes in leaf angle and grain length. Arrows and bars indicate positive and negative regulation, respectively. Solid and dashed lines indicate direct regulation and possible feed-forward regulation, respectively.

2.6. OsbHLH079 Might Indirectly Regulate Expressions of OsBZR1, and ILI1

As OsBZR1 directly regulates the expression of *ILI1*, and *DLT* [23], it is possible that (1) OsbHLH079 modulates the expression and/or activity of OsBZR1, (2) OsbHLH079 directly regulates other genes because OsBZR1 has a limited function, in which the effect of overexpressed OsBZR1 occurs only when the binding site of negative regulators, such as 14-3-3 or GSK2, is mutated in plants [23,63,76], or (3) OsbHLH079 directly regulates downstream genes of *OsBZR1*, such as *ILI1*, and *DLT*, and expression of *OsBZR1* is increased by a feedback regulatory loop.

To examine the possible roles of OsbHLH079 in the transcription of *OsBZR1*, *ILI1*, and *DLT* in rice, we investigated the promoter sequences of *OsBZR1*, *ILI1*, and *DLT* (−2000 to −1 from the ATG). It revealed that *OsBZR1* or *ILI1* did not contain the G-box sequence (CACGTG; a putative binding site of bHLH-type transcription factors), but only one CACGTG sequence in the promoter region (-989 to −984 from ATG) of *DLT*. These findings suggest that OsbHLH079 regulates *OsBZR1* and *ILI1* indirectly, although we cannot exclude the possibility that OsbHLH079 binds to the promoter regions of *OsBZR1* and *ILI1*.

2.7. OsbHLH079 Might Directly Regulate the Cell Expansion-Associated Genes

PHYTOCHROME INTERACTING FACTOR LIKE1 (*OsPIL1*) functions as a key regulator of internode elongation [77]. *OsPIL1*-overexpressing rice plants (*Ubi::OsPIL1*) formed elongated internodes via larger cells through direct regulation of its downstream genes, such as *OsEXPA4* and *1-ACC OXIDASE*, via binding to the G-box element. Both OsPIL1 and OsbHLH079 are bHLH-type transcription factors. Thus, it can be speculated that OsbHLH079 directly binds to the promoter regions of cell expansion-related genes, such as *OsEXPAs* and *OsXTHs*. The expression of cell expansion-related genes was increased to a much greater extent in the *osbhlh079-D* mutant compared to the increased expression of BR signaling-associated genes (*OsBZR1*, and *ILI1*) (Figure 5c, Figure 7b). The increased expression of BR-related genes in *osbhlh079-D* and *OsbHLH079*-overexpressing plants may have caused a significant change in the expression of the downstream genes, but we could not exclude the possibility that OsbHLH079 directly regulates the cell expansion-associated genes.

2.8. OsbHLH079 Increases Grain Length by Altering TGW6 Expression

In rice, several genes and QTLs, including *GS3*, *GS5*, *GW2*, *GW5*, *GW8*, *TGW6*, *GW6a*, *qGL3*, and *BG1*, affect grain size by regulating cell number, and *GS2/GL2*, *GL7*, and *PGL1* affect grain size by influencing cell size [33–43]. GW2 is a RING-type E3 ubiquitin ligase and acts as a negative regulator of cell division [33]. GW2-overexpressing transgenic rice showed a reduced grain-width phenotype. Increased expression of *GW2* could be one of the reasons for the slender-grain phenotype of the *osbhlh079-D* mutant (Figure S1). *TGW6* encodes an IAA-glucose hydrolase, and loss of function of *TGW6* results in enhanced grain weight, resulting in increased grain yield [38]. *RNAi-TGW6* rice plants showed an increased-grain-length phenotype, which could be one of the reasons for the long-grain phenotype of the *osbhlh079-D* mutant (Figure S1). It is possible that the regulatory function of OsbHLH079 is associated with not only the BR signaling pathway, but also the auxin-related pathway, which is closely involved in the control of grain size and grain yield [38].

2.9. OsbHLH079 is an Ortholog of Arabidopsis CRYPTOCHROME-INTERACTING bHLH 1

In the genome-wide analysis, 167 *bHLH* genes in rice and 162 *bHLH* genes in *Arabidopsis* were identified and analyzed by amino acid sequence-based alignments. As a result, they were divided into 25 subfamilies [45]. *OsbHLH079* belongs to the C group with 24 genes in rice and 23 genes in *Arabidopsis*. The C group contains all the *Arabidopsis CRYPTOCHROME-INTERACTING bHLH* (*AtCIB*) genes (*AtCIB1*, *AtCIB2*, *AtCIB3*, *AtCIB4*, and *AtCIB5*), and *OsbHLH079* was annotated to be an *AtCIB1*-like gene [45]. In *Arabidopsis*, CIB1 interacts with cryptochrome 2 (CRY2), and this complex affects various developmental processes, such as hypocotyl elongation, flowering time, stomata opening, hypocotyl bending, programmed cell death, plastid development, and silique elongation [35,78–85]. In soybean, GmCIB1 binds to the E-box (CANNTG) motif and acts as a transcriptional activator to regulate leaf senescence-associated genes [86]. In rice, OsbHLH079 might be an ortholog of CIB1, and the bHLH domain was a highly conserved among amino acid sequences of AtCIB1, GmCIB1, and OsbHLH079 (Figure S2). Additionally, the *osbhlh079-D* mutant flowered earlier than WT under long-day and short-day conditions (Figure S3), which provides more insight into additional regulatory functions of OsbHLH079 in rice growth and development, similar to AtCIB1.

3. Materials and Methods

3.1. Plant Materials and Growth Conditions

The enhancer-trap T-DNA insertion mutant of *OsbHLH079* (LOC_Os02g47660; PFG_3A-01275; designated as *osbhlh079-D*) in rice was isolated from the Korean *japonica* cultivar 'Dongjin', and obtained from the Rice Functional Genomic Express Database [51]. For phenotypic characterization, rice plants were grown under natural long day (NLD) conditions in the paddy field (37°N latitude, Suwon, Republic of Korea). Rice was also grown in growth chambers under long-day (14.5 h light, 30 °C/9.5 h dark, 24 °C) conditions or short-day (10 h light, 30 °C/14 h dark, 24 °C) conditions with 60% relative humidity. The light sources used in the artificial growth chamber were light-emitting diodes (LEDs), and the average photon flux density was around 500 μmol m^{-2} s^{-1}.

3.2. Vector Construction and Rice Transformation

To generate the *35S::OsbHLH079*, and *35S::RNAi-OsbHLH079* transgenic rice plants, the full-length cDNA of *OsbHLH079*, and the partial cDNA fragment of *OsbHLH079* were amplified from the first-strand cDNA obtained from leaves of WT by reverse-transcription polymerase chain reaction (RT-PCR) using gene-specific primers (Table S1), and subcloned into pCR8/GW/TOPO (Invitrogen, USA). After confirming the sequences, the full-length cDNA of *OsbHLH079*, and the partial cDNA fragment of *OsbHLH079* were transferred into the pMDC32 Gateway binary vector [87], and the pANDA vector [88], respectively, by LR reaction using Gateway LR Clonase II Enzyme Mix (Invitrogen, USA). The resulting constructs, *35S::OsbHLH079*, and *35S::RNAi-OsbHLH079*, were transformed into *Agrobacterium tumefaciens* strain LBA4404, and then introduced into calli generated from mature embryos of WT through *Agrobacterium*-mediated transformation, respectively [89].

3.3. RNA Extraction and Reverse Transcription-Quantitative PCR (RT-qPCR) Analysis

Total RNA was extracted from 2-cm lamina joint segments or other tissues using the MG Total RNA Extraction Kit (Macrogen, Seoul, Republic of Korea) according to the manufacturer's instructions. First-strand cDNAs were synthesized from 2 µg of total RNA using oligo(dT)$_{15}$ primers and M-MLV reverse transcriptase (Promega, USA). The relative transcript levels of each gene were measured by quantitative PCR (qPCR) using gene-specific primers, and rice *Ubiquitin5* (*UBQ5*) was used as an internal control (Table S1) [90]. GoTaq qPCR Master Mix (Promega, USA) was used in a 20 µl total reaction volume, and quantitative PCR was performed using a LightCycler 480 (Roche, Switzerland). qPCR conditions were 95 °C for 2 min, and then 45 cycles of 95 °C for 10 s and 60 °C for 1 min.

3.4. Subcellular Localization of OsbHLH079

To investigate the subcellular localization of OsbHLH079, the *35S::YFP-OsbHLH079* construct was prepared. The full-length coding sequence of *OsbHLH079* was amplified with gene-specific primers (Table S1), and fused with *YFP* in the pEarleyGate 104 (pEG104) vector through LR reactions using Gateway LR Clonase II Enzyme Mix (Invitrogen, USA). The resultant construct, *35S::YFP-OsbHLH079*, and the *35S::YFP* construct were introduced into onion epidermal cells using a DNA particle delivery system (Biolistic PDS-1000/He; Bio-Rad, Hercules, CA, USA), respectively. The transformed onion epidermal cells were incubated on Murashige and Skoog phytoagar medium (pH 5.7) under dark conditions at 25 °C for 18 h, and then onion nuclei were stained with 300 nM 4′,6-diamidino-2-phenylindole (DAPI; Invitrogen, USA) in phosphate-buffered saline for 5 min. YFP and DAPI fluorescence were observed using a confocal laser scanning microscope (SP8X, Leica, Germany) with excitation wavelengths of 458 and 405 nm and emission wavelengths of 514 and 488 nm for YFP and DAPI, respectively.

3.5. Transactivation Activity Assay

Transactivation activity assay was performed as previously described with some modifications [91]. The full-length coding sequence of *OsbHLH079* was amplified by PCR and fused with the yeast GAL4 activation domain in the pGADT7 vector (Biosciences Clontech, Palo Alto, CA, USA), or with the yeast GAL4 DNA binding domain in the pGBKT7 vector (Biosciences Clontech, Palo Alto, CA, USA), respectively. Then, the yeast strain AH109 was co-transformed with a pair of plasmids, and plated on each medium, as shown in Figure 4c. The yeast β-galactosidase liquid assay was carried out according to the Yeast Protocols Handbook (Clontech) using chlorophenol red-β-D-galactopyranoside (CPRG, Roche Biochemical) as the substrate.

3.6. Scanning Electron Microscopy

Scanning electron microscopy was conducted as previously described with some modifications [92]. The lamina joints of flag leaves of WT and *osbhlh079-D* at heading stage grown under NLD conditions in the paddy field were excised and sectioned longitudinally as previously described [56]. The samples were fixed with modified Karnovsky's fixative (2% paraformaldehyde, 2% glutaraldehyde, and 50 mM sodium cacodylate buffer, pH 7.2) at 4 °C for 24 h, and washed with 50 mM sodium cacodylate buffer (pH 7.2) three times at 4 °C for 10 min each. Next, the samples were post-fixed at 4 °C for 2 h with 1% osmium tetroxide in 50 mM sodium cacodylate buffer (pH 7.2), and washed twice with distilled water at room temperature, followed by dehydration with a gradient series of ethanol. After dehydration, the samples were processed as follows: dried in liquid CO_2 using a critical point dryer

(EM CPD300, Leica, Germany), and coated with platinum using a sputter coater (EM ACE200, Leica, Austria). The processed samples were observed by scanning electron microscope (AURIGA, Carl Zeiss, Germany).

3.7. BR-Induced Lamina Joint Inclination Assay

The BR-induced lamina joint inclination assay was performed as previously described with some modifications [93]. Sterilized seeds of WT and the *osbhlh079-D* mutant were grown on Murashige and Skoog (MS) medium in an artificial growth chamber at 30 °C under dark conditions for 10 days. Then, the 2-cm lamina joint segments of WT and *osbhlh079-D* were excised, and incubated on distilled water containing various concentrations of 24-epibrassinolide (BL), an active form of brassinosteroid (Sigma), at 30 °C in darkness for 48 h. The angle between lamina and sheath was measured using ImageJ software [94].

3.8. Gene Information

Sequence data from this article can be found in the National Center for Biotechnology Information (NCBI): *OsbHLH079*, Os02g0705500; *UBQ5*, Os01g0328400; *OsCDC6*, Os01g0856000; *OsMCM3*, Os05g0476200; *OsE2F1*, Os02g0537500; *OsCYCA3;1*, Os03g0607600; *OsEXPA3*, Os05g0276500; *OsEXPA4*, Os05g0477600; *OsEXPA5*, Os02g0744200; *OsEXPA6*, Os03g0336400; *OsEXPA7*, Os03g0822000; *OsXTH2*, Os11g0539200; *OsXTH28*, Os03g0239000; *D2*, Os01g0197100; *D11*, Os04g0469800; *BRD1*, Os03g0602300; *OsBRI1*, Os01g0718300; *OsBAK1*, Os08g0174700; *OsBSK3*, Os04g0684200; *GSK2*, Os05g0207500; *BU1*, Os06g0226500; *OsBZR1*, Os07g0580500; *ILI1*, Os04g0641700; *DLT*, Os06g0127800; *GS2*, Os02g0701300; *GS3*, Os03g0407400; *GS5*, Os05g0158500; *GS6*, Os06g0127800; *GW2*, Os02g0244100; *GW6a*, Os06g0650300; *GLW7*, Os07g0505200; *GW8*, Os08g0531600; *TGW6*, Os06g0623700; *GL7*, Os07g0603300; *PGL1*, Os03g0171300; *PGL2*, Os02g0747900; *qGL3*, Os03g0646900; *RGA1*, Os05g0333200; *RGB1*, Os03g0669200; *SRS5*, Os11g0247300; *TH1*, Os02g0811000; *BG1*, Os03g0175800; *DEP1*, Os09g0441900; *OsMAPK6*, Os06g0154500.

4. Conclusions

We found that OsbHLH079 increases leaf angle and grain length in rice. Rice overexpressing *OsbHLH079* (*osbhlh079-D* and *35S::OsbHLH079*) showed wider leaf angle and longer grain length, and RNA-mediated knockdown lines of *OsbHLH079* (*35S::RNAi-OsbHLH079*) exhibited narrower leaf angle and shorter grain length compared to those of WT. Our data also revealed that OsbHLH079 enhances BR signaling by modulating the expression levels of BR signaling-related genes (*OsBZR1*, *ILI1*, and *DLT*) which lead to increases in leaf angle and grain length. This study opens up the possibility to improve grain yield per unit area in rice by controlling plant architecture and seed shape.

Supplementary Materials: Supplementary materials can be found at http://www.mdpi.com/1422-0067/21/6/2090/s1.

Author Contributions: B.-D.L. and N.-C.P. designed and supervised the research. H.S., S.-H.K., and J.-H.L. performed the experiments and analyzed data. S.-H.K. and S.-J.L. performed the revision experiments. G.A. developed *osbhlh079-D* mutant and provided critical advice. H.S., S.-H.K., B.-D.L., and N.-C.P. wrote the manuscript. All authors have read and agreed to the published version of the manuscript.

Funding: This research was supported by grants from the Next-Generation BioGreen21 Program (PJ013656 to B.-D.L.), Rural Development Administration, and the Basic Science Research Program through the National Research Foundation (NRF) of Korea funded by the Ministry of Education (NRF-2017R1A2B3003310 to N.-C.P.), Republic of Korea.

Conflicts of Interest: The authors have no potential conflicts of interest.

Abbreviations

bHLH	basic Helix-Loop-Helix
BL	24-epibrassinolide
BR	Brassinosteroid
DAPI	4′,6-diamidino-2-phenylindole
GL	Grain length
GS	Grain size
GW	Grain width
LD	Long day
NLD	Natural long day
QTL	Quantitative trait loci
SD	Short day
WT	Wild-type

References

1. Sakamoto, T.; Morinaka, Y.; Ohnishi, T.; Sunohara, H.; Fujioka, S.; Ueguchi-Tanaka, M.; Mizutani, M.; Sakata, K.; Takatsuto, S.; Yoshida, S.; et al. Erect leaves caused by brassinosteroid deficiency increase biomass production and grain yield in rice. *Nat. Biotechnol.* **2006**, *24*, 105–109. [CrossRef]
2. Yoon, J.; Cho, L.H.; Lee, S.; Pasriga, R.; Tun, W.; Yang, J.; Yoon, H.; Jeong, H.J.; Jeon, J.S.; An, G. Chromatin interacting factor OsVIL2 is required for outgrowth of axillary buds in rice. *Mol. Cells* **2019**, *42*, 858–868.
3. Sinclair, T.R.; Sheehy, J.E. Erect leaves and photosynthesis in rice. *Science* **1999**, *283*, 1455. [CrossRef]
4. Nakamura, A.; Fujioka, S.; Takatsuto, S.; Tsujimoto, M.; Kitano, H.; Yoshida, S.; Asami, T.; Nakano, T. Involvement of C-22-hydroxylated brassinosteroids in auxin-induced lamina joint bending in rice. *Plant Cell Physiol.* **2009**, *50*, 1627–1635. [CrossRef] [PubMed]
5. Wada, K.; Marumo, S.; Ikekawa, N.; Morisaki, M.; Mori, K. Brassinolide and homobrassinolide promotion of lamina inclination of rice seedling. *Plant Cell Physiol.* **1981**, *22*, 323–325.
6. Bajguz, A. Brassinosteroids—Occurrence and chemical structures in plants. In *Brassinosteroids: A Class of Plant Hormone*; Hayat, S., Ahmad, A., Eds.; Springer: Dordrecht, The Netherlands, 2011.
7. Szekeres, M.; Nemeth, K.; Koncz-Kalman, Z.; Mathur, J.; Kauschmann, A.; Altmann, T.; Redei, G.P.; Nagy, F.; Schell, J.; Koncz, C. Brassinosteroids rescue the deficiency of CYP90, a cytochrome P450, controlling cell elongation and de-etiolation in Arabidopsis. *Cell* **1996**, *85*, 171–182. [CrossRef]
8. Clouse, S.D.; Sasse, J.M. Brassinosteroids: Essential regulators of plant growth and development. *Annu. Rev. Plant Physiol. Plant Mol. Biol.* **1998**, *49*, 427–451. [CrossRef]
9. Fujioka, S.; Yokota, T. Biosynthesis and metabolism of brassinosteroids. *Annu. Rev. Plant Biol.* **2003**, *54*, 137–164. [CrossRef]
10. Nakashita, H.; Yasuda, M.; Nitta, T.; Asami, T.; Fujioka, S.; Arai, Y.; Sekimata, K.; Takatsuto, S.; Yamaguchi, I.; Yoshida, S. Brassinosteroid functions in a broad range of disease resistance in tobacco and rice. *Plant J.* **2003**, *33*, 887–898. [CrossRef]
11. Wu, C.Y.; Trieu, A.; Radhakrishnan, P.; Kwok, S.F.; Harris, S.; Zhang, K.; Wang, J.; Wan, J.; Zhai, H.; Takatsuto, S.; et al. Brassinosteroids regulate grain filling in rice. *Plant Cell* **2008**, *20*, 2130–2145. [CrossRef]
12. Gudesblat, G.E.; Russinova, E. Plants grow on brassinosteroids. *Curr. Opin. Plant Biol.* **2011**, *14*, 530–537. [CrossRef]
13. Kim, S.; Moon, J.; Roh, J.; Kim, S.K. Castasterone can be biosynthesized from 28-homodolichosterone in Arabidopsis thaliana. *J. Plant Biol.* **2018**, *61*, 330–335. [CrossRef]
14. Yamamuro, C.; Ihara, Y.; Wu, X.; Noguchi, T.; Fujioka, S.; Takatsuto, S.; Ashikari, M.; Kitano, H.; Matsuoka, M. Loss of function of a rice brassinosteroid insensitive1 homolog prevents internode elongation and bending of the lamina joint. *Plant Cell* **2000**, *12*, 1591–1606. [CrossRef] [PubMed]
15. Hong, Z.; Ueguchi-Tanaka, M.; Umemura, K.; Uozu, S.; Fujioka, S.; Takatsuto, S.; Yoshida, S.; Ashikari, M.; Kitano, H.; Matsuoka, M. A rice brassinosteroid-deficient mutant, ebisu dwarf (d2), is caused by a loss of function of a new member of cytochrome P450. *Plant Cell* **2003**, *15*, 2900–2910. [CrossRef]

16. Tanabe, S.; Ashikari, M.; Fujioka, S.; Takatsuto, S.; Yoshida, S.; Yano, M.; Yoshimura, A.; Kitano, H.; Matsuoka, M.; Fujisawa, Y.; et al. A novel cytochrome P450 is implicated in brassinosteroid biosynthesis via the characterization of a rice dwarf mutant, dwarf11, with reduced seed length. *Plant Cell* **2005**, *17*, 776–790. [CrossRef] [PubMed]

17. Li, Z.K.; Paterson, A.H.; Pinson, S.R.M.; Khush, G.S. A major gene, Ta1 and QTLs affecting tiller and leaf anlgles in rice. *Rice Genet. Newsl.* **1998**, *15*, 154–156.

18. Li, Z.; Paterson, A.H.; Pinson, S.R.M.; Stansel, J.W. RFLP facilitated analysis of tiller and leaf angles in rice. *Euphytica* **1999**, *109*, 79–84. [CrossRef]

19. Zhao, S.Q.; Hu, J.; Guo, L.B.; Qian, Q.; Xue, H.W. Rice leaf inclination2, a VIN3-like protein, regulates leaf angle through modulating cell division of the collar. *Cell Res.* **2010**, *20*, 935–947. [CrossRef]

20. Ning, J.; Zhang, B.; Wang, N.; Zhou, Y.; Xiong, L. Increased leaf angle1, a Raf-like MAPKKK that interacts with a nuclear protein family, regulates mechanical tissue formation in the Lamina joint of rice. *Plant Cell* **2011**, *23*, 4334–4347. [CrossRef]

21. Feng, Z.; Wu, C.; Wang, C.; Roh, J.; Zhang, L.; Chen, J.; Zhang, S.; Zhang, H.; Yang, C.; Hu, J.; et al. SLG controls grain size and leaf angle by modulating brassinosteroid homeostasis in rice. *J. Exp. Bot.* **2016**, *67*, 4241–4253. [CrossRef]

22. Morinaka, Y.; Sakamoto, T.; Inukai, Y.; Agetsuma, M.; Kitano, H.; Ashikari, M.; Matsuoka, M. Morphological alteration caused by brassinosteroid insensitivity increases the biomass and grain production of rice. *Plant Physiol.* **2006**, *141*, 924–931. [CrossRef] [PubMed]

23. Zhang, C.; Bai, M.Y.; Chong, K. Brassinosteroid-mediated regulation of agronomic traits in rice. *Plant Cell Rep.* **2014**, *33*, 683–696. [CrossRef] [PubMed]

24. Li, J.; Nam, K.H. Regulation of brassinosteroid signaling by a GSK3/SHAGGY-like kinase. *Science* **2002**, *295*, 1299–1301. [PubMed]

25. Kinoshita, T.; Cano-Delgado, A.; Seto, H.; Hiranuma, S.; Fujioka, S.; Yoshida, S.; Chory, J. Binding of brassinosteroids to the extracellular domain of plant receptor kinase BRI1. *Nature* **2005**, *433*, 167–171. [CrossRef] [PubMed]

26. He, J.X.; Gendron, J.M.; Sun, Y.; Gampala, S.S.; Gendron, N.; Sun, C.Q.; Wang, Z.Y. BZR1 is a transcriptional repressor with dual roles in brassinosteroid homeostasis and growth responses. *Science* **2005**, *307*, 1634–1638. [CrossRef] [PubMed]

27. Yin, Y.; Vafeados, D.; Tao, Y.; Yoshida, S.; Asami, T.; Chory, J. A new class of transcription factors mediates brassinosteroid-regulated gene expression in Arabidopsis. *Cell* **2005**, *120*, 249–259. [CrossRef]

28. Sun, Y.; Fan, X.Y.; Cao, D.M.; Tang, W.; He, K.; Zhu, J.Y.; He, J.X.; Bai, M.Y.; Zhu, S.; Oh, E.; et al. Integration of brassinosteroid signal transduction with the transcription network for plant growth regulation in Arabidopsis. *Dev. Cell* **2010**, *19*, 765–777. [CrossRef]

29. Wang, L.; Xu, Y.Y.; Li, J.; Powell, R.A.; Xu, Z.H.; Chong, K. Transgenic rice plants ectopically expressing AtBAK1 are semi-dwarfed and hypersensitive to 24-epibrassinolide. *J. Plant Physiol.* **2007**, *164*, 655–664. [CrossRef]

30. Li, D.; Wang, L.; Wang, M.; Xu, Y.Y.; Luo, W.; Liu, Y.J.; Xu, Z.H.; Li, J.; Chong, K. Engineering OsBAK1 gene as a molecular tool to improve rice architecture for high yield. *Plant Biotechnol. J.* **2009**, *7*, 791–806. [CrossRef]

31. Potter, C.J.; Xu, T. Mechanisms of size control. *Curr. Opin. Genet. Dev.* **2001**, *11*, 279–286. [CrossRef]

32. Sugimoto-Shirasu, K.; Roberts, K. "Big it up": Endoreduplication and cell-size control in plants. *Curr. Opin. Plant. Biol.* **2003**, *6*, 544–553. [CrossRef] [PubMed]

33. Song, X.J.; Huang, W.; Shi, M.; Zhu, M.Z.; Lin, H.X. A QTL for rice grain width and weight encodes a previously unknown RING-type E3 ubiquitin ligase. *Nat. Genet.* **2007**, *39*, 623–630. [CrossRef] [PubMed]

34. Weng, J.; Gu, S.; Wan, X.; Gao, H.; Guo, T.; Su, N.; Lei, C.; Zhang, X.; Cheng, Z.; Guo, X.; et al. Isolation and initial characterization of GW5, a major QTL associated with rice grain width and weight. *Cell Res.* **2008**, *18*, 1199–1209. [CrossRef] [PubMed]

35. Mao, J.; Zhang, Y.C.; Sang, Y.; Li, Q.H.; Yang, H.Q. A role for Arabidopsis cryptochromes and COP1 in the regulation of stomatal opening. *Proc. Natl. Acad. Sci. USA* **2005**, *102*, 12270–12275. [CrossRef] [PubMed]

36. Li, Y.; Fan, C.; Xing, Y.; Jiang, Y.; Luo, L.; Sun, L.; Shao, D.; Xu, C.; Li, X.; Xiao, J.; et al. Natural variation in GS5 plays an important role in regulating grain size and yield in rice. *Nat. Genet.* **2011**, *43*, 1266–1269. [CrossRef] [PubMed]

37. Zhang, X.; Wang, J.; Huang, J.; Lan, H.; Wang, C.; Yin, C.; Wu, Y.; Tang, H.; Qian, Q.; Li, J.; et al. Rare allele of OsPPKL1 associated with grain length causes extra-large grain and a significant yield increase in rice. *Proc. Natl. Acad. Sci. USA* **2012**, *109*, 21534–21539. [CrossRef]

38. Ishimaru, K.; Hirotsu, N.; Madoka, Y.; Murakami, N.; Hara, N.; Onodera, H.; Kashiwagi, T.; Ujiie, K.; Shimizu, B.; Onishi, A.; et al. Loss of function of the IAA-glucose hydrolase gene TGW6 enhances rice grain weight and increases yield. *Nat. Genet.* **2013**, *45*, 707–711. [CrossRef]

39. Liu, L.; Tong, H.; Xiao, Y.; Che, R.; Xu, F.; Hu, B.; Liang, C.; Chu, J.; Li, J.; Chu, C. Activation of Big Grain1 significantly improves grain size by regulating auxin transport in rice. *Proc. Natl. Acad. Sci. USA* **2015**, *112*, 11102–11107. [CrossRef]

40. Wang, Y.; Xiong, G.; Hu, J.; Jiang, L.; Yu, H.; Xu, J.; Fang, Y.; Zeng, L.; Xu, E.; Xu, J.; et al. Copy number variation at the GL7 locus contributes to grain size diversity in rice. *Nat. Genet.* **2015**, *47*, 944–948. [CrossRef]

41. Zhang, S.; Wang, S.; Xu, Y.; Yu, C.; Shen, C.; Qian, Q.; Geisler, M.; de Jiang, A.; Qi, Y. The auxin response factor, OsARF19, controls rice leaf angles through positively regulating OsGH3-5 and OsBRI1. *Plant Cell Environ.* **2015**, *38*, 638–654. [CrossRef]

42. Heang, D.; Sassa, H. Antagonistic actions of HLH/bHLH proteins are involved in grain length and weight in rice. *PLoS ONE* **2012**, *7*, e31325. [CrossRef] [PubMed]

43. Duan, P.; Ni, S.; Wang, J.; Zhang, B.; Xu, R.; Wang, Y.; Chen, H.; Zhu, X.; Li, Y. Regulation of OsGRF4 by OsmiR396 controls grain size and yield in rice. *Nat. Plants* **2015**, *2*, 15203. [CrossRef]

44. Ledent, V.; Vervoort, M. The basic helix-loop-helix protein family: Comparative genomics and phylogenetic analysis. *Genome Res.* **2001**, *11*, 754–770. [CrossRef] [PubMed]

45. Li, X.; Duan, X.; Jiang, H.; Sun, Y.; Tang, Y.; Yuan, Z.; Guo, J.; Liang, W.; Chen, L.; Yin, J.; et al. Genome-wide analysis of basic/helix-loop-helix transcription factor family in rice and Arabidopsis. *Plant Physiol.* **2006**, *141*, 1167–1184. [CrossRef] [PubMed]

46. Massari, M.E.; Murre, C. Helix-loop-helix proteins: Regulators of transcription in eucaryotic organisms. *Mol. Cell Biol.* **2000**, *20*, 429–440. [CrossRef]

47. Zhang, L.Y.; Bai, M.Y.; Wu, J.; Zhu, J.Y.; Wang, H.; Zhang, Z.; Wang, W.; Sun, Y.; Zhao, J.; Sun, X.; et al. Antagonistic HLH/bHLH transcription factors mediate brassinosteroid regulation of cell elongation and plant development in rice and Arabidopsis. *Plant Cell* **2009**, *21*, 3767–3780. [CrossRef]

48. Tanaka, A.; Nakagawa, H.; Tomita, C.; Shimatani, Z.; Ohtake, M.; Nomura, T.; Jiang, C.J.; Dubouzet, J.G.; Kikuchi, S.; Sekimoto, H.; et al. BRASSINOSTEROID UPREGULATED1, encoding a helix-loop-helix protein, is a novel gene involved in brassinosteroid signaling and controls bending of the lamina joint in rice. *Plant Physiol.* **2009**, *151*, 669–680. [CrossRef]

49. Jang, S.; An, G.; Li, H.Y. Rice leaf angle and grain size are affected by the OsBUL1 transcriptional activator complex. *Plant Physiol.* **2017**, *173*, 688–702. [CrossRef]

50. Yang, X.; Ren, Y.; Cai, Y.; Niu, M.; Feng, Z.; Jing, R.; Mou, C.; Liu, X.; Xiao, L.; Zhang, X.; et al. Overexpression of OsbHLH107, a member of the basic helix-loop-helix transcription factor family, enhances grain size in rice (*Oryza sativa* L.). *Rice* **2018**, *11*, 41. [CrossRef]

51. Rice Functional Genomic Express Database. Available online: http://signal.salk.edu/cgi-bin/RiceGE (accessed on 21 July 2017).

52. NCBI-BLASTP Program. Available online: https://blast.ncbi.nlm.nih.gov/Blast.cgi?PROGRAM=blastp&PAGE_TYPE=BlastSearch&LINK_LOC=blasthome (accessed on 20 June 2018).

53. Cao, H.; Chen, S. Brassinosteroid-induced rice lamina joint inclination and its relation to indole-3-acetic acid and ethylene. *Plant Growth Regul.* **1995**, *16*, 189–196. [CrossRef]

54. Zhao, S.Q.; Xiang, J.J.; Xue, H.W. Studies on the rice LEAF INCLINATION1 (LC1), an IAA-amido synthetase, reveal the effects of auxin in leaf inclination control. *Mol. Plant* **2013**, *6*, 174–187. [CrossRef] [PubMed]

55. Sun, S.; Chen, D.; Li, X.; Qiao, S.; Shi, C.; Li, C.; Shen, H.; Wang, X. Brassinosteroid signaling regulates leaf erectness in Oryza sativa via the control of a specific U-type cyclin and cell proliferation. *Dev. Cell* **2015**, *34*, 220–228. [CrossRef] [PubMed]

56. Zhou, L.J.; Xiao, L.T.; Xue, H.W. Dynamic cytology and transcriptional regulation of rice lamina joint development. *Plant Physiol.* **2017**, *174*, 1728–1746. [CrossRef] [PubMed]

57. Tang, Y.; Liu, H.; Guo, S.; Wang, B.; Li, Z.; Chong, K.; Xu, Y. OsmiR396d affects gibberellin and brassinosteroid signaling to regulate plant architecture in rice. *Plant Physiol.* **2018**, *17*, 946–959. [CrossRef]

58. Wang, W.; Li, G.; Zhao, J.; Chu, H.; Lin, W.; Zhang, D.; Wang, Z.; Liang, W. DWARF TILLER1, a WUSCHEL-related homeobox transcription factor, is required for tiller growth in rice. *PLoS Genet.* **2014**, *10*, e1004154. [CrossRef] [PubMed]

59. Yokoyama, R.; Rose, J.K.; Nishitani, K. A surprising diversity and abundance of xyloglucan endotransglucosylase/hydrolases in rice. Classification and expression analysis. *Plant Physiol.* **2004**, *134*, 1088–1099. [CrossRef] [PubMed]

60. Choi, D.; Cho, H.T.; Lee, Y. Expansins: Expanding importance in plant growth and development. *Physiol. Plant* **2006**, *126*, 511–518. [CrossRef]

61. Hong, Z.; Ueguchi-Tanaka, M.; Shimizu-Sato, S.; Inukai, Y.; Fujioka, S.; Shimada, Y.; Takatsuto, S.; Agetsuma, M.; Yoshida, S.; Watanabe, Y.; et al. Loss-of-function of a rice brassinosteroid biosynthetic enzyme, C-6 oxidase, prevents the organized arrangement and polar elongation of cells in the leaves and stem. *Plant J.* **2002**, *32*, 495–508. [CrossRef]

62. Sakamoto, T.; Matsuoka, M. Characterization of CONSTITUTIVE PHOTOMORPHOGENESIS AND DWARFISM homologs in rice (*Oryza sativa* L.). *J. Plant Growth Regul.* **2006**, *25*, 245–251. [CrossRef]

63. Bai, M.Y.; Zhang, L.Y.; Gampala, S.S.; Zhu, S.W.; Song, W.Y.; Chong, K.; Wang, Z.Y. Functions of OsBZR1 and 14-3-3 proteins in brassinosteroid signaling in rice. *Proc. Natl. Acad. Sci. USA* **2007**, *104*, 13839–13844. [CrossRef]

64. Zurek, D.M.; Clouse, S.D. Molecular cloning and characterization of a brassinosteroid-regulated gene from elongating soybean (*Glycine max* L.) epicotyls. *Plant Physiol.* **1994**, *104*, 161–170. [CrossRef] [PubMed]

65. Uozu, S.; Tanaka-Ueguchi, M.; Kitano, H.; Hattori, K.; Matsuoka, M. Characterization of XET-related genes of rice. *Plant Physiol.* **2000**, *122*, 853–860. [CrossRef] [PubMed]

66. Liu, Y.; Liu, D.; Zhang, H.; Gao, H.; Guo, X.; Wang, D.; Zhang, X.; Zhang, A. The α-and β-expansin and xyloglucan endotransglucosylase/hydrolase gene families of wheat: Molecular cloning, gene expression, and EST data mining. *Genomics* **2007**, *90*, 516–529. [CrossRef]

67. Genovesi, V.; Fornalé, S.; Fry, S.C.; Ruel, K.; Ferrer, P.; Encina, A.; Sonbol, F.M.; Bosch, J.; Puigdomenech, P.; Rigau, J. ZmXTH1, a new xyloglucan endotransglucosylase/hydrolase in maize, affects cell wall structure and composition in Arabidopsis thaliana. *J. Exp. Bot.* **2008**, *59*, 875–889. [CrossRef] [PubMed]

68. Kozuka, T.; Kobayashi, J.; Horiguchi, G.; Demura, T.; Sakakibara, H.; Tsukaya, H.; Nagatani, A. Involvement of auxin and brassinosteroid in the regulation of petiole elongation under the shade. *Plant Physiol.* **2010**, *153*, 1608–1618. [CrossRef] [PubMed]

69. Abuqamar, S.; Ajeb, S.; Sham, A.; Enan, M.R.; Iratni, R. A mutation in the expansin-like A2 gene enhances resistance to necrotrophic fungi and hypersensitivity to abiotic stress in Arabidopsis thaliana. *Mol. Plant Pathol.* **2013**, *14*, 813–827. [CrossRef] [PubMed]

70. Rao, X.; Dixon, R.A. Brassinosteroid mediated cell wall remodeling in grasses under abiotic stress. *Front. Plant Sci.* **2017**, *8*, 806. [CrossRef]

71. Tong, H.; Liu, L.; Jin, Y.; Du, L.; Yin, Y.; Qian, Q.; Zhu, L.; Chu, C. DWARF AND LOW-TILLERING acts as a direct downstream target of a GSK3/SHAGGY-Like Kinase to medate brassinosteroid responses in rice. *Plant Cell* **2012**, *24*, 2562–2577. [CrossRef]

72. Tong, H.; Jin, Y.; Liu, W.; Li, F.; Fang, J.; Yin, Y.; Qian, Q.; Zhu, L.; Chu, C. DWARF AND LOW-TILLERING, a new member of the GRAS family, plays positive roles in brassinosteroid signaling in rice. *Plant J.* **2009**, *58*, 803–816. [CrossRef]

73. Bai, M.Y.; Shang, J.X.; Oh, E.; Fan, M.; Bai, Y.; Zentella, R.; Sun, T.P.; Wang, Z.Y. Brassinosteroid, gibberellin and phytochrome impinge on a common transcription module in Arabidopsis. *Nat. Cell Biol.* **2012**, *14*, 810–817. [CrossRef]

74. Zhu, J.Y.; Sae-Seaw, J.; Wang, Z.Y. Brassinosteroid signalling. *Development* **2013**, *140*, 1615–1620. [CrossRef] [PubMed]

75. Zhang, B.; Wang, X.; Zhao, Z.; Wang, R.; Huang, X.; Zhu, Y.; Yuan, L.; Wang, Y.; Wang, Y.; Xu, X.; et al. OsBRI1 activates BR signaling by preventing binding between the TPR and kinase domains of OsBSK3 via phosphorylation. *Plant Physiol.* **2016**, *170*, 1149–1161. [CrossRef] [PubMed]

76. Wang, Z.Y.; Nakano, T.; Gendron, J.; He, J.; Chen, M.; Vafeados, D.; Yang, Y.; Fujika, S.; Yoshida, S.; Asami, T.; et al. Nuclear-localized BZR1 mediates brassinosteroid-induced growth and feedback suppression of brassinosteroid biosynthesis. *Dev. Cell* **2002**, *2*, 505–513. [CrossRef]

77. Todaka, D.; Nakashima, K.; Maruyama, K.; Kidokoro, S.; Osakabe, Y.; Ito, Y.; Matsukura, S.; Fujita, Y.; Yoshiwara, K.; Ohme-Takagi, M.; et al. Rice phytochrome-interacting factor-like protein OsPIL1 functions as a key regulator of internode elongation and induces a morphological response to drought stress. *Proc. Natl. Acad. Sci. USA* **2012**, *109*, 15947–15952. [CrossRef] [PubMed]

78. El-Din El-Assal, S.; Alonso-Blanco, C.; Peeters, A.J.; Raz, V.; Koornneef, M. A QTL for flowering time in Arabidopsis reveals a novel allele of CRY2. *Nat. Genet.* **2001**, *29*, 435–440. [CrossRef]

79. Ahmad, M.; Grancher, N.; Heil, M.; Black, R.C.; Giovani, B.; Galland, P.; Lardemer, D. Action spectrum for cryptochrome-dependent hypocotyl growth inhibition in Arabidopsis. *Plant Physiol.* **2002**, *129*, 774–785. [CrossRef]

80. Ohgishi, M.; Saji, K.; Okada, K.; Sakai, T. Functional analysis of each blue light receptor, cry1, cry2, phot1, and phot2, by using combinatorial multiple mutants in Arabidopsis. *Proc. Natl. Acad. Sci. USA* **2004**, *101*, 2223–2228. [CrossRef]

81. Usami, T.; Mochizuki, N.; Kondo, M.; Nishimura, M.; Nagatani, A. Cryptochromes and phytochromes synergistically regulate Arabidopsis root greening under blue light. *Plant Cell Physiol.* **2004**, *45*, 1798–1808. [CrossRef]

82. Danon, A.; Coll, N.S.; Apel, K. Cryptochrome-1-dependent execution of programmed cell death induced by singlet oxygen in Arabidopsis thaliana. *Proc. Natl. Acad. Sci. USA* **2006**, *103*, 17036–17041. [CrossRef]

83. Liu, H.; Yu, X.; Li, K.; Klejnot, J.; Yang, H.; Lisiero, D.; Lin, C. Photoexcited CRY2 interacts with CIB1 to regulate transcription and floral initiation in Arabidopsis. *Science* **2008**, *322*, 1535–1539. [CrossRef]

84. Kennedy, M.J.; Hughes, R.M.; Peteya, L.A.; Schwartz, J.W.; Ehlers, M.D.; Tucker, C.L. Rapid blue-light-mediated induction of protein interactions in living cells. *Nat. Methods* **2010**, *7*, 973–975. [CrossRef] [PubMed]

85. Idevall-Hagren, O.; Dickson, E.J.; Hille, B.; Toomre, D.K.; De Camilli, P. Optogenetic control of phosphoinositide metabolism. *Proc. Natl. Acad. Sci. USA* **2012**, *109*, E2316–E2323. [CrossRef]

86. Meng, Y.; Li, H.; Wang, Q.; Liu, B.; Lin, C. Blue light-dependent interaction between cryptochrome2 and CIB1 regulates transcription and leaf senescence in soybean. *Plant Cell* **2013**, *25*, 4405–4420. [CrossRef]

87. Curtis, M.D.; Grossniklaus, U. A gateway cloning vector set for high-throughput functional analysis of genes in planta. *Plant Physiol.* **2003**, *133*, 462–469. [CrossRef] [PubMed]

88. Miki, D.; Shimamoto, K. Simple RNAi vectors for stable and transient suppression of gene function in rice. *Plant Cell Physiol.* **2004**, *45*, 490–495. [CrossRef] [PubMed]

89. Jeon, J.S.; Lee, S.; Jung, K.H.; Jun, S.H.; Jeong, D.H.; Lee, J.; Kim, C.; Jang, S.; Yang, K.; Nam, J.; et al. T-DNA insertional mutagenesis for functional genomics in rice. *Plant J.* **2000**, *22*, 561–570. [CrossRef]

90. Jain, M.; Nijhawan, A.; Tyagi, A.K.; Khurana, J.P. Validation of housekeeping genes as internal control for studying gene expression in rice by quantitative real-time PCR. *Biochem. Biophys. Res. Commun.* **2006**, *345*, 646–651. [CrossRef]

91. Wu, T.; Zhang, M.; Zhang, H.; Huang, K.; Chen, M.; Chen, C.; Yang, X.; Li, Z.; Chen, H.; Ma, Z.; et al. Identification and characterization of EDT1 conferring drought tolerance in rice. *J. Plant Biol.* **2019**, *62*, 39–47. [CrossRef]

92. Lee, D.W.; Lee, S.K.; Rahman, M.M.; Kim, Y.J.; Zhang, D.; Jeon, J.S. The role of rice vacuolar invertase2 in seed size control. *Mol. Cells* **2019**, *42*, 711–720.

93. Zhang, C.; Xu, Y.; Guo, S.; Zhu, J.; Huan, Q.; Liu, H.; Wang, L.; Luo, G.; Wang, X.; Chong, K. Dynamics of brassinosteroid response modulated by negative regulator LIC in rice. *PLoS Genet.* **2012**, *8*, e1002686. [CrossRef]

94. ImageJ Software. Available online: https://imagej.nih.gov/ij/ (accessed on 18 May 2019).

International Journal of
Molecular Sciences

Article

InDel Marker Based Estimation of Multi-Gene Allele Contribution and Genetic Variations for Grain Size and Weight in Rice (*Oryza sativa* L.)

Sadia Gull [1], Zulqarnain Haider [2], Houwen Gu [1], Rana Ahsan Raza Khan [2], Jun Miao [1], Tan Wenchen [1], Saleem Uddin [3], Irshad Ahmad [4] and Guohua Liang [1,*]

[1] Jiangsu Key Laboratory of Crop Genetics and Physiology/Co-Innovation Center for Modern Production Technology of Grain Crops, Key Laboratory of Plant Functional Genomics of the Ministry of Education, Yangzhou University, Yangzhou 225009, China

[2] Rice Breeding and Genetics Section, Rice Research Institute, Kala Shah Kaku, Lahore, Pakistan

[3] Key Laboratory of Genetics and Breeding in Forest Trees and Ornamental Plants, College of Biological Sciences and Technology, Beijing Advanced Innovation Center for Tree Breeding by Molecular Design, National Engineering Laboratory for Tree Breeding , Beijing Forestry University, Beijing 100083, China

[4] Joint International Laboratory of Agriculture and Agri-Product Safety, Yangzhou University, Yangzhou 225009, China

* Correspondence: ricegb@yzu.edu.cn; Tel.: +86-514-87972138

Received: 27 July 2019; Accepted: 24 September 2019; Published: 28 September 2019

Abstract: The market success of any rice cultivar is exceedingly dependent on its grain appearance, as well as its grain yield, which define its demand by consumers as well as growers. The present study was undertaken to explore the contribution of nine major genes, *qPE9~1*, *GW2*, *SLG7*, *GW5*, *GS3*, *GS7*, *GW8*, *GS5*, and *GS2*, in regulating four size and weight related traits, i.e., grain length (GL), grain width (GW), grain thickness (GT), and thousand grain weight (TGW) in 204 diverse rice germplasms using Insertion/Deletion (InDel) markers. The studied germplasm displayed wide-ranging variability in the four studied traits. Except for three genes, all six genes showed considerable association with these traits with varying strengths. Whole germplasm of 204 genotypes could be categorized into three major clusters with different grain sizes and weights that could be utilized in rice breeding programs where grain appearance and weight are under consideration. The study revealed that TGW was 24.9% influenced by GL, 37.4% influenced by GW, and 49.1% influenced by GT. Hence, assuming the trend of trait selection, i.e., GT > GW > GL, for improving TGW in the rice yield enhancement programs. The InDel markers successfully identified a total of 38 alleles, out of which 27 alleles were major and were found in more than 20 genotypes. GL was associated with four genes (*GS3*, *GS7*, *GW8*, and *GS2*). GT was also found to be regulated by four different genes (*GS3*, *GS7*, *GW8*, and *GS2*) out of the nine studied genes. GW was found to be under the control of three studied genes (*GW5*, *GW8*, and *GS2*), whereas TGW was found to be under the influence of four genes (*SLG7*, *GW5*, *GW8*, and *GS5*) in the germplasm under study. The Unweighted Pair Group Method with Arithmetic means (UPGMA) tree based on the studied InDel marker loci segregated the whole germplasm into three distinct clusters with dissimilar grain sizes and weights. A two-dimensional scatter plot constructed using Principal Coordinate Analysis (PCoA) based on InDel markers further separated the 204 rice germplasms into four sub-populations with prominent demarcations of extra-long, long, medium, and short grain type germplasms that can be utilized in breeding programs accordingly. The present study could help rice breeders to select a suitable InDel marker and in formulation of breeding strategies for improving grain appearance, as well as weight, to develop rice varieties to compete international market demands with higher yield returns. This study also confirms the efficient application of InDel markers in studying diverse types of rice germplasm, allelic frequencies, multiple-gene allele contributions, marker-trait associations, and genetic variations that can be explored further.

Int. J. Mol. Sci. **2019**, *20*, 4824

Keywords: rice (*Oryza sativa* L.), grain size and weight; Insertion/Deletion (InDel) markers; multi-gene allele contributions; genetic variation; rice germplasm

---

1. Introduction

Rice is the most consumable food commodity in the world and is used as a staple by more than 50% of the world's population [1]. Rice is also the third-highest produced agricultural commodity after sugarcane and maize [2]. It is a highly valuable grain crop with regard to human nutrition, as well as caloric intake, which provides more than one-fifth of the calories being consumed by humans worldwide [3]. The rapid increase in the human population is further boosting its demand. In some countries, rice is the only staple food, whereas in some other countries, rice is consumed as a traditional dish, as well as an important ingredient in different dishes. In the international market, quality of rice grain is highly indicative of its price, which also reflects its sale-ability. Therefore, both yield and grain quality are equally important parameters for varietal improvement in rice breeding programs.

Commercial success of a modern rice cultivar is highly dependent on grain size related traits (e.g., grain length, grain width, and thickness) for quality and grain weight associated traits (most importantly thousand grain weight) for grain yield [4]. Grain size related traits determine the rice's final market value as defined by consumer preferences, which are a combination of grain size, length, and thickness (Figure 1). Some consumers prefer long grains, while some prefer short and bold grains. Likewise, grain yield is determined by three major components, including grain weight, panicles per plant, and grains per panicle. Among these, the most associated trait is grain weight which is determined as the 1000 grain weight. Therefore, grain length, width, and thickness along with the 1000 grain weight, are central benchmarks for breeding grain appearance, as well as yield improvement, in rice. However, due to the quantitatively inherited nature of these traits, breeders hardly rely on phenotypes for their improvement [5]. Therefore, the use of genetic markers is considered superior to phenotyping [6], because such markers are not affected by the environment and are more efficient and reliable compared to phenotypic data.

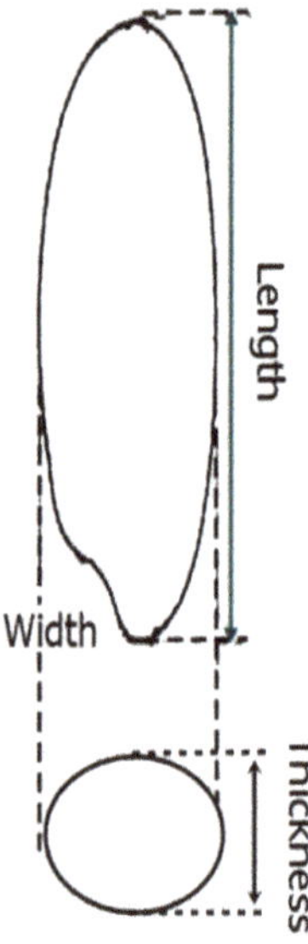

Figure 1. Longitudinal and radial-pattern cross section diagrams of rice grain showing the grain length (GL), grain width (GW), and grain thickness (GT).

A number of Quantitative Trait Loci (QTLs) for grain appearance and weight have already been investigated and reported by different scientists [7–20]. More often, grain length, thickness, and width are regarded as determinants of grain appearance whereas the 1000 grain weight determines grain weight and eventually grain yield. As reported, these traits are under the control of several or many genes and are highly influenced by environmental factors. So far, several major QTLs influencing grain appearance and grain weight have already been characterized and investigated by many researchers. These major genes/QTLs include *qPE9~1* [17], *GW2* [12], *SLG7* [21,22], *GW5* [23–26], *GS3* [27–29], *GS7* [30], *GW8* [31], *GS5* [32], and *GS2* [33–36].

GS2 (*GRAIN SIZE 2*), also reported as *GL2* (*GRAIN LENGTH 2*) or *PT2* (*PANICLE TRAIT 2*), is a rare allele directly controlling two important grain size related traits, including grain width and grain length in rice. *GS2* is found to encode a transcriptional regulator protein named Growth-Regulating Factor 4 (OsGRF4), which is then targeted by OsmiR396, which is a microRNA causing termination of the OsGRF4 function. Several studies have shown that a 2 bp substitution mutation in *GS2* disturbs the binding of OsmiR396 on OsGRF4, resulting in its overexpression, which in turns increases cell enlargement and enhances cell division in grains, causing longer and wider rice grains [33–36]. *GS3* (GRAIN SIZE 3) was among the first reported genes to have minor effects on grain thickness and width. In published studies on *GS3*, the *GS3* was demonstrated to be a negative regulator of grain size, and its encoded putative transmembrane protein contains a plant-specific organ size regulation (OSR) domain as a negative regulatory motif, whose function is inhibited by its tumor necrosis factor receptor/nerve growth factor receptor (TNFR/NGFR) family, cysteine-rich domain and von Willebrand factor type C (VWFC) domain. All four domains have been reported to regulate cell divisions in the upper epidermis of the glume inside the rice seed, causing minor effects on the cell size [29].

GS5 (*GRAIN SIZE 5*) has been reported by many researchers who described this gene as a regulator of grain filling and weight. *GS5* promotes cell division in rice seed and, to some extent, elongation of the cells located in the lemma and palea [32]. The encoded protein of *GS5* (i.e., putative serine carboxypeptidase) executes its function as a positive regulator of a subset of the transition genes (G1-to-S) of cell cycle, causing increased cell divisions and resulting in enhanced grain filling and grain weight. Likewise, the *GW2* (*GRAIN WIDTH 2*) gene encodes a protein (RING-type) that has E3 ubiquitin ligase activity, which degrades the ubiquitin–proteasome pathway. *GW2* negatively regulates cell division by suppressing its substrate(s) to proteasomes for regulated proteolysis. The absence or loss of the *GW2* function via its mutation causes enhanced milk filling in grains and enlarged endosperm cells, resulting in a wider spikelet hull [12].

GW5 (*GRAIN WIDTH 5*), also reported as *SW5* (*SEED WIDTH 5*) and *GSE5*, was investigated by many researchers [23–26], who discovered that *GW5* is negatively associated with rice grain width and weight. Later, it was revealed that *GW5* actually encodes a calmodulin-binding protein, and *GW5* physically interacts with calmodulin AsCaM1-1, which is responsible for grain width in rice. The deletion of *GW5* or its mutations result in wider grains, indicating its negative effects on grain width. Likewise, *SLG7* (*GRAIN LENGTH 7*), also known as *GW7* (*GRAIN WIDTH 7*), has been identified to encode a TONNEAU1-recruiting motif protein responsible for increased cellular division in the longitudinal direction and reduced cell division in the transverse direction [37]. This gene was found to be responsible for grain appearance by altering cell divisions, thereby having significant effects on regulating grain weight, as well.

GW8 (*GRAIN WIDTH 8*) has been reported as a positive regulator of cell proliferation and has a positive association with seed width and seed weight [31]. It encodes SQUAMOSA promoter-binding protein-like 16 (AsSPL16), that was discovered to regulate the expression of several genes involved in G1-to-S transition, similar to the regulatory role of the *GS5* gene [31,32]. It was revealed that a higher expression of the *GW8* gene promoted cell division and grain filling, resulting in increased grain width and a higher grain yield.

GS7 (*GRAIN SHAPE 7*), a robust QTL known to regulate grain shape, has been reported [30] to control the grain length, roundness (thickness), and area (size) in rice. Likewise, another gene, *qPE9~1*, also known as *DEP1 (DENSE AND ERECT PANICLE 1)*, encodes a G protein γ subunit found to be involved in the regulation of erect panicles, grains per panicle, nitrogen uptake, and stress tolerance through a G protein signal pathway [17,38]. In another study, the protein was also found to regulate plant architecture, grain size, and grain yield in rice. The qPE9–1 protein contains an N-terminal G gamma-like (GGL) domain, a putative transmembrane domain, and a C-terminal cysteine-rich domain [39]. Overexpression of protein qPE9–1 has been found to be responsible for increased grain size and yield in rice.

In the past few years, PCR based InDel markers have gained popularity in diversity studies because of their reproducibility, ease of use, and co-dominant inheritance [40]. InDel markers have been extensively utilized as powerful phylogenetic markers for mapping and other genetic studies in different crops [41–48]. Here, based on deletion insertion polymorphisms (DIPs), InDel markers were deployed successfully to study marker trait association and genetic variations. InDels are becoming more famous, as their genotyping requires a low start-up cost, and because they are efficient, relatively simple, and applicable to a wide range of species for which expressed sequence tag (EST) collections are available. Therefore, the leading goals of this study were to assess the efficacy of InDels to (1) estimate the population structure, allelic frequencies, and genetic variation in diverse germplasms comprising 204 rice genotypes; (2) sort the germplasms based on the distribution of the InDel marker loci; (3) assess the allele based contribution of the target genes and their association with individual traits; (4) and engagement of InDel markers to understand the genetics of traits for efficient breeding [49,50].

2. Results

2.1. Descriptive Statistics and Phenotypic Variability for Rice Grain Size and Weight

Descriptive statistics (Table 1) were determined for all four studied traits, i.e., grain length (GL), grain thickness (GT), grain width (GW), and thousand grain weight (TGW), to elaborate the phenotypic variations of the respective traits in 204 rice germplasms. Collected data of whole studied germplasm for each trait is given in Table S1 (as Supplementary Material) The average values (Mean ± Standard Error) of 204 rice genotypes for GL, GW, GT, and TGW were observed to be 8.162 ± 0.065 mm, 2.932 ± 0.019 mm, 2.156 ± 0.012 mm, and 25.858 ± 0.199 g, respectively. The germplasm consisting of 204 rice genotypes showed an appreciable range for the estimated GL: 4.640 (ranging from 6.01 to 10.65 mm). Conversely, GW and GT showed lower range values (i.e., 1.800 and 1.000, respectively), ranging from 2.05 to 3.85 mm and 1.86 to 2.86 mm, respectively. Likewise, TGW was found to have an substantial range of 20 (ranging from 17 g to 37 g), depicting a wide range of variation also suggested by the higher value of variance (i.e., 8.070 in the studied germplasm). On other hand, the variance for GL (0.864) was recorded to be higher than the variance of GW and GT (having a variance of 0.073 and 0.027, respectively) (Table 1). The coefficient of variation (CV%) for all the studied traits (i.e., GL, GW, GT, and TGW) was 11.4%, 9.2%, 7.7%, and 11%, respectively. Kurtosis and skewness both symbolize the modes of gene action [51], and estimate the gene numbers controlling the trait [52], respectively. Estimated values of the skewness and kurtosis for all the studied traits are given in Table 1. Skewness was observed for GL, GT, and TGW, with values of 0.514, −0.148, 1.075, and 1.501, while the estimated kurtosis values were −0.511, 1.092, 1.965, and 1.590, respectively (Table 1).

Figure 2 shows the score plot showing phenotypic variability within the germplasm on a biplot using principal component analysis (PCA), with the first two components representing the maximum proportion (PC1 = 56.8%, PC2 = 31.1%) of the total variation. This shows that sufficient phenotypic variation is present in the germplasm to study genetic variation in the germplasm [53].

Table 1. Descriptive statistics for grain length (GL), grain width (GW), grain thickness (GT), and thousand grain weight (TGW) of germplasms containing two hundred and four (204) rice germplasms.

Parameters	GL (mm)	GW (mm)	GT (mm)	TGW (g)
Minimum	6.010	2.050	1.860	17.0
Maximum	10.650	3.850	2.860	37.0
Range	4.640	1.800	1.000	20.0
Arithmetic Mean	8.162	2.932	2.156	25.858
Standard Error of Arithmetic Mean	0.065	0.019	0.012	0.199
Standard Deviation	0.929	0.270	0.165	2.841
Variance	0.864	0.073	0.027	8.070
Coefficient of Variation (CV)	0.114	0.092	0.077	0.110
Skewness (G1)	0.514	−0.148	1.075	1.501
Kurtosis (G2)	−0.511	1.092	1.965	1.591

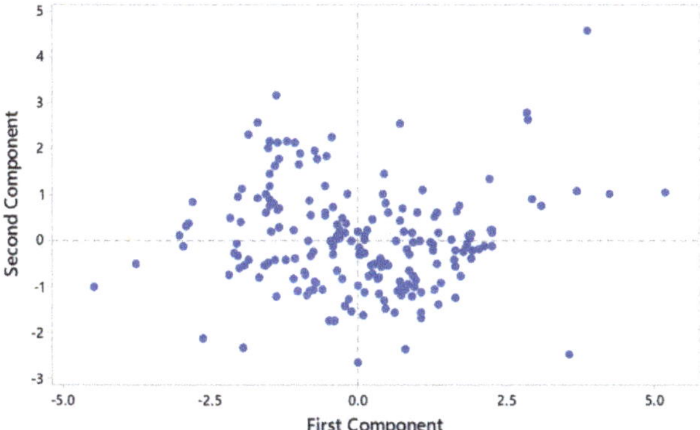

Figure 2. Score plot showing variability within the germplasms of 204 rice germplasms on a biplot using principal component analysis, with the first two components representing the maximum proportion (87.9%) of the total phenotypic variation for the studied traits.

Based on clustering, 204 germplasms were classified into three distinct clusters (I, II, III), as depicted in Figure 3. The major cluster, i.e., Cluster I, consisted of 182 genotypes. Cluster II consisted of 8 genotypes, whereas Cluster III had only 14 genotypes. Cluster I was further subdivided into Cluster IA and Cluster IB for simplification. Cluster IA consisted of 78 entries, and Cluster IB contained 104 entries of germplasm. The average grain length of Cluster I was calculated to be 8.01 mm, whereas the average GW and GT were 2.96 mm and 2.16 mm, respectively. The thousand grain weight (TGW) of this group was 25.63 g. Cluster II showed the highest values for the average GL (8.83 mm), indicating that the entries of this group had the maximum grain length. This group also had the maximum GT (2.27 mm), GW (3.02 mm), and the heaviest grain, as indicated by its TGW value (i.e., 30.43 g). Cluster III contained genotypes with a medium grain length (8.25 mm), but their GW (2.71 mm), GT (2.03 mm), and TGW (21.70 g) were the lowest among the groups [54].

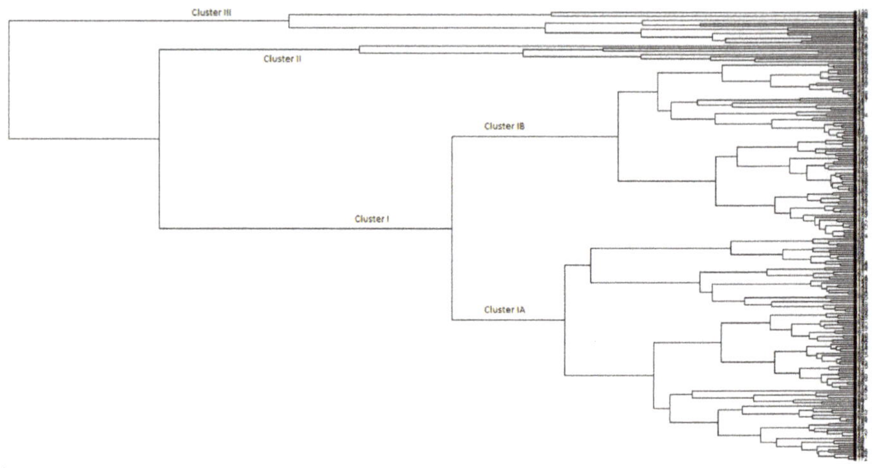

Figure 3. UPGMA Dendrogram showing variability in 204 rice germplasms in three distinct clusters for the studied traits estimated on the similarity index using the Euclidean distances between the groups.

2.2. Factor Analysis and Phenotypic Correlation among the Four Traits for Rice Grain Size and Weight

The Pearson correlation coefficients (r) were estimated for pair-wise analysis among the studied traits (i.e., GL, GW, GT, and TGW) using diverse germplasms comprising 204 rice germplasms (Table 2). A correlation analysis between GL and GW showed a highly significant ($p \leq 0.01$) but negative association (r = −0.604 **) between these two traits. Such a negative linear relationship has also been reported by other researchers [20,55]. GL and GT were also negatively correlated (r = −0.398 *), which was also significant ($p \leq 0.05$) and strong. In the case of TGW, GL showed a significant ($p \leq 0.05$) and positive association (r = 0.249 *), depicting the linear relationship between these two traits. A highly significant ($p \leq 0.01$) and positive relationship (r = 0.686 **) was also observed between GW and GT, indicating that GW was 68.6% positively influenced by GT in the studied rice germplasm. Likewise, GW showed a significant ($p \leq 0.05$) and positive association (r = 0.374 *) with TGW. GT was also found to have a highly significant ($p \leq 0.01$) and positive association (r = 0.491 **) with TGW. These results further revealed that TGW was 24.9% influenced by GL, 37.4% influenced by GW, and 49.1% influenced by GT. Hence, we assume the trend of trait selection (i.e., GT > GW> GL) for improving TGW in yield enhancement programs. The same results were also confirmed using Factor analysis in the Minitab software by producing a biplot for all the traits to further study the pattern of association among the traits. The results confirmed the importance of GT followed by GW and GL in improving TGW and paddy yield. Figure 4 shows a two-dimensional (2D) representation of the associations among the traits [56–58].

Table 2. Correlation coefficient analysis among the four studied traits for grain size and weight.

Traits	GL	GW	GT
Grain length (GL)	1.000		
Grain width (GW)	−0.604 **	1.000	
Grain thickness (GT)	−0.398 *	0.686 **	1.000
Thousand grain weight (TGW)	0.249 *	0.374 *	0.491 **

** highly significant at $p \leq 0.01$; * significant at $p \leq 0.05$.

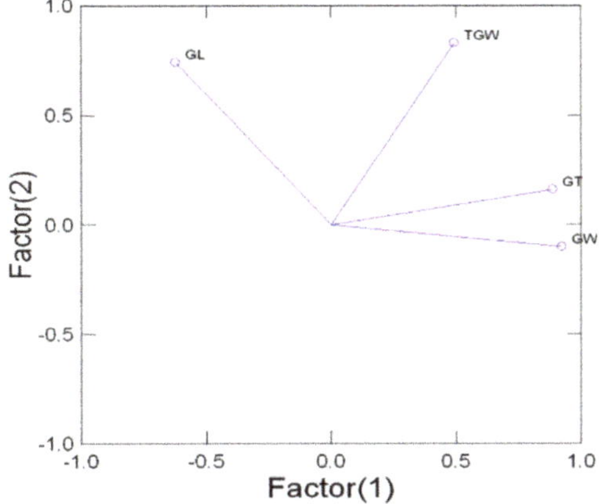

Figure 4. Factor analysis of the four grain size and weight related traits in rice germplasm comprising 204 genotypes; the first two factors represent more than 85% of the total variation. Traits with a positive association have vectors with an acute angle (<90) between their vectors and are closer to one another, while traits with negative associations have an obtuse (>90) angle between their vectors and are found far from one another in the biplot.

2.3. InDel Based Estimation of the Allelic Distribution of Fifteen Genes in the Determination of Grain Size and Weight

A number of genes and their alleles in different combinations have been found to be involved in determining the ultimate size, shape, and weight of rice grains (also detected by previous researchers) [7–20]. An amplification profile of 14 Insertion/Deletion (InDel) markers of nine studied genes related to grain size and weight in 204 rice germplasms is given in Figure 5. A functional insertion–deletion (InDel) marker (sequences of reverse and forward primers are given in Table 8) was used to determine the allelic frequency of the *qPE9~1* gene in the germplasm. The InDel marker produced 270 bp and 350 bp fragments, which corresponded to the A- and B-allele, respectively (Figure 5). The results (Table 3) showed that A- and B-alleles were distributed in the germplasm with frequencies of 0.8922 and 0.1078 (i.e., ~89% and ~11% of the whole germplasm) in 364 and 44 genotypes, respectively. The frequencies of alleles in the genotypes are given in Table 3. The germplasm with an A-allele of this gene was observed to have a lower grain length (7.94 ± 0.920 mm), whereas the B-allele was found to control grain length (Table 4) in rice grains, as suggested by the longer grain length (8.24 ± 0.961 mm) in the germplasm possessing the allele. Entries with lower grain widths (2.95 ± 0.275 mm) retained the A-allele, whereas the B-allele was retained by germplasm with higher values of gain width (3.05 ± 0.209 mm). For the *GW2* gene, another InDel marker (Table 8) was used to determine its allelic distribution in the germplasm. The A-allele (labelled for the fragment at 500 bp) was present in 290 rice genotypes (with 0.7108 allelic frequency), whereas the B-allele (labelled for the fragment at 520 bp) was present in 116 entries, with an allelic frequency of 0.2843 (Table 3). The A-allele was observed in the germplasm having grains with a lower grain width (2.95 ± 0.273 mm), whereas the B-allele was found in the germplasm with more grain width (3.02 ± 0.253 mm). However, the results for the gene associations with any traits were non-significant.

The InDel marker *SLG7*-InDel (Table 8) was used to determine the allelic distribution of *SLG7* in the 204 rice germplasms. Two alleles (A and B, designated for bands at 450 bp and 500 bp, respectively) successfully divided all the germplasms into two groups: the A-allele group (308 entries) and the B-allele group (100 entries), with allelic frequencies estimates of 0.7549 (~75% of the total population) and 0.2451 (~24% of the total population), respectively (Table 3). It was further shown that both the alleles have no effect in regulating grain shape (i.e., GL, GT, and GW) (Table 4). The results depicted in Table 4 show that the gene is significantly ($p \leq 0.05$) associated with TGW. Two alleles of the *GW5* gene, the A- and B-alleles (DNA fragments at 450 bp and 500 bp, respectively), were distinguished by another insertion–deletion (InDel) marker, *GW5*-InDel (Table 8), used to investigate the allelic frequencies of both alleles in the target germplasms of 204 rice lines. Both alleles were found in germplasm with different frequencies (Table 3). The A-allele was present in 294 genotypes, and the B-allele was found in 114 genotypes, with estimated frequencies of 0.7206 and 0.2794, respectively. Notably, the gene was found to have a significant ($p \leq 0.05$) contribution in controlling GW and TGW. Two alleles (the A-allele labelled for the DNA fragment at 650 bp and the B-allele at 750 bp) were observed for the *GS3* gene by using an InDel marker, *GS3*-InDel (Table 8), which distinguished all the germplasms into groups, with A- and B-alleles in 204 germplasms with the frequencies of 69.6% and 30.4%, respectively, in 284 and 124 rice genotypes (Table 3). The gene was also found to have a significant ($p \leq 0.05$) association (Table 4) with GL and GT, as indicated in Table 4, which shows that the A-allele of this gene was found in the germplasm with a less average GL (7.90 ± 0.926 mm) and thicker (2.14 ± 0.166 mm) grains, while the B-allele was present in genotypes with longer grains (8.16 ± 0.917 mm) and a low average GT (2.11 ± 0.159). The *GS7*-InDel marker (Table 8) was used to determine allelic frequencies in the subject population for the *GS7* gene. This InDel marker divided all germplasms into three groups: a group containing A-alleles and B-alleles only and a third one possessing both A- and B-alleles, which were distributed among the germplasms with different frequencies (0.4412, 0.3088, and 0.2451 respectively), present in 90, 63, and 50 entries, respectively (Table 3). It was further observed that the A-allele (DNA band labelled at 200 bp) was present in 56.4% of the total germplasm, whereas the B-allele (at 250 bp) was present in 43% of the total germplasm under study (Figure 5). Interestingly, both the A- and B-alleles showed a significantly higher mean grain length (8.19 ± 0921 and 8.08 ± 0.828 respectively) separately. However, in combination (AB), both alleles showed a lower mean grain length (7.57 ± 0.974). These studies, based on 204 germplasms, also showed that the gene has significant ($p \leq 0.05$) associations with two grain size traits (i.e., GL and GT), indicating its contribution in regulating grain size in rice.

An InDel marker, *GW8*-InDel (Table 8), was applied to distinguish all germplasms into two groups carrying A- and B-alleles, as designated by DNA fragments/bands at 350 bp and 450 bp, respectively (Figure 5). The A-allele was found to be present in 29.9% (122 genotypes) and the B-allele was present in the remaining 62.8% (256 genotypes) of the total germplasms (Table 3). Remarkably, the B-allele had a significantly higher mean grain length (8.41 ± 0.890), a significantly lower grain width (2.92 ± 0.259), and a lower grain thickness (2.09 ± 0.118) compared to the A-allele, which showed a lower mean grain length (7.34 ± 0.578), a significantly higher grain width (3.05 ± 0.203), and a higher thickness (2.27 ± 0.133) in the studied 204 germplasms. Based on the *GW8*-InDel marker loci in the *GW8* gene, the gene was found to be strongly associated ($p \leq 0.01$) with three grain size traits (i.e., GL, GW, and GT) (Table 4). However, no allelic associations were observed in the case of TGW in this study. Likewise, another InDel marker, *GW8*-InDel1A was used for the same germplasm. It also separated the whole germplasm into two groups carrying A- (350 bp) and B-alleles (450 bp), with a frequency of 31.1% (127 genotypes) and 66.4% (271 genotypes), respectively. The results show that the B-allele has a significantly higher average gain length (8.40 ± 0.901) and a lower grain width (2.91 ± 0.260) and grain thickness (2.09 ± 0.118) compared to the A-allele, which showed a lower average grain length (7.42 ± 0.667), a significantly higher grain width (3.06 ± 0.231), and an average grain thickness (2.27 ± 0.169) in the studied genotypes. TGW showed no changes due to these alleles. A third InDel marker, *GW8*-InDel2B, was also used for further studies, as it separated the whole germplasm into A-allele (at 270 bp) with a lower frequency (46%) and the B-allele (at 300 bp) with frequency (51.9%) that was found in 94 and 106 genotypes out of the 204 genotypes (Figure 5). These results show that B-allele has a significantly higher grain length (8.35 ± 0.852), a lower grain width (2.92 ± 0.274), and a lower grain thickness (2.10 ± 0.129) compared to the A-allele, which showed a lower average grain length (7.64 ± 0.965) and a higher average grain width (3.00 ± 0.238) in the studied genotypes. Unlike other InDel markers, *GW8*-InDel2B also distinguished the germplasm into two groups for the thousand grain weight, with the A-allele having a significantly higher TGW (26.0 ± 2.824) and the B-allele with a lower average TGW (25.0 ± 2.611). For all three InDel markers, the same results were observed for GL, GT, and GW, showing a very strong association ($p \leq 0.01$), whereas, in the case of TGW, only one marker (*GW8*-InDel2B) showed a strong and highly significant ($p \leq 0.01$) association, as depicted in Table 4. A-alleles of this gene should be considered while improving the GW, GT, and TGW in rice, whereas the B-alleles should be considered when breeding for longer grains.

Another QTL *GS5* has already been reported to regulate grain size in rice via grain filling and weight [32]. In this study, two InDel markers, *GS5*-InDel1A and *GS5*-InDel1B (Table 8), were applied to determine the contribution of the affective alleles for the *GS5* gene in the germplasm. The InDel1A marker yielded two alleles, A (500 bp) and B (550 bp), with a frequency of 181 (44.3%) and 215 (52.7%), respectively (Table 3). The other InDel marker, *GS5*-InDel2B, also yielded two alleles, A (500 bp) and B (550 bp), with a frequency of 101 (24.5%) and 295 (72.3%), respectively. Both these markers had no observable association with the grain size traits in this study, except the thousand grain weight (TGW). The B-alleles of both InDel markers (i.e., *GS5*-InDel1A and *GS5*-InDel1B) were found to control the heavier grains (Table 4), as indicated by the significantly higher average TGW values (27.0 ± 2.535 and 26.2 ± 2.899, respectively) of the genotypes possessing B-alleles, whereas the germplasm with the other allele (A-allele) showed a lower average TGW for both markers (24.0 ± 3.180 and 23.0 ± 2.729, respectively). Likewise, in the case of the *GS5*-InDel1A makers, the A-allele was found to be present in 90 germplasms with a lower average grain length (7.79 ± 0.921 mm), and B-allele was found in 107 genotypes with a longer grain length (8.16 ± 0.925 mm). For the marker *GS5*-InDel2B, the A-allele was present in only seven genotypes with shorter grains (7.40 ± 0.900 mm) and in the B-allele in the larger portion of germplasm (185 genotypes) with a longer grain length (8.24 ± 0.934 mm), as represented in Table 4. The study further revealed that there is a strong ($p \leq 0.05$) association of the *GS5* gene for both the InDel markers with TGW.

For the *GS2* gene, three InDel markers (i.e., *GS2*-InDel, *GS2*-InDel1A, and *GS2*-InDel2B (Table 8)) were used to determine the different alleles for the *GS2* gene in the germplasm of 204 entries. Two alleles, A and B, were determined for the DNA fragments at 400 bp and 500 bp, respectively, which distinguished the whole germplasm into two groups: one having 16% (65 entries) germplasm of the total (with A-alleles) and other having 83% of the total (339 entries) (with B-alleles). However, no significant associations were detected for these alleles with regards to grain size and weight related traits, suggesting the inefficiency of this marker in distinguishing this gene in the rice germplasm under study. The other InDel marker, *GS2*-InDel1A, distinguished germplasm into two groups: one with A-alleles (designated for the band at 400 bp), comprising 52.7%, and one with B-alleles (at 500 bp), comprising 45.3% of the total germplasm. Genotypes possessing B-alleles showed a significantly higher average grain length (8.44 ± 0.908), a significantly lower grain width (2.84 ± 0.275), and a lower grain thickness (2.09 ± 0.126); whereas genotypes with A-alleles showed a significantly lower average grain length (7.66 ± 0.820), as well as wider and thicker grains compared to the rest of the germplasm. No associations were observed for the TGW marker in the present study. The third marker, *GS2*-InDel2B, also distinguished A- and B-alleles, labelled for the bands at 250 bp and 300 bp, respectively, and present in 24.7% and 72.3% of the total germplasm (Figure 5). This marker separated the genotypes into two groups: one with genotypes possessing A-alleles with a significantly higher average grain length (9.32 ± 0.951), a significantly lower grain width (2.65 ± 0.303), and a lower grain thickness (2.08 ± 0.153), and the other group consisting of genotypes with B-alleles with a significantly lower average grain length (7.81 ± 0.742) and wider and thicker grains with a higher average grain width (3.01 ± 0.201) compared to the rest of the germplasm (Table 3). This gene was found to have a strong and highly significant ($p \leq 0.01$) association with GL, GT, and GW, thereby revealing its importance in regulating all these traits in rice grains (Table 4). *GS2* showed no association with TGW, indicating that this gene is responsible only for regulating the grain size in the studied germplasm. These results showed that, for longer grains, the B-allele should be taken into consideration for the InDel1A marker, and the A-allele should be considered for the InDel2B marker.

Table 3. Fragment lengths on electrophoresis gel, expected frequency estimates, allelic variance, and standard deviation (SD) estimates of each allele of the 14 InDel loci.

Marker	Allele	Band Length (bp)	Genotype	Expected Frequency	Variance	SD
qPE9~1-InDel	A	270	364	0.8922	0.0005	0.0217
	B	350	44	0.1078	0.0005	0.0217
GW2-InDel	A	500	290	0.7108	0.0010	0.0317
	B	520	116	0.2843	0.0010	0.0316
SLG7-InDel	A	450	308	0.7549	0.0009	0.0301
	B	500	100	0.2451	0.0009	0.0301
GW5-InDel	A	450	294	0.7206	0.0010	0.0314
	B	500	114	0.2794	0.0010	0.0314
GS3-InDel	A	650	284	0.6961	0.0010	0.0322
	B	750	124	0.3039	0.0010	0.0322
GS7-InDel	A	200	230	0.5637	0.0009	0.0301
	B	250	176	0.4314	0.0009	0.0300
GW8-InDel	A	350	122	0.2990	0.0010	0.0321
	B	450	256	0.6275	0.0011	0.0339
GW8-InDel1A	A	350	127	0.3113	0.0010	0.0323
	B	450	271	0.6642	0.0011	0.0330
GW8-InDel2B	A	270	188	0.4608	0.0012	0.0349
	B	300	212	0.5196	0.0012	0.0350
GS5-InDel1A	A	500	181	0.4436	0.0012	0.0347
	B	530	215	0.5270	0.0012	0.0349
GS5-InDel2B	A	500	14	0.0343	0.0002	0.0127
	B	550	370	0.9069	0.0004	0.0203
GS2-InDel	A	400	65	0.1593	0.0007	0.0255
	B	500	339	0.8309	0.0007	0.0261
GS2-InDel1A	A	400	215	0.5270	0.0012	0.0349
	B	450	185	0.4534	0.0012	0.0348
GS2-InDel2B	A	250	101	0.2475	0.0009	0.0301
	B	300	295	0.7230	0.0010	0.0312

Table 4. Allelic variations of the grain size and weight regulating genes in 204 rice germplasms, based on InDel marker loci.

Markers	Alleles	Germplasm	Grain Length (mm)		Grain Width (mm)		Grain Thickness (mm)		Thousand Grain Weight (g)	
			Mean ± SD	*p* Value	Mean ± SD	*p* Value	Mean ± SD	*p* Value	Mean ± SD	*p* Value
qPE9~-InDel	A	182	7.94 ± 0.920 [b]	0.32	2.95 ± 0.275 [b]	0.49	2.13 ± 0.167 [a]	0.86	26.0 ± 2.837 [a]	0.32
	B	22	8.24 ± 0.961 [a]		3.06 ± 0.209 [a]		2.12 ± 0.149 [a]		26.5 ± 2.672 [a]	
GW2-InDel	A	145	7.98 ± 0.934 [a]	0.56	2.95 ± 0.273 [b]	0.17	2.13 ± 0.167 [a]	0.47	26.0 ± 2.984 [a]	0.52
	B	58	8.03 ± 0.907 [a]		3.02 ± 0.253 [a]		2.12 ± 0.160 [a]		26.0 ± 2.441 [a]	
SLG7-InDel	A	154	7.97 ± 0.933 [a]	0.89	2.95 ± 0.268 [a]	0.98	2.12 ± 0.160 [a]	0.23	26.0 ±2.857 [a]	0.027 *
	B	50	8.03 ± 0.907 [a]		2.99 ± 0.273 [a]		2.16 ± 0.178 [a]		26.0 ± 2.695 [a]	
GW5-InDel	A	147	7.98 ± 0.894 [a]	0.93	2.95 ± 0.261 [a]	0.047 *	2.13 ± 0.163 [a]	0.29	25.0 ± 2.847 [b]	0.031 *
	B	57	7.93 ± 1.008 [a]		3.00 ± 0.285 [a]		2.13 ± 0.170 [a]		26.0 ± 2.708 [a]	
GS3-InDel	A	142	7.90 ± 0.926 [a]	0.019 *	2.97 ± 0.279 [a]	0.44	2.14 ± 0.166 [a]	0.012 *	26.0 ± 2.729 [a]	0.61
	B	62	8.16 ± 0.917 [a]		2.95 ± 0.244 [a]		2.11 ± 0.159 [a]		25.5 ± 3.054 [a]	
GS7-InDel	A	90	8.19 ± 0.921 [a]	0.022 *	2.95 ± 0.299 [a]	0.56	2.10 ± 0.162 [b]	0.0194 *	26.0 ±2.889 [a]	0.96
	B	63	8.08 ± 0.828 [a]		2.97 ± 0.190 [a]		2.12 ± 0.162 [b]		26.0 ± 2.504 [a]	
	AB	50	7.57 ± 0.974 [b]		2.95 ± 0.293 [a]		2.20 ± 0.159 [a]		25.0 ± 3.116 [b]	
GW8-InDel	A	60	7.34 ± 0.578 [b]	≤0.0001 *	3.05 ± 0.203 [b]	≤0.0001 *	2.27 ± 0.133 [a]	≤0.0001 *	25.0 ± 2.694 [b]	0.06
	B	128	8.41 ± 0.890 [a]		2.92 ± 0.259 [a]		2.08 ± 0.118 [b]		26.0 ± 2.583 [a]	
GW8-InDel1A	A	63	7.42 ± 0.667 [b]	≤0.0001 *	3.06 ± 0.231 [b]	≤0.0001 *	2.27 ± 0.169 [a]	≤0.0001 *	26.0 ± 3.309 [a]	0.0620
	B	135	8.40 ± 0.901 [a]		2.91 ± 0.260 [a]		2.09 ± 0.118 [b]		26.0 ± 2.615 [a]	
GW8-InDel2B	A	94	7.64 ±0.965 [b]	0.0045 *	3.00 ± 0.238 [a]	0.003 *	2.19 ± 0.158 [a]	≤0.0001 *	26.0 ± 2.824 [a]	≤0.0001 *
	B	106	8.35 ± 0.852 [a]		2.92 ± 0.274 [b]		2.10 ± 0.129 [b]		25.0 ± 2.611 [b]	
GS5-Ir-Del1A	A	90	7.79 ± 0.921 [b]	0.12	2.97 ± 0.259 [a]	0.75	2.13 ± 0.165 [a]	0.22	24.0 ± 3.180 [b]	0.016 *
	B	107	8.16 ± 0.925 [a]		2.94 ± 0.272 [a]		2.12 ± 0.150 [a]		27.0 ± 2.535 [a]	
GS5-Ir-Del2B	A	7	7.40 ± 0.900 [b]	0.87	2.78 ± 0.261 [b]	0.33	2.14 ± 0.161 [a]	0.44	23.0 ± 2.729 [b]	0.038 *
	B	185	8.24 ± 0.934 [a]		2.99 ± 0.265 [a]		2.21 ± 0.161 [b]		26.2 ± 2.899 [a]	
GS2-InDel	A	32	7.57 ± 1.035 [a]	0.96	3.03 ± 0.314 [a]	0.99	2.22 ± 0.180 [a]	0.18	27.0 ± 3.258 [a]	0.58
	B	169	7.98 ± 0.900 [a]		2.94 ± 0.261 [b]		2.12 ± 0.161 [b]		25.0 ± 2.729 [b]	
GS2-Ir-Del1A	A	107	7.66 ± 0.820 [b]	≤0.0001 *	3.01 ± 0.246 [a]	0.0005 *	2.19 ± 0.180 [a]	≤0.0001 *	26.0 ± 2.943 [a]	0.89
	B	92	8.44 ± 0.908 [a]		2.84 ± 0.275 [b]		2.09 ± 0.126 [b]		26.0 ± 2.754 [a]	
GS2-Ir-Del2B	A	50	9.32 ± 0.951 [a]	≤0.0001 *	2.65 ± 0.303 [b]	≤0.0001 *	2.08 ± 0.153 [b]	0.0074 *	26.0 ± 3.136 [a]	0.12
	B	147	7.81 ± 0.742 [b]		3.01 ± 0.201 [a]		2.16 ± 0.155 [a]		25.0 ± 2.761 [b]	

Data present the mean values ± standard deviation; ANOVA test for significant at the level of *p* level <0.05 and <0.01. * significant at $p \leq 0.05$. [a], [b] and [c] were ranked by Duncan's test.

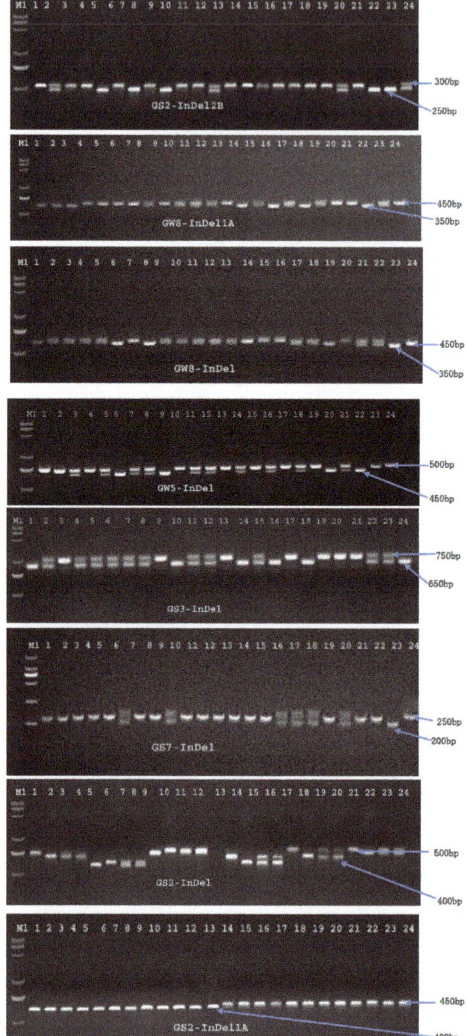

Figure 5. *Cont.*

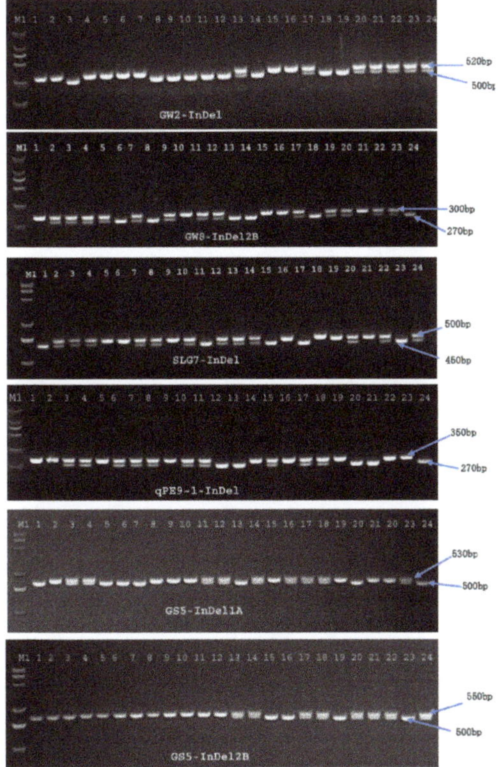

Figure 5. Amplification profile of the 14 Insertion/Deletion (InDel) markers of nine genes related to grain size and weight in 204 rice germplasms. M1 represents a 2000 base pair (bp) DNA ladder, 1–24 represent the rice germplasm. The size of the DNA fragment is depicted on the right side of the gel picture.

2.4. Favorable Alleles of Studied Genes for Gene Pyramiding

In the present study, a total of 38 alleles were identified. Out of these, only 27 alleles were major alleles (genotype frequency ≥0.2). Alleles present in less than 20 genotypes were considered minor alleles. Major alleles were further classified into favorable and non-favorable categories. A total of seven alleles were found to be favorable/beneficial for improving the grain length (>8 mm), having a better TGW, followed by GT and GW, respectively (Table 5). The individual role of the nine studied genes for improving grain size (GL, GW, and GT) and weight (TGW) were estimated, and the combined impact of these favorable alleles was comprehensively analyzed. ANOVA was used to test the difference between the germplasms possessing favorable alleles and non-favorable alleles in the studied germplasms (Table 5). The frequency of the favorable alleles was recorded to be in the range of 24.7% to 66.4%, with the highest for the gene *GW8* (with marker InDel1A), followed by *GW8* (InDel), *GS7* (InDel), *GW8* (InDel2B), *GS2* (InDel1A), *GS3* (InDel), and *GS2* (InDel2B), as depicted in Figure S1. All the favorable alleles, except *GS3* (InDel) and *GW8* (InDel2B), were observed to contribute to the grain length (GL > 8 mm) and thousand grain weight (Table 5). As GL and TGW are positively correlated, these genotypes may be utilized in pyramiding the target genes for longer and heavier grains to simultaneously improve the yield and quality. Average GL of the cumulated FAs were found to be higher compared with the average GL of the cumulated N-FAs depicted in Table 5.

Table 5. Pyramiding of the favorable alleles of the studied genes for Grain Length (GL).

Genes	Markers	Alleles	Grain Length (mm)		Grain Width (mm)		Grain Thickness (mm)		Thousand Grain Weight (g)	
			FA (Mean ± SD)	N-FA (Mean ± SD)	FA (Mean ± SD)	N-FA (Mean ± SD)	FA (Mean ± SD)	N-FA (Mean ± SD)	FA (Mean ± SD)	N-FA (Mean ± SD)
GS3	GS3-InDel	A	-	7.90 ± 0.926 [a]	2.97 ± 0.279 [a]	-	-	-	-	-
		B	8.16 ± 0.917 [a]	-	-	2.95 ± 0.244 [a]	-	-	-	25.5 ± 3.054 [a]
GS7	GS7-InDel	A	8.19 ± 0.921 [a]	-	-	-	-	-	26.0 ± 2.889 [a]	-
		B	8.08 ± 0.828 [a]	-	-	2.97 ± 0.190 [a]	-	2.12 ±0.162 [b]	-	26.0 ± 2.504 [a]
	GW8-InDel	A	-	7.34 ± 0.578 [b]	-	3.05 ± 0.203 [b]	-	2.27 ± 0.133 [a]	-	25.0 ± 2.694 [b]
		B	8.41 ± 0.890 [a]	-	-	-	-	-	26.0 ± 2.583 [a]	-
GW8	GW8-InDel1A	A	-	7.42 ± 0.667 [b]	-	3.06 ± 0.231 [b]	-	2.27 ± 0.169 [a]	-	26.0 ±3.309 [a]
		B	8.40 ± 0.901 [a]	-	-	-	-	-	26.0 ± 2.615 [a]	-
	GW8-InDel2B	A	-	7.64 ± 0.965 [b]	-	3.00 ± 0.238 [a]	-	2.19 ± 0.158 [a]	-	26.0 ±2.824 [a]
		B	8.35 ± 0.852 [a]	-	-	-	-	-	-	-
GS2	GS2-InDel1A	A	-	7.66 ± 0.820 [b]	-	3.01 ± 0.246 [a]	-	2.19 ± 0.180 [a]	-	26.0 ±2.943 [a]
		B	8.44 ± 0.908 [a]	-	-	-	-	-	26.0 ± 2.754 [a]	-
	GS2-InDel2B	A	9.32 ± 0.951 [a]	-	-	-	-	-	26.0 ± 3.136 [a]	-
		B	-	7.81 ± 0.742 [b]	-	3.01 ± 0.201 [a]	-	2.16 ± 0.155 [a]	-	25.0 ±2.761 [b]
	Mean ± SD		8.47 ± 0.39 [a]	7.62 ± 0.22 [b]	2.97 ± 0.279 [a]	3.01 ± 0.030 [a]	-	2.20 ± 0.060	26.0 ± 0.00 [a]	25.6 ±0.51 [b]

FA and N-FA symbolize favorable and non-favorable alleles, respectively; a and b represent significantly (p value < 0.01) different mean values. Sign (-) symbolizes no FA/N-FA allele under this category

2.5. InDel Polymorphism and Assessment of the Genetic Relationship among the 204 Rice Germplasms

To study the genetic relatedness among the studied germplasm consisting of 204 rice genotypes, fourteen (14) developed Insertion/Deletion (InDel) markers from the nine grain shape, size, and weight related genes were used. Most of the InDel markers amplified 2 bands / alleles, as depicted in Figure 5. Four markers (i.e., *qPE9~1*-InDel, *SLG7*-InDel, *GW5*-InDel, and *GS3*-InDel) showed two alleles per locus, whereas the rest of the markers showed three allele types for their respective genes (Table 5). The major allele frequency ranged from 0.5196 to 0.9069, with an average value of 0.6903. The studied markers also revealed higher gene diversity (D) ranging from 0.1730–0.5246, with an average of 0.4073, indicating that the genotypes in the studied germplasm possess a considerable range of gene variations that can be further exploited. This statement is further strengthened by the higher values of the polymorphism information content (PIC) values, which have an average value of 0.3310 (0.1653–0.4359). The maximum PIC value was observed for the marker *GW8*-InDel (0.4359), followed by *GS5*-InDel1A (0.4145), *GW8*-InDel2B (0.4023), and *GS2*-InDel1A (0.4018), as mentioned in Table 6. The Average value of the observed heterozygosity was very low (0.0193), indicating that the rice germplasm under study was mostly homozygous and had uniform lines. Among all the studied markers, *GS7*-InDel (for *GS7* gene) showed considerable heterozygosity in its results compared to the other markers. The study showed that InDel markers (i.e., *GW8*-InDel, *GW8*-InDel2B, *GS5*-InDel1A, and *GS2*-InDel1A) of the *GW8*, *GS5*, and *GS2* genes are highly informative regarding these traits and can be used to understand the genetic variations in germplasm, whereas the rest of the markers are moderate to slightly informative for these traits in this study (Table 6).

Table 6. Estimation of the number of alleles per locus, the major allele frequency, gene diversity (D), expected heterozygosity, and the polymorphism information content (PIC) values in the 204 rice germplasms.

Marker	A	p_{ma}	D	H_{Exp}	PIC
qPE9~1-InDel	2	0.8922	0.1924	0.0000	0.1739
GW2-InDel	3	0.7108	0.4139	0.0000	0.3322
SLG7-InDel	2	0.7549	0.3700	0.0000	0.3016
GW5-InDel	2	0.7206	0.4027	0.0000	0.3216
GS3-InDel	2	0.6961	0.4231	0.0000	0.3336
GS7-InDel	3	0.5637	0.4961	0.2451	0.3778
GW8-InDel	3	0.6275	0.5115	0.0000	0.4359
GW8-InDel1A	3	0.6642	0.4613	0.0049	0.3752
GW8-InDel2B	3	0.5196	0.5173	0.0000	0.4023
GS5-InDel1A	3	0.5270	0.5246	0.0049	0.4145
GS5-InDel2B	3	0.9069	0.1730	0.0000	0.1653
GS2-InDel	3	0.8309	0.2842	0.0049	0.2490
GS2-InDel1A	3	0.5270	0.5163	0.0049	0.4018
GS2-InDel2B	3	0.7230	0.4151	0.0049	0.3500
Mean	3	0.6903	0.4073	0.0193	0.3310

A = alleles per locus; p_{ma} = major allele frequency; D = gene diversity; H_{Exp} = expected heterozygosity; PIC = polymorphism information content.

In order to understand genetic variations and distinctiveness in the selected germplasm, an un-weighted neighbor joining tree was constructed (bootstraps value of 10,000) based on a dissimilarity index calculated from allelic data in the bit data format (0 and 1 indicating the absence and presence of alleles, respectively) using the Otsuka–Ochiai coefficient [47,48]. The Formula for calculating the coefficients was

$$dij = 1 - \frac{a}{\sqrt{(a+b)(a+c)}}.$$

Based on cluster analysis, 204 germplasms were classified into six distinct clusters, encircled separately, as illustrated by the UPGMA tree in Figure 6. The three major and three minor clusters showed distinct grain characteristics. Major clusters were designated as Cluster I, Cluster II, and Cluster III. Cluster I consisted of 80 genotypes that can be further divided into two sub-clusters, IA and IB. Sub-cluster IA consists of 35 genotypes, having the highest average grain length (GL) of 9.58 mm. The average grain width (GW) and average grain thickness (GT) were 2.70 mm and 2.09 mm, respectively, whereas the thousand grain weight (TGW) of this sub-cluster was 27.07 g. Likewise, sub-cluster IB consists of 43 genotypes, with an average GL of 8.78 mm, an average GW of 2.76, an average GT of 2.07, and an average TGW of 25.52 g. The germplasm in cluster I contains extra-long grain rice genotypes that can be utilized for breeding long grain rice varieties. The major Cluster II, consisting of 35 genotypes, comprises the germplasms with an average GL, GW, GT, and TGW of 8.05 mm, 2.95 mm, 2.14 mm, and 25.36 g, respectively. This cluster contains germplasms with long and bolder grains that can be utilized in breeding programs, where grain length along with grain width and thickness are under consideration. Major Cluster III included 74 germplasms, with a mean GL of 7.38, a mean GW of 3.09 mm, a mean GT of 2.29 mm, and a mean TGW of 25.81 g, illustrating that this group consisted of rice germplasms with shorter and bolder grains. This germplasm can be utilized for short and thicker grains with a higher grain weight. The other three minor clusters contained 15 germplasms, as depicted in Figure 6.

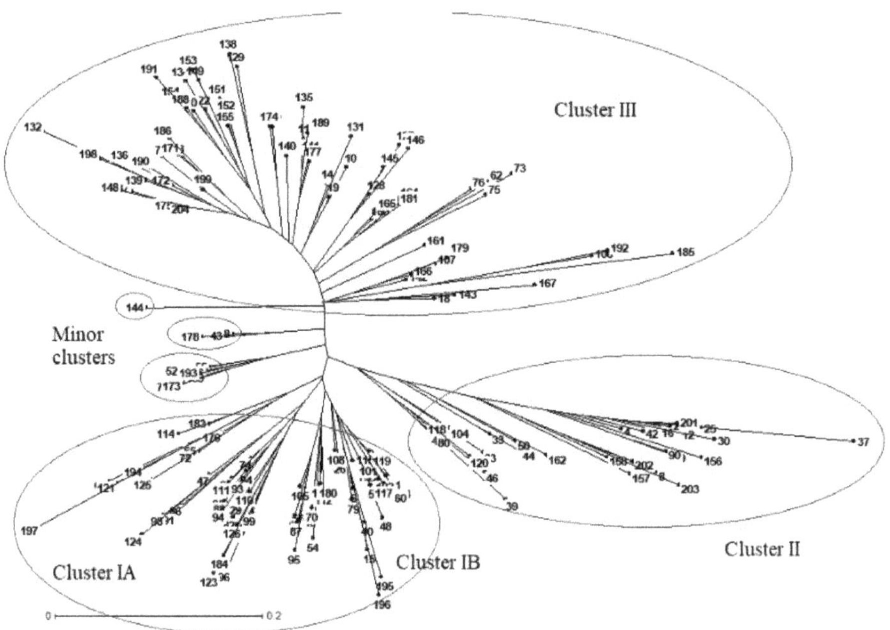

Figure 6. InDel marker based genetic relationship among the 204 rice germplasm entries, estimated using an un-weighted neighbor joining tree (bootstraps value of 10,000).

2.6. Estimation of Population Genetics for Grain Size and Weight Based on Fourteen InDel Markers

To estimate the genetic relationship among the populations of the 204 rice genotypes in the germplasm, the total rice germplasm was divided into four sub-populations according to grain type. Extra-long grains with an average grain length (GL) of more than 9.5 mm comprising 23 rice germplasms, long grains with average AGL in the range of 8.5–9.5 mm comprising 48 rice germplasms, medium grains comprising 78 rice germplasms with an average GL between 7.5–8.5 mm, and a short grain type containing 55 germplasms with an average GL below 7.5 mm. Principal Coordinate Analysis (PCoA) was used to establish the genetic relationship of all the germplasms based on fourteen

InDel markers, as depicted in Figure 7. A scatter plot was constructed with two first coordinates that collectively explain 57.3% of the total genetic variation and separate the germplasm into four types of distinct clusters according to their grain types. These results are also in agreement with the results depicted in UPGMA and the dendrogram.

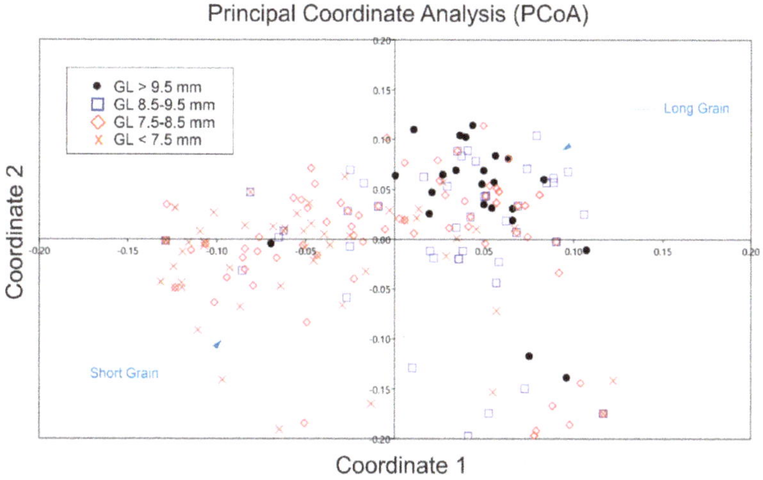

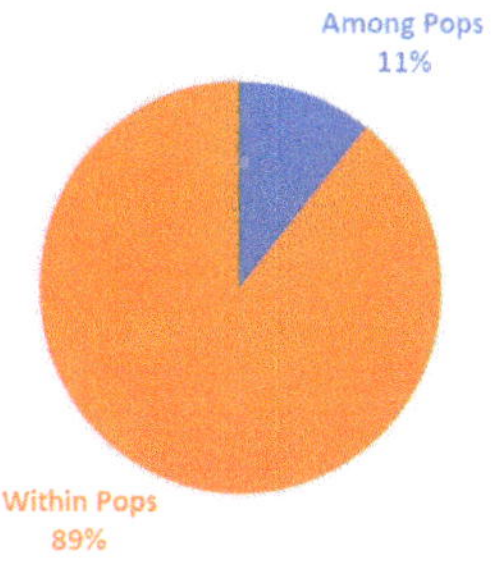

Principal Coordinate Analysis (PCoA)

Figure 7. Principal Coordinate Analysis (PCoA) of 204 rice germplasms based on 14 InDel markers. A 2D scatter plot distributed the 204 germplasms based on InDel markers into four distinct groups according to the average grain length. Extra-long grains (with AGL > 9.5 mm) are shifted towards the right side, whereas short grain (AGL < 7.5 mm) germplasms shifted towards the left side. Medium grain germplasm is dispersed over the central region of the biplot.

These four sub-populations were further investigated based on the InDel marker binary data in order to separate the total molecular variance into variance within and between sub-populations. The study demonstrated that the total molecular variance was partitioned into two, of which the maximum variance (89%) was observed within the population, and the minimum was found between the populations (11%), as demonstrated in Figure 8. The results extracted from AMOVA analysis (Table 7) clearly showed that there was a maximum (Φ_{PT} = 0.449) and significant ($p \leq 0.001$) genetic between the short grain and extra-long grain germplasms / sub-populations, while the lowest (Φ_{PT} = 0.035) genetic differentiation was observed between the long grain and medium grain sub-populations.

Figure 8. Percentages of molecular variation within and among populations.

Table 7. Pair-wise population Φ_{PT} value estimates among the four sub-populations of 204 rice germplasms.

	Extra-Long Grain	Long Grain	Medium Grain
Long grain	0.097		
Medium grain	0.212	0.035	
Short grain	0.449	0.264	0.129

Φ_{PT} values given below the diagonal; calculated for 10,000 permutations.

3. Discussion

Before this study, the size and weight related traits of rice grains, including grain length, grain width, grain thickness, and thousand grain weight, had not been explored at the same time using a large and diverse germplasm group with the help of InDel markers. New Generation Sequencing (NGS) tools have yielded advanced, cheaper, and more efficient methods for developing such markers. This investigatory research efficaciously shows the capable utilization of deletion / insertion variations (DIV) that naturally accrue in the rice genome. InDel polymorphisms are the second most abundant (after SNPs) forms of genetic variations in animals and plants, with great diversity. Moreover, previous studies only showed the contributions of the reported genes in regulating these grain traits separately [11,15,16,19,20,59,60]. However, recent studies have shown the modes of InDel based allelic contributions in the expression of grain size, as well as weight related traits, the allelic combination of all the InDel marker loci involved in the final texture of the rice grains, and the potential of each loci in regulating these traits in rice. This research can assist in selecting or deselecting genes for rapid breeding strategies. The correlation coefficient values suggest that the thousand grain weight (TGW) was positively influenced by the other studied grain size traits but with a different aptitude. The TGW can be improved by using all three studied grain traits—most importantly GT, which contributes 49%, followed by GW and GL, which contribute 37.4% and 24.9% of the final grain weight, respectively. These findings are in agreement with those of previous studies [19,54,61,62], which showed that grain weight is significantly correlated with grain size. The present study further showed that the thickness of rice grains was most strongly correlated with grain weight, followed by width, whereas the length of the grains was the least associated, suggesting that selecting for grain thickness is more fruitful for heavier grains (Table 2, Figure 4).

In the present study, two hundred and four rice genotypes were used to investigate alleles in nine different genes that regulate size and weight in rice grains, using InDel markers. In the past few years, PCR based InDel markers have gained popularity in variation studies because of their reproducibility, easy to use nature, and co-dominant inheritance [40]. Dendrograms were used to separate the germplasms according to their grain size and weight and divide the InDel marker data into distinguishable clusters that could be used for breeding preferential grain appearances and weights. Results of the genetic diversity (D) analysis and InDel based polymorphism information content (PIC) values (Table 6) indicate that the InDel markers (*GW8*-InDel, *GS5*-InDel1A, *GW8*-InDel2B, *GS2*-InDel1A) are highly informative (D ≥ 0.5; PIC ≥ 0.4) for the studied traits, whereas the rest of the markers were found to be moderately (D 0.3–0.5; PIC 0.2–0.4) to slightly/less (D ≤ 0.3; PIC ≤ 0.2) informative. These InDel markers show potential to be efficiently used to study the genetic variations (DIVs) in rice germplasm. Only two InDel markers showed very low values (gene diversity ≤0.3; PIC ≤ 0.2) for D and PIC, indicating that the deployed markers were fairly informative [20]. Furthermore, this investigatory research also showed the capability of InDel markers to distinguish a diverse rice germplasm into distinctive groups (Figure 6) with different combinations of grain lengths, widths, thicknesses, and weights that could be used to breed desirable genotypes with better potential for higher market value rice grains and heavier grains for a better yield.

This study successfully identified 25 InDel marker derived loci highly associated ($p \leq 0.05$) with grain size and weight in rice (Table 4). A total of 38 alleles were identified, out of which 27 alleles were major and were found in more than 20 genotypes. In the case of GL, five markers (*GW8*-InDel, *GW8*-InDel1A, *GW8*-InDel2B, *GS2*-InDel1A, and *GS2*-InDel2B) corresponding to two genes (*GW8* and *GS2*) were found to have a highly significant association with GL at $p \leq 0.01$. Similarly, two markers (*GS3*-InDel and *GS5*-InDel1A) corresponding to two genes (*GS3* and *GS5*) were found to have a significant association with GL at $p \leq 0.05$. In the case of GT, five markers (*GW8*-InDel, *GW8*-InDel1A, *GW8*-InDel2B, *GS2*-InDel1A, and *GS2*-InDel2B), corresponding to two genes (*GW8* and *GS2*), were found to have a significant association with GT (at $p \leq 0.01$). Similarly, two markers (*GS3*-InDel and *GS7*-InDel), corresponding to genes *GS3* and *GS7*, were found to have a significant association with GT (at $p \leq 0.05$).

For GW, five markers (*GW8*-InDel, *GW8*-InDel1A, *GW8*-InDel2B, *GS2*-InDel1A, and *GS2*-InDel2B), corresponding to genes *GW8* and *GS2*, respectively, were found to have a significant association with GW at $p \leq 0.01$. Similarly, one marker (*GW5*-InDel), corresponding to gene *GW5*, was found to have a significant association with GW at $p \leq 0.1$. For TGW, one marker (*GW8*-InDel2B), corresponding to gene *GW8* was found to have a significant association with GW at $p \leq 0.01$. Similarly, three markers (*GW5*-InDel, *GS5*-InDel1A, and *GS5*-InDel2B), corresponding to two genes (*GW8* and *GS5*) were found to have a significant association with GW at $p \leq 0.05$.

The *SLG7* gene is known to regulate the grain size in rice via increased cell division, longitudinally resulting in longer grains [21]. In the present study, the *SLG7*-InDel marker showed a significant ($p \leq 0.05$) association with the thousand grain weight (TGW), which is also in agreement with the results of other researchers. This gene encodes the TONNEAU1-recruiting motif protein, which was found by many researchers [37] to be responsible for grain appearance by altering cell divisions, thus having significant effects in regulating grain weight, as well. Notably, GW5 gene was found to have a significant contribution in controlling GW and TGW ($p \leq 0.05$), also revealed by previous studies [23–26], which showed that this gene encodes a calmodulin-binding protein and *GW5* physically interacts with calmodulin AsCaM1-1, which is responsible for grain width in rice. Recent studies have also demonstrated that this gene is responsible for regulating TGW (a significant correlation with $p \leq 0.05$), as TGW is directly and highly associated with GW (Table 4), thereby confirming its utilization in grain yield improving objectives in rice breeding programs. Previous studies also identified this gene for controlling seed width and weight in rice [23,24,63].

Previous studies showed that *GS3* was among the first reported genes to have minor effects on grain thickness and width. The domains on its encoded protein have been reported to regulate cell divisions in the upper epidermis of the glume inside the rice seed, causing minor effects on cell size [29]. In the present study, the *GS3*-InDel marker was found to be significantly associated ($p \leq 0.05$) with grain length and grain thickness, which is consistent with previous reports [20,28,31]. Another InDel marker for the *GS7* gene (*GS7*-InDel) was also found to be significantly ($p \leq 0.05$) related with grain length and thickness. Our studies showed that the investigated alleles for both genes *GS3* and *GS7* affectively regulated grain length and thickness in the rice (Table 3). Previous studies [20,30] also reported that the germplasm carrying different alleles of the *GS3* gene with different allele combinations of *GS7* produced different grain lengths and thicknesses. Shao et al. [30] also reported that *GS7* is a strong QTL known to regulate grain size and controls grain length, roundness (thickness), and area (size) in rice. Ngangkham et al. [20] also found this gene to be associated with GL and GT, thereby playing a significant role in regulating grain size. For *GW2* gene, the results for gene associations with any trait were non-significant, but in previous studies, the gene was found to control the grain width in rice grains [12]. Ngangkham et al. [20] also found no association of this gene with any of these traits using STS (Sequence-Tagged Sites) markers, thus emphasizing the ineffectuality of the markers used. This might be due to inter- and/or intra-allelic interactions that may be subjected to further studies.

Among all the studied genes for grain size and weight, *GW8* was detected to represent a highly significant ($p \leq 0.01$) association with all the grain size related traits, thereby suggesting its great importance in regulating grain size in rice. The *GW8* bearing genotypes were reported to have a higher grain length and grain length-width ratio [20]. The scanning results of the electron microscopy analysis of the lemma in *GW8*.1 carrying NILs showed that the inner epidermal cell length was higher than the lines without this gene, indicating that *GW8*.1 might be responsible for regulating cell elongation [64]. *GW8* (OsSPL16) encodes a protein that is positively associated with cell proliferation [31]. Its higher expression promoted cell division and grain filling, consequentially increasing grain width and yield in rice. Another study also suggested that *GW8* suppresses the expression of the GW7 gene and plays a significant role in controlling grain size [21]. In the present study, *GW8* was determined to regulate grain length, width, and thickness. All three InDel markers successfully distinguished the germplasm into two alleles: the A-allele, which is responsible for shorter, thicker, and wider grains, and the B-allele, which carries genotypes possessing longer but narrower grains. This is due to the fact that grain length (GL) is negatively correlated with grain width (GW) and grain thickness (GT), as depicted by correlation analysis in Table 2. Based on these results, the B-allele carrying germplasm may be selected to breed longer grains, and the A-alleles may be screened for broader, shorter, and thicker grains. However, all three traits (i.e., GL, GT, and GW) contributed to the thousand grain weight (as suggested by the positive correlation between GL, GT, and GW with thousand grain weight), assuming that both alleles contribute to an increased yield. Two markers (i.e., *GW8*-InDel and *GW8*-InDel1A) were shown to have highly significant ($p \leq 0.0001$) associations with GL, GT, and GW, thus indicating the ample potential of InDel markers in variation studies and genome-wide association studies. The third marker (i.e., *GW8*-InDel2B) for the *GW8* gene was also identified to have a highly significant ($p \leq 0.01$) association with GL and GW (Table 5). However, unlike the other two InDel markers for the *GW8* gene, *GW8*-InDel2B also showed a highly significant ($p \leq 0.0001$) relationship with GT and the thousand grain weight (Table 4), indicating its potential to be used for all four studied traits to improve the grain size and grain weight in rice.

Two InDel based markers were used for the *GS5* gene, and both of these markers showed a significant ($p \leq 0.05$) association with only the thousand grain weight. These findings are partially inconsistent with other studies [32,65], which suggested that the *GS5* gene is associated with grain width and grain weight in rice. Previously, Lee et al. [63] attained three types of alleles by applying the markers generated from the promoter-region of the *GS5* gene, thus demonstrating the relatedness of this gene with grain weight. However, in another study, this gene was reported to participate in the regulation of grain length and grain width [63]. This might be due to higher genetic and/or allelic interactions with other genes/alleles that must be studied more comprehensively. This gene has been reported to have significant importance in regulating grain yield, as concluded by Li et al. [32], who showed that this gene encoded proteins—i.e., the putative serine carboxypeptidase executes its function as a positive regulator of a subset of the transition genes (G1-to-S) of the cell cycle, thereby causing increased cell divisions and resulting in enhanced grain filling and grain weight.

The Present study further explored the previously reported gene *GS2* to be highly associated ($p \leq 0.001$) with all three-grain size related traits. Out of the three markers, two markers for this gene (including *GS2*-InDel1A and *GS2*-InDel2B) showed the potential for GL, GT, and GW, whereas the marker *GS2*-InDel showed no association with any trait (Table 4). For both markers 1A and 1B for this gene, the germplasm was separated into two groups carrying A- and B-alleles with different grain size traits. In the case of InDel1A, the A-allele was associated ($p \leq 0.001$) with a shorter grain length (7.66 ± 0.820 mm) with thicker and wider grains, whereas its B-allele had a germplasm with a longer (8.44 ± 0.908 mm) grain length and narrower and slander grains (Table 5). Conversely, for the other marker, *GS2*-InDel2B, the A-allele was detected to relate to the germplasm with the longest grain lengths (9.32 ± 0.951 mm) and narrowest grains (the GW average is 2.65 ± 0.303 mm) in the whole germplasm, whereas the genotypes carrying the B-allele possessed shorter grains (7.81 ± 0.742 mm) and wider grains (3.01 ± 201 mm), as depicted in Table 4. This finding suggests that these InDel markers can further be investigated to breed for >9 mm grain lengths. Previous studies also showed that this gene directly controls two important grain size related traits, including grain width and grain length in rice. Researchers showed that its overexpression increased cell enlargement and enhanced cell division in the grain, thus producing longer and wider rice grains [33–36].

This study further investigated the favorable alleles in the studied germplasm to improve the grain length (>8 mm) with heavier grains. Identifying the beneficial alleles of the target traits is one of the most important prerequisites to improve modern cultivars via introgressions of favorable alleles from a vast gene pool using marker assisted selection approaches. This investigatory research discovered seven favorable alleles for grain length that can be utilized to improve grain size, while keeping in mind the recent criterion for longer grains with improved grain sizes and weights.

This research explored 7 genes and 11 InDel marker associations with grain size and weight related traits in rice. The present study further showed that InDel markers may be used efficiently in research investigations related to genetic variations, genome-wide association studies, germplasm genetic characterization, gene mapping, and other studies to further develop the ease and efficiency of breeding procedures and create more desirable varieties to cope with climate change and food security risks.

4. Materials and Methods

4.1. Rice Material Collection and Phenotyping

The plant material of this experiment included 204 rice germplasms, which included lines with a wide range of grain size related traits. Plant material was received from the Key Laboratory of Crop Genetics and Breeding section, Yangzhou University, Jiangsu province, P.R. China, comprising rice lines with higher yields. The field experiment was carried out in the experimental fields of the research capacity in Yangzhou University (E, N) during the normal rice season (from May to November) in 2018. Sowing of the material was carried out in the nursery for the grain cell, and then the 30 day old nursery was transplanted into well prepared fields following standard agronomic practices in order to assure full expression of the traits. Plant to Plant distance was maintained at 10 cm, whereas row to row distance was maintained at 15 cm. For grain trait phenotyping, data were collected during the stage of plant maturity from the middle plants in the central row of each entry. Grain size related traits (i.e., grain length (GL), grain width (GW), and grain thickness (GT)) were measured in (millimeters) as the average from ten completely mature and filled grains using digital Vernier calipers. Likewise, data for grain weight related traits (i.e., thousand grain weight (TGW)) were measured for 250 carefully counted, filled, and fully mature rice grains using an electronic digital weighing balance, and the values were multiplied by a factor of 4 to obtain the 1000 grain weight.

4.2. DNA Extraction and PCR

DNA was extracted from fresh rice leaves according to the method described in [66], with minor modifications. PCR (polymerase chain reaction) was conducted, and the products were separated by agarose gel electrophoresis [67]. A set of newly developed insertion/deletion (InDel) markers was selected as the primary markers, as found in the open rice genome sequence library (http: //www.ncbi.nlm.nih.gov). Amplified DNA products were analyzed on 3% agarose gels stained with ethidium bromide and photographed with a UVP system.

4.3. Marker Genotyping

The germplasm comprising 204 rice germplasms was mined for the presence of 9 grain size and weight related genes (i.e., *qPE9~1*, *GW2*, *SLG7*, *GW5*, *GS3*, *GS7*, *GW8*, *GS5*, and *GS2*) using the 14 InDel markers given in Table 8, along with detailed information of the primer pairs for each marker. All the markers were scored visually as 1 for their presence and 0 for their absence.

Table 8. Detailed information for the 14 molecular markers for the 9 grain size and weight related genes.

S.N.	Genes	Markers	Forward (5′ to 3′)	Reverse (5′ to 3′)	Tm(°C)	Types
1	*qPE9~1*	*qPE9~1*-InDel	AGTGGTGCCTATAACTCTGC	AGCAAAGAGGACGTCATACT	55	InDel
2	*GW2*	*GW2*-InDel	GCAATGCAAAAGCATATGGC	AAAGCCAAAGATGCACACAG	53	InDel
3	*SLG7*	*SLG7*-InDel	TGGTTTCGATTAGGTTCCTCT	AAAACGCCGTTTAGCTATCC	52	InDel
4	*GW5*	*GW5*-InDel	GAACTACATGTCCAACACGC	CTCCACCACCACCACCTC	58	InDel
5	*GS3*	*GS3*-InDel	ATGACCACGTCGATCATCAA	ACTCCACCTGCAGATTTCTT	53	InDel
6	*GS7*	*GS7*-InDel	TGGTCAAATCATGGGCTAAT	TATTATTGTGCCTGCGATCC	51	InDel
7		*GW8*-InDel	AAAAGAGACAGCCACGGAAT	TCTTGAGATCCCACTCCATG	54	InDel
8	*GW8*	*GW8*-InDel1A	AAAAGAGACAGCCACGGAAT	TCTTGAGATCCCACTCCATG	55	InDel
9		*GW8*-InDel2B	TTTCAGTGTCCTCCTGTCTG	ACCACTAAACCAGGTGCTAC	56	InDel
10	*GS5*	*GS5*-InDel1A	TGACGCCGTTGACTTTTTGA	GAATCCGGCGTTGATTTCGA	55	InDel
11		*GS5*-InDel2B	AGGTGTTGTTCGAAACTCACG	TGAAAATTTGGATATTCGTGGCA	54	InDel
12		*GS2*-InDel	CCCACCGCATGATACATCTA	ATGTGGGAATTTCTAGCCCC	55	InDel
13	*GS2*	*GS2*-InDel1A	GCCGCGGTCTTTAGTAATGG	CCTTCTCGTGTCGGGCTC	59	InDel
14		*GS2*-InDel2B	AAATTGCAGCCGACCGTAG	AAATTGTGACGAGTCCCAGC	55	InDel

4.4. Statistical Analyses

Descriptive statistics and factorial analyses for the trait association of GL, GT, GW, and TGW were conducted using the statistical package SYSTAT, version 13.1. Principal Component Analysis (PCA) for testimation of the phenotypic variability within the germplasm was performed using Minitab software, version 18. Correlation coefficients among the grain size and weight related traits were determined using the computer-based software, IBM-SPSS, version 25.

For allele scoring, the genetic variations, expected heterozygosity, and polymorphism information content (PIC) values of the 14 markers were calculated using binary data in the POWER MARKER software, version 3.25 [68]. The presence of alleles was scored as 1, whereas the absence of alleles was scored as 0 to generate a binary matrix for each genotype. An un-weighted neighbour joining UPGMA tree was built based on the marker data using the DARwin 6 software [69]. Another UPGMA dendrogram (to categorize different groups based on phenotypic data) was constructed using the PAST version 3.25 software. A scatter plot was constructed using the PAST version 3.25 software to disperse the germplasm further into the sub-populations based on the InDel marker data [63,70]. The Principal Coordinate Analysis (PCoA) analysis and Analysis of Molecular Variance (AMOVA) were conducted from the binary data of the InDel markers using GenAlEx version 6.502. The binary data of the genetic markers were constructed according to grain size, which was used to separate the total molecular variance between and within the groups. Φ_{PT} was also calculated using GenAlEx v. 6.502.

5. Conclusions

Collectively, it can be concluded that allelic variations of the nine genes are extensively distributed in the studied germplasm. Further, we observed that most of these alleles are significantly associated with variations in one or more of the studied traits related to grain size and grain weight in rice. The results also suggest that several genotypes have similar grain size characteristics and share particular allele and/or allele combinations of the nine key genes examined in this study. The examinations of allelic contributions from different genes in regulating grain size and weight related traits undertaken in the study will further strengthen our understanding of the complex mechanisms involved in rice grain appearance and the grain weight to be utilized in rice breeding programs. The identified genes and their linked InDel markers could be highly informative in pyramided breeding strategies and selecting parental lines for developing rice varieties according to consumer needs.

Supplementary Materials: Supplementary materials can be found at http://www.mdpi.com/1422-0067/20/19/4824/s1.

Author Contributions: S.G., H.G. and Z.H. analyzed genotype data for InDel marker development/ Estimation; Z.H. and R.A.R.K. analyzed all statistical analyses; H.G., J.M. and T.W. conducted phenotype experiment; G.L., S.G., Z.H., S.U. and I.A. helped conceive the basic idea, gave suggestions, G.L., Z.H. and S.G. arranged the outline of the content in the manuscript and prepared a draft article; corrected the entire article, and improved on the prospects in breeding programs. All authors read and approved the research article.

Acknowledgments: This work was financially supported by grants from the National Natural Science Foundation of China (31771350, 31471458) and the Key Science and Technology Support Program of Jiangsu Province (BE2018336) the Priority Academic Program Development of Jiangsu Higher Education Institutions (PAPD).

Conflicts of Interest: The authors declare no conflict of interest.

References

1. Ngangkham, U.; Parida, S.K.; De, S.; Kumar, K.A.R.; Singh, A.K.; Singh, N.K.; Mohapatra, T. Genic markers for wild abortive (WA) cytoplasm based male sterility and its fertility restoration in rice. *Mol. Breed.* **2010**, *26*, 275–292. [CrossRef]

2. Food, U. Agriculture Organization Corporate Statistical Database [Faostat] (2017). Production/Crops, Persimmons, Food and Agriculture Organization of the United Nations: Division of Statistics. Available online: http://faostat3.fao.org/browse/Q/QC/E (accessed on 19 April 2016).

3. Thornton, P. The emergence of agriculture: Bruce D. Smith. Scientific American Library, distributed by W. H. Freeman & Co. Ltd, New York and Oxford, 1995. 231 pp. ISBN 0 7167 5055 4. *Agric. Sci.* **1996**, *51*, 496–497.

4. Redona, E.; Mackill, D. Quantitative trait locus analysis for rice panicle and grain characteristics. *Theor. Appl. Genet.* **1998**, *96*, 957–963. [CrossRef]

5. McKenzie, K.; Rutger, J. Genetic Analysis of Amylose Content, Alkali Spreading Score, and Grain Dimensions in Rice 1. *Crop Sci.* **1983**, *23*, 306–313. [CrossRef]

6. Melchinger, A.; Messmer, M.; Lee, M.; Woodman, W.; Lamkey, K. Diversity and relationships among US maize inbreds revealed by restriction fragment length polymorphisms. *Crop Sci.* **1991**, *31*, 669–678. [CrossRef]

7. Lin, H.; Qian, H.; Zhuang, J.; Lu, J.; Min, S.; Xiong, Z.; Huang, N.; Zheng, K. RFLP mapping of QTLs for yield and related characters in rice (*Oryza sativa* L.). *Theor. Appl. Genet.* **1996**, *92*, 920–927. [CrossRef]

8. Hua, J.; Xing, Y.; Xu, C.; Sun, X.; Yu, S.; Zhang, Q. Genetic dissection of an elite rice hybrid revealed that heterozygotes are not always advantageous for performance. *Genetics* **2002**, *162*, 1885–1895. [PubMed]

9. Aluko, G.; Martinez, C.; Tohme, J.; Castano, C.; Bergman, C.; Oard, J. QTL mapping of grain quality traits from the interspecific cross Oryza sativa× O. glaberrima. *Theor. Appl. Genet.* **2004**, *109*, 630–639. [CrossRef] [PubMed]

10. Agrama, H.; Eizenga, G.; Yan, W. Association mapping of yield and its components in rice cultivars. *Mol. Breed.* **2007**, *19*, 341–356. [CrossRef]

11. Bai, X.; Luo, L.; Yan, W.; Kovi, M.R.; Zhan, W.; Xing, Y. Genetic dissection of rice grain shape using a recombinant inbred line population derived from two contrasting parents and fine mapping a pleiotropic quantitative trait locus qGL7. *BMC Genet.* **2010**, *11*, 16. [CrossRef]

12. Song, X.-J.; Huang, W.; Shi, M.; Zhu, M.-Z.; Lin, H.-X. A QTL for rice grain width and weight encodes a previously unknown RING-type E3 ubiquitin ligase. *Nat. Genet.* **2007**, *39*, 623–630. [CrossRef] [PubMed]

13. Li, J.; Xiao, J.; Grandillo, S.; Jiang, L.; Wan, Y.; Deng, Q.; Yuan, L.; McCouch, S.R. QTL detection for rice grain quality traits using an interspecific backcross population derived from cultivated Asian (*O. sativa* L.) and African (*O. glaberrima* S.) rice. *Genome* **2004**, *47*, 697–704. [CrossRef] [PubMed]

14. Wang, L.; Wang, A.; Huang, X.; Zhao, Q.; Dong, G.; Qian, Q.; Sang, T.; Han, B. Mapping 49 quantitative trait loci at high resolution through sequencing-based genotyping of rice recombinant inbred lines. *Theor. Appl. Genet.* **2011**, *122*, 327–340. [CrossRef] [PubMed]

15. Thi, T.G.T.; Dang, X.; Liu, Q.; Zhao, K.; Wang, H.; Hong, D. Association analysis of rice grain traits with SSR markers. *Chin. J. Rice Sci.* **2014**, *28*, 243–257.

16. Dang, X.; Thi, T.G.T.; Edzesi, W.M.; Liang, L.; Liu, Q.; Liu, E.; Wang, Y.; Qiang, S.; Liu, L.; Hong, D. Population genetic structure of Oryza sativa in East and Southeast Asia and the discovery of elite alleles for grain traits. *Sci. Rep.* **2015**, *5*, 11254. [CrossRef] [PubMed]

17. Huang, X.; Qian, Q.; Liu, Z.; Sun, H.; He, S.; Luo, D.; Xia, G.; Chu, C.; Li, J.; Fu, X. Natural variation at the DEP1 locus enhances grain yield in rice. *Nat. Genet.* **2009**, *41*, 494–497. [CrossRef]

18. Xing, Y.; Tan, Y.; Hua, J.; Sun, X.; Xu, C.; Zhang, Q. Characterization of the main effects, epistatic effects and their environmental interactions of QTLs on the genetic basis of yield traits in rice. *Theor. Appl. Genet.* **2002**, *105*, 248–257. [CrossRef]

19. Edzesi, W.M.; Dang, X.; Liang, L.; Liu, E.; Zaid, I.U.; Hong, D. Genetic diversity and elite allele mining for grain traits in rice (*Oryza sativa* L.) by association mapping. *Front. Plant Sci.* **2016**, *7*, 787. [CrossRef]

20. Ngangkham, U.; Samantaray, S.; Yadav, M.K.; Kumar, A.; Chidambaranathan, P.; Katara, J.L. Effect of multiple allelic combinations of genes on regulating grain size in rice. *PLoS ONE* **2018**, *13*, e0190684. [CrossRef]

21. Wang, S.; Li, S.; Liu, Q.; Wu, K.; Zhang, J.; Wang, S.; Wang, Y.; Chen, X.; Zhang, Y.; Gao, C. The OsSPL16-GW7 regulatory module determines grain shape and simultaneously improves rice yield and grain quality. *Nat. Genet.* **2015**, *47*, 949–954. [CrossRef]

22. Zhou, Y.; Miao, J.; Gu, H.; Peng, X.; Leburu, M.; Yuan, F.; Gu, H.; Gao, Y.; Tao, Y.; Zhu, J. Natural diversity in *SLG7* regulate grain shape in rice. *Genetics* **2015**, *201*, 1591–1599. [CrossRef]

23. Shomura, A.; Izawa, T.; Ebana, K.; Ebitani, T.; Kanegae, H.; Konishi, S.; Yano, M. Deletion in a gene associated with grain size increased yields during rice domestication. *Nat. Genet.* **2008**, *40*, 1023–1028. [CrossRef] [PubMed]

24. Weng, J.; Gu, S.; Wan, X.; Gao, H.; Guo, T.; Su, N.; Lei, C.; Zhang, X.; Cheng, Z.; Guo, X. Isolation and initial characterization of *GW5*, a major QTL associated with rice grain width and weight. *Cell Res.* **2008**, *18*, 1199–1209. [CrossRef] [PubMed]

25. Duan, P.; Xu, J.; Zeng, D.; Zhang, B.; Geng, M.; Zhang, G.; Huang, K.; Huang, L.; Xu, R.; Ge, S. Natural variation in the promoter of GSE5 contributes to grain size diversity in rice. *Mol. Plant* **2017**, *10*, 685–694. [CrossRef] [PubMed]

26. Liu, J.; Chen, J.; Zheng, X.; Wu, F.; Lin, Q.; Heng, Y.; Tian, P.; Cheng, Z.; Yu, X.; Zhou, K. GW5 acts in the brassinosteroid signalling pathway to regulate grain width and weight in rice. *Nat. Plants* **2017**, *3*, 17043. [CrossRef] [PubMed]

27. Fan, C.; Xing, Y.; Mao, H.; Lu, T.; Han, B.; Xu, C.; Li, X.; Zhang, Q. *GS3*, a major QTL for grain length and weight and minor QTL for grain width and thickness in rice, encodes a putative transmembrane protein. *Theor. Appl. Genet.* **2006**, *112*, 1164–1171. [CrossRef]

28. Mao, H.; Sun, S.; Yao, J.; Wang, C.; Yu, S.; Xu, C.; Li, X.; Zhang, Q. Linking differential domain functions of the *GS3* protein to natural variation of grain size in rice. *Proc. Natl. Acad. Sci. USA* **2010**, *107*, 19579–19584. [CrossRef] [PubMed]

29. Takano-Kai, N.; Jiang, H.; Powell, A.; McCouch, S.; Takamure, I.; Furuya, N.; Doi, K.; Yoshimura, A. Multiple and independent origins of short seeded alleles of *GS3* in rice. *Breed. Sci.* **2013**, *63*, 77–85. [CrossRef] [PubMed]

30. Shao, G.; Wei, X.; Chen, M.; Tang, S.; Luo, J.; Jiao, G.; Xie, L.; Hu, P. Allelic variation for a candidate gene for *GS7*, responsible for grain shape in rice. *Theor. Appl. Genet.* **2012**, *125*, 1303–1312. [CrossRef]

31. Wang, S.; Wu, K.; Yuan, Q.; Liu, X.; Liu, Z.; Lin, X.; Zeng, R.; Zhu, H.; Dong, G.; Qian, Q. Control of grain size, shape and quality by OsSPL16 in rice. *Nat. Genet.* **2012**, *44*, 950–954. [CrossRef]

32. Li, Y.; Fan, C.; Xing, Y.; Jiang, Y.; Luo, L.; Sun, L.; Shao, D.; Xu, C.; Li, X.; Xiao, J. Natural variation in *GS5* plays an important role in regulating grain size and yield in rice. *Nat. Genet.* **2011**, *43*, 1266–1269. [CrossRef] [PubMed]

33. Che, R.; Tong, H.; Shi, B.; Liu, Y.; Fang, S.; Liu, D.; Xiao, Y.; Hu, B.; Liu, L.; Wang, H. Control of grain size and rice yield by GL2-mediated brassinosteroid responses. *Nat. Plants* **2016**, *2*, 15195. [CrossRef] [PubMed]

34. Duan, P.; Ni, S.; Wang, J.; Zhang, B.; Xu, R.; Wang, Y.; Chen, H.; Zhu, X.; Li, Y. Regulation of OsGRF4 by OsmiR396 controls grain size and yield in rice. *Nat. Plants* **2016**, *2*, 15203. [CrossRef] [PubMed]

35. Hu, J.; Wang, Y.; Fang, Y.; Zeng, L.; Xu, J.; Yu, H.; Shi, Z.; Pan, J.; Zhang, D.; Kang, S. A rare allele of *GS2* enhances grain size and grain yield in rice. *Mol. Plant* **2015**, *8*, 1455–1465. [CrossRef] [PubMed]

36. Sun, P.; Zhang, W.; Wang, Y.; He, Q.; Shu, F.; Liu, H.; Wang, J.; Wang, J.; Yuan, L.; Deng, H. OsGRF4 controls grain shape, panicle length and seed shattering in rice. *J. Integr. Plant Biol.* **2016**, *58*, 836–847. [CrossRef] [PubMed]

37. Wang, Y.; Xiong, G.; Hu, J.; Jiang, L.; Yu, H.; Xu, J.; Fang, Y.; Zeng, L.; Xu, E.; Xu, J. Copy number variation at the GL7 locus contributes to grain size variations in rice. *Nat. Genet.* **2015**, *47*, 944–948. [CrossRef] [PubMed]

38. Xu, H.; Zhao, M.; Zhang, Q.; Xu, Z.; Xu, Q. The DENSE AND ERECT PANICLE 1 (DEP1) gene offering the potential in the breeding of high-yielding rice. *Breed. Sci.* **2016**, *66*, 659–667. [CrossRef]

39. Li, X.; Tao, Q.; Miao, J.; Yang, Z.; Gu, M.; Liang, G.; Zhou, Y. Evaluation of differential qPE9-1/DEP1 protein domains in rice grain length and weight variation. *Rice* **2019**, *12*, 5. [CrossRef]

40. Vasemägi, A.; Gross, R.; Palm, D.; Paaver, T.; Primmer, C.R. Discovery and application of insertion-deletion (INDEL) polymorphisms for QTL mapping of early life-history traits in Atlantic salmon. *BMC Genom.* **2010**, *11*, 156. [CrossRef]

41. Li, W.; Cheng, J.; Wu, Z.; Qin, C.; Tan, S.; Tang, X.; Cui, J.; Zhang, L.; Hu, K. An InDel-based linkage map of hot pepper (Capsicum annuum). *Mol. Breed.* **2015**, *32–35*. [CrossRef]

42. Lister, R.; Gregory, B.D.; Ecker, J.R. Next is now: New technologies for sequencing of genomes, transcriptomes, and beyond. *Curr. Opin. Plant Biol.* **2009**, *12*, 107–118. [CrossRef] [PubMed]

43. Liu, B.; Wang, Y.; Zhai, W.; Deng, J.; Wang, H.; Cui, Y.; Cheng, F.; Wang, X.; Wu, J. Development of INDEL markers for *Brassica rapa* based on whole-genome re-sequencing. *Theor. Appl. Genet.* **2013**, *126*, 231–239. [CrossRef] [PubMed]

44. Moghaddam, S.M.; Song, Q.; Mamidi, S.; Schmutz, J.; Lee, R.; Cregan, P.; Osorno, J.M.; McClean, P.E. Developing market class specific InDel markers from next generation sequence data in *Phaseolus vulgaris* L. *Front. Plant Sci.* **2014**, *5*, 185. [CrossRef] [PubMed]

45. Păcurar, D.I.; Păcurar, M.L.; Street, N.; Bussell, J.D.; Pop, T.I.; Gutierrez, L.; Bellini, C. A collection of INDEL markers for map-based cloning in seven Arabidopsis germplasm. *J. Exp. Bot.* **2012**, *63*, 2491–2501. [CrossRef]

46. Schneeberger, K.; Weigel, D. Fast-forward genetics enabled by new sequencing technologies. *Trends Plant Sci.* **2011**, *16*, 282–288. [CrossRef]

47. Wu, K.; Yang, M.; Liu, H.; Tao, Y.; Mei, J.; Zhao, Y. Genetic analysis and molecular characterization of Chinese sesame (*Sesamum indicum* L.) cultivars using insertion-deletion (InDel) and simple sequence repeat (SSR) markers. *BMC Genet.* **2014**, *15*, 35. [CrossRef] [PubMed]

48. Yamaki, S.; Ohyanagi, H.; Yamasaki, M.; Eiguchi, M.; Miyabayashi, T.; Kubo, T.; Kurata, N.; Nonomura, K.-I. Development of INDEL markers to discriminate all genome types rapidly in the genus *Oryza*. *Breed. Sci.* **2013**, *63*, 246–254. [CrossRef]

49. Wu, D.-H.; Wu, H.-P.; Wang, C.-S.; Tseng, H.-Y.; Hwu, K.-K. Genome-wide InDel marker system for application in rice breeding and mapping studies. *Euphytica* **2013**, *192*, 131–143. [CrossRef]

50. Zeng, Y.; Wen, Z.; Ma, L.; Ji, Z.; Li, X.; Yang, C. Development of 1047 insertion-deletion markers for rice genetic studies and breeding. *Genet. Mol. Res.* **2013**, *12*, 5226–5235. [CrossRef]

51. Fisher, R.; Immer, F.; Tedin, O. The genetical interpretation of statistics of the third degree in the study of quantitative inheritance. *Genetics* **1932**, *17*, 107–124.

52. Robson, D.S. Applications of the k 4 Statistic to Genetic Variance Component Analyses. *Biometrics* **1956**, *12*, 433–444. [CrossRef]

53. Nachimuthu, V.V.; Robin, S.; Sudhakar, D.; Raveendran, M.; Rajeswari, S.; Manonmani, S. Evaluation of rice genetic diversity and variability in a population panel by principal component analysis. *Indian J. Sci. Technol.* **2014**, *7*, 1555–1562.

54. Romesburg, C. *Cluster Analysis for Researchers*; Lulu.com: Morrisville, North Carolina, 2004.

55. Haider, Z.; Farooq, U.; Naseem, I.; Zia, S.; Alamgeer, M. Impact of drought stress on some grain quality traits in rice (*Oryza sativa*). *Agric. Res.* **2015**, *4*, 132–138. [CrossRef]

56. Schneider, A.; Hommel, G.; Blettner, M. Linear regression analysis: Part 14 of a series on evaluation of scientific publications. *Dtsch. Ärztebl. Int.* **2010**, *107*, 776–782. [PubMed]

57. Seo, H.; Yang, G.; Kim, N.; Kim, H.; Kim, M. *SPSS (PASW) Regression Analysis*; Hannarae Publishing Co.: Seoul, Korea, 2009.

58. Ochiai, A. Zoogeographic studies on the soleoid fishes found in Japan and its neighbouring regions. *Bull. Jpn. Soc. Sci. Fish.* **1957**, *22*, 526–530. [CrossRef]

59. Mogga, M.; Sibiya, J.; Shimelis, H.; Lamo, J.; Yao, N. Diversity analysis and genome-wide association studies of grain shape and eating quality traits in rice (*Oryza sativa* L.) using DArT markers. *PLoS ONE* **2018**, *13*, e0198012. [CrossRef] [PubMed]

60. Li, N.; Xu, R.; Duan, P.; Li, Y. Control of grain size in rice. *Plant Reprod.* **2018**, *31*, 237–251. [CrossRef]

61. Huang, R.; Jiang, L.; Zheng, J.; Wang, T.; Wang, H.; Huang, Y.; Hong, Z. Genetic bases of rice grain shape: So many genes, so little known. *Trends Plant Sci.* **2013**, *18*, 218–226. [CrossRef]

62. Tan, Y.-F.; Xing, Y.-Z.; Li, J.-X.; Yu, S.-B.; Xu, C.-G.; Zhang, Q. Genetic bases of appearance quality of rice grains in Shanyou 63, an elite rice hybrid. *Theor. Appl. Genet.* **2000**, *101*, 823–829. [CrossRef]
63. Lee, C.-M.; Park, J.; Kim, B.; Seo, J.; Lee, G.; Jang, S.; Koh, H.-J. Influence of multi-gene allele combinations on grain size of rice and development of a regression equation model to predict grain parameters. *Rice* **2015**, *8*, 33. [CrossRef]
64. Kang, Y.-J.; Shim, K.-C.; Lee, H.-S.; Jeon, Y.-A.; Kim, S.-H.; Kang, J.-W.; Yun, Y.-T.; Park, I.-K.; Ahn, S.-N. Fine mapping and candidate gene analysis of the quantitative trait locus *GW8.* 1 associated with grain length in rice. *Genes Genom.* **2018**, *40*, 389–397. [CrossRef] [PubMed]
65. Xu, C.; Liu, Y.; Li, Y.; Xu, X.; Xu, C.; Li, X.; Xiao, J.; Zhang, Q. Differential expression of *GS5* regulates grain size in rice. *J. Exp. Bot.* **2015**, *66*, 2611–2623. [CrossRef] [PubMed]
66. Murray, M.; Thompson, W.F. Rapid isolation of high molecular weight plant DNA. *Nucleic Acids Res.* **1980**, *8*, 4321–4326. [CrossRef] [PubMed]
67. Panaud, O.; Chen, X.; McCouch, S. Development of microsatellite markers and characterization of simple sequence length polymorphism (SSLP) in rice (*Oryza sativa* L.). *Mol. Gen. Genet.* **1996**, *252*, 597–607. [CrossRef] [PubMed]
68. Liu, K.; Muse, S.V. PowerMarker: An integrated analysis environment for genetic marker analysis. *Bioinformatics* **2005**, *21*, 2128–2129. [CrossRef]
69. Perrier, X. DARwin Software (2006). Available online: http://darwin.cirad.fr/darwin (accessed on 11 July 2014).
70. Lestari, P.; Ham, T.-H.; Lee, H.-H.; Woo, M.-O.; Jiang, W.; Chu, S.-H.; Kwon, S.-W.; Ma, K.; Lee, J.-H.; Cho, Y.-C. PCR marker-based evaluation of the eating quality of japonica rice (*Oryza sativa* L.). *J. Agric. Food Chem.* **2009**, *57*, 2754–2762. [CrossRef]

International Journal of
Molecular Sciences

Article

Systems Metabolic Alteration in a Semi-Dwarf Rice Mutant Induced by *OsCYP96B4* Gene Mutation

Limiao Jiang [1,2,*,†], Rengasamy Ramamoorthy [3,†], Srinivasan Ramachandran [4] and Prakash P. Kumar [3,5,*]

1   Department of Epidemiology and Biostatistics, MOE Key Lab of Environment and Health, School of Public Health, Tongji Medical College, Huazhong University of Science and Technology, 13 Hangkong Road, Wuhan 430030, China
2   Department of Diagnostic Radiology, Yong Loo Lin School of Medicine, National University of Singapore, 5 Lower Kent Ridge Road, Singapore 119074, Singapore
3   Department of Biological Sciences, Faculty of Science, National University of Singapore, Singapore 117543, Singapore; ram10375@u.nus.edu
4   Temasek Life Sciences Laboratory, 1 Research Link National University of Singapore, Singapore 117604, Singapore; sri@tll.org.sg
5   NUS Environmental Research Institute (NERI), 5A Engineering Drive 1, Singapore 117411, Singapore
*   Correspondence: limiao_jiang@hust.edu.cn (L.J.); dbskumar@nus.edu.sg (P.P.K.)
†   The authors contributed equally.

Received: 26 November 2019; Accepted: 8 March 2020; Published: 11 March 2020

Abstract: Dwarfism and semi-dwarfism are among the most valuable agronomic traits in crop breeding, which were adopted by the "Green Revolution". Previously, we reported a novel semi-dwarf rice mutant (*oscyp96b4*) derived from the insertion of a single copy of *Dissociator (Ds)* transposon into the gene *OsCYP96B4*. However, the systems metabolic effect of the mutation is not well understood, which is important for understanding the gene function and developing new semi-dwarf mutants. Here, the metabolic phenotypes in the semi-dwarf mutant (M) and ectopic expression (ECE) rice line were compared to the wild-type (WT) rice, by using nuclear magnetic resonance (NMR) metabolomics and quantitative real-time polymerase chain reaction (qRT-PCR). Compared with WT, ECE of the *OsCYP96B4* gene resulted in significant increase of γ-aminobutyrate (GABA), glutamine, and alanine, but significant decrease of glutamate, aromatic and branched-chain amino acids, and some other amino acids. The ECE caused significant increase of monosaccharides (glucose, fructose), but significant decrease of disaccharide (sucrose); induced significant changes of metabolites involved in choline metabolism (phosphocholine, ethanolamine) and nucleotide metabolism (adenosine, adenosine monophosphate, uridine). These metabolic profile alterations were accompanied with changes in the gene expression levels of some related enzymes, involved in GABA shunt, glutamate and glutamine metabolism, choline metabolism, sucrose metabolism, glycolysis/gluconeogenesis pathway, tricarboxylic acid (TCA) cycle, nucleotide metabolism, and shikimate-mediated secondary metabolism. The semi-dwarf mutant showed corresponding but less pronounced changes, especially in the gene expression levels. It indicates that *OsCYP96B4* gene mutation in rice causes significant alteration in amino acid metabolism, carbohydrate metabolism, nucleotide metabolism, and shikimate-mediated secondary metabolism. The present study will provide essential information for the *OsCYP96B4* gene function analysis and may serve as valuable reference data for the development of new semi-dwarf mutants.

Keywords: rice; dwarfism; *OsCYP96B4*; metabolomics; NMR; qRT-PCR

1. Introduction

Rice (*Oryza sativa* L.) is an important staple food for more than half of the global population [1]. Its sustainable production is critical to the world's food security and the health of the ever-increasing global population. Although nitrogen fertilization is essential for improving grain yield, the increasing usage rate will lead to increased plant height [2,3]. The resulting taller plants are prone to lodging caused by wind/rain, which will increase the difficulty of harvest, promote pre-harvest germination and fungal contamination, and ultimately lead to significant reduction in grain yield and quality [2,4].

As adopted by the "Green Revolution" [2], dwarfism and semi-dwarfism are among the most valuable agronomic traits in crop breeding [5,6]. The semi-dwarfism in rice enhances their lodging resistance (e.g., to wind and rain), improves harvest index (i.e., grain/straw ratio) and enhances biomass production [2]. Therefore, a wide range of studies have been done on the development of rice semi-dwarf mutants [3,6–8], which can be broadly categorized into phytohormone-dependent (e.g., brassinosteroids- or gibberellins-related) and phytohormone-independent mutants [6]. Previously, we reported a novel phytohormone-independent semi-dwarf rice mutant derived from the insertion of a single copy of transposon *Dissociator (Ds)* into the gene *OsCYP96B4* (*Oryza sativa* CytochromeP450 *96B4*), which resulted in defects in cell elongation and pollen germination [5]. As one of the largest protein-encoding gene families in plants, the cytochrome P450 (CYP) superfamily plays important roles in plant growth, development, and responses to biotic and abiotic stresses [9]. CYP96B subfamily belongs to the CYP96 family of the CYP86 clan. Recently, the functions of the CYP96B subfamily have been gradually revealed. CYP96B5 hydroxylates alkanes to primary alcohols and is involved in rice leaf cuticular wax synthesis [10]. *OsCYP96B4* is involved in secondary cell wall formation in rice [11], associated with the growth and drought stress responses in rice [12], and may be an important regulator of plant growth that affects plant height in rice [6]. However, to the best of our knowledge, currently, the systems metabolic effect of the *OsCYP96B4* gene mutation in rice is unclear, which is important for understanding the gene function and developing new semi-dwarf mutants.

Metabolomics is useful in characterizing the systems metabolic changes of biological systems to genetic modification or environmental stimuli [13–16]. Nuclear magnetic resonance (NMR) and mass spectrometry (MS) are the two dominant high-throughput analytical platforms employed in metabolomics, possessing their own advantages and providing complementary metabolic information [17]. For NMR analysis, it shows very high reproducibility, requires minimal sample preparation, detects in a non-selective manner and favors unambiguous metabolite identification [17]. Metabolomics has been widely applied in evaluating the metabolic response of plants to gene manipulation and biotic/abiotic stresses, e.g., the insertion of a moss Na$^+$ transporter gene in rice and barley [18], the influence of *SUB1A* gene during submergence stress in rice [19], *Tri5* gene deletion in *Fusarium graminearum* [20], antisense thioredoxin s (*anti-trx-s*) introduction to wheat for pre-harvest sprouting resistance [21], *Fusarium oxysporum* inoculation in chickpea roots [22], brown planthopper infestation in pest-susceptible and -resistant rice plants [23]. Metabolomics has also been used to characterize the metabolic phenotypes of plant dwarf and semi-dwarf mutants, e.g., BW312 barley (*Hordeum vulgare*) semi-dwarf mutant [24], dwarf banana (*Musa* spp.; Musaceae) variants [25], tomato (*Solanum lycopersicum* L.) dwarf cultivar Micro-Tom [26], a dwarf genotype of soybean named MiniMax [27], and dwarfed *tcd2* (*totally cyanide deficient 2*) mutants of *Sorghum bicolor* (L.) Moench [28]. Such studies indicate that there are significant metabolic profile alterations in the dwarf/semi-dwarf mutant plants. However, to the best of our knowledge, currently, there is no metabolomics study on the dwarf rice.

In the present study, we investigated the differences in the metabolome and related gene expression levels among the *oscyp96b4* semi-dwarf mutant (M), ectopic expression (ECE), and wild-type (WT) rice using NMR-based metabolomics and quantitative real-time polymerase chain reaction (qRT-PCR). We aimed to (1) characterize the metabolic phenotypes of the mutant rice and (2) elucidate the systems metabolic effect of the mutation by integrating the metabolomics and gene expression data. The analysis provides essential information on the *OsCYP96B4* gene function and may serve as reference data for the development of new semi-dwarf mutants.

2. Results

2.1. Metabolite Profiling in Wild-Type and Mutant Rice

In the ^{1}H NMR spectra of rice extracts, a total of 42 abundant metabolites were assigned (Table S1) according to 2D NMR experiments and literature reports [29]. Typical 1D ^{1}H NMR spectra of WT, M, and ECE rice extracts were shown in Figure 1. The NMR spectra contain signals mainly from amino acids and derivatives, choline metabolism-related metabolites, carbohydrate metabolites, tricarboxylic acid (TCA) cycle intermediates, nucleotide metabolites, N-methylnicotinate, nicotinamide mononucleotide, lipid, formate, methylamine, mono-methyl phosphate, and methanol.

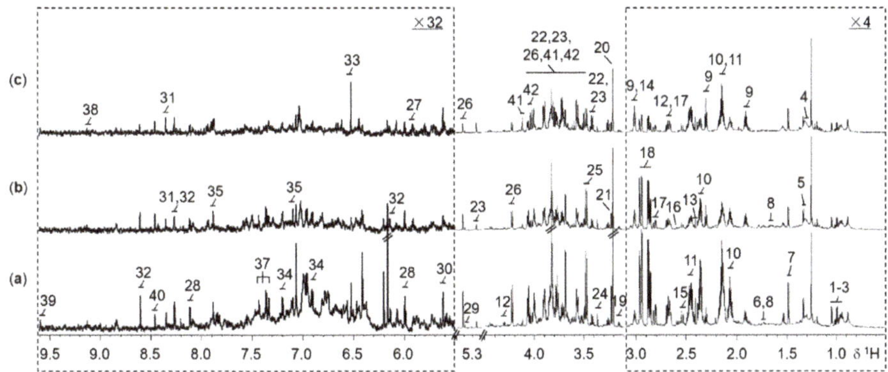

Figure 1. Representative 800 MHz ^{1}H NMR spectra for the extracts of 2-week-old seedling shoots (without roots) from (**a**) the wild type, (**b**) the *oscyp96b4* semi-dwarf mutant, and (**c**) the *OsCYP96B4* ectopic expression rice lines. The dotted regions δ0.5–3.1 and δ5.5–9.61 were vertically expanded 4 and 32 times respectively, compared to the region δ3.1-4.5. Metabolite keys: 1, isoleucine; 2, leucine; 3, valine; 4, lipid; 5, threonine; 6, lysine; 7, alanine; 8, arginine; 9, γ-aminobutyrate; 10, glutamate; 11, glutamine; 12, malate; 13, succinate; 14, 2-oxoglutarate; 15, citrate; 16, methylamine; 17, aspartate; 18, asparagine; 19, ethanolamine; 20, choline; 21, phosphocholine; 22, β-glucose; 23, α-glucose; 24, methanol; 25, mono-methyl phosphate; 26, sucrose; 27, uridine; 28, uridine 5′-monophosphate (UMP); 29, allantoin; 30, uridine diphosphate glucuronic acid (UDP glucuronate); 31, adenosine; 32, adenosine monophosphate (AMP); 33, fumarate; 34, tyrosine; 35, histidine; 36, tryptophan; 37, phenylalanine; 38, N-methylnicotinate (trigonelline, NMNA); 39, nicotinamide mononucleotide (nicotinamide ribotide, NMN); 40, formate; 41, β-D-fructopyranose; 42, β-D-fructofuranose.

2.2. OsCYP96B4 Gene Mutation Induced Metabolic Changes in Mutant Rice

Clear differentiation in the metabolic profiles among the three rice phenotypes was shown in the PCA scores plots (Figure S1) and the corresponding OPLS-DA scores plots (Figure 2). The OPLS-DA model quality was evaluated by the sevenfold internal cross-validation, with R^2(X) = 0.368, R^2(Y) = 0.988, Q^2 = 0.940 for the comparison between WT and M (Figure 2a), R^2(X) = 0.537, R^2(Y) = 0.985, Q^2 = 0.959 for the comparison between WT and ECE (Figure 2b), and R^2(X) = 0.464, R^2(Y) = 0.972, Q^2 = 0.915 for the comparison between M and ECE (Figure 2c). The model validities were demonstrated by CV-ANOVA of OPLS-DA models (Figure 2) and permutation tests on the corresponding PLS-DA models (Figure S2). Detailed metabolic profile difference among the three phenotypes were shown in the color-coded loadings plots of OPLS-DA models (Figure 2) and further summarized in Figure 3 and Table S2.

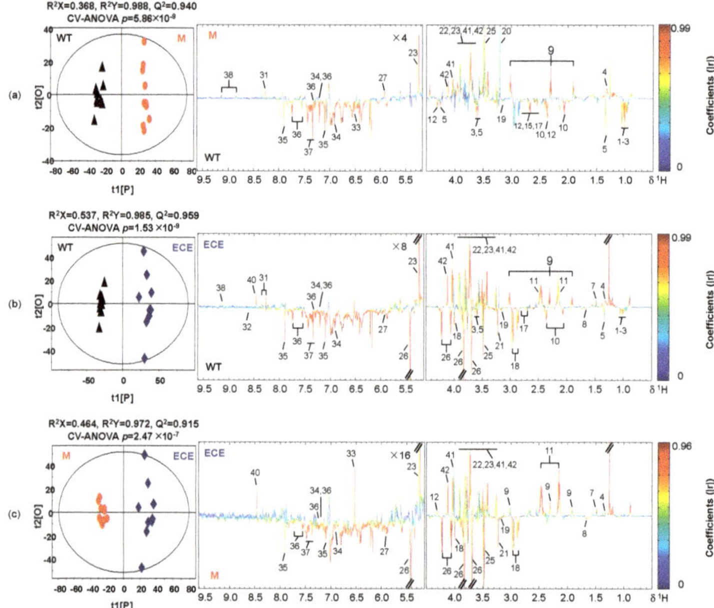

Figure 2. Cross-validated OPLS-DA scores plots (**left**) and the corresponding loadings plots (**right**) derived from the comparison of ^{1}H NMR spectra for (**a**) the wild type (WT, ▲) and the *oscyp96b4* semi-dwarf mutant (M, •), (**b**) the wild type (WT, ▲) and the *OsCYP96B4* ectopic expression (ECE, ◆), (**c**) the *oscyp96b4* semi-dwarf mutant (M, •) and the *OsCYP96B4* ectopic expression (ECE, ◆). Metabolite keys are shown in Figure 1 and Table S1.

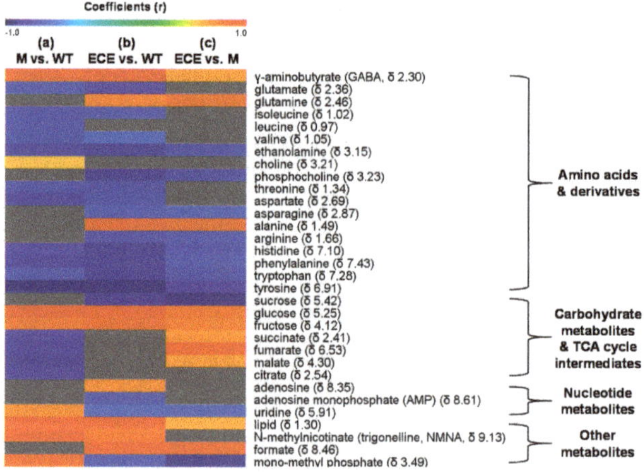

Figure 3. Heat map showing metabolites with significant level changes ($p < 0.05$) for (**a**) the *oscyp96b4* semi-dwarf mutant (M) vs. the wild type (WT), (**b**) the *OsCYP96B4* ectopic expression (ECE) vs. the wild type (WT), and (**c**) the *OsCYP96B4* ectopic expression (ECE) vs. the *oscyp96b4* semi-dwarf mutant (M). It was color-coded with the Pearson correlation coefficients from the corresponding OPLS-DA models, where a warm color (e.g., red) indicates significant increase of metabolites in M (**a**) or ECE (**b,c**) as compared to the counterpart, a cool color (e.g., blue), indicating significant decrease, and the grey color indicates no significant difference.

Compared to WT, M showed higher levels of γ-aminobutyrate, choline, carbohydrates (glucose, fructose), uridine, N-methylnicotinate, lipid, and mono-methyl phosphate, but lower levels of most amino acids (isoleucine, leucine, valine, threonine, glutamate, aspartate, tyrosine, histidine, tryptophan, phenylalanine), ethanolamine, and TCA cycle intermediates (malate, succinate, citrate, and fumarate) ($p < 0.05$, Figures 2a and 3a, Table S2). ECE presented more alanine, γ-aminobutyrate, glutamine, carbohydrates (glucose, fructose), adenosine, N-methylnicotinate, lipid, and formate, but less amino acids (isoleucine, valine, threonine, arginine, glutamate, aspartate, asparagine, tyrosine, histidine, tryptophan, phenylalanine), choline metabolites (ethanolamine, phosphocholine), sucrose, nucleotide metabolites (uridine, adenosine monophosphate), and mono-methyl phosphate than WT ($p < 0.05$, Figures 2b and 3b, Table S2). Compared with M, there were increases of γ-aminobutyrate, glutamine, alanine, carbohydrates (glucose, fructose), TCA cycle intermediates (succinate, fumarate and malate), lipid, and formate, but decreases of amino acids (asparagine, arginine, histidine, phenylalanine, tyrosine, tryptophan), choline metabolites (ethanolamine, phosphocholine), sucrose, uridine, and mono-methyl phosphate in ECE ($p < 0.05$, Figures 2c and 3c, Table S2).

2.3. Gene Expression Analysis

The qRT-PCR analysis was performed to acquire supporting information for the aforementioned metabolic changes induced by *OsCYP96B4* gene mutation. Compared with WT, significant alterations ($p < 0.05$) in M or ECE were observed for the expression levels of key enzyme-encoding genes, responsible for the regulation of GABA shunt, glutamate/glutamine metabolism, choline metabolism, carbohydrate metabolism, nucleotide metabolism, and secondary metabolism (Figure 4).

Along with the metabolite level changes, *OsCYP96B4* gene mutation induced significant alterations in the expression levels of key genes in GABA shunt and glutamate/glutamine metabolism, including the down-regulation of glutamate decarboxylase 4 gene (*GAD4*) in M and the up-regulation of glutamate synthase 2 gene (*GOGAT2*) and glutamine synthetase gene (*GS*) in ECE. *OsCYP96B4* gene mutation also resulted in significant changes of gene expression related to choline metabolism, i.e., down-regulation of betaine aldehyde dehydrogenase 2 gene (*BADH2*) in M and up-regulation of phosphoethanolamine/phosphocholine phosphatase gene (*PHOSPHO1*) in ECE. In addition, significant alterations were also observed for the gene expression pertaining to the metabolism of other amino acids, including the down-regulation of aspartate aminotransferase gene (*AAT*) and LL-diaminopimelate aminotransferase gene (*LL-DAP-AT*) in M, along with the up-regulation of *AAT* and aspartate kinase gene (*AK*) and down-regulation of alanine transaminase gene (*ALT*) and *LL-DAP-AT* in ECE (Figure 4a).

There were significant changes in the gene expression involved in carbohydrate metabolism, including significant down-regulation of genes encoding sucrose synthase 1 (*SUS1*), fructose-bisphosphate aldolase (*FBA*), glyceraldehyde 3-phosphate dehydrogenase (*GAPDH*), 2,3-bisphosphoglycerate-independent phosphoglycerate mutase (*iPGM*) and aconitase (*ACO*) in M, along with the up-regulation of genes encoding hexokinase-8 (*HXK8*), phosphoglucomutase (*PGM*), and succinate dehydrogenase (*SDH*) and down-regulation of genes encoding malate dehydrogenase (*MDH*) in ECE (Figure 4b).

Significant changes also occurred in the gene expression related to nucleotide metabolism, i.e., the up-regulation of adenine phosphoribosyltransferase gene (*APRT*) in ECE, and down-regulation of allantoinase (*ALN*) gene in both M and ECE (Figure 4c). In addition to the significant alterations of gene expression in primary metabolism, *OsCYP96B4* gene mutation also led to significant changes of gene expression in secondary metabolism, i.e., the down-regulation of tyrosine aminotransferase gene (*TAT*), phosphoribosylanthranilate isomerase gene (*PRAI*), indole-3-glycerol phosphate synthase gene (*IGPS*), anthranilate synthase beta subunit 2 gene (*ASB2*) in M, and the down-regulation of *PRAI* in ECE (Figure 4c).

For the comparison between ECE and M, the gene expression alterations were nearly the same as the comparison between ECE and WT, except for the significant increase of *ACO* and no significant change of *PRAI* in ECE when compared with M (at the significance level of 0.05, Figure 4a–c).

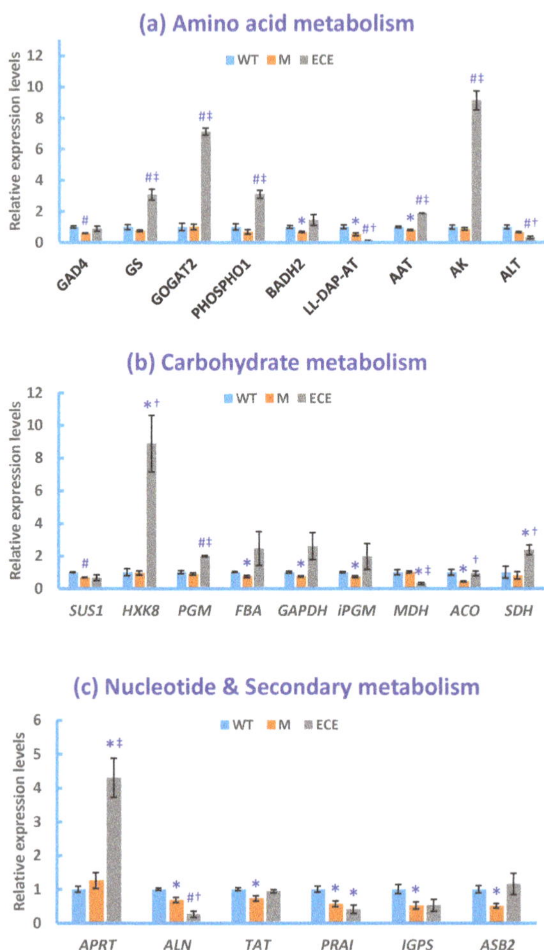

Figure 4. Gene expression levels in the wild-type (WT, blue bars), the *oscyp96b4* semi-dwarf mutant (M, orange bars), and the *OsCYP96B4* ectopic expression (ECE, grey bars) rice lines measured by qRT-PCR. Data shown are means ± SE of three biological replicates each with three technical replicates (* $p < 0.05$, #$P < 0.01$, as compared to WT; † $p < 0.05$, ‡ $p < 0.01$, as compared to M).

3. Discussion

The aforementioned metabolomics and gene expression data showed that *OsCYP96B4* gene mutation resulted in comprehensive metabolic responses in rice plants. The responses were summarized in Figure 5, mainly including alterations in amino acid metabolism, carbohydrate metabolism, nucleotide metabolism, and secondary metabolism. In general, such changes were more comprehensive in ECE than in M when compared with WT, especially on the gene expression levels. Moreover, significant differences were observed in the metabolic and gene expression levels between ECE and M. These differences may be related to the developmental disparities in the two mutants. The *oscyp96b4* mutant displayed a semi-dwarf phenotype, but with the development of panicles and seeds (i.e., complete mature plant formation). However, the ECE plants remained severely dwarf, without any panicle formation, and normally died after about 4 weeks in the pot. Taken together, the results shed light on the *OsCYP96B4* gene function and may be associated with the plant phenotype (i.e., dwarfism).

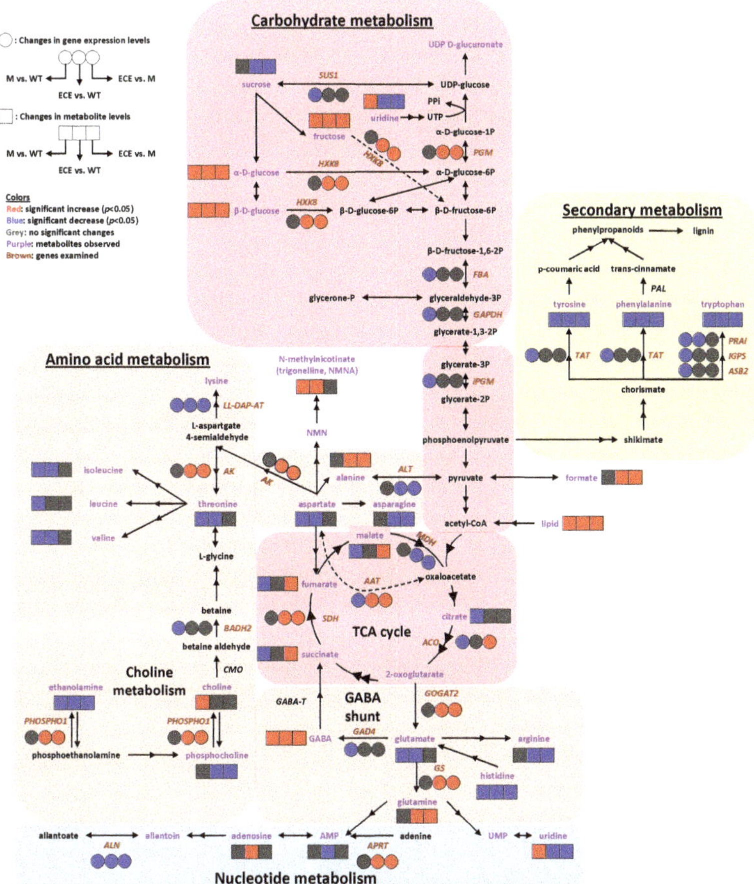

Figure 5. Systems metabolic reprogramming induced by *OsCYP96B4* gene mutation in rice. Symbols: boxes (□) represent metabolite levels; circles (○) represent gene expression levels; both the solid and dashed arrows indicate the enzyme catalyzed metabolic reactions, where the dashed arrows were intentionally used to prevent the intersection between solid ones. Colors: red indicates significant increase ($p < 0.05$), blue indicates significant decrease ($p < 0.05$), grey indicates no significant change; purple letters denote identified metabolites; italic brown letters denote genes with examined transcript levels. Metabolite abbreviations: AMP, adenosine monophosphate; GABA, γ-aminobutyrate; NMN, nicotinamide mononucleotide; NMNA, N-methylnicotinate; PPi, diphosphate; UDP, uridine diphosphate; UMP, uridine 5′-monophosphate; UTP, uridine triphosphate. Gene abbreviations: *SUS1*, sucrose synthase 1; *PGM*, phosphoglucomutase; *HXK8*, hexokinase-8; *FBA*, fructose-bisphosphate aldolase; *GAPDH*, glyceraldehyde 3-phosphate dehydrogenase; *iPGM*, 2,3-bisphosphoglycerate-independent phosphoglycerate mutase; *SDH*, succinate dehydrogenase; *MDH*, malate dehydrogenase; *ACO*, aconitase; *PAL*, phenylalanine ammonia-lyase; *TAT*, tyrosine aminotransferase; *PRAI*, phosphoribosylanthranilate isomerase; *IGPS*, indole-3-glycerol phosphate synthase; *ASB2*, anthranilate synthase beta subunit 2; *LL-DAP-AT*, LL-diaminopimelate aminotransferase; *AK*, aspartate kinase; *ALT*, alanine transaminase; *AAT*, aspartate aminotransferase; *BADH2*, betaine aldehyde dehydrogenase 2; *CMO*, choline monooxygenase; *PHOSPHO1*, phosphoethanolamine/phosphocholine phosphatase; *GABA-T*, GABA transaminase; *GAD4*, glutamate decarboxylase 4; *GOGAT2*, glutamine:2-oxoglutarate amidotransferase 2; *GS*, glutamine synthetase; *APRT*, adenine phosphoribosyltransferase; *ALN*, allantoinase.

3.1. Amino Acid Metabolism

The GABA shunt is comprised of three steps, including the α-decarboxylation of glutamate to GABA by glutamate decarboxylase (*GAD*), the conversion of GABA to succinic semialdehyde by GABA transaminase (*GABA-T*), and the oxidization of succinic semialdehyde to succinate catalyzed by succinic semialdehyde dehydrogenase (*SSADH*) [30,31]. There are reported connections between GABA shunt and plant dwarfism. The suppression of *GABA-T* induces prominent GABA accumulation, dwarfism, and infertility in the tomato [32]. Rice plants overexpressing *OsGAD2ΔC* have dwarf phenotypes [33]. In the present study, there was significant increase of GABA and decrease of glutamate levels in both M and ECE. Besides, the succinate level and the *GAD4* gene expression was down-regulated in M. Although both glutamate and *GAD4* were down-regulated in M, other metabolic pathways (e.g., arginine and proline metabolism) may contribute to the increase of GABA level. It indicates that *OsCYP96B4* functions affect rice GABA shunt, which may be connected with the dwarfism phenotype [32,33].

The glutamine synthetase/glutamine:2-oxoglutarate amidotransferase (i.e., GS/GOGAT) cycle plays a regulatory role in the nitrogen assimilation process in plants [34]. While glutamine synthetase (*GS*) catalyzes the conversion of glutamate and ammonia into glutamine, glutamate synthase (i.e., *GOGAT*, or glutamine:2-oxoglutarate amidotransferase) catalyzes the formation of glutamate from glutamine and 2-oxoglutarate. Here, the significant decrease of glutamate was observed in both M and ECE, together with the up-regulated *GS*, *GOGAT2* gene expression, and glutamine level in ECE. It suggests the function of *OsCYP96B4* on rice glutamate and glutamine metabolism, along with GS/GOGAT cycle, which may be connected with the nitrogen assimilation process [34] and the dwarfism phenotype.

Isoleucine, leucine, and valine are three branched-chain amino acids (BCAAs). Here, significant decrease of isoleucine and valine were observed in both M and ECE, along with the significantly reduced leucine level in M. As both valine and isoleucine possess the glucogenic property, their level decrease and the increase of glucose concentration may indicate the possible gluconeogenesis process in the mutant rice. Although at present there are no relevant reports, it suggests the effect of *OsCYP96B4* mutation on rice branched-chain amino acid metabolism, which may be connected with the dwarfism phenotype.

Phosphoethanolamine/phosphocholine phosphatase (*PHOSPHO1*) catalyzes the conversion of phosphoethanolamine to ethanolamine and the conversion of phosphocholine to choline. Choline can be further converted to betaine by the action of choline monooxygenase (*CMO*) and betaine aldehyde dehydrogenase 2 (*BADH2*). It was shown that simultaneous expression of *Spinacia oleracea* chloroplast *CMO* and *BADH* genes contribute to dwarfism in transgenic *Lolium perenne* [35]. In the current study, there were significantly decreased ethanolamine level and *BADH2* gene expression level, along with significantly increased choline level in M. While, there were significantly decreased ethanolamine and phosphocholine levels and significantly increased *PHOSPHO1* gene expression level in ECE. These observations suggest the influence of *OsCYP96B4* function on rice choline metabolism and dwarfism [35].

In addition, *OsCYP96B4* mutation showed potential effects on the metabolism of other amino acids, including the significant decrease of threonine, aspartate, histidine levels, and *LL-DAP-AT* gene expression in both M and ECE, the down-regulation of *AAT* expression level in M, and the significantly decreased levels of asparagine, arginine, and *ALT* expression along with the significantly increased levels of alanine and *AK* expression in ECE.

3.2. Carbohydrate Metabolism

Sucrose synthase (*SUS*) is a key enzyme for the regulation of carbon partitioning in plants by providing UDP-glucose as a substrate for the biosynthesis of cellulose and other polysaccharides [36]. It has been shown that AtCesA8::SUS3 transgenic rice plants exhibited largely improved biomass saccharification and lodging resistance by reducing cellulose crystallinity and increasing cell wall thickness [36]. *OsIDD2* overexpression leads to severely dwarfed rice plants and *OsIDD2* negatively regulates the transcription of genes involved in lignin biosynthesis, cinnamyl alcohol dehydrogenase 2 and 3 (*CAD2* and *3*), and sucrose metabolism sucrose synthase 5 (*SUS5*) [37]. In the present study, although there were significant increases of glucose, fructose, and uridine levels in M, the significant decrease of *SUS1* gene expression may lead to the unchanged level of sucrose. While for ECE, the significantly increased levels of glucose, fructose, and *HXK8* and *PGM* gene expression and the decreased uridine level may result in the decrease of sucrose level, in spite of the unchanged *SUS1* gene expression. It suggests the influence of *OsCYP96B4* on sucrose metabolism, which may be connected with dwarfism [37]. Besides, the significantly decreased *FBA*, *GAPDH*, and *iPGM* gene expression levels in M further supported the effect of *OsCYP96B4* gene mutation on rice carbohydrate metabolism.

The TCA cycle is carried out through a variety of interconnected enzymatic reactions, e.g., the conversion of succinate to fumarate by succinate dehydrogenase (*SDH*), the oxidation of malate to oxaloacetate via malate dehydrogenase (*MDH*), and the transformation of citrate to *cis*-aconitate and then isocitrate by aconitase (*ACO*). *OsAPX2* (rice ascorbate peroxidase 2) knock-out leads to shoot dwarfing and up-regulation of enzymes linked to glycolysis and TCA cycle in rice flag leaves [38]. *Arabidopsis thaliana* mutants lacking plastidial NAD-dependent MDH (*pdnad-mdh*) are embryo-lethal, and constitutive silencing (*miR-mdh-1*) leads to a dwarfed phenotype [39]. In the present study, there were significant decreases of TCA cycle intermediates (i.e., succinate, fumarate, malate, and citrate) and *ACO* gene expression level in M, along with the significant up-regulation of *SDH* and the down-regulation of *MDH* gene expression levels in ECE. It is inferred that the perturbation in TCA cycle is related to *OsCYP96B4* gene function and may be connected with the rice dwarfism phenotype [38,39].

3.3. Nucleotide Metabolism

As an important intermediate in the purine metabolism, adenosine monophosphate (AMP) is formed from adenine by adenine phosphoribosyltransferase (*APRT*) [40] and is interconvertible with adenosine through hydroxylation and phosphorylation reactions. Allantoin, an intermediary metabolite of purine catabolism, is hydrolyzed to allantoate under the catalysis of allantoinase (*ALN*) [41]. Similarly, uridine and UMP are two interconvertible intermediates in the pyrimidine metabolism. *APRT* was shown to be potentially involved in thermo-sensitive genic male sterility (TGMS) in the rice line 'Annong S-1' [42] and associated with growth retardation and male sterility in Arabidopsis [43]. In the current study, there was significant increase of adenosine and *APRT* gene expression, but significant decrease of AMP, uridine, and *ALN* gene expression level in ECE, along with significant increase of uridine and the significant decrease of *ALN* gene expression level in M. It suggests the potential function of *OsCYP96B4* on rice nucleotide metabolism, which may be associated with the rice dwarfism phenotype. However, to the best of our knowledge, currently, there are no reports on such association.

3.4. Secondary Metabolism

The shikimate and aromatic amino acids (AAA) biosynthesis pathways represent a link between primary and secondary metabolism in plants [44]. The catabolism of AAAs results in a variety of secondary metabolites, e.g., phenylpropanoids and flavonoids from phenylalanine, tocochromanols and phenylpropanoids from tyrosine, and indole-containing metabolites from tryptophan [45]. Cytochrome P450 monooxygenases (*CYP450s*) play important roles in the biosynthesis of plant secondary metabolites, including phenylpropanoids, terpenes, and alkaloids [46]. As one of the major phenylpropanoid pathway end-products, lignin is a major component of the secondary cell wall and is vital for providing mechanical strength to reduce lodging stress in plants [47,48]. Phenylalanine ammonia-lyase (*PAL*) catalyzes the conversion of phenylalanine to *trans*-cinnamic acid and plays an important role in the biosynthetic pathway of lignin [49]. In the dwarf rice mutant Fukei 71, elevated levels of p-coumaric acid (PCA), ferulic acid (FA), and *PAL* were observed in the abnormal parenchyma tissue, indicating that the abnormal activation of phenylpropanoid pathway leads to the biosynthesis of polysaccharide-linked

FA and PCA [50]. The AAAs metabolism is also connected with dwarfism. The overexpression of a rice tyrosine decarboxylase (*TyDC*) leads to tyramine accumulation in rice cells and causes a dwarf phenotype via reduced cell division [51]. The suppression of serotonin N-acetyltransferase 2 (*SNAT2*) leads to melatonin-deficient rice with a semidwarf phenotype [52]. Our present study showed that there was significant decrease in levels of phenylalanine, tyrosine, and tryptophan in both M and ECE, along with the significant decrease of *TAT*, *PRAI*, *IGPS*, and *ASB2* expression levels in M and the significant decrease of *PRAI* expression level in ECE. It may indicate the declined synthesis of phenylalanine, tyrosine, and tryptophan and reflect the function of *OsCYP96B4* on shikimate-mediated secondary metabolism in rice, which may be possibly connected with the dwarfism phenotype [51,52].

3.5. Other Metabolism

In our previous study [5], GC-MS lipid profiling was performed on non-polar extracts of rice leaves from WT and M. Three classes of lipid were analyzed, i.e., plant glycolipids, the common membrane phospholipid classes, and the minor membrane lipid metabolites. Compared with WT, there were no significant alteration in total monogalactosyldiacylglycerol (MGDG) or total digalactosyldiacylglycerol (DGDG), but significant increase of MGDG 34:6, and significant decreases of DGDG 34:1, DGDG 36:2, total phosphatidylglycerol (PG), PG 36:2, total lysoPG and lysoPG 16:1 in M. While, in the present study, untargeted NMR metabolic profiling was performed on the polar extraction of the whole 2-week-old seedling shoots without roots. The accumulation of total lipids, with a wider species coverage than our previous study [5], was observed in both M and ECE. Though the difference in plant material, extraction method, and lipid coverage may lead to different results, both studies suggested that *OsCYP96B4* might be involved in rice plant lipid metabolism. The accumulation of lipids observed here may indicate the reduced conversion of lipids to other metabolites or the increased synthesis of lipids from other metabolites. The significant elevation of N-methylnicotinate level in both M and ECE may indicate the potential effect of this mutation on vitamin B3 metabolism. In addition, the significant increase of formate concentration suggests altered formate metabolism in ECE.

4. Materials and Methods

4.1. Plant Materials

The *Japonica* rice (*Oryza sativa* ssp. *japonica* cv. Nipponbare) was used as the wild-type (WT) rice plant for the current study. The *oscyp96b4* semi-dwarf (M) was a *Ds* insertion mutant, and the *OsCYP96B4* ectopic expression (ECE) lines showing the most severe dwarf phenotype were generated constitutively expressing the *OsCYP96B4* gene in the Nipponbare background. The detailed descriptions about the mutant and ectopic expression lines can be found in our previous report [5]. For each genotype of rice plants (i.e., WT, M, and ECE, Figure S3), 19 seedlings were grown in a Phytatray™ II (Sigma-Aldrich, St. Louis, MO, USA) containing half strength Murashige and Skoog medium [53] (Sigma-Aldrich, St. Louis, MO, USA) in a plant growth room maintained at 28 °C with 12 h dark and 12 h light. For metabolites extraction, 10 whole 2-week-old seedling shoots without roots were used for each genotype. While for qRT-PCR, 3 independent biological replicates were used, where each replicate was made up of RNA extracted from 3 seedlings pooled together for better representation. Each plant sample was collected individually into a 2 mL Eppendorf tube, weighed, immediately snap-frozen in liquid nitrogen and stored at −80 °C until further analyses.

4.2. Metabolome Analysis

4.2.1. Chemicals

Analytical grade methanol, $K_2HPO_4 \cdot 3H_2O$, $NaH_2PO_4 \cdot 2H_2O$ and sodium azide (NaN_3) were purchased from Sigma-Aldrich (St. Louis, MO, USA). Sodium 3-trimethylsilyl [2,2,3,3–$^2H_4$]-propionate (TSP) and deuterium oxide (D_2O, 99.9% D) were purchased from Cambridge Isotope Laboratories, Inc.

(Tewksbury, MA, USA). The phosphate buffer for NMR analysis was prepared by dissolving K_2HPO_4 and NaH_2PO_4 in water (0.15 M, pH 7.44, K_2HPO_4/NaH_2PO_4 = 4:1) containing 0.001% TSP, 0.1% NaN_3, and 50% D_2O [54].

4.2.2. Plant Extraction and Sample Preparation

For each of the three groups (i.e., WT, M, and ECE), 10 biological replicates of rice plants were used for the extraction and subsequent NMR-based metabolomics analysis. Each rice plant was extracted individually using a method modified from previous reports [23,54]. Firstly, the rice sample-containing tube was snap-frozen in liquid nitrogen and the rice tissue was then swiftly ground into fine powder using a pre-cooled pestle. Pre-cooled methanol/water (v/v = 2/1, −20 °C) was added into the homogenized sample at a ratio of 600 µL per 100 mg powder. Afterwards, the mixture was further homogenized using a 2010 Geno/Grinder® (SPEX Sample Prep, Metuchen, NJ, USA) at 1300 rpm for 90 s. Then the homogenate mixture was sonicated in an ice bath with 10 cycles of 30 s sonication and 30 s break. Following centrifugation (14,489× *g*, 10 min, 4 °C), the supernatant was collected, and the remaining pellets were further treated twice using the same procedure. Three supernatants were combined and lyophilized after removal of methanol *in vacuo* (CentriVap Centrifugal Vacuum Concentrators, Labconco, MO, USA). Each dried extract was reconstituted into 600 µL phosphate buffer prepared as previously described. After a final centrifugation, 500 µL supernatant was transferred into a 5 mm NMR tube for NMR analysis.

4.2.3. NMR Spectroscopy

All NMR spectra were acquired on a Bruker Avance 800 MHz NMR spectrometer (800.15 MHz for proton frequency) at 302 K using a 5 mm cryoprobe (Bruker Biospin, Rheinstetten, Germany) [55]. A one-dimensional (1D) ^{1}H NMR spectrum was acquired for each sample in a random order using the first increment of the gradient selected NOESY pulse sequence (recycle delay$-G_1-90°-t_1-90°-t_m-G_2-90°-$acquisition) with water presaturation during both the recycle delay (2 s) and mixing time (t_m, 80 ms) [56]. For each spectrum, a total of 64 transients were collected into 32,768 data points over a spectral width of 16,025 Hz with a 90° pulse length adjusted to around 11 µs. For selected samples, a variety of two-dimensional (2D) NMR spectra [54] were acquired for the purpose of resonance assignment, including ^{1}H$-^1$H Correlation Spectroscopy (COSY), ^{1}H$-^1$H Total Correlation Spectroscopy (TOCSY), ^{1}H J-Resolved Spectroscopy (JRES), ^{1}H$-^{13}$C Heteronuclear Single Quantum Correlation Spectroscopy (HSQC), and ^{1}H$-^{13}$C Heteronuclear Multiple Bond Correlation Spectroscopy (HMBC).

After Fourier transformation with 1 Hz exponential line broadening and 65,536 data points zero-filling, each 1D spectrum was manually corrected for phase and baseline distortions and referenced to TSP (δ 0.00) using Topspin 2.0 (Bruker Biospin, Germany). The spectra regions δ 0.500−9.610 were segmented into discrete bins of 0.003 ppm width using AMIX (V3.9.15, Bruker Biospin, Germany) [57]. Spectra regions with imperfect water suppression (δ 4.500−5.170) or residual methanol signal (δ 3.356−3.370) were discarded. The intersample chemical-shift variations for some metabolites were manually corrected to prevent the possible adverse effect on subsequent data analysis [58].

4.2.4. Multivariate Data Analysis

The integral of each included bin was normalized to the summed integral of all included bins. The normalized data was then utilized for multivariate data analysis using SIMCA-P$^+$ (V11.0 and 13.0, Umetrics AB, Umea, Sweden). Initially, for the overview of data distribution and detection of possible outliers, Principal Component Analysis (PCA) was performed on mean-centered data using two principal components. Then, Projection to Latent Structure Discriminant Analysis (PLS-DA) and Orthogonal Projection to Latent Structure Discriminant Analysis (OPLS-DA) were performed on unit-variance scaled data by using grouping information as Y-matrix [59]. Two PLS components were calculated for PLS-DA models, while one PLS and one orthogonal component were used for

OPLS-DA models. Both supervised models were validated using a 7-fold cross-validation method [59]. Further assessments of model quality were also performed, including a permutation test with 200 permutations for PLS models [60] and ANOVA of the cross-validated residuals (CV-ANOVA) tests for OPLS-DA models [61].

The OPLS-DA models were interpreted as back-transformed and color-coded correlation coefficients loadings plots [62] (MATLAB 7.0, The Mathworks Inc., Natick, MA, USA), where the colors indicate the significance of differentiating metabolites, with a warm color (e.g., red) being more significant than a cool color (e.g., blue). The cutoffs for correlation coefficients were chosen on the basis of discrimination significance ($p < 0.05$), e.g., a cutoff value (|r|) of 0.602 was corresponding to the sample number (n) of 10. Differentiating metabolites were also summarized in a heat map, color-coded with the Pearson correlation coefficients from the OPLS-DA models (MeV version 4.9.0).

4.3. RNA Extraction and qRT-PCR Analysis

For RNA extraction and subsequent qRT-PCR analysis, 3 biological replicates were used for each of the three genotypes (i.e., WT, M, and ECE). Each replicate consisted of 3 seedlings pooled together before RNA extraction, and the seedlings were from the same batch used for the metabolomics analysis. Moreover, each qRT-PCR reaction was performed with 3 technical replicates. Total RNA was extracted using Qiagen RNeasy Mini Kit (Cat No. 74904) from 2-week-old seedling shoots without roots according to the manufacturer's instructions. The RNA quality was determined by a Nanodrop ND-1000 spectrophotometer (Nanodrop Technologies), and RNA samples with A260/A280 ratios between 1.9 and 2.1 were selected for further analysis. One microgram total RNA was reverse transcribed using Maxima First-Strand cDNA Synthesis Kit (ThermoFisher Scientific, Waltham, MA, USA, Cat No. K1671), as per the manufacturer's instructions. The qRT-PCR analysis was carried out using selected gene-specific primer pairs (Table S3) on *BIO-RAD* CFX384 Real-Time system with denaturation at 95 °C for 10 min, followed by 40 cycles of denaturation at 95 °C for 15s and annealing/extension at 60 °C for 1 min. The amplification of an *ACTIN* gene (*OsACT1*) was used as an internal control to normalize the data. Melting curve analyses were performed to confirm the amplicon specificity [63]. The values were expressed as the average of three independent biological samples, each averaged from its three technical replicates. The relative expression levels were calculated using the $2^{-\Delta\Delta Ct}$ method [64]. Differentially expressed transcripts were derived from two-sided unpaired t-tests, with a p value of less than 0.05 considered to be statistically significant. The data analysis and charting (with reference to WT) was performed using Microsoft Excel 2016 (Microsoft, Redmond, WA, USA).

4.4. Metabolic Pathway Analysis

Based on the differential metabolites and transcripts, the comprehensive metabolic responses in rice plants induced by *OsCYP96B4* gene mutation were mapped onto relevant metabolic pathways, with reference to the KEGG pathway database [65].

5. Conclusions

In the present study, the combination of NMR-based metabolomics and qRT-PCR analyses revealed that there were systems alteration in the metabolic phenotypes of semi-dwarf mutant (M) and ectopic expression (ECE) rice lines in comparison with the wild-type (WT) rice, as a result of the *OsCYP96B4* gene mutation. Such changes included the significant effect on amino acid metabolism (e.g., GABA shunt, glutamate and glutamine metabolism, branched-chain amino acid metabolism, choline metabolism), carbohydrate metabolism (e.g., sucrose metabolism, TCA cycle), nucleotide metabolism, and shikimate-mediated secondary metabolism. The present findings provide useful information on understanding the *OsCYP96B4* gene function possibly pertaining to the metabolism and dwarfism, which may be helpful for the development of valuable new semi-dwarf plant mutants in the future.

Supplementary Materials: The following are available online at http://www.mdpi.com/1422-0067/21/6/1924/s1, Figure S1: PCA scores plots derived from ^{1}H NMR spectra of rice plant extracts from different groups; Figure S2: Permutation test results (with 200 permutations) for PLS-DA models (with 2 components) derived from ^{1}H NMR spectra of rice plant extracts from different groups; Figure S3: Representative phenotypes of the 2-week-old (A) *oscyp96b4* semi-dwarf mutant (M), (B) *OsCYP96B4* ectopic expression (ECE), and (C) wild-type (WT) rice plants; Table S1: ^{1}H and ^{13}C NMR assignment for metabolites in rice plant extracts; Table S2: OPLS-DA loadings correlation coefficients; Table S3: Primers for quantitative real-time PCR analysis on selected genes.

Author Contributions: Conceptualization, L.J., R.R. and P.P.K.; Data curation, L.J., R.R. and S.R.; Formal analysis, L.J., R.R. and S.R.; Funding acquisition, L.J. and P.P.K.; Investigation, L.J., R.R. and S.R.; Methodology, L.J. and R.R.; Project administration, L.J. and P.P.K.; Resources, L.J., S.R. and P.P.K.; Software, L.J. and R.R.; Supervision, L.J. and P.P.K.; Validation, L.J. and R.R.; Visualization, L.J. and R.R.; Writing—original draft, L.J.; Writing—review & editing, L.J., R.R., S.R. and P.P.K. All authors have read and agreed to the published version of the manuscript.

Funding: This research was funded by National University of Singapore (R-180-000-016-733; NRF-CRP7-2010-02), Huazhong University of Science and Technology (513-3004513113), and the Chinese Academy of Sciences (KJZD-EW-G20-01).

Acknowledgments: We acknowledge the help of Siyi Guo and Toshiro Ito (Temasek Life Sciences Laboratory) for the use of Geno/Grinder®. Limiao Jiang gratefully acknowledges Huiru Tang (Fudan University) for the kind help on facilitating this collaborative study and thanks Caixiang Liu (Wuhan Institute of Physics and Mathematics) for the useful discussion.

Conflicts of Interest: The authors declare no conflict of interest. The funders had no role in the design of the study; in the collection, analyses, or interpretation of data; in the writing of the manuscript, or in the decision to publish the results.

Abbreviations

Ds	Dissociator
ECE	Ectopic expression
GABA	γ-aminobutyrate
M	Mutant
NMR	Nuclear magnetic resonance
OPLS-DA	Orthogonal projection to latent structure discriminant analysis
OsCYP96B4	*Oryza sativa* Cytochrome P450 96B4
PCA	Principal component analysis
PLS-DA	Projection to latent structure discriminant analysis
qRT-PCR	Quantitative real-time polymerase chain reaction
TCA	Tricarboxylic acid
WT	Wild type

References

1. Yu, J.; Hu, S.N.; Wang, J.; Wong, G.K.S.; Li, S.G.; Liu, B.; Deng, Y.J.; Dai, L.; Zhou, Y.; Zhang, X.Q.; et al. A draft sequence of the rice genome (Oryza sativa L. ssp indica). *Science* **2002**, *296*, 79–92. [CrossRef] [PubMed]

2. Khush, G.S. Green revolution: Preparing for the 21st century. *Genome* **1999**, *42*, 646–655. [CrossRef] [PubMed]

3. Morinaka, Y.; Sakamoto, T.; Inukai, Y.; Agetsuma, M.; Kitano, H.; Ashikari, M.; Matsuoka, M. Morphological alteration caused by brassinosteroid insensitivity increases the biomass and grain production of rice. *Plant Physiol.* **2006**, *141*, 924–931. [CrossRef] [PubMed]

4. Berry, P.M.; Sterling, M.; Spink, J.H.; Baker, C.J.; Sylvester-Bradley, R.; Mooney, S.J.; Tams, A.R.; Ennos, A.R. Understanding and reducing lodging in cereals. *Adv. Agron.* **2004**, *84*, 217–271.

5. Ramamoorthy, R.; Jiang, S.Y.; Ramachandran, S. Oryza sativa Cytochrome P450 Family Member OsCYP96B4 Reduces Plant Height in a Transcript Dosage Dependent Manner. *PLoS ONE* **2011**, *6*, e28069. [CrossRef]

6. Zhang, J.; Liu, X.Q.; Li, S.Y.; Cheng, Z.K.; Li, C.Y. The Rice Semi-Dwarf Mutant sd37, Caused by a Mutation in CYP96B4, Plays an Important Role in the Fine-Tuning of Plant Growth. *PLoS ONE* **2014**, *9*, e88068. [CrossRef]

7. Asano, K.; Hirano, K.; Ueguchi-Tanaka, M.; Angeles-Shim, R.B.; Komura, T.; Satoh, H.; Kitano, H.; Matsuoka, M.; Ashikari, M. Isolation and characterization of dominant dwarf mutants, Slr1-d, in rice. *Mol. Genet. Genomics* **2009**, *281*, 223–231. [CrossRef]

8.   Tong, J.P.; Han, Z.S.; Han, A.N.; Liu, X.J.; Zhang, S.Y.; Fu, B.Y.; Hu, J.; Su, J.P.; Li, S.Q.; Wang, S.J.; et al. Sdt97: A Point Mutation in the 5 Untranslated Region Confers Semidwarfism in Rice. *G3-Genes Genomes Genet.* **2016**, *6*, 1491–1502. [CrossRef]

9.   Xu, J.; Wang, X.Y.; Guo, W.Z. The cytochrome P450 superfamily: Key players in plant development and defense. *J. Integr. Agric.* **2015**, *14*, 1673–1686. [CrossRef]

10. Zhang, D.; Yang, H.F.; Wang, X.C.; Qiu, Y.J.; Tian, L.H.; Qi, X.Q.; Qu, L. Cytochrome P450 family member CYP96B5 hydroxylates alkanes to primary alcohols and is involved in rice leaf cuticular wax synthesis. *New Phytol.* **2020**, *225*, 2094–2107. [CrossRef]

11. Wang, X.L.; Cheng, Z.J.; Zhao, Z.C.; Gan, L.; Qin, R.Z.; Zhou, K.N.; Ma, W.W.; Zhang, B.C.; Wang, J.L.; Zhai, H.Q.; et al. BRITTLE SHEATH1 encoding OsCYP96B4 is involved in secondary cell wall formation in rice. *Plant Cell Rep.* **2016**, *35*, 745–755. [CrossRef] [PubMed]

12. Tamiru, M.; Undan, J.R.; Takagi, H.; Abe, A.; Yoshida, K.; Undan, J.Q.; Natsume, S.; Uemura, A.; Saitoh, H.; Matsumura, H.; et al. A cytochrome P450, OsDSS1, is involved in growth and drought stress responses in rice (Oryza sativa L.). *Plant Mol. Biol.* **2015**, *88*, 85–99. [CrossRef]

13. Nicholson, J.K.; Lindon, J.C.; Holmes, E. 'Metabonomics': Understanding the metabolic responses of living systems to pathophysiological stimuli via multivariate statistical analysis of biological NMR spectroscopic data. *Xenobiotica* **1999**, *29*, 1181–1189. [CrossRef] [PubMed]

14. Fiehn, O. Metabolomics-the link between genotypes and phenotypes. *Plant Mol. Biol.* **2002**, *48*, 155–171. [CrossRef] [PubMed]

15. Nicholson, J.K.; Connelly, J.; Lindon, J.C.; Holmes, E. Metabonomics: A platform for studying drug toxicity and gene function. *Nat. Rev. Drug Discovery* **2002**, *1*, 153–161. [CrossRef] [PubMed]

16. Fiehn, O.; Kopka, J.; Dormann, P.; Altmann, T.; Trethewey, R.N.; Willmitzer, L. Metabolite profiling for plant functional genomics. *Nat. Biotechnol.* **2000**, *18*, 1157–1161. [CrossRef]

17. Tang, H.R.; Wang, Y.L. Metabonomics: A revolution in progress. *Prog. Biochem. Biophys.* **2006**, *33*, 401–417.

18. Jacobs, A.; Lunde, C.; Bacic, A.; Tester, M.; Roessner, U. The impact of constitutive heterologous expression of a moss Na+ transporter on the metabolomes of rice and barley. *Metabolomics* **2007**, *3*, 307–317. [CrossRef]

19. Barding, G.A.; Fukao, T.; Beni, S.; Bailey-Serres, J.; Larive, C.K. Differential Metabolic Regulation Governed by the Rice SUB1A Gene during Submergence Stress and Identification of Alanylglycine by H-1 NMR Spectroscopy. *J. Proteome Res.* **2012**, *11*, 320–330. [CrossRef]

20. Chen, F.F.; Zhang, J.T.; Song, X.S.; Yang, J.; Li, H.P.; Tang, H.R.; Liao, Y.C. Combined metabonomic and quantitative real-time PCR analyses reveal systems metabolic changes of *Fusarium graminearum* induced by *Tri5* gene deletion. *J. Proteome Res.* **2011**, *10*, 2273–2285. [CrossRef]

21. Liu, C.X.; Ding, F.; Hao, F.H.; Yu, M.; Lei, H.H.; Wu, X.Y.; Zhao, Z.X.; Guo, H.X.; Yin, J.; Wang, Y.L.; et al. Reprogramming of Seed Metabolism Facilitates Pre-harvest Sprouting Resistance of Wheat. *Sci. Rep.* **2016**, *6*, 20593. [CrossRef] [PubMed]

22. Kumar, Y.; Zhang, L.M.; Panigrahi, P.; Dholakia, B.B.; Dewangan, V.; Chavan, S.G.; Kunjir, S.M.; Wu, X.Y.; Li, N.; Rajmohanan, P.R.; et al. Fusarium oxysporum mediates systems metabolic reprogramming of chickpea roots as revealed by a combination of proteomics and metabolomics. *Plant Biotechnol. J.* **2016**, *14*, 1589–1603. [CrossRef] [PubMed]

23. Liu, C.X.; Hao, F.H.; Hu, J.; Zhang, W.L.; Wan, L.L.; Zhu, L.L.; Tang, H.R.; He, G.C. Revealing different systems responses to brown planthopper infestation for pest susceptible and resistant rice plants with the combined metabonomic and gene-expression analysis. *J. Proteome Res.* **2010**, *9*, 6774–6785. [CrossRef] [PubMed]

24. Villette, C.; Zumsteg, J.; Schaller, H.; Heintz, D. Non-targeted metabolic profiling of BW312 Hordeum vulgare semi dwarf mutant using UHPLC coupled to QTOF high resolution mass spectrometry. *Sci. Rep.* **2018**, *8*, 13178. [CrossRef]

25. Cevallos-Cevallos, J.M.; Jines, C.; Mariduena-Zavala, M.G.; Molina-Miranda, M.J.; Ochoa, D.E.; Flores-Cedeno, J.A. GC-MS metabolite profiling for specific detection of dwarf somaclonal variation in banana plants. *Appl. Plant Sci.* **2018**, *6*, e1194. [CrossRef]

26. Flores, P.; Hernandez, V.; Hellin, P.; Fenoll, J.; Cava, J.; Mestre, T.; Martinez, V. Metabolite profile of the tomato dwarf cultivar Micro-Tom and comparative response to saline and nutritional stresses with regard to a commercial cultivar. *J. Sci. Food Agric.* **2016**, *96*, 1562–1570. [CrossRef]

27. John, K.M.M.; Khan, F.; Luthria, D.L.; Matthews, B.; Garrett, W.M.; Natarajan, S. Proteomic and metabolomic analysis of minimax and Williams 82 soybeans grown under two different conditions. *J. Food Biochem.* **2017**, *41*, e12404. [CrossRef]

28. Blomstedt, C.K.; O'Donnell, N.H.; Bjarnholt, N.; Neale, A.D.; Hamill, J.D.; Moller, B.L.; Gleadow, R.M. Metabolic consequences of knocking out UGT85B1, the gene encoding the glucosyltransferase required for synthesis of dhurrin in Sorghum bicolor (L. Moench). *Plant Cell Physiol.* **2016**, *57*, 373–386. [CrossRef]

29. Fan, W.M.T. Metabolite profiling by one- and two-dimensional NMR analysis of complex mixtures. *Prog. Nucl. Magn. Reson. Spectrosc.* **1996**, *28*, 161–219. [CrossRef]

30. Shelp, B.J.; Bown, A.W.; McLean, M.D. Metabolism and functions of gamma-aminobutyric acid. *Trends Plant Sci.* **1999**, *4*, 446–452. [CrossRef]

31. Bouche, N.; Fait, A.; Bouchez, D.; Moller, S.G.; Fromm, H. Mitochondrial succinic-semialdehyde dehydrogenase of the gamma-aminobutyrate shunt is required to restrict levels of reactive oxygen intermediates in plants. *Proc. Natl. Acad. Sci. USA* **2003**, *100*, 6843–6848. [CrossRef] [PubMed]

32. Koike, S.; Matsukura, C.; Takayama, M.; Asamizu, E.; Ezura, H. Suppression of gamma-Aminobutyric Acid (GABA) Transaminases Induces Prominent GABA Accumulation, Dwarfism and Infertility in the Tomato (Solanum lycopersicum L.). *Plant Cell Physiol.* **2013**, *54*, 793–807. [CrossRef] [PubMed]

33. Akama, K.; Takaiwa, F. C-terminal extension of rice glutamate decarboxylase (OsGAD2) functions as an autoinhibitory domain and overexpression of a truncated mutant results in the accumulation of extremely high levels of GABA in plant cells. *J. Exp. Bot.* **2007**, *58*, 2699–2707. [CrossRef] [PubMed]

34. Yang, X.L.; Nian, J.Q.; Xie, Q.J.; Feng, J.; Zhang, F.X.; Jing, H.W.; Zhang, J.; Dong, G.J.; Liang, Y.; Peng, J.L.; et al. Rice Ferredoxin-Dependent Glutamate Synthase Regulates Nitrogen-Carbon Metabolomes and Is Genetically Differentiated between japonica and indica Subspecies. *Mol. Plant* **2016**, *9*, 1520–1534. [CrossRef] [PubMed]

35. Bao, Y.X.; Zhao, R.; Li, F.F.; Tang, W.; Han, L.B. Simultaneous Expression of Spinacia oleracea Chloroplast Choline Monooxygenase (CMO) and Betaine Aldehyde Dehydrogenase (BADH) Genes Contribute to Dwarfism in Transgenic Lolium perenne. *Plant Mol. Biol. Rep.* **2011**, *29*, 379–388. [CrossRef]

36. Fan, C.F.; Feng, S.Q.; Huang, J.F.; Wang, Y.T.; Wu, L.M.; Li, X.K.; Wang, L.Q.; Tu, Y.Y.; Xia, T.; Li, J.Y.; et al. AtCesA8-driven OsSUS3 expression leads to largely enhanced biomass saccharification and lodging resistance by distinctively altering lignocellulose features in rice. *Biotechnol. Biofuels* **2017**, *10*, 221. [CrossRef]

37. Huang, P.; Yoshida, H.; Yano, K.; Kinoshita, S.; Kawai, K.; Koketsu, E.; Hattori, M.; Takehara, S.; Huang, J.; Hirano, K.; et al. OsIDD2, a zinc finger and INDETERMINATE DOMAIN protein, regulates secondary cell wall formation. *J. Integr. Plant Biol.* **2018**, *60*, 130–143. [CrossRef]

38. Wu, B.M.; Li, L.; Qiu, T.H.; Zhang, X.; Cui, S.X. Cytosolic APX2 is a pleiotropic protein involved in H2O2 homeostasis, chloroplast protection, plant architecture and fertility maintenance. *Plant Cell Rep.* **2018**, *37*, 833–848. [CrossRef]

39. Beeler, S.; Liu, H.C.; Stadler, M.; Schreier, T.; Eicke, S.; Lue, W.L.; Truernit, E.; Zeeman, S.C.; Chen, J.; Kotting, O. Plastidial NAD-Dependent Malate Dehydrogenase Is Critical for Embryo Development and Heterotrophic Metabolism in Arabidopsis. *Plant Physiol.* **2014**, *164*, 1175–1190. [CrossRef]

40. Itai, R.; Suzuki, K.; Yamaguchi, H.; Nakanishi, H.; Nishizawa, N.K.; Yoshimura, E.; Mori, S. Induced activity of adenine phosphoribosyltransferase (APRT) in iron-deficient barley roots: A possible role for phytosiderophore production. *J. Exp. Bot.* **2000**, *51*, 1179–1188. [CrossRef]

41. Casartelli, A.; Melino, V.J.; Baumann, U.; Riboni, M.; Suchecki, R.; Jayasinghe, N.S.; Mendis, H.; Watanabe, M.; Erban, A.; Zuther, E.; et al. Opposite fates of the purine metabolite allantoin under water and nitrogen limitations in bread wheat. *Plant Mol.Biol.* **2019**, *99*, 477–497. [CrossRef] [PubMed]

42. Zhou, C.J.; Li, J.; Zou, J.C.; Liang, F.S.; Ye, C.J.; Jin, D.M.; Weng, M.L.; Wang, B. Cloning and characterization of a second form of the rice adenine phosphoribosyl transferase gene (OsAPT2) and its association with TGMS. *Plant Mol.Biol.* **2006**, *60*, 365–376. [CrossRef] [PubMed]

43. Moffatt, B.; Somerville, C. Positive selection for male-sterile mutants of Arabidopsis lacking adenine phosphoribosyl transferase activity. *Plant Physiol.* **1988**, *86*, 1150–1154. [CrossRef] [PubMed]

44. Tzin, V.; Galili, G. The Biosynthetic Pathways for Shikimate and Aromatic Amino Acids in Arabidopsis thaliana. *Arabidopsis Book* **2010**, *8*, e0132. [CrossRef]

45. Tzin, V.; Galili, G. New Insights into the Shikimate and Aromatic Amino Acids Biosynthesis Pathways in Plants. *Mol. Plant* **2010**, *3*, 956–972. [CrossRef]

46. Yang, D.H.; Chung, B.Y.; Kim, J.S.; Kim, J.H.; Yun, P.Y.; Lee, Y.K.; Lim, Y.P.; Lee, M.C. cDNA cloning and sequence analysis of the rice cinnamate-4-hydroxylase gene, a cytochrome P450-dependent monooxygenase involved in the general phenylpropanoid pathway. *J. Plant Biol.* **2005**, *48*, 311–318. [CrossRef]

47. Kim, J.I.; Dolan, W.L.; Anderson, N.A.; Chapple, C. Indole Glucosinolate Biosynthesis Limits Phenylpropanoid Accumulation in Arabidopsis thaliana. *Plant Cell* **2015**, *27*, 1529–1546. [CrossRef]

48. Ahmad, I.; Kamran, M.; Ali, S.; Bilegjargal, B.; Cai, T.; Ahmad, S.; Meng, X.P.; Su, W.N.; Liu, T.N.; Han, Q.F. Uniconazole application strategies to improve lignin biosynthesis, lodging resistance and production of maize in semiarid regions. *Field Crop. Res.* **2018**, *222*, 66–77. [CrossRef]

49. Wang, C.; Hu, D.; Liu, X.B.; She, H.Z.; Ruan, R.W.; Yang, H.; Yi, Z.L.; Wu, D.Q. Effects of uniconazole on the lignin metabolism and lodging resistance of culm in common buckwheat (Fagopyrum esculentum M.). *Field Crop. Res.* **2015**, *180*, 46–53. [CrossRef]

50. Nishikubo, N.; Araki, T.; Kajita, S.; Kuroda, K.; Kitano, H.; Katayama, Y. Specific accumulation of polysaccharide-linked hydroxycinnamoyl esters in the cell walls of irregularly shaped and collapsed internode parenchyma cells of the dwarf rice mutant Fukei 71. *Plant Cell Physiol.* **2000**, *41*, 776–784. [CrossRef]

51. Kim, Y.S.; Park, S.; Kang, K.; Lee, K.; Back, K. Tyramine accumulation in rice cells caused a dwarf phenotype via reduced cell division. *Planta* **2011**, *233*, 251–260. [CrossRef] [PubMed]

52. Lee, K.; Back, K. Melatonin-deficient rice plants show a common semidwarf phenotype either dependent or independent of brassinosteroid biosynthesis. *J. Pineal Res.* **2019**, *66*, e12537. [CrossRef] [PubMed]

53. Murashige, T.; Skoog, F. A revised medium for rapid growth and bio assays with tobacco tissue cultures. *Physiol. Plant.* **1962**, *15*, 473–497. [CrossRef]

54. Jiang, L.M.; Huang, J.; Wang, Y.L.; Tang, H.R. Metabonomic analysis reveals the CCl_4-induced systems alterations for multiple rat organs. *J. Proteome Res.* **2012**, *11*, 3848–3859. [CrossRef] [PubMed]

55. Jiang, L.; Lee, S.C.; Ng, T.C. Pharmacometabonomics Analysis Reveals Serum Formate and Acetate Potentially Associated with Varying Response to Gemcitabine-Carboplatin Chemotherapy in Metastatic Breast Cancer Patients. *J. Proteome Res.* **2018**, *17*, 1248–1257. [CrossRef] [PubMed]

56. Jiang, L.M.; Zhao, X.J.; Huang, C.Y.; Lei, H.H.; Tang, H.R.; Wang, Y.L. Dynamic changes in metabolic profiles of rats subchronically exposed to mequindox. *Mol. Biosyst.* **2014**, *10*, 2914–2922. [CrossRef]

57. Jiang, L.; Wang, J.; Li, R.; Fang, Z.-M.; Zhu, X.-H.; Yi, X.; Lan, H.; Wei, X.; Jiang, D.-S. Disturbed energy and amino acid metabolism with their diagnostic potential in mitral valve disease revealed by untargeted plasma metabolic profiling. *Metabolomics* **2019**, *15*, 57. [CrossRef]

58. Jiang, L.M.; Huang, J.; Wang, Y.L.; Tang, H.R. Eliminating the dication-induced intersample chemical-shift variations for NMR-based biofluid metabonomic analysis. *Analyst* **2012**, *137*, 4209–4219. [CrossRef]

59. Trygg, J.; Wold, S. Orthogonal projections to latent structures (O-PLS). *J. Chemom.* **2002**, *16*, 119–128. [CrossRef]

60. Lindgren, F.; Hansen, B.; Karcher, W.; Sjostrom, M.; Eriksson, L. Model validation by permutation tests: Applications to variable selection. *J. Chemom.* **1996**, *10*, 521–532. [CrossRef]

61. Eriksson, L.; Trygg, J.; Wold, S. CV-ANOVA for significance testing of PLS and OPLS® models. *J. Chemom.* **2008**, *22*, 594–600. [CrossRef]

62. Cloarec, O.; Dumas, M.E.; Trygg, J.; Craig, A.; Barton, R.H.; Lindon, J.C.; Nicholson, J.K.; Holmes, E. Evaluation of the orthogonal projection on latent structure model limitations caused by chemical shift variability and improved visualization of biomarker changes in ^{1}H NMR spectroscopic metabonomic studies. *Anal. Chem.* **2005**, *77*, 517–526. [CrossRef] [PubMed]

63. Pryor, R.J.; Wittwer, C.T. Real-time polymerase chain reaction and melting curve analysis. In *Methods in Molecular Biology*; Lo, Y.M.D., Chiu, R.W.K., Chan, K.C.A., Eds.; 999 Riverview Dr, Ste 208; Humana Press Inc.: Totowa, NJ, USA, 2006; Volume 336, pp. 19–32.

64. Livak, K.J.; Schmittgen, T.D. Analysis of relative gene expression data using real-time quantitative PCR and the $2^{-\Delta\Delta Ct}$ method. *Methods* **2001**, *25*, 402–408. [CrossRef] [PubMed]
65. Kanehisa, M.; Goto, S. KEGG: Kyoto Encyclopedia of Genes and Genomes. *Nucleic Acids Res.* **2000**, *28*, 27–30. [CrossRef]

International Journal of
Molecular Sciences

Article

Genome Sequence and QTL Analyses Using Backcross Recombinant Inbred Lines (BILs) and BILF₁ Lines Uncover Multiple Heterosis-related Loci

Yahui Yu [1,2,3,†], Mengmeng Zhu [1,†], Yue Cui [1,†], Yu Liu [2,†], Zhenyu Li [2], Nan Jiang [4], Zhengjin Xu [1], Quan Xu [1,*] and Guomin Sui [3,*]

[1] Rice Research Institute, Shenyang Agricultural University, Shenyang 110866, China; yyh3655@126.com (Y.Y.); ZMM895486619@126.com (M.Z.); CY544549733@126.com (Y.C.); xuzhengjin@126.com (Z.X.)

[2] Liaoning Institute of Saline-Alkali Land Utilization, Panjin 124010, China; lyyyh8685@126.com (Y.L.); ZL17004745@126.com (Z.L.)

[3] Liaoning Academy of Agricultural Sciences, Shenyang 110866, China

[4] Shenyang Research and Development Service Center of Modern Agriculture, Shenyang 110034, China; jnan0423@163.com

* Correspondence: kobexu34@syau.edu.cn (Q.X.); guomin666@126.com (G.S.)

† These authors contributed equally to this work.

Received: 5 December 2019; Accepted: 23 January 2020; Published: 25 January 2020

Abstract: Heterosis is an interesting topic for both breeders and biologists due to its practical importance and scientific significance. Cultivated rice (*Oryza sativa* L.) consists of two subspecies, *indica* and *japonica*, and hybrid rice is the predominant form of *indica* rice in China. However, the molecular mechanism underlying heterosis in *japonica* remains unclear. The present study determined the genome sequence and conducted quantitative trait locus (QTL) analysis using backcross recombinant inbred lines (BILs) and BILF₁ lines to uncover the heterosis-related loci for rice yield increase under a *japonica* genetic background. The BIL population was derived from an admixture variety Habataki and *japonica* variety Sasanishiki cross to improve the genetic diversity but maintain the genetic background close to *japonica*. The results showed that heterosis in F₁ mainly involved grain number per panicle. The BILF₁s showed an increase in grain number per panicle but a decrease in plant height compared with the BILs. Genetic analysis then identified eight QTLs for heterosis in the BILF₁s; four QTLs were detected exclusively in the BILF₁ population only, presenting a mode of dominance or super-dominance in the heterozygotes. An additional four loci overlapped with QTLs detected in the BIL population, and we found that *Grains Height Date 7* (*Ghd7*) was correlated in days to heading in both BILs and BILF₁s. The admixture genetic background of Habataki was also determined by subspecies-specific single nucleotide polymorphisms (SNPs). This investigation highlights the importance of high-throughput sequencing to elucidate the molecular mechanism of heterosis and provides useful germplasms for the application of heterosis in *japonica* rice production.

Keywords: rice; heterosis; yield components; high-throughput sequence

---

1. Introduction

Food security is a major global problem as the competition for arable land between food and energy crops, and population growth, continue to increase. Improving crop productivity has been the key focus of national and international efforts in breeding crops such as maize and rice. The vigor of *indica* rice hybrids often exhibits phenotypes that surpass their parents in terms of growth and fertility, which is also known as heterosis. In crop production, successful agronomic exploitation of yield heterosis has been achieved in the past decades. Three major competing but non-mutually-exclusive hypotheses (dominance, overdominance, and epistasis) have been proposed to explain heterosis at the genetic level [1–6]. However, the progress in elucidating the molecular mechanism underlying crop heterosis has lagged.

In the past decades, heterosis has become a priority research target for both breeders and scientists based on its practical importance and scientific significance. Heterosis in crop breeding was first applied to hybrid maize in the 1930s, and the three-line method or cytoplasmic male sterility (CMS) system contributed to the commercialization of hybrid rice in the 1970s. Crop heterosis has been extensively applied to rice and maize production, significantly improving global yield compared with traditional inbred lines [7]. Currently, the planting area of hybrid rice accounts for 50–60% of the total rice planting area, and about 80% of *indica* rice is hybrid and is mainly planted in southern China. A study used recombinant inbred lines (RIL) derived from a cross between PA64s and 93-11, and RIL backcross F_1 populations were analyzed to elucidate the molecular mechanism of heterosis among *indica* and *javanica* varieties [8]. A mega sequence project for 10,0742 F_2 lines revealed the genomic architecture of heterosis for yield traits in rice using the cross combination of *Oryza sativa* subspecies (ssp.) *indica–indica* (three-line system), *indica–indica* (two-line system), and *O. sativa* ssp. *indica–O. sativa* ssp. *japonica* crosses [9]. Nevertheless, research on heterosis in *japonica* is limited. Almost all *japonica* varieties are conventional rice, and the hybrid *japonica* accounts for less than 3% among total *japonica*, which is mainly distributed in northern China and the middle range of China. The disappearance of hybrid *japonica* rice is due to the lack of genetic diversity among *japonica* cultivars.

In this study, we used a backcross RIL derived from the cross between Habataki (an admixture variety between the *indica* and *japonica* variety) and *japonica* variety Sasanishiki, and F_1 plants were backcrossed to Sasanishiki once before inbreeding. The backcross improved genetic diversity through the introgression of the *indica* pedigree while maintaining the population under the *japonica* genetic background. After inbreeding for 10 generations, we obtained 85 lines of the backcross recombinant inbred line (BIL) population. Then, we crossed all of the 85 lines to Sasanishiki again to generate the $BILF_1$ population. The present study used genome sequencing and quantitative trait locus (QTL) analysis of BILs and $BILF_1$s to identify heterosis-related loci for yield increase.

2. Results

2.1. Population Sequencing and Linkage Map Construction

We used a strategy of sequencing-based map construction to conduct QTL mapping for the BIL derived from the cross between Sasanishiki and Habataki (Figure 1A). We sequenced a segregating population of Sasanishiki and Habataki BILs together with parental lines on an Illumina HiSeq2500 platform. A total of 224.02 GB of raw data were generated for all of the BILs, with approximately 6.29-fold depth for each BIL, 22.94 GB for Habataki (51.00-fold), and 23.24 GB for Sasanishiki (56.00-fold). We aligned the sequence data to the reference genome (Os-Nipponbare-Reference-IRGSP-1.0) using SOAP2 software [10,11]. A total of 1,947,668 single-nucleotide polymorphisms (SNPs) with homozygous genotypes between both parents were identified using the SOAPsnp software [12]. These SNPs were used as potential markers in the subsequent analysis. The SNP markers localized to highly repetitive regions, and those with low genotyping scores were removed to avoid ambiguity in linkage map construction. We used an effective imputation model, k-nearest neighbor algorithm to impute the missing genotypes of each RIL caused by low-coverage sequencing [13]. Finally, we used 1,591,495 high-quality polymorphic SNP markers to construct a recombinant bin map (Figure 1B). Subsequently, a recombinant bin map was constructed, and the map contained 3652 recombinant blocks, with the average genetic length of 0.44 cM. Then, we determined the introgression rate of the Habataki pedigree using the data of bin map. The introgression rate showed a normal distribution that indicated that the population is ideal for the subsequent survey (Figure 1C). We analyzed the correlation between the introgression rate of Habataki and the important yield-related traits. The results showed that the introgression rate was significantly negatively correlated to 1000-grain weight (TGW) (Figure 1D).

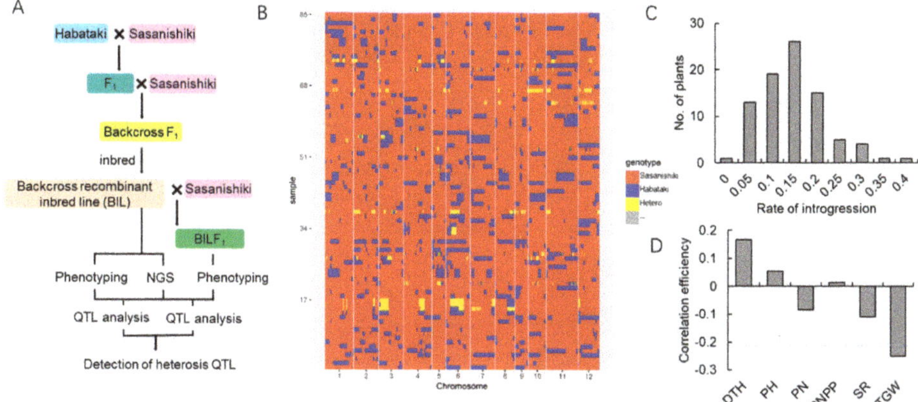

Figure 1. Schematic overview of the parent backcross recombinant inbred line (BIL) system construction and the map of genome-wide graphic genotypes. (**A**) The technology roadmap using genome sequence and quantitative trait locus (QTL) analysis of BILs and BILF$_1$s to uncover heterosis-related loci for yield increase. (**B**) Graphic genotypes of 85 BILs were identified by a sliding window approach along each chromosome. Various colors represent different genotypes. (**C**) The introgression rate of the Habataki pedigree among the BILs. (**D**) The correlation between the introgression rate of the Habataki pedigree and yield-related traits. The dotted lines indicate significance at the 5% level.

2.2. Grain Yield Heterosis is the Result of Hybrid Vigor in Grain Number Per Panicle

To obtain ideal agronomic characteristics of heterosis, the paternal line itself is usually an excellent inbred variety with superior agronomic performance. In addition, the hybrid F_1 plant should surpass its paternal parent, particularly the traits related to yield components. Interestingly, the F_1 plants did not exhibit greater plant height compared with the parental lines in the present study (Figure 2A). We surveyed the yield components of the parental lines and F_1 plants (Figure 2). The panicle length of the F_1 plants was significantly longer than that of Sasanishiki and Habataki (Figure 2B). Sasanishiki exhibited a short round grain shape, whereas the F_1 and Habataki exhibited a slender grain shape. However, the 1000-grain weight of Habataki was significantly lower than that of the F_1 plants, whereas the 1000-grain weight of Sasanishiki was similar to that of the F_1 plants. The results showed that the F_1 plants exhibited an advantage in grain number per panicle and panicle number compared with Sasanishiki and Habataki. A similar setting rate was observed in F_1, Sasanishiki, and Habataki. The slenderest grain shape of Habataki had the lowest 1000-grain weight in the parent line and F_1 plants, whereas Sasanishiki had similar 1000-grain weight as the F_1 plants. Taken together, the advantage in panicle number and grain number per panicle makes the F_1 plants exhibit heterosis in grain yield per plant.

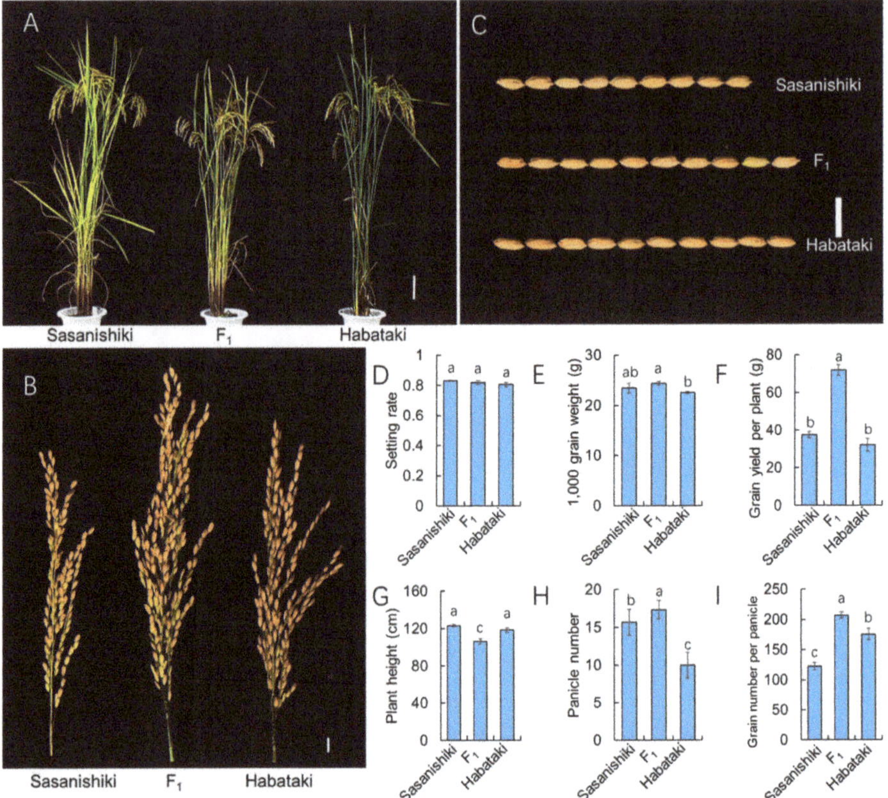

Figure 2. Heterosis in F_1 plants. (**A**) The plant architecture of F_1 plant and parental lines. Scale bar = 10 cm. (**B**) The panicle of F_1 plants and parental lines. Scale bar = 1 cm. (**C**) Grains of F_1 plants and parental lines. Scale bar = 10 cm. (**D–I**) the yield-related traits of F_1 plants and parental lines.

2.3. Yield Components and Other Important Agronomic Traits of BILs and BILF$_1$s

To assess the distribution pattern of yield-related traits between BILs and BILF$_1$s, we surveyed the yield components of BILs and BILF$_1$s in the paddy field of Shenyang Agricultural University in 2018. The normal distribution and transgressive segregation were observed in all of the traits surveyed in both BIL and BILF$_1$s. These results indicate that all of these agronomic traits were controlled by multiple genes. The plant height of BILF$_1$s mainly ranged from 105 to 120 cm, whereas that of BILs was from 120 to 135 cm. These results indicate that the heterozygous genotype of the F_1 plants reduces the plant height, based on the fact that the F_1 plants were shorter than Sasanishiki and Habataki (Figure 1A). In addition, the grain number per panicle of BILs ranged from 80 to 180, whereas that of the BILF$_1$s ranged from 130 to 260. These results suggest that the heterozygous genotype of the F_1 plants increases the grain number per panicle, which also coincides with the result that the F_1 plants have significantly more grains per panicle than Sasanishiki and Habataki (Figure 2). Similar distributions of days to heading, panicle number, and 1000-grain weight were observed in both BILs and BILF$_1$s. However, the variation range of setting rate in BILs was larger than in the BILF$_1$s (Figure 3).

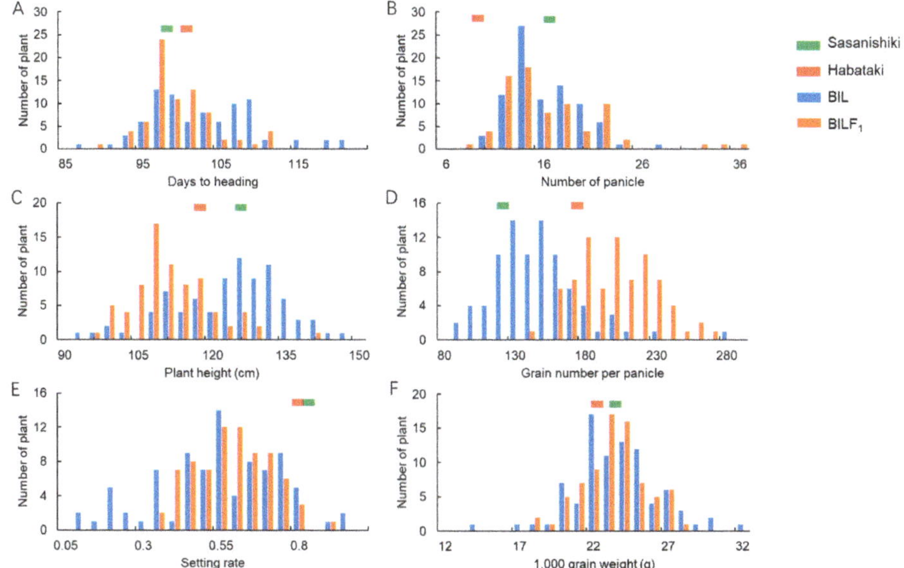

Figure 3. The distribution of yield-related traits in BILs and BILs. (**A–F**) The distribution of days to heading, number of panicles, plant height, grain number per panicle, setting rate, and 1000-grain weight in BILs and BILF$_1$s.

2.4. QTL Detection and Analysis Using the BILs and BILF$_1$ Population

To elucidate the genetic mechanism underlying heterosis of yield traits, we primarily focused on the days to heading (DTH), plant height (PH), panicle number (PN), grain number per panicle (GNPP), setting rate (SR), and 1000-grain weight (TGW). A molecular linkage map with 3652 bins was constructed based on sequence variations in Sasanishiki and Habataki and those among the BILs using Highmaps software. In the BIL population, 10 QTLs for all of the traits were mapped independently to rice chromosomes 1, 4, 5, 7, 9, 10, and 12 (Figure 4). To detect QTLs for heterosis that was associated with the effects of the heterozygous on the genetic background of Sasanishiki, we used the BILF$_1$ to conduct QTL analysis. Eight loci for the respective phenotypes were detected on chromosomes 1, 3, 6, 7, 9, and 10. Among the eight QTLs, four of these loci overlapped with QTLs detected in the BIL population, and four QTLs were detected independently in the BILF$_1$ population. These results presented a mode of dominance or super-dominance in the heterozygote. Among the QTLs detected in both BILs and BILF$_{1S}$, the cluster at the short arm of chromosome 10 corresponded to the days to heading in BILF$_1$s and the grain number per panicle in both BILs and BILF$_1$s. In addition, a cluster at chromosome 7 was related to days to heading in both BILs and BILF$_1$s. As heterosis was mainly observed in the grain number per panicle, we subsequently summarized the grain number per panicle of each genotype for the four QTLs detected in the BIL and BILF$_{1s}$ (Figure 5). The results showed that in the *Gn1* and *Ghd7* loci, the heterozygous genotype and the homozygous Habataki genotype plants had significant advantage in grain number per panicle compared with the homozygous Sasanishiki genotype plants. In the *GNNP3* locus, the homozygous Sasanishiki and homozygous Habataki genotype plants had similar grain number per panicle but significantly fewer than the heterozygous genotype plants. In the *Gn3* locus, the heterozygous genotype plants showed a significant increase of grain number per panicle compared with homozygous Habataki genotype plants, whereas the homozygous Habataki genotype plants had a significantly higher grain number per panicle than the homozygous Sasanishiki genotype plants.

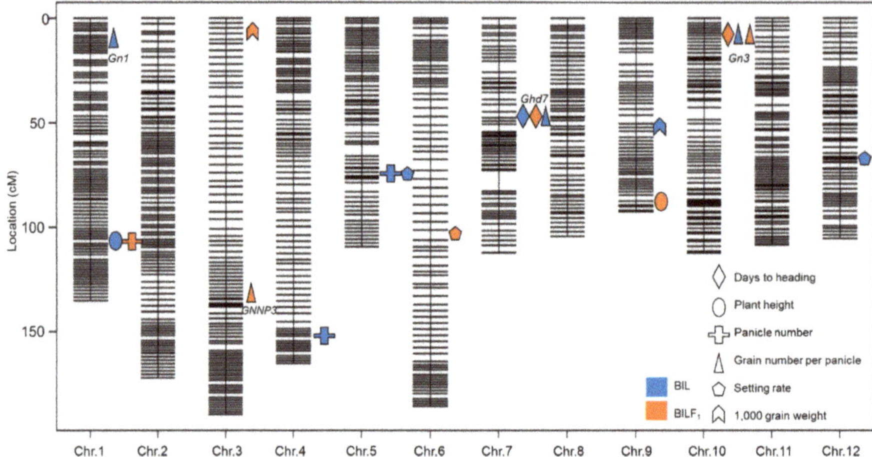

Figure 4. The position of quantitative trait locuses (QTLs) for yield-related traits and heterosis QTLs in BILs and BILF$_1$s. Different colors represent the QTLs detected in BILs and BILF$_1$s.

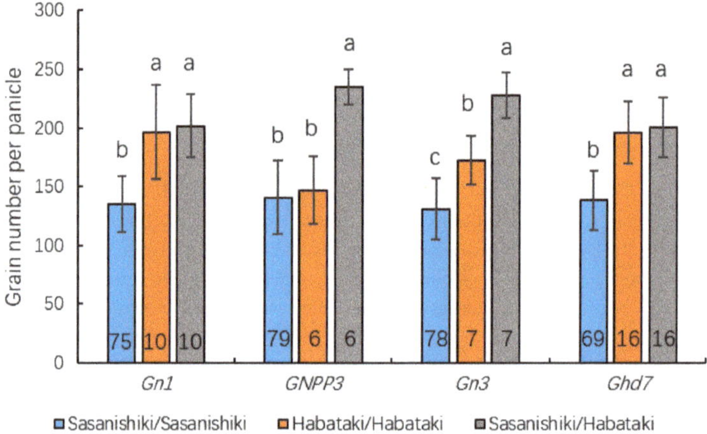

Figure 5. Grain number per panicle of each genotype for the four QTLs detected in BIL and BILF$_{1s}$. The number inside the column represents the number of plants for each genotype. All data are given as mean ± s.e.m.,different letters indicate the significant differences at the 5% level.

2.5. The Indica/Japonica Pedigree Analysis of Sasanishiki and Habataki

Breeding history indicates that Habataki is an admixture line between *indica* and *japonica*. Thus we conducted an analysis of the *indica/japonica* pedigree of Sasanishiki and Habataki. The *indica/japonica* pedigree was defined using the subspecies-specific SNPs. The subspecies-specific SNPs were those of the same type in all *japonica*, but not in *indica*, which is based on the divergence of the 517 rice landraces [14]. In total, 100,529 subspecies-specific SNPs were selected. We matched the 1,947,668 SNPs between Sasanishiki and Habataki to 100,529 subspecies-specific SNPs, and 81,690 SNPs were merged. The 81,690 SNPs were then used to analyze the *indica/japonica* pedigree of Sasanishiki and Habataki. The results showed that there are 621 *japonica*-type SNPs in the genome of Habataki, which indicated that 0.76% of *japonica* genomic introgression involved Habataki. Meanwhile, there were only 81 *indica*-type SNPs in Sasanishiki, indicating that only 0.01% *indica* pedigree introgression occurred in the Sasanishiki genome. The distribution of subspecies-specific SNPs and *indica/japonica* genomic introgression is shown in Figure 6. Agronomic-traits-related QTLs were located out of the *indica/japonica* genomic introgression, which confirmed that the heterosis QTL originated from the difference between *indica* and *japonica*.

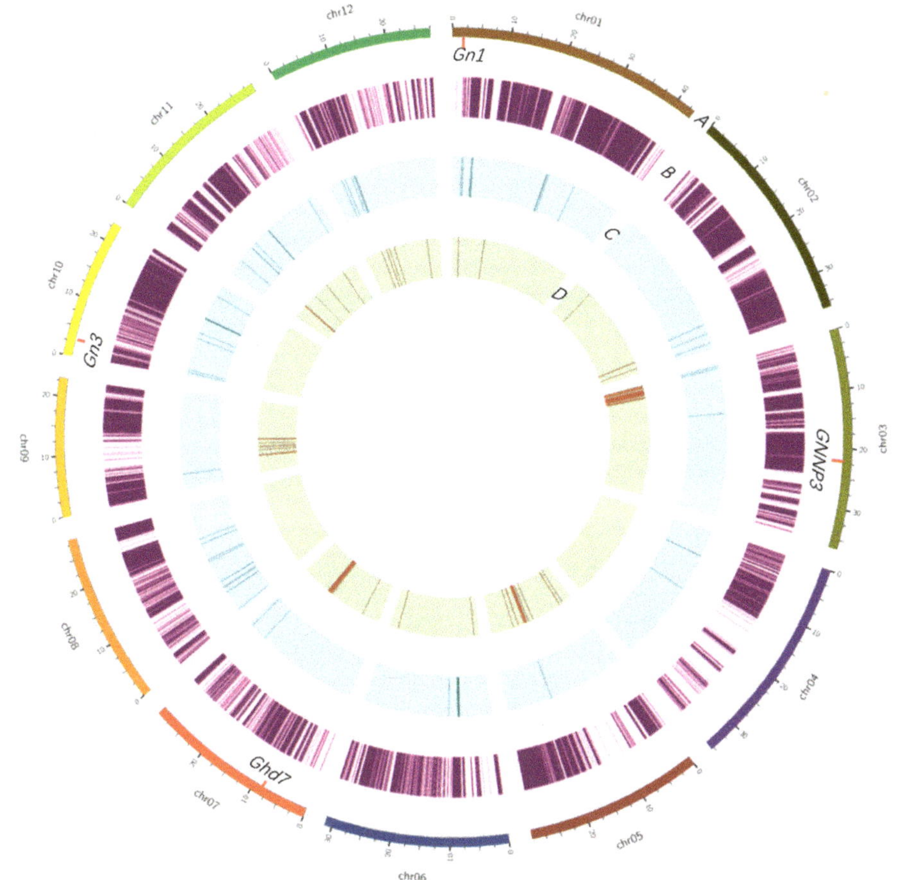

Figure 6. Overview of the subspecies-specific SNPs and introgression in Sasanishiki and Habataki. Tracks from the outer to inner circles indicate the following. (**A**) The chromosome length. The loci of grain number per panicle are indicated on the inside of the circle. (**B**) The distribution of 81,690 subspecies-specific SNPs in the genome. (**C**) The introgression of *japonica*-type SNPs into the genome of Habataki. (**D**) The introgression of *indica*-type SNPs in the genome of Sasnishiki.

3. Discussion

Since its generation in 1973, hybrid rice has predominated *indica* rice production in China. Numbers of elite combinations have been developed and released as commercial varieties, with yields roughly 20% higher than their inbred counterparts [15]. Recent molecular research has investigated the number of hybrid combinations to construct a model system for studying the molecular mechanism of heterosis for three- and two-line hybrids. A study using the yield components data and an ultra-high-density SNP bin map of an immortalized F₂ population derived from the cross between Zhenshan97 and Minghui63 demonstrated that the relative contributions of the genetic components vary with traits. The results indicate that overdominance/pseudo-overdominance are most important to the heterosis of grain number per panicle, 1000-grain weight, and grain yield per plant. In heterosis of panicle number and 1000-grain weight, the dominance × dominance interaction is important. Among these yield-related traits, single-locus dominance has relatively small contributions [16]. An integrated analysis using the RIL population and RILBCF$_1$ population derived from the cross between PA64S (which has a mix genetic background of *indica* and *javanica*) and 93-11 showed that heterosis was mainly detected in grain number per panicle and panicle number [8]. Huang et al. sequenced 10,074 F$_2$ lines derived from 17 representative hybrid combinations, and they found that a small number of genomic loci from female parents explain a large proportion of the yield advantage of hybrids over their male parents [9]. Taken together, these studies improve our understanding of heterosis. However, research on heterosis between *japonica* and *japonica* crosses is limited. The genetic diversity of *japonica* was not as high as that of *indica*, making heterosis among *japonica* inconspicuous compared with that among *indica*. Moreover, several factors have limited the utilization of heterosis between *indica* and *japonica*, which includes sterility. Thus, we used a BIL population derived from an *indica* and *japonica* cross to improve the genetic diversity but maintain the genetic background close to *japonica*. The results of the present study showed that heterosis between Sasanishiki and Habataki mainly came from a complicated quantitative and components-specific phenotype. Here, we clearly demonstrate yield heterosis, mainly by the outperformance of grain number per panicle and panicle number. Yield components survey also showed that the grain number per panicle of BILF$_1$ population was significantly higher than that of the BIL population. Our previous study demonstrated that the introgression of *indica* pedigree in the *japonica* genome contributed to the increase of rice production in northern China [17], and thus the present study confirmed that the *indica* pedigree could increase the grain number per panicle in *japonica*.

Using high-throughput sequencing, we conducted QTL mapping of the BIL and BILF$_1$ populations, and the number of QTLs that are responsible for yield and yield-related heterosis was determined. A total of 10 QTLs for all of the traits were mapped independently in BIL population (Figure 4), and eight loci for the respective phenotypes were detected. Among the eight QTLs, four of these loci overlapped with QTLs detected in the BIL population, and the remaining four QTLs were detected exclusively in the BILF$_1$ population, indicating a mode of dominance or super-dominance in the heterozygote. At the gene level, Gao et al. demonstrated that a heading-time-regulated gene *Day to Heading 8 (DTH8)* is a candidate locus for yield heterosis in Liang-you-pei 9 (LYP9) [18], and Li et al. and Huang et al. confirmed that *DTH8* corresponds to yield heterosis [8,9]. These results suggest that the heading time gene strongly participates in yield heterosis in hybrid rice. The present study detected that the heading-time gene *Ghd7* is also a candidate locus for both yield heterosis and heading time regulation. A study using the BILs derived from the cross between Habataki and Koshihikari identified five QTLs (*Gn1*–*Gn5*) that were related to grain number per panicle on chromosome 1, 4, 10, and 12 [19]. The present study confirmed that *Gn1* and *Gn3* correspond to grain number per panicle. Moreover, *Gn3*, on the short arm of chromosome 10, also corresponded to grain number per panicle in BILF$_1$s. Thus, this QTL may be further assessed in terms of heterosis.

4. Materials and Methods

4.1. Plant Materials

The parental line Sasanishiki is typical *japonica* varieties, and the parental line Habataki is an admixture variety with both *indica* and *japonica* pedigree. Sasanishiki was crossed to Habataki, and the F$_1$ plant was crossed to Sasanishiki, and then inbred over 10 generations by single-seed descent to generate a population containing 85 BILs. The BILs were then backcrossed to the maternal parent Sasanishiki to obtain 85 BILF$_1$s. All materials are maintained at the Rice Research Institute of Shenyang Agricultural University (Shenyang, China). All plants were planted in random block design under standard agricultural management practice in the paddy field of Shenyang Agricultural University (N41°, E123°). All lines were planted with three biological replicates A total of 300 plants in a 16 m^2

plot were characterized for each line. Plant height, panicle number, grain number per panicle, setting rate, and 1000-grain weight were surveyed in the field in 2018. Both BILs and BILF$_1$s were used in QTL analysis.

4.2. DNA Extraction and QTL Analysis

Young leaves of each line were collected two weeks after transplant. The CTAB method was used in extracting the high-quality genomic DNA. The sequencing libraries were constructed on an Illumina HiSeq2500 platform (Illumina, Inc.; San Diego, CA, USA) according to the manufacturer's instructions. We aligned the sequence data to the reference genome (Nipponbare, http://rapdb.dna.affrc.go.jp/download/irgsp1.html/) using SOAP2 software [11]. To construct the genetic linkage map for QTL analysis, we combined the cosegregating SNP/InDel into bins via HighMap software [20]. A map containing 3652 bins and 1592.12 cM in length was constructed with an average of 304 bins on each chromosome. Twelve linkage groups corresponded to the 12 rice chromosomes. We observed the full collinearity between the genetic map and the rice genome, and the minimum value of the Spearman coefficient for chromosomes was 0.962 (Chr. 7). The HighMap software was used to construct a linkage map. The software constructs high-quality linkage map according to the maximum likelihood estimation method. We used the R/qtl (version: 1.44-9) software to conduct QTL analysis via a composite interval mapping (CIM) model. The significance thresholds were determined by 1000 permutations. The percentage of phenotypic variance calculation explained by each QTL was obtained according to the population variance within the mapping population. The details of the QTL analysis were described in our previous studies [21].

Author Contributions: Q.X., Z.X., and G.S. designed this study and contributed to the original concept of the project. Y.Y., M.Z., Y.C., and Y.L. performed most of the experiments. Z.L. and N.J. participated in crossing B.I.L. and Sasanishiki. Q.X. wrote the manuscript. All authors have read and agreed to the published version of the manuscript.

Funding: The National Key R&D Program of China (2017YFD0100500), and the Liaoning BaiQianWan Talents Program (2016921039) supported this study.

References

1. Fiévet, J.B.; Thibault, N.; Christine, D.; de Dominique, V. Heterosis Is a Systemic Property Emerging From Non-linear Genotype-Phenotype Relationships: Evidence From in Vitro Genetics and Computer Simulations. *Front. Genet.* **2018**, *9*, 159. [CrossRef] [PubMed]

2. Bruce, B.A. The Mendelian Theory of Heredity and the Augmentation of Vigor. *Science* **1910**, *32*, 627–628. [CrossRef] [PubMed]

3. Jones, D.F. Dominance of Linked Factors as a Means of Accounting for Heterosis. *Proc. Natl. Acad. Sci. USA* **1917**, *3*, 310–312. [CrossRef] [PubMed]

4. Shull, G.H. The Genotypes of Maize. *Am. Nat.* **1911**, *45*, 234–252. [CrossRef]

5. East, E.M. Heterosis. *Genetics* **1936**, *21*, 375–397. [PubMed]

6. Minvielle, F. Dominance is not necessary for heterosis: A two-locus model. *Genet. Res.* **1987**, *49*, 245. [CrossRef]

7. Xu, Y. *Developing Marker-Assisted Selection Strategies for Breeding Hybrid Rice*; John Wiley & Sons, Ltd.: Hoboken, NJ, USA, 2010.

8. Li, D.; Huang, Z.; Song, S.; Xin, Y.; Mao, D.; Lv, Q.; Zhou, M.; Tian, D.; Tang, M.; Wu, Q. Integrated analysis of phenome, genome, and transcriptome of hybrid rice uncovered multiple heterosis-related loci for yield increase. *Proc. Natl. Acad. Sci. USA* **2016**, *113*, E6026–E6035. [CrossRef] [PubMed]

9. Huang, X.; Yang, S.; Gong, J.; Zhao, Q.; Feng, Q.; Zhan, Q.; Zhao, Y.; Li, W.; Cheng, B.; Xia, J. Genomic architecture of heterosis for yield traits in rice. *Nature* **2016**, *537*, 629–633. [CrossRef] [PubMed]

10. Kawahara, Y.; Bastide, M.D.L.; Hamilton, J.P.; Kanamori, H.; Mccombie, W.R.; Shu, O.; Schwartz, D.C.; Tanaka, T.; Wu, J.; Zhou, S. Improvement of the Oryza sativa Nipponbare reference genome using next generation sequence and optical map data. *Rice* **2013**, *6*, 4. [CrossRef] [PubMed]

11. Li, R.; Yu, C.; Li, Y.; Lam, T.W.; Yiu, S.M.; Kristiansen, K.; Wang, J. SOAP2: An improved ultrafast tool for short read alignment. *Bioinformatics* **2009**, *25*, 1966–1967. [CrossRef] [PubMed]
12. Li, R.; Li, Y.; Fang, X.; Yang, H.; Wang, J.; Kristiansen, K.; Wang, J. SNP detection for massively parallel whole-genome resequencing. *Genome Res.* **2009**, *19*, 1124. [CrossRef] [PubMed]
13. Huang, X.; Wei, X.; Sang, T.; Zhao, Q.; Feng, Q.; Zhao, Y.; Li, C.; Zhu, C.; Lu, T.; Zhang, Z. Genome-wide association studies of 14 agronomic traits in rice landraces. *Nat. Genet.* **2010**, *42*, 961–967. [CrossRef] [PubMed]
14. Huang, X.; Zhao, Y.; Wei, X.; Li, C.; Wang, A.; Zhao, Q.; Li, W.; Guo, Y.; Deng, L.; Zhu, C. Genome-wide association study of flowering time and grain yield traits in a worldwide collection of rice germplasm. *Nat. Genet.* **2012**, *44*, 32. [CrossRef] [PubMed]
15. Pan, X.W.; Qiang, H.E.; Zhang, W.H.; Shu, F.; Xing, J.J.; Sun, P.Y.; Deng, H.F.; University, H.A. Reflection of Rice Development in Yangtze River Basin under the New Situation. *Hybrid Rice* **2015**, *30*, 1–5.
16. Zhou, G.; Chen, Y.; Yao, W.; Zhang, C.; Xie, W.; Hua, J.; Xing, Y.; Xiao, J.; Zhang, Q. Genetic composition of yield heterosis in an elite rice hybrid. *Proc. Natl. Acad. Sci. USA* **2012**, *109*, 15847–15852. [CrossRef] [PubMed]
17. Sun, J.; Liu, D.; Wang, J.-Y.; Ma, D.-R.; Tang, L.; Gao, H.; Xu, Z.-J.; Chen, W.-F. The contribution of intersubspecific hybridization to the breeding of super-high-yielding japonica rice in northeast China. *Theor. Appl. Genet.* **2012**, *125*, 1149–1157. [CrossRef] [PubMed]
18. Gao, Z.-Y.; Zhao, S.-C.; He, W.-M.; Guo, L.-B.; Peng, Y.-L.; Wang, J.-J.; Guo, X.-S.; Zhang, X.-M.; Rao, Y.-C.; Zhang, C. Dissecting yield-associated loci in super hybrid rice by resequencing recombinant inbred lines and improving parental genome sequences. *Proc. Natl. Acad. Sci. USA* **2013**, *110*, 14492–14497. [CrossRef] [PubMed]
19. Ashikari, M.; Sakakibara, H.; Lin, S.; Yamamoto, T.; Takashi, T.; Nishimura, A.; Angeles, E.R.; Qian, Q.; Kitano, H.; Matsuoka, M. Cytokinin oxidase regulates rice grain production. *Science* **2005**, *309*, 741–745. [CrossRef] [PubMed]
20. Liu, D.; Ma, C.; Hong, W.; Huang, L.; Liu, M.; Liu, H.; Zeng, H.; Deng, D.; Xin, H.; Song, J. Construction and Analysis of High-Density Linkage Map Using High-Throughput Sequencing Data. *PLoS ONE* **2014**, *9*, e98855. [CrossRef] [PubMed]
21. Li, X.; Wu, L.; Wang, J.; Sun, J.; Xia, X.; Geng, X.; Wang, X.; Xu, Z.; Xu, Q. Genome sequencing of rice subspecies and genetic analysis of recombinant lines reveals regional yield- and quality-associated loci. *BMC Biol.* **2018**, *16*, 102. [CrossRef] [PubMed]

International Journal of *Molecular Sciences*

MDPI

Article

Transcriptomic and Co-Expression Network Profiling of Shoot Apical Meristem Reveal Contrasting Response to Nitrogen Rate between *Indica* and *Japonica* Rice Subspecies

Xiaoxiang Zhang [1,2,†], Juan Zhou [1,†], Niansheng Huang [2], Lanjing Mo [1], Minjia Lv [1], Yingbo Gao [1], Chen Chen [3], Shuangyi Yin [1], Jing Ju [4], Guichun Dong [1], Yong Zhou [1], Zefeng Yang [1], Aihong Li [2], Yulong Wang [1], Jianye Huang [1,*] and Youli Yao [1,*]

[1] Jiangsu Key Laboratory of Crop Genetics and Physiology/Co-Innovation Center for Modern Production Technology of Grain Crops, Yangzhou University, Yangzhou 225009, China; zhangxiaoxiangyzu@126.com (X.Z.); juanzhou@yzu.edu.cn (J.Z.); m2013325@163.com (L.M.); lvminjia@126.com (M.L.); yingbogao_yzu@163.com (Y.G.); shuangyiyin@126.com (S.Y.); gcdong@yzu.edu.cn (G.D.); zhouyong@yzu.edu.cn (Y.Z.); zfyang@yzu.edu.cn (Z.Y.); ylwang@yzu.edu.cn (Y.W.)

[2] Lixiahe Agricultural Research Institute of Jiangsu Province, Yangzhou 225007, China; jsyzhns@163.com (N.H.); yzlah@126.com (A.L.)

[3] Zhenjiang Agricultural Research Institute of Jiangsu Province, Jurong 212400, China; Chen852235436@163.com

[4] College of Environmental Science and Engineering, Yangzhou University, Yangzhou 225000, China; jujing@yzu.edu.cn

[*] Correspondence: jyhuang@yzu.edu.cn (J.H.); yaoyl@yzu.edu.cn (Y.Y.); Tel.: +86-136-0527-8988 (J.H.); +86-183-6282-9071 (Y.Y.)

[†] These authors contributed equally to this work.

Received: 6 October 2019; Accepted: 23 November 2019; Published: 25 November 2019

Abstract: Reducing nitrogen (N) input is a key measure to achieve a sustainable rice production in China, especially in Jiangsu Province. Tiller is the basis for achieving panicle number that plays as a major factor in the yield determination. In actual production, excessive N is often applied in order to produce enough tillers in the early stages. Understanding how N regulates tillering in rice plants is critical to generate an integrative management to reduce N use and reaching tiller number target. Aiming at this objective, we utilized RNA sequencing and weighted gene co-expression network analysis (WGCNA) to compare the transcriptomes surrounding the shoot apical meristem of *indica* (Yangdao6, YD6) and *japonica* (Nipponbare, NPB) rice subspecies. Our results showed that N rate influenced tiller number in a different pattern between the two varieties, with NPB being more sensitive to N enrichment, and YD6 being more tolerant to high N rate. Tiller number was positively related to N content in leaf, culm and root tissue, but negatively related to the soluble carbohydrate content, regardless of variety. Transcriptomic comparisons revealed that for YD6 when N rate enrichment from low (LN) to medium (MN), it caused 115 DEGs (LN vs. MN), from MN to high level (HN) triggered 162 DEGs (MN vs. HN), but direct comparison of low with high N rate showed a 511 DEGs (LN vs. HN). These numbers of DEG in NPB were 87 (LN vs. MN), 40 (MN vs. HN), and 148 (LN vs. HN). These differences indicate that continual N enrichment led to a bumpy change at the transcription level. For the reported sixty-five genes which affect tillering, thirty-six showed decent expression in SAM at tiller starting phase, among them only nineteen being significantly influenced by N level, and two genes showed significant interaction between N rate and variety. Gene ontology analysis revealed that the majority of the common DEGs are involved in general stress responses, stimulus responses, and hormonal signaling process. WGCNA network identified twenty-two co-expressing gene modules and ten candidate hubgenes for each module.

Several genes associated with tillering and N rate fall on the related modules. These indicate that there are more genes participating in tillering regulation in response to N enrichment.

Keywords: shoot apical meristem; transcriptomic analysis; co-expression network; tiller; nitrogen rate; rice (*Oryza sativa* L.)

---

1. Introduction

As one of the three most important cereal crops cultivated for thousands of years, rice (*Oryza sativa* L.) provides staple foods for nearly half of the world's population. Keeping its production apace with the increasing demand is critical to food security. The most gain in rice yield is attributable to increased N fertilizer addition to the paddy field. N is an essential macronutrient for completing rice plant growth and development, therefore, usually a limit to its production [1]. Inarguably, N is the most effective fertilizer in promoting crop growth and increasing crop yield [1]. However, the magic effects of N fertilization deceptively lead to excessive application, which gravely contributes to more direct ammonium gas loss, N run-off, water eutrophication, nitrous oxide greenhouse gas emissions and soil acidification [1–3]. Consequentially, excessive N fertilization has become a fundamental environmental issue and a health problem. Improving N use efficiency is believed to be the ultimate solution to mitigate these problems. Ideal N application rate and proper timing are the keys to reaching a balance of yield gain and N use efficiency [2,3]. *Indica* and *japonica* rice are two major subspecies in Asia, and reportedly being different in response to N enrichment [4,5].

Rice tiller begins at the leaf sheath auxiliary of the bottom nodes of a host culm, close to shoot apical meristem (SAM) at the early growth stage. Usually, rice tiller first appears when the 4th true leaf emerges. Tiller number at 8-9th week after germination essentially represents a variety's tiller production feature [6]. Tillers provide more opportunities to the development of more panicles. Among the agronomic yield traits of rice, panicle number is the one that being determined at the earliest stage. More importantly, panicle number largely associates with the spikelet number and filled grain number per panicle. Therefore, reaching a suitable number of tillers timely is a benchmark in setting the cornerstone for achieving an ideal grain yield in rice production.

Meanwhile, as a monocot fully sequenced species, rice provides a great model to decode the molecular secrets of tiller regulation. Recent progress in rice molecular genetics reveals that more than 65 genes at different stages engage in tiller number regulation [7–14]. However, how these genes concert in tiller control remains to be an enigma. High throughput RNA sequencing (RNA-Seq) lends us an effective tool to discover the transcriptomic profile of every active gene [15]. Though much transcriptome data exists in rice, none targets at the tissues near the SAM region, especially in response to N enrichment.

In the present study, we validated the differential tillering responses to N rates of *indica* (Yangdao6, YD6) and *japonica* (Nipponbare, NPB) rice subspecies, performed RNA sequencing and weighted gene co-expression network analysis (WGCNA) to compare their transcriptomes in the surrounding tissues of SAM. The objectives of this study is to compare their differential transcriptomic responses to N rate, and to reveal the concerted molecular network in tiller regulation in rice. The results would expand our understanding of how complicated an important agronomic trait is controlled.

2. Results

2.1. N Enrichment Promotes Tillers in a Different Pattern between NPB and YD6

Rice tiller depends much on N availability, and hence a substantial portion of N fertilizer is applied before seeding or transplanting in field production. In this study, we enriched N supply at two rates, 9 (MN) and 18 (HN) g N m^{-2}, which were comparable to the average and high range of top-dressing level in the local field production practice at this growth stage. Results showed that enriching N rate to MN level promoted tillering significantly at the 4th leaf emerging stage in both NPB and YD6 (Figure 1). However, further increasing N input to HN could not further enhance tillering at this stage. On the contrary, HN even suppressed tillering at the 4th leaf stage, compared with MN. At the 6th leaf emerging stage, MN, and HN treatments produced the same number of tillers in NPB, both significantly higher than that of LN (CK); but the tiller number was not significantly different among LN, MN and HN at the 8th leaf stage in NPB. The tiller number in YD6 was consistently HN > MN > LN at the 6th and 8th leaf stage. These indicate that for producing more tillers at an early stage, varieties like NPB did not require much N enrichment if any, whereas varieties like YD6 required a higher N rate, albeit preferably in a mild rate. Excessive N enrichment did not promote tiller as wished, irrespective of varieties.

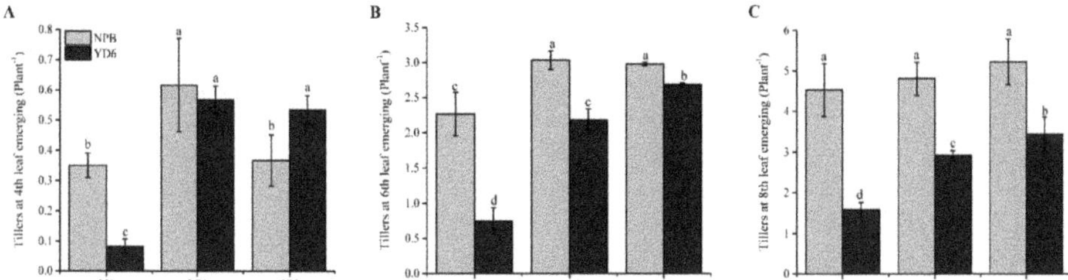

Figure 1. Number of tillers under different N rates in NPB and YD6. (**A**) Tillers at fourth leaf stage; (**B**) Tillers at sixth leaf stage; (**C**) Tillers at eighth leaf stage. The mean and standard deviation (SD) were from three biological replicates, different lowercase letters represent significant differences by Stduent's *t* test between the treatments at $p \leq 0.05$.

2.2. Tiller Number Is Related with N and Carbohydrate Content

Analysis of the N and carbohydrate content of the tissues revealed that MN enrichment increased N content in all tissues (root, sheath, and leaf) at the 4th leaf emergence stage, irrespective of varieties (Figure 2). However, further enrichment of N rate from MN to HN did not alter the N content at this stage. At the 6th leaf stage, the N content showed consistently to be HN > MN > LN (CK), irrespective of tissues or varieties. Yet, the gap of N content diminished at the 8th leaf emerging stage. These indicate that MN could enhance N absorption at the 4th leaf emergence stage, but more N enrichment could not promote N absorption further. The differential gap in N content disappeared as the plant grew for two more leaf-age. It is worthy to notice that the N content dropped faster in YD6 than in NPB from 4th to 6th leaf stage, corroborating that YD6 requires more N input to maintain a similar N content than NPB. This suggests different N management strategy should be applied to different variety for keep N content for tillering.

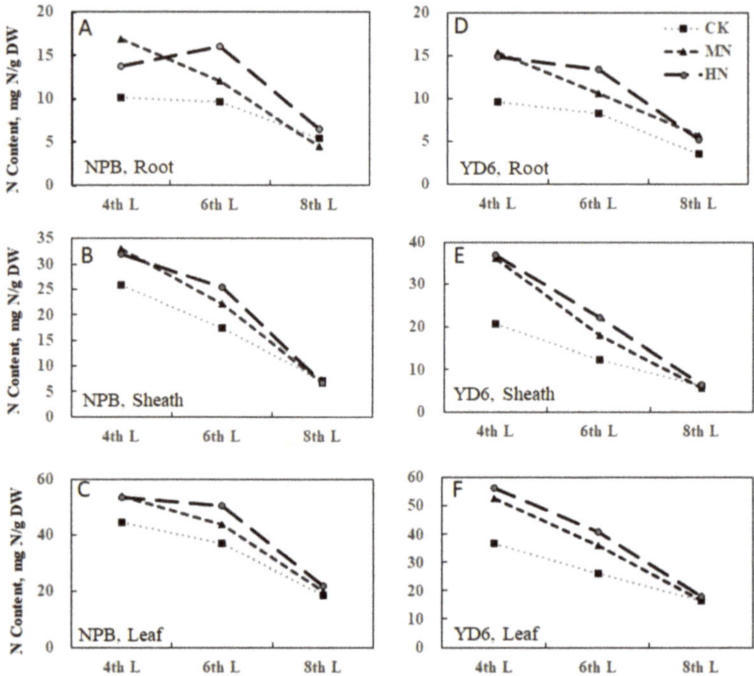

Figure 2. Nitrogen (N) content in root, sheath and leaf organs in different stage of NPB and YD6. N content in root of NPB (**A**) and YD6 (**D**); N content in sheath of NPB (**B**) and YD6 (**E**); N content in leaf of NPB (**C**) and YD6 (**F**).

Water soluble carbohydrate (WSC) (i.e., soluble sugar) content was almost comparable between root and sheath, both being much lower than that in leaf (Figure 3A,C). MN enrichment treatment lowered WSC drastically, whereas HN caused slightly further decrease. Similarly, starch content reduced as N rate increased. Yet, the highest starch content was consistently observed in the sheath, with root being higher than that in leaf (Figure 3B,D), irrespective of growth stages. The correlation between N content and starch, soluble sugar are not significant, irrespective of varieties (Figure 3E,F).

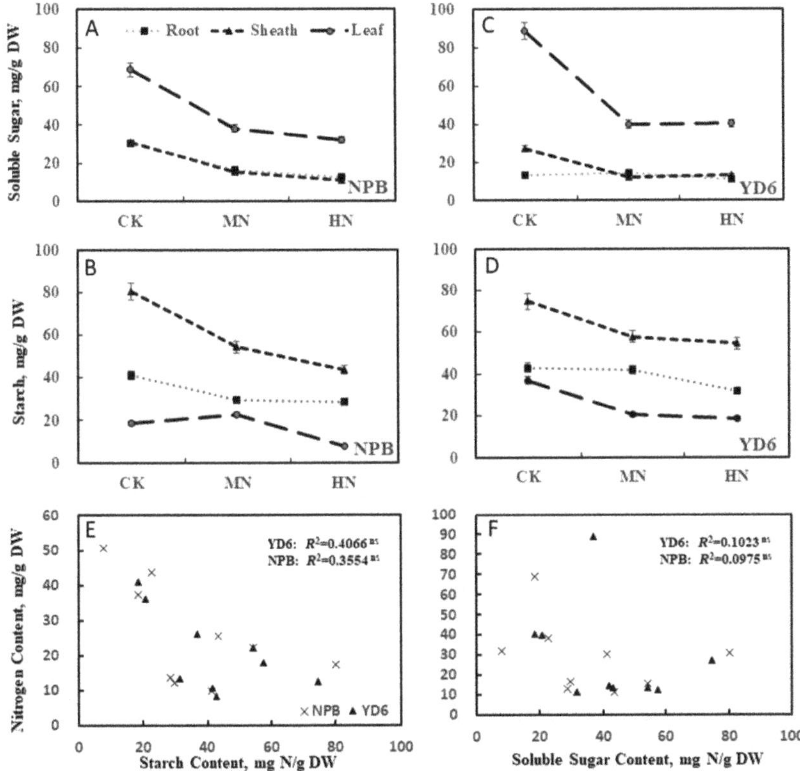

Figure 3. Soluble sugar and starch under different N rates in NPB and YD6. Soluble sugar in NPB (**A**) and YD6 (**C**); Starch in NPB (**B**) and YD6 (**D**); Correlation between N content and starch (**E**), N content and soluble sugar (**F**).

The tiller number at the 4th leaf stage was significantly correlated with the tiller number at the 6th leaf stage, but not significantly correlated with those at the 8th leaf emerging stage, respectively (Figure S1A–C), suggesting the association was diluted along with the growth. The tiller number at the 6th leaf stage was positively correlated with the N content in the root ($R^2 = 0.6303$ ns), sheath ($R^2 = 0.8783$ **) and leaf ($R^2 = 0.0.8719$ **) at this stage (Figure S1D–F). It displayed a negative correlation with WSC in the sheath ($R^2 = 0.3970$ ns) and leaf ($R^2 = 0.7956$ **) (Figure S1G–I), and also negative correlation with the starch content in the root ($R^2 = 0.6677$ *), sheath ($R^2 = 0.4848$ ns) and leaf ($R^2 = 0.7036$ *) (Figure S1J–L). These generally confirm that the tiller number correlated positively with the N content, but negatively with WSC and starch content.

2.3. RNA-Seq Data Quality and Assembly

To reveal the transcriptomic change responsive to N rates, cDNA libraries from tissues surrounding SAM were constructed and subjected to RNA-Seq analysis. A total of 624 million raw reads were generated from two subspecies *japonica* NPB and *indica* YD6 in combination with three levels of N rate (Table S1). After removal of adaptor sequences, short reads, low quality, and rice ribosome RNA reads, clean reads ratio proved to range from 91.6% to 97.1%, indicating that the sequencing reads were qualified for further analysis.

These clean reads were mapped to the NPB reference genome, the mapped ratio ranged from 97.3% to 98.1% in NPB samples, and from 90.9% to 93.5% in YD6. Since the most recent physical map covers about 97% of the NPB genome [16], our mapped ratio of the transcriptomes in SAM tissues in NPB was generally consistent with it. These also indicate that the alignments were successful. As expected, the transcriptome from YD6 was genetically more heterogeneous to NPB reference genome. Even when trying to map the clean reads of YD6 to a reference genome of 9311, a sister line of YD6, the successfully mapped ratio for YD6 ranged from 93.5% to 95.9%. For comparison consistency, we used the physical genomic map of NPB as a common reference genome in all of the following analysis.

Principal component analysis (PCA) projects the whole transcriptome profile onto a few variables to reflect the distribution across samples, which are often clustered in two or three-dimensional space [17]. PCA revealed that transcriptome profile between YD6 and NPB were more distant than that among N rates (Figure S2A), reflecting that the variety defined the most range of the differences. The variability of LN and MN were less than that of HN in NPB, while MN was less variable than that of LN and HN in YD6. Yet, the general variability in NPB was much less than that of YD6. These suggest that the transcriptomic response to N rate was more variable in YD6 than in NPB, the transcriptomic variability caused by N enrichment also depends on the plant's existing N rate condition, and the transcriptomic differences between varieties were more distant than their responses to N rate.

The correlation analysis of gene expression between the biological replicates reflects the experimental and sequencing reproducibility, whereas the correlation between the groups reflects the variety and N rate effects. The correlation coefficients were determined by the FPKM values of the individual genes to verify the general reproducibility and variability among replicates and groups, respectively (Figure S2B). The correlation coefficient R value between the biological replicates ranged from 0.96–0.97 in NPB and 0.95-0.96 in YD6, respectively, exhibiting a remarkably high consistency. This meant this set of RNA-Seq results had very high reproducibility in replicates. Meanwhile, the R value between NPB and YD6 at corresponding N rate ranged from 0.89–0.92, displaying certain consistency. Yet, these values were lower than the R between the biological replicates, indicating more variability between varieties, as expected. Furthermore, the heat map between the groups is consistent (Figure S2C,D). These ensure that the RNA-Seq generated a highly reproducible expression data among the replicates and a proper variability across the comparison groups, which corroborated with PCA results as well as phenotypic performance.

2.4. Validation of Selected DEG Confirms RNA-Seq Data Reliability

Twenty-one DEGs of different FPKM levels were selected for validation by qRT-PCR (Table S2). The quantification data were used to correlate with the FPKM value of each treatment. Among them, twenty genes showed a positive linear correlation between the FPKM and qRT-PCR data, with the correlation coefficient R^2 ranging from 0.6785 * to 0.9721 ** (Figure 4). One gene could not be amplified due to the primers failure. This affirmative validation result indicates that the transcriptome profiling generally reflected the virtual transcript expression differences in the experiment.

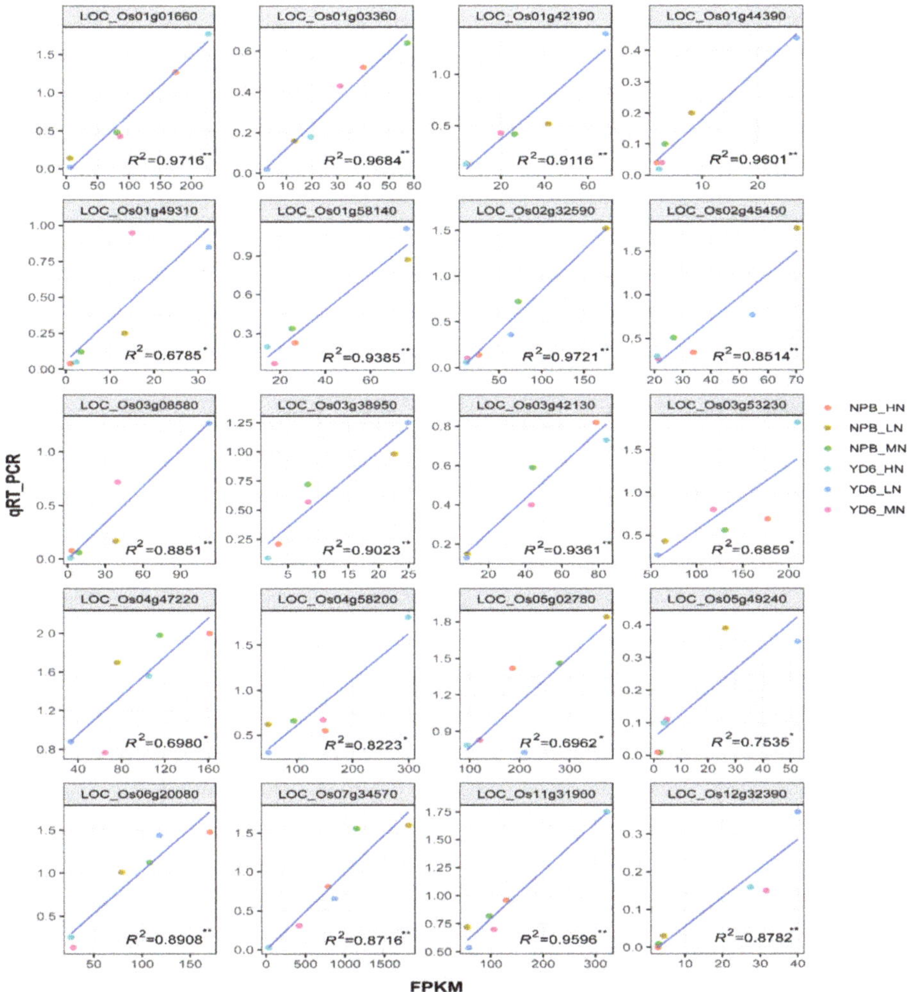

Figure 4. Correlation of qRT-PCR validation and RNA-seq results in 20 representative genes. * and ** represent significance at $p \leq 0.05$ and $p \leq 0.01$, respectively.

2.5. Differentially Expressed Genes (DEGs) in Response to N Rate

To identify the DEG in response to N rate, Cuffdiff [18] program was used in comparing the FPKM of the treatment vs. control according to the following criteria: fold change (FC) |log2| ≥ 1.0 and false discovery rate (FDR, q-value) ≤ 0.05. The total DEG in response to N enrichment in NPB were 87 (65 down-regulated and 22 up-regulated), 40 (12 down and 28 up) and 148 (69 down and 79 up) between LN and MN, MN and HN, LN and HN comparisons, respectively (Figure 5A,D). Among them, merely four genes were consistently present in every comparison. The corresponding numbers of DEG were 115 (58 down and 57 up), 162 (80 down and 82 up) and 511 (239 down and 272 up) in YD6, and only 15 genes were commonly present in these comparisons (Figure 5B,E). The number of DEG indicate that more drastic transcriptome changes from LN to HN, rather than from LN to MN, or MN to HN, irrespective of varieties.

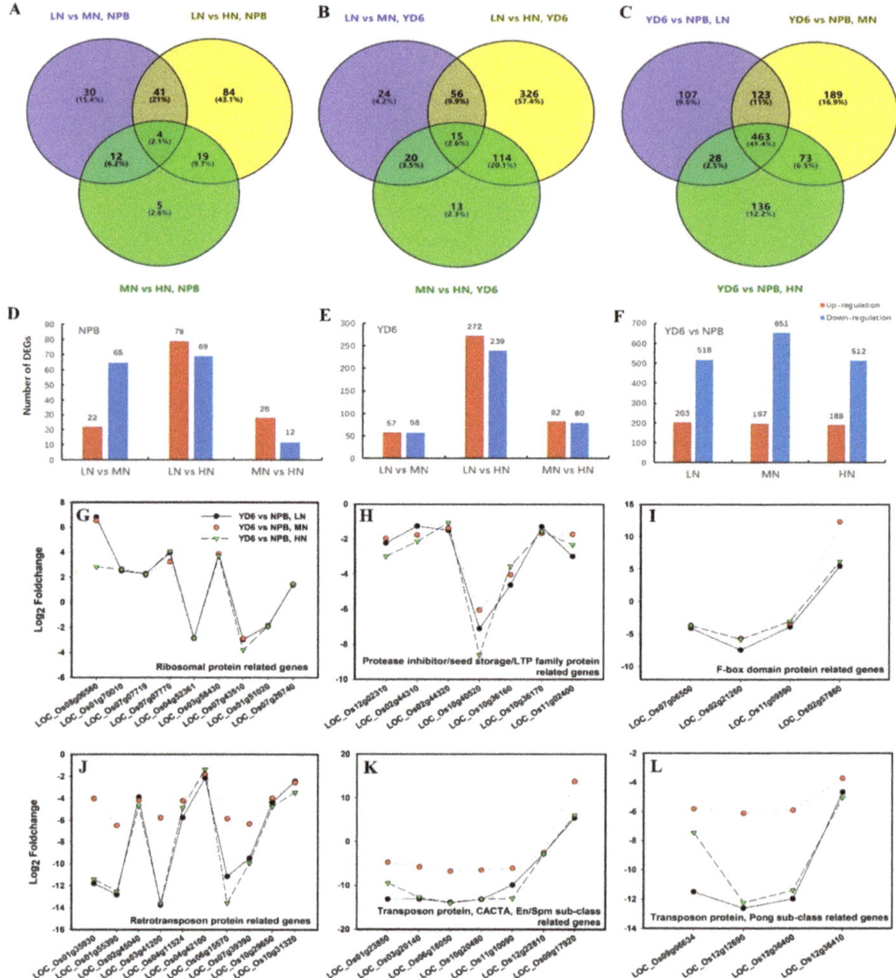

Figure 5. Profiles of DEGs between the comparisons. (**A,D**) DEGs in response to N rate in NPB; (**B,E**) DEGs in response to N rate in YD6; (**C,F**) DEGs between the varieties; (**G–L**) Fold changes of DEGs related to ribosomal protein (**G**), Protease inhibitor/seed storage/LTP family protein (**H**), F-box domain protein (**I**), retrotransposon protein, transposon protein (**J**), CACTA, En/Spm sub-class protein, transposon protein (**K**), Pong sub-class protein (**L**).

Apparently, the number of DEG in response to N enrichment was much more in the *indica* variety YD6 than in the *japonica* variety NPB. Surprisingly, of those DEG commonly presented in response to N rate in NPB and YD6, only one common gene was found (LOC_Os01g01660), which putatively encodes an isoflavone reductase-like gene and confers tolerance to reactive oxygen species. This scarce overlapping in the DEGs suggested that the transcriptome responses to N enrichment were very much dependent on the different genetic background. Combination of different sets of background genes may totally alter the transcription profile in response to certain N enrichment.

On the contrary, comparison between NPB and YD6 at each respective N rate revealed that the number of DEG commonly presented at each N level was a majority of 463 genes (Figure 5C,F), much more than those numbers of distinctive DEGs at each N rate. Among these DEGs, we listed several specific types of genes (Figure 5F–L). This corroborates that the varietal set of DEG was a core determinant in shaping their response to various N enrichment.

2.6. Gene Ontology (GO) Analysis and KEGG Clustering

Web-based String (version 10.5) program was employed to define the classification of N-responsive DEG involved in the three domains of the biological process, cellular component and molecular functions, and their KEGG pathway clustering (Figures S3–S5). In the comparison of LN vs. HN in NPB, the number of DEG in the category of biological process was in the order of the cellular process (41), metabolic process (34), single-organism process (32), response to stimulus (30) and biological regulation (22). Of the DEG fell into the category of cellular component, the order was the cell (44), cell part (44), organelle (35) and membrane (14). Of the DEG that fell into the category of molecular function, the order was the binding activity (23), catalytic activity (16), transporter activity (four), and transcription factor (TF) activity (four, Figure S3C). Similar orders in the number of DEG were also revealed in LN vs. HN comparison in YD6 as well as in other N rate comparisons, and the comparisons between varieties at the respective N rate (Figures S3 and S4, Tables S3–S6). Based on the number of DEGs, more genes are concentrated in the biological process, followed by cellular components, and finally in molecular function. Apparently, these biological procedures were the major changes in the tissues near SAM in response to N rate.

KEGG clustering of the DEG revealed that the first seven categories of common pathways between NPB and YD6 responsive to N rate were the metabolic pathway, ribosome, plant-pathogen interaction, microbial metabolism in diverse environments, plant hormone signal transduction, antenna proteins in photosynthesis and protein processing in the endoplasmic reticulum (Figure S5). The major differential pathways between NPB and YD6 were the glutathione metabolism, protein export, and selenocompound metabolism, which were unique to NPB, and 13 pathways unique to YD6 including the biosynthesis of secondary metabolites, photosynthesis, phenylpropanoid biosynthesis and phenylalanine metabolism. These indicate that the common basic pathways (metabolic and ribosome) were important for both varieties' response to N rate, and many unique pathways were involved in the procedures as well.

2.7. Alternative Splicing Transcripts and Novel Genes

Alternative splicing (AS) creates more versatile regulating RNAs and translatable mRNAs, thus a more enriched diversity in transcriptome and possible code proteins [19]. In this experiment, a total of 68,385 AS events, derived from 12,044 annotated genes, had been revealed (Figure 6A,B). Among them, the retained intron was the most common event type (34.4%–34.9%), followed by the alternative 3′ splicing sites (31.3%–31.7%), alternative 5′ splicing sites (17.6%–17.8%), skipped exon (12.7%–12.8%), a mix of two or more different types of AS (complex, 2.6%) and mutually exclusive exons (0.8%). The number of AS events between the varieties and among N rates did not show any statistically significant difference. However, the AS events relative to clean reads in all category type was consistently 1–4% higher in NPB than that in YD6, which probably reflected the varietal differences.

Totally 352 novel genes were found in the RNA-seq data. Among them, 74 and 174 genes were unique to NPB and YD6, respectively; 104 genes were common in the two varieties (Figure 6C). Novel genes responsive to LN were 58 and 89 specific to NPB and YD6 respectively; 55 were common in the two varieties (Figure 6D). These corresponding novel transcripts numbers were 48, 102 and 60 in MN, 48, 93 and 52 in HN, respectively (Figure 6E,F). These results suggest that deep sequencing may reveal more tissue-specific transcripts. Comparison of the novel gene number among N rates within a variety also revealed similar trends, though the numbers were always bigger in YD6 than that in NPB (Figure 6G,H). The novel genes revealed were always more in the *indica* variety YD6 than the *japonica* variety NPB, probably partially attributable to that the transcript annotation in the *japonica* subspecies was more fully covered in the reference genome.

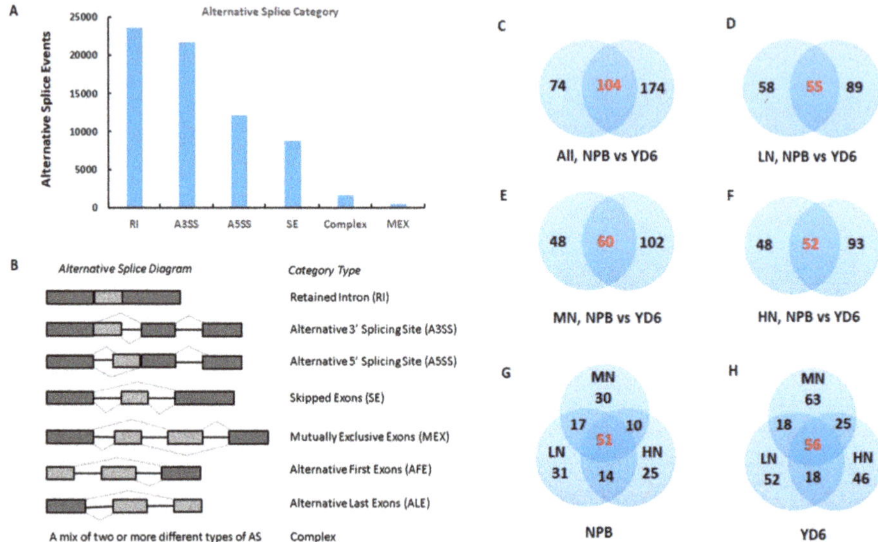

Figure 6. Profiles of alternative splice (AS) and novel genes. (**A,B**) Events and types of AS; (**C–H**) Novel genes between varieties in general (**C**); at LN (**D**); at MN (**E**); at HN (**F**); Novel genes in response to N rate in NPB (**G**); Novel genes in response to N rate in YD6 (**H**).

2.8. Identifition of Weighted Gene Co-Expression Network

To obtain candidate key genes associated with the phenotypical traits, WGCNA [20] was used to distinguish the specific genes that are related to the traits, including tiller number, N content in leaf and in stem, dry weight of root, stem and leaf etc. After removing the genes with low FPKM levels, 21,700 genes were retained for the further WGCNAs analysis. This analysis identified twenty-two distinct gene co-expression modules (labeled in different colors) shown in the dendrogram (Figure 7A), and a total of eleven modules (brown, green, darkred, red, blue, magenta, pink, tan, lightyellow, turquoise and purple) were significantly related to the above-mentioned various phenotypical traits, respectively (Figure 7B). The eigengene expression in blue and green modules (M10 and M3) are shown in Figure 7C,D, respectively. The tiller number was positively most correlated with the eigengene expression levels of the lightyellow module (M18), and the correlation coefficient (r) was 0.61 ($p = 0.04$). Gene ontology (GO) annotation analysis reveals that the cellular components of this module genes mostly fall into extracellular region, yet molecular functions and biological processes are not significantly enriched (Table S7). The tiller number was negatively most correlated with the green module, with $r = -0.71$ ($p = 0.01$). GO annotation shows that the biological processes of the green module genes fall into biological regulation, regulation of cellular process, transport, establishment of localization, localization, regulation of transcription, etc. The molecular function fall into transporter activity, protein binding, transmembrane transporter activity, active transmembrane transporter activity, and transcription regulator activity. The cellular component of this module fall into membrane, integral to membrane, intrinsic to membrane, and cell part (Table S8).

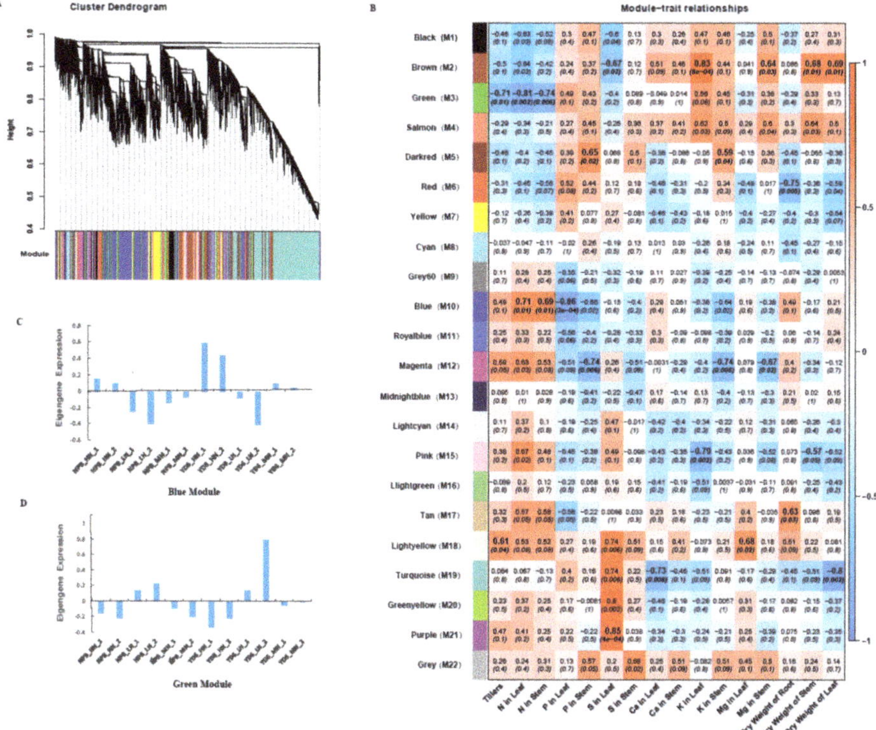

Figure 7. Weighted gene co-expression network analysis (WGCNAs) of gene expressions and the related traits. (**A**) Hierarchical cluster tree showing 22 co-expression modules identified by WGCNAs. Each branch in the tree represents an individual gene; (**B**) Matrix of module – trait correlation: A total of 22 modules shown on the left and the relevance color scale from −1 to 1 is on the right. The numbers in parentheses represent the significance (*p*), and the numbers above represent the correlation coefficient (*r*). The eigengene expression profile in blue module (**C**) and green module (**D**).

The scales of N content in stem and in leaf were highly positively correlated with the eigengene expression in the blue module (M10), with *r* at 0.71 (*p* = 0.01) and 0.69 (*p* = 0.01), respectively. GO annotation shows that the biological processes of the blue module genes fall into the translation, gene expression, cellular protein metabolic process, small molecule metabolic process, protein folding, protein metabolic process, cellular ketone metabolic process, carboxylic acid metabolic process etc. The molecular functions of the blue module are the structural constituent of ribosome, structural molecule activity, translation factor activity, nucleic acid binding, pyrophosphatase activity, and GTP binding. Their cellular components are cytoplasm, macromolecular and ribonucleoprotein complex, ribosome, and cytoplasmic part (Table S9). And the scale of N content in stem and in leaf were all negatively correlated with the green module (*r* = −0.81, *p* = 0.002) and (*r* = −0.74, *p* = 0.006), consistent with the negative correlation module of the scale of tillers. In addition, the modules related to other traits and their GO annotations are shown in Figure 7B and Tables S10–S14. Interestingly, modules positively related with N content are in a negatively association with phosphorus content, and vice versa for the positively related modules. Their associations with sulphur, calcium and potassium content are in a similar pattern but not that consistent. These reveal that the phenotypic traits are more closely associated with certain modules of genes than other modules.

2.9. Construction of the Gene Co-Expression Networks and Identification of Candidate Hub Genes

After screening for the highest significant correlated modules, we selected two modules (one positive correlation module and one negative module) to construct the gene expression networks (Figure 8, Figures S6–S8). Since the scale of N in stem and N in leaf, dry weight in stem and in leaf share the same selected modules, totally seven most relevant modules (light yellow, green, blue, tan, red, brown and pink module) were chosen for the further analysis. The association degree of eigengenes within the module was used to construct the gene expression network (Figures 8A,B, S5A,B, and S6A,B). The hub genes are selected by K_{ME} values through their closest connections in the gene network. The top 20 node genes of each module were screened to generate a network map (Figures 8C,D, S5C,D, and S6C,D).

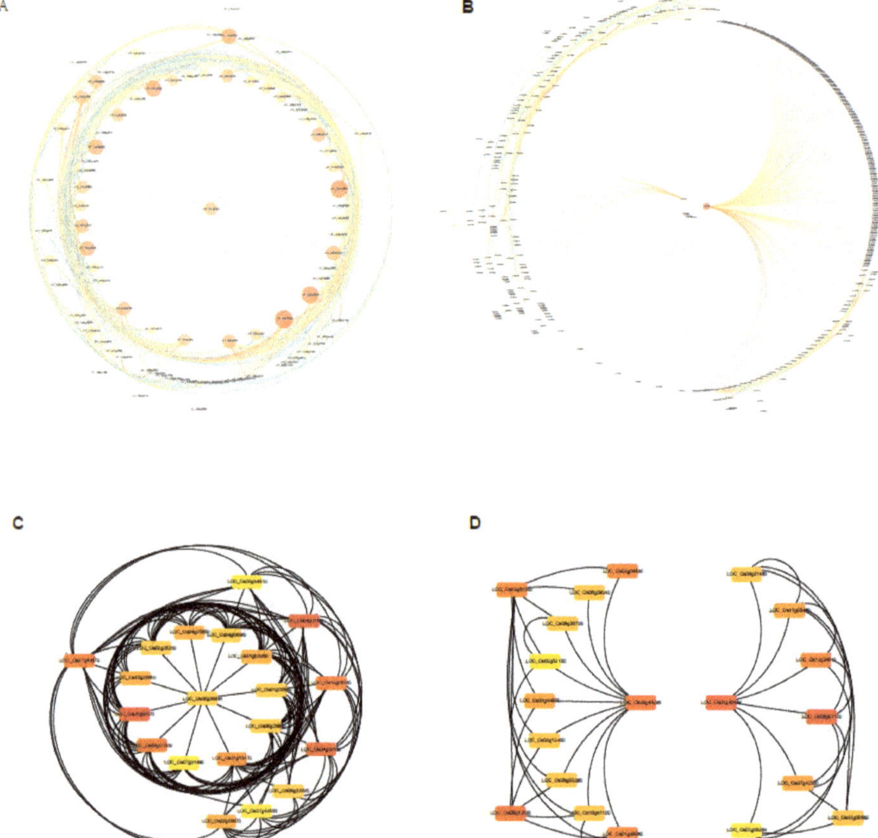

Figure 8. Co-expression network analysis of tiller number related modules. (**A,B**) Gene co-expression networks of positively correlated lightyellow module (**A**) and negatively correlated green module (**B**) visualized using Cytoscape software platform. The circle size of and color depth indicate the degree of connectivity; (**C,D**) The correlation networks of top 20 nodes in lightyellow module (**C**) and green module (**D**). The color depth represents the number of associated nodes.

To screen for candidate hub genes, the top ten connected genes in the specific modules were picked out by their K_{ME} value (Table 1, Table S15). The tiller number relate positively with the lightyellow module (M18) and negatively with the green module (M3), respectively (Table 1). In the light yellow module, the top ten genes encode HEV3—Hevein family protein precursor, BBTI5, BBTI8, VP15, ribonuclease T2 family, pyridoxal-dependent decarboxylase protein, cysteine-rich receptor-like protein kinase, and kinesin motor domain containing protein. The correlation network of the light yellow module is shown in Figure 9A. Genes encoding the aforementioned proteins were identified as key candidate hub genes for the light yellow module (M18) (Figure 8C). In the green module (M3), the top ten genes encode MATE efflux protein, ethylene-responsive protein, protein kinase, GDSL-like lipase/acylhydrolase, aspartic proteinase oryzasin-1 precursor, eukaryotic translation initiation factor 1A, thaumatin, tetraspanin family protein, AGAP002737-PA, and RFC5. Genes encoding these proteins were identified as candidate hub genes for the green module (Figure 8D). Similarly, multiple functional proteins were enriched in this module, indicating that these regulatory networks may play a pivotal role in regulation of the scale of tillers.

Table 1. Candidate hub genes related to the tiller numbers, N rate in leaf and in stem.

Gene Name	Description	K_{ME} Value
Light yellow module (M18) with positive correlation associated with the tillers		
LOC_Os04g09390	HEV3 - Hevein family protein precursor, expressed	0.97
LOC_Os01g03680	BBTI8 - Bowman-Birk type bran trypsin inhibitor precursor, expressed	0.95
LOC_Os08g28880	patatin, putative, expressed	0.94
LOC_Os10g08780	expressed protein	0.94
LOC_Os01g03360	BBTI5 - Bowman-Birk type bran trypsin inhibitor precursor, expressed	0.94
LOC_Os02g35200	VP15, putative, expressed	0.88
LOC_Os07g43670	ribonuclease T2 family domain containing protein, expressed	0.85
LOC_Os04g01690	pyridoxal-dependent decarboxylase protein, putative, expressed	0.85
LOC_Os04g25650	cysteine-rich receptor-like protein kinase, putative, expressed	0.85
LOC_Os05g02670	kinesin motor domain containing protein, putative, expressed	−0.80
Green module (M3) with negative correlation associated with the tillers, N rate in leaf and stem		
LOC_Os10g13940	MATE efflux protein, putative, expressed	0.98
LOC_Os07g28890	ethylene-responsive protein related, putative, expressed	0.98
LOC_Os01g12720	protein kinase domain containing protein, expressed	0.98
LOC_Os07g47210	GDSL-like lipase/acylhydrolase, putative, expressed	0.97
LOC_Os01g18630	aspartic proteinase oryzasin-1 precursor, putative, expressed	0.95
LOC_Os02g19770	eukaryotic translation initiation factor 1A, putative, expressed	0.94
LOC_Os09g36580	thaumatin, putative, expressed	0.93
LOC_Os05g03530	tetraspanin family protein, putative, expressed	0.90
LOC_Os04g07280	AGAP002737-PA, putative, expressed	0.90
LOC_Os02g53500	RFC5 - Putative clamp loader of PCNA, replication factor C subunit 5, expressed	−0.92
Blue module (M10) with positive correlation associated with the N rate in leaf and stem		
LOC_Os08g44290	RNA recognition motif containing protein, putative, expressed	0.98
LOC_Os03g61260	ribosomal L18p/L5e family protein, putative, expressed	0.98
LOC_Os03g04530	cytochrome P450, putative, expressed	0.95
LOC_Os09g36700	ribonuclease T2 family domain containing protein, expressed	0.94
LOC_Os02g52150	heat shock 22 kDa protein, mitochondrial precursor, putative, expressed	0.91
LOC_Os03g07960	expressed protein	0.88
LOC_Os10g41100	CCT motif family protein, expressed	0.84
LOC_Os11g34880	NB-ARC domain containing protein, expressed	−0.70
LOC_Os11g05480	transcription factor, putative, expressed	−0.76
LOC_Os01g40499	S-locus lectin protein kinase family protein, putative, expressed	−0.82

The N content in stem and in leaf relate positively with the blue module and negatively to the green module, respectively (Table 1). In the blue module, the top ten genes encode RNA recognition motif containing protein, ribosomal L18p/L5e family protein, cytochrome P450, ribonuclease T2 family, heat shock 22 kDa protein, CCT motif family protein, NB-ARC domain containing protein, and transcription factor etc. The correlation network of the blue module is shown in Figure S5A. This indicates that these genes may play a crucial role in regulation of the scale of N in stem and in leaf. Interestingly, the negative module falls into the same module with the scale of tillers. In addition, the co-expression network and candidate hub genes in other modules are shown in Figures S6–S8 and Table S15.

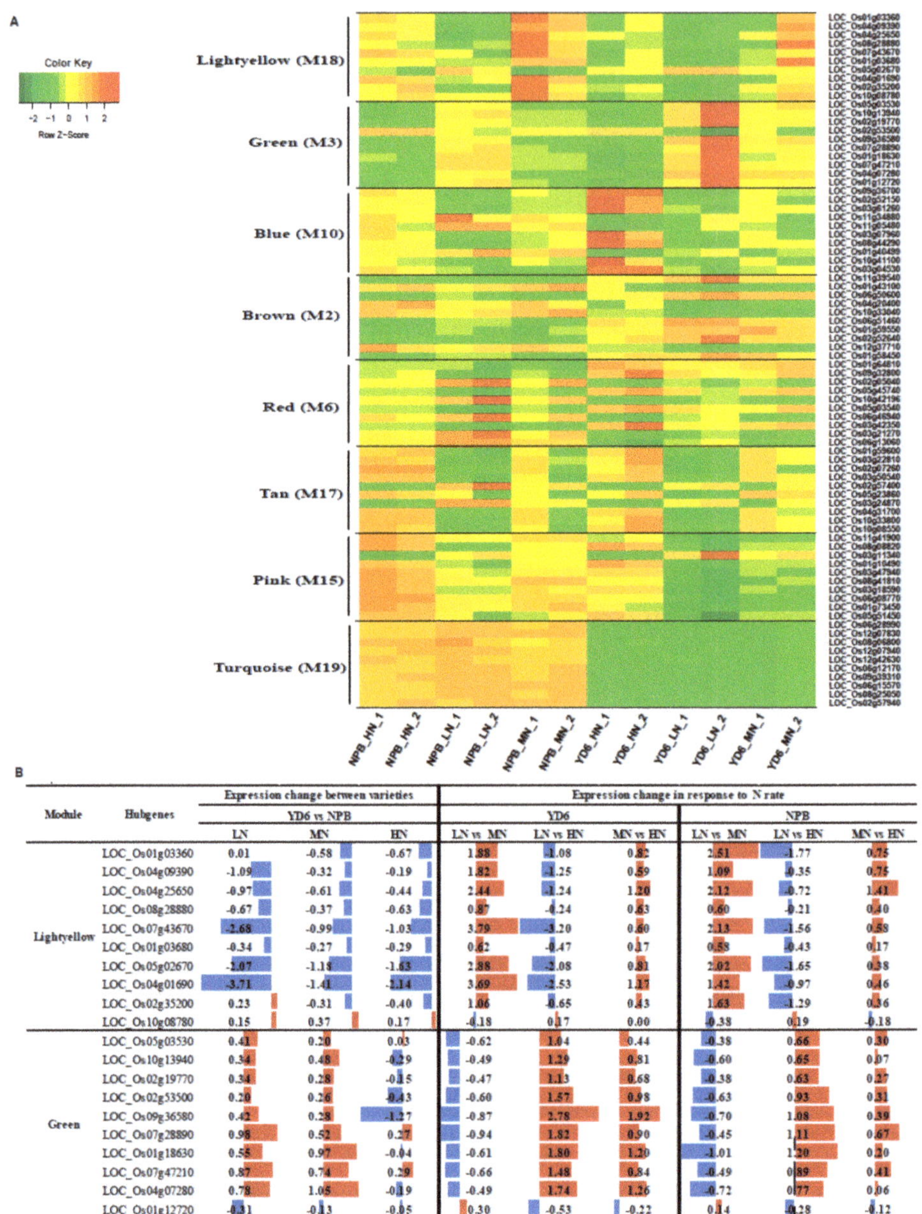

Module	Hubgenes	Expression change between varieties			Expression change in response to N rate					
		YD6 vs NPB			YD6			NPB		
		LN	MN	HN	LN vs MN	LN vs HN	MN vs HN	LN vs MN	LN vs HN	MN vs HN
Lightyellow	LOC_Os01g03360	0.01	-0.58	-0.67	1.88	-1.08	0.82	2.51	-1.77	0.75
	LOC_Os04g09390	-1.09	-0.32	-0.19	1.82	-1.25	0.59	1.09	-0.35	0.75
	LOC_Os04g25650	-0.97	-0.61	-0.44	2.44	-1.24	1.20	2.12	-0.72	1.41
	LOC_Os08g28880	-0.67	-0.37	-0.63	0.87	-0.24	0.63	0.60	-0.21	0.40
	LOC_Os07g43670	-2.68	-0.99	-1.03	3.79	-3.20	0.60	2.13	-1.56	0.58
	LOC_Os01g03680	-0.34	-0.27	-0.29	0.62	-0.47	0.17	0.58	-0.43	0.17
	LOC_Os05g02670	-2.07	-1.18	-1.63	2.88	-2.08	0.81	2.02	-1.65	0.38
	LOC_Os04g01690	-3.71	-1.41	-2.14	3.69	-2.53	1.17	1.42	-0.97	0.46
	LOC_Os02g35200	0.23	-0.31	-0.40	1.06	-0.65	0.43	1.63	-1.29	0.36
	LOC_Os10g08780	0.15	0.37	0.17	-0.18	0.17	0.00	-0.38	0.19	-0.18
Green	LOC_Os05g03530	0.41	0.20	0.03	-0.62	1.04	0.44	-0.38	0.66	0.30
	LOC_Os10g13940	0.34	0.48	-0.29	-0.49	1.29	0.81	-0.60	0.65	0.07
	LOC_Os02g19770	0.34	0.28	-0.15	-0.47	1.13	0.68	-0.38	0.63	0.27
	LOC_Os02g53500	0.20	0.26	-0.43	-0.60	1.57	0.98	-0.63	0.93	0.31
	LOC_Os09g36580	0.42	0.28	-1.27	-0.87	2.78	1.92	-0.70	1.08	0.39
	LOC_Os07g28890	0.98	0.52	0.27	-0.94	1.82	0.90	-0.45	1.11	0.67
	LOC_Os01g18630	0.55	0.97	-0.04	-0.61	1.80	1.20	-1.01	1.20	0.20
	LOC_Os07g47210	0.87	0.74	0.29	-0.66	1.48	0.84	-0.49	0.89	0.41
	LOC_Os04g07280	0.78	1.05	-0.19	-0.49	1.74	1.26	-0.72	0.77	0.06
	LOC_Os01g12720	-0.31	-0.13	-0.05	0.30	-0.53	-0.22	0.14	-0.28	-0.12

Figure 9. Gene expression of top ten node hub genes of eight related modules. (**A**) Heatmap showing the expression profiles of the top ten node hub genes in each module; (**B**) Fold changes (log2) of the top ten node hub genes in lightyellow and green module. Red and blue color represent scale of up-and down-regulation, respectively.

It is noteworthy that we find four hub genes overlap with DEGs (between varieties) in the turquoise module (Table S15), which negatively correlated with the scale of dry weight in leaf. These four genes are LOC_Os12g07830, LOC_Os06g15570, LOC_Os08g25050 and LOC_Os06g12170. The results suggest that these genes in the regulatory network may play a core role in differential dry weight accumulation between varieties.

The heatmap and changes in the expression levels of the candidate hub genes from the selected modules are shown in Figures 9 and S9. The expression level of these hub genes are contrastingly different between the varieties, indicating that the hub genes may play a crucial role in their differential response to N rate between the varieties.

2.10. Expression Profiles of Tiller Related Genes and Their Network

Many genes are previously reported to impact upon tiller numbers in their mutants [9]. To reveal their transcriptional responses to N rate, we extracted their FPKM value from the transcriptomes, and subjected it to statistical comparison (Table 2). Totally of 65 genes were selected to screen for their transcription levels. Among the 36 decently expressed genes (FPKM > 0.1), 27 showed significant varietal difference ($p \leq 0.05$), 18 showed to be responsive to N rate, and three showed significant interaction between the variety and N rate. Of these significantly influenced tiller genes, only six of the 27 genes in the varietal comparison and nine of 18 genes in the N rate comparisons displayed more than twofold change in at least one comparison. Apparently, the expression level of these genes was more determined by a variety than to be affected by the N rate. A protein and protein interaction network (PPI) constructed from known connections seems to illustrate their complexity more (Figure 10). However, the change did not need to surpass a threshold of twofold to produce a significant impact on the phenotypes such as the tiller number.

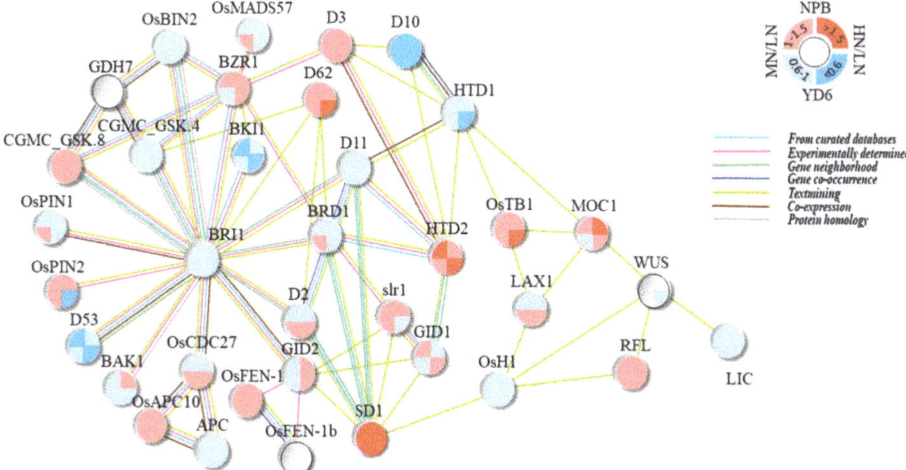

Figure 10. The graphical visualization of protein and protein interaction (PPI) network show the relationship of tiller genes. The gene expression in their fold change are represented by the up and down half of the circle, for NPB and YD6, respectively; MN/LN denotes fold change of MN treatment to LN, HN/LN for HN treatment to LN, represented by the left and right half of the circle, respectively; red and pink mean induction (up regulation), blue and pale blue for suppression, and white for not detected; line colors denote that their connections are generated from curated database (blue), experimentally determined (purple), gene neighborhood (green), textmining (yellow green), co-expression (brown), or protein homology (light purple).

Meanwhile, through co-expression network analysis, these tiller related genes belong to different modules, and *D10/OsCCD8*, *HTD2/D14/D88/qPPB3*, and *NRR/CRCT* genes fall in the green module, which negatively controls the occurrence of the tiller number (Table 2). Surprisingly, these tiller genes are not part of the candidate hub genes. This suggests that these tiller related genes are not at the nodes

Int. J. Mol. Sci. **2019**, *20*, 5922

of the gene expression network, which may locate downstream of certain pathways, and changes of hub genes will regulate their expression accordingly. In addition, we found that other tiller-related genes belong to different network regulation modules, showing the complexity and diversity of the gene regulation at the early stage of tillering. It seems that many genes have other functions that are not yet discovered during this growth stage. For example, *TAC*, *LAX* and *IPA1* fell in the turquoise module are negatively correlated with dry weight in the leaf during this period.

Table 2. Fold change of the expression level change of tiller genes, their significance, and present module.

Module	Gene symbol	Expression change between varieties (YD6 vs NPB) LN	MN	HN	Expression change in response to N rate — YD6 LN vs MN	LN vs HN	MN vs HN	NPB LN vs MN	LN vs HN	MN vs HN	P value Variety	N rate	Variety x N
Blue	*OsAPC10*	0.11	0.21	0.37	0.23	-0.54	-0.30	0.15	-0.29	-0.12	0.017	0.008	0.286
Blue	*DLT/OsGRAS-32/D62/GS6*	0.02	0.06	0.41	0.25	-0.80	-0.54	0.24	-0.42	-0.17	0.029	0.002	0.134
Blue	*OsEXP4*	-0.37	0.10	0.24	0.84	-1.10	-0.25	0.40	-0.49	-0.08	0.607	0.078	0.636
Blue	*OsFEN-1*	0.18	0.19	0.13	0.02	-0.04	-0.01	0.03	-0.09	-0.05	0.019	0.353	0.904
Blue	*THIS1*	-0.09	-0.01	0.22	0.49	-0.46	-0.36	0.04	-0.16	-0.11	0.424	0.016	0.140
Blue	*OsHAP2E/OsNF-YA2*	-0.26	-0.36	-0.67	-0.51	1.09	0.60	-0.38	0.67	0.30	0.001	0.000	0.711
Blue	*TAD1/TE/OsCCS52A*	-0.29	-0.18	-0.33	-0.13	0.19	0.07	-0.21	0.15	-0.05	0.010	0.107	0.622
Green	*NRR/CRCT*	0.23	0.22	-0.30	-0.68	1.22	0.56	-0.63	0.67	0.05	0.541	0.004	0.365
Green	*HTD2/D14/D88/qPPB3*	0.22	-0.27	0.58	0.57	-0.85	-0.27	1.09	-0.50	0.60	0.043	0.002	0.047
Green	*D10/OsCCD8*	0.32	0.36	0.54	-0.85	0.94	0.10	-1.36	1.15	-0.20	0.012	0.002	0.654
Brown	*BAK1/SERK1/BISERK1*	0.45	0.50	0.15	-0.04	0.23	0.20	-0.06	-0.08	-0.13	0.000	0.174	0.035
Brown	*OsCCD7/htd1/sd-t/dit1*	1.17	0.98	0.71	-0.34	0.73	0.40	-0.12	0.26	0.16	0.000	0.021	0.082
Brown	*OsGSK3*	0.21	0.23	0.07	-0.10	0.21	0.12	-0.09	0.06	-0.01	0.006	0.020	0.341
Brown	*BKI1*	0.30	1.13	-0.17	-0.03	1.12	1.10	-0.83	0.65	-0.17	0.041	0.050	0.180
Brown	*sd1/OsGA20ox2/qSD1-2*	-0.36	-0.94	-0.26	0.82	-1.27	-0.43	1.44	-1.18	0.27	0.067	0.008	0.259
Brown	*OsTB1/FC1*	-1.09	-0.74	-0.42	0.51	-0.74	-0.22	0.19	-0.09	0.12	0.004	0.173	0.532
Brown	*RFL/APO2*	-0.52	-0.54	-0.36	0.16	-0.63	-0.46	0.21	-0.49	-0.26	0.029	0.049	0.905
Red	*MOC1*	0.79	0.50	0.35	-0.05	-0.23	-0.27	0.27	-0.68	0.39	0.006	0.162	0.621
Pink	*CGMC_GSK.8*	-0.12	-0.03	-0.15	0.16	-0.20	-0.03	0.10	-0.23	-0.12	0.000	0.000	0.018
Pink	*D63*	0.07	0.09	0.11	-0.04	0.11	0.07	-0.04	0.15	0.12	0.194	0.063	0.601
Pink	*OsH1/Oskn1*	0.34	0.24	-0.02	-0.33	0.64	0.32	-0.20	0.27	0.09	0.012	0.004	0.270
Pink	*DEP1/DN1/qPE9-1/qNGR9*	-0.39	-0.32	0.21	0.15	-0.88	-0.72	0.11	-0.29	-0.17	0.546	0.035	0.208
Turquoise	*GID2/OsGF14e/14-3-3*	0.25	0.18	0.33	-0.07	-0.17	-0.23	0.03	-0.10	-0.06	0.001	0.034	0.240
Turquoise	*OsPIN2*	-0.53	-0.46	-1.39	0.32	0.73	1.05	0.28	-0.15	0.14	0.014	0.235	0.377
Turquoise	*OsPIN1*	0.37	0.56	0.30	0.00	0.11	0.11	-0.16	0.02	-0.13	0.000	0.322	0.226
Turquoise	*brd1/OsDWARF/OsBR60x*	0.33	0.39	0.31	-0.01	0.09	0.10	-0.03	0.07	0.05	0.019	0.783	0.937
Turquoise	*d11/CPB1/zg4/cl4*	0.40	0.43	0.58	-0.35	0.09	-0.25	-0.36	0.37	0.02	0.000	0.002	0.068
Turquoise	*sdt/OsmiR156h*	1.15	1.20	1.21	0.02	0.06	0.10	0.00	0.12	0.13	0.000	0.459	0.891
Turquoise	*TAC1*	0.24	0.40	0.78	0.00	-0.31	-0.30	-0.14	0.22	0.09	0.019	0.667	0.192
Turquoise	*LAX1*	-1.22	-1.09	-0.36	0.05	-0.58	-0.51	-0.05	0.27	0.24	0.005	0.968	0.325
Turquoise	*qPN1*	-0.46	-0.67	-0.22	0.39	0.01	-0.36	-0.15	0.25	0.11	0.003	0.086	0.340
Turquoise	*OsSPL14/IPA1/WFP*	0.33	0.13	0.43	-0.09	-0.03	-0.11	0.14	0.05	0.21	0.004	0.871	0.514
Turquoise	*BRI1/D61*	-0.26	-0.18	-0.17	-0.08	0.00	-0.08	-0.13	0.07	-0.04	0.002	0.079	0.512
/	*OMTN2/OsNAC2/Osnil1*	-0.41	-0.12	-0.15	0.15	-0.43	-0.28	-0.11	-0.19	-0.29	0.084	0.031	0.459
/	*OsNAC2*	-0.20	-0.29	-0.08	-0.53	0.44	-0.07	-0.40	0.55	0.17	0.129	0.017	0.718
/	*PROG1*	1.53	3.19	1.48	1.79	-0.95	0.84	-0.85	-0.02	-0.85	0.053	0.427	0.278

Note: The expression changes are in the form of log2, and significances at their respective probability level. Red and blue color represent scale of up-and down-regulation, respectively.

2.11. Expression Profiles of No Apical Meristem Family Genes, Carbohydrate, and N Metabolism and Transport-Related Genes

No apical meristem family genes (NAM) are critical for meristem maintenance and floral development [21]. Tillers contain newly branched meristems and therein arise perspective panicles. Rice genome contains 90 putative NAM genes. Among the 76 NAM genes detected in the transcriptome at this stage, 27 had an extremely low expression (FPKM ≤ 0.1). Among the 49 decently expressed NAM genes (FPKM > 0.1), 18 displayed significant varietal differences, 12 being responsive to the N rate, and 10 showed an interaction between the variety and N rate (Figure S10). Among those significantly changed, only 11 of the 18 genes in the varietal comparison and six of the 12 genes responsive to the N rate displayed a twofold or higher change at least once. Apparently, more NAM genes were defined by the variety than to be responsive to the N rate.

Carbohydrates provide basic carbon substrates for multiple primary and secondary metabolic pathways, and are subject to the impact of N availability [22,23]. Among the totally 555 carbohydrate metabolism-related genes we screened for in the transcriptome data, 29 were not detected in the

SAM at this stage, and 45 showed an extremely low expression (FPKM < 0.1). Of those 482 decently expressed genes, 200 genes did not show significant ($p > 0.05$) differences in any comparison; 212 showed significant varietal differences ($p \leq 0.05$, Figure S11); 128 being responsive to the N rates; and 43 showed significant interaction between the variety and N rate. This corroborates with that more genes were expressing differentially between the varieties than those in response to the N rate. All these specific pathway profile analysis points toward that the expression profile differences were more likely to be determined by a variety rather than the N enrichment.

N enrichment directly affects the N availability, absorption, synthesis and transport of the related amino acids and the downstream N metabolism [2,24]. We screened for 210 genes pertaining to ammonium, nitrate, nitrite, glutamate, glutamine, asparagine, tryptophan, methionine, aspartate, proline, glycine, cysteine and NADH metabolism. We detected 199 genes expressed in the SAM, with 178 showed a decent expression (FPKM > 0.1). Among them, 97 showed significant differences between the varieties; 69 showed responsive to the N rate; 28 displayed significant interaction between the variety and N rate (Figure S12). Comparably, much more genes expressed differentially between the varieties than those being responsive to the N rate. At the same time, we found that the genes related to N metabolism are in different modules and play different roles during this period (Figure 10). The N transporter proteins, *OsNRT1* and *OsNRT2.3/OsNRT2.3a/OsNRT2.3b*, are in the brown module which positively relate to dry weight of stem and leaf (Table 3). Interestingly, an ammonium transporter protein-coding gene *OsAMT1.2* locates in the turquoise module, which negatively regulates the dry weight of leaf. These results show that the N related genes in rice plant might play very different roles in the absorption, function and distribution of N in response to N rate as well as in tillering regulation.

Table 3. Fold change of the expression level of N metabolism and transporter genes between the varieties and their responses to the N rate.

Module	Gene symbol	Accession number	Expression change between varieties			Expression change in response to N rate						P value		
			YD6 vs NPB			YD6			NPB			Variety	N rate	Variety x N
			LN	MN	HN	LN vs MN	LN vs HN	MN vs HN	LN vs MN	LN vs HN	MN vs HN			
Green		LOC_Os05g04220	-0.06	0.11	-0.07	0.43	-0.27	0.18	0.29	-0.29	0.02	0.471	0.000	0.017
Green	OsNPF2.4	LOC_Os03g48180	0.18	0.05	-0.29	-0.31	0.72	0.42	-0.15	0.24	0.10	0.535	0.119	0.348
Brown	TOND1	LOC_Os12g43440	-1.38	0.02	-2.53	0.84	-0.01	0.84	-0.53	-1.17	-1.69	0.196	0.532	0.411
Brown		LOC_Os09g07920	0.50	0.28	0.10	-0.27	0.43	0.17	-0.02	0.02	0.02	0.010	0.275	0.124
Brown	OsNRT1	LOC_Os03g13274	0.49	0.81	0.22	-0.09	0.79	0.72	-0.38	0.51	0.14	0.004	0.007	0.145
Brown	OsNRT2.3	LOC_Os01g50820	1.20	1.53	1.62	-0.55	0.74	0.21	-0.84	1.15	0.32	0.001	0.049	0.522
Tan		LOC_Os01g65000	-0.77	-0.92	-0.99	-0.39	1.31	0.94	-0.21	1.09	0.89	0.027	0.028	0.856
Turquoise		LOC_Os01g54530	0.97	0.80	0.68	-0.30	0.28	-0.01	-0.11	-0.02	-0.11	0.000	0.005	0.026
Turquoise		LOC_Os12g01420	-1.88	-1.77	-2.18	0.00	0.29	0.30	-0.08	-0.02	-0.08	0.004	0.961	0.905
Turquoise		LOC_Os03g62200	-2.53	-1.89	-2.69	0.98	-0.27	0.72	0.37	-0.44	-0.06	0.001	0.497	0.669
Turquoise	OsAMT1.2	LOC_Os02g40730	0.31	1.67	1.51	1.27	-0.80	0.49	-0.05	0.39	0.36	0.013	0.214	0.200

Note: The expression changes are in the form of log2, and significances at their respective probability level. Red and blue color represent scale of up-and down-regulation, respectively.

3. Discussion

SAM differentiates all organs, and initiates new branches/tillers for more panicles. Understanding the molecular mechanism regulating SAM is of great importance to improve N use efficiency in many crop species, especially in rice. There are so many transcriptomic researches in rice plant. However, surprisingly, none is dealing with this specific important tissue. NPB and 9311 (sister line of YD6) are the two representative *japonica* and *indica* rice varieties being deeply sequenced and several mutant collections are derived from them. In this study, we utilized the transcriptome and co-expression network to analyze the different response to N rate between *indica* and *japonica* rice (YD6 and NPB) at the early stage of tillering occurrence. This study also reveals that the hub genes, tiller genes, NAM genes, and N related genes in each module are playing specific roles.

3.1. Reducing N Input to Low or Moderate Rate Is Still Good to Promote Enough Tillers

Indica and *japonica* rice subspecies possess much differences in morphological, physiological and cultivation characteristics especially in some important agronomic traits such as tillering. Tiller feature of a specific variety is the product of its genetic background, cultivation practices and environmental conditions. Among the later factors, N availability acts a key role [25,26]. Tiller is the basis for achieving panicle number which plays as a major factor in yield determination. In actual production, excessive N is usually applied aiming to produce enough number of tillers in the early growth stages. Therefore, to reduce N without yield penalty lies much in achieving enough tiller number at a minimal N requirement. Our results suggest that, even for varieties like NPB and YD6 being very different in their sensitivity to N enrichment, they both consistently show that a mild N enrichment can enhance tillering; however, excessive N enrichment will not promote more tillers as wished, instead it even suppresses tillering for a short term effect. Yet, for a more N tolerant variety like YD6, a frequent mild N enrichment may be necessary to boost tillering. Therefore, for reducing N input to a paddy field, cutting N rate to avoid heavy topdressing and switching to a low to moderate N enrichment can serve the same effect on promoting tillers. Meanwhile, at a reduced N rate, it can significantly cut off N run off as well as raise N use efficiency [25].

Meanwhile, we found the genes related to N uptake, transport, and N metabolism mainly fall in the blue, brown and turquoise modules through co-expression network in this study (Figure S12). Among them, the brown module correlated positively with the dry weight accumulation. *OsNRT1* encodes a low-affinity nitrate transporter and belongs to the constitutive expression of the outermost layer of roots, epidermis and root hairs [27]. This gene is not only homologous to the *CHL1* (*AtNRT1*) gene in Arabidopsis, but also to the polypeptide transporter that widely present in plants, animals, fungi, bacteria etc. *OsNRT2.3a* plays an important role in the long-distance transport of nitrate from root to shoot in the low-nitrate supply condition [28]. *OsNRT2.3a* is a rice vascular-specific, NRT2 family high-affinity nitrate transporter. *OsNAR2.1* can interact with *OsNRT2.1/2.2* and *OsNRT2.3a* to promote nitrate uptake by rice roots at different nitrate supply levels [29]. *OsNRT2.3b* can enhance the buffering capacity of rice to pH status, increase the absorption of N, iron and phosphorus, improve the effective utilization of N, and is very important for plants to adapt to different forms of N sources [30]. In our study, though *OsNRT1*, *OsNRT2.3* and *OsNPF2.4* showed similar response pattern to N rate, *OsAMT1.2* showed an even different response pattern to N rate between varieties. In previous transcriptome and co-expression network analysis, *NRT* and *AMT* play an important role in N uptake and utilization efficiency in *Brassica juncea* cultivars and N transporters regulate some aspects (shoot or leaf) of the coordination of N and C metabolism in *Arabidopsis* [31,32]. Interestingly, these genes have a close positive regulation of dry weight of stem and leaf in this study, which suggests that these genes may share common roles in regulating N transportation and biomass. These results indicate that these genes have a promoting effect on the growth and development of the above-ground organs. Although it does not belong to hub genes, it plays a role in the transport and absorption of N under the control of upstream hub genes.

3.2. Most of the Tiller Genes Are Not Drastically Respond to N Rate

Recent progress in molecular genetics have deciphered that *MOC1* [9], *LAX1* and *LAX2* [12], *OsFC1/OsTB1* [7,8], *APC/C*$^{TAD1/TE}$ [11,12], and more than 60 other genes [13,14] are involved in tiller number control in rice. However, it is almost unknown which set of genes are involved in tiller regulation under general cultivation conditions, especially which ones are responsible to N rate. In an attempt to answer such a question, we turn to RNA-seq approach to compare the transcriptomic changes in the tissue around the SAM, where the tillers emerge.

Our RNA-seq data point to that drastic N enrichment caused more dramatic transcription change rather than mild N rate (Figure S12). The top categories of those DEGs fall into the metabolic and hormonal/signal transduction pathways. However, all changes of the tiller related genes did not reach the two-fold bar to be a DEG (Table 2). Furthermore, half of these tiller genes were showing opposite change between the two varieties in response to the N rate (Table 2). These suggest us that drastic N rate may trigger a dramatic global transcription shift, but those changes may not necessarily directly relate to tillering. More importantly, change in a gene expression level does not have to surpass the two-fold bar to impact on tiller number.

In addition, previous studies have shown that the tiller genes are mainly analyzed in a certain genetic background. Moreover, the interaction network or regulatory mechanisms of tiller genes at specific stages are not well understood, especially in the early stages of tillering occurrence around the SAM in rice. In this study, we use GWCNA methods to find co-expressed gene modules and explore the relationship between gene networks and phenotypes, as well as core genes in the network. Interestingly, we found that several tiller genes (*HTD2/D14/D88/qPPB3, D10/OsCCD8 and NRR/CRT*) fall in the green module, which is negatively correlated with tiller regulation (Figure 7B). *HTD2/D14*, which is a component of the stragolactones (SLs) signaling pathway, encodes an esterase that inhibits rice branching and negatively regulates rice tiller number [33]. *D88*, function through the MAX/RMS/D pathway, is expressed in most tissues of rice, including leaves, stems and roots and ultimately affect rice plant type by regulating cell growth and organ development. Mutations in *D88* affected the expression of genes involved in tiller formation, including *HTD1, OSH1, D10* and other genes were significantly up-regulated in *d88* mutants [34]. *D10* is a rice ortholog of MAX4/RMS1/DAD1, which encodes a carotenoid cleavage dioxygenase and is involved in the biosynthesis of levodolactone/levylactone derivative SLs [35]. *D10/OsCCD8* is involved in the synthesis of rice aboveground branching inhibitors, and transcription of *D10* may be a key step in regulating the branching inhibition pathway. The interaction among *D10*, auxin and cytokinin affects lateral bud elongation in rice [36]. *NRR/CCRT* can regulate the structure of rice roots, so that they can better absorb a large number of nutrients and play a negative regulation role in the growth of rice roots. It can respond to the level of photo-contracted compounds and coordinate the expression of genes related to starch synthesis [37,38]. Although *HTD2/D88* and *D10/OsCCD8* genes belong to the negative regulatory module and they are not key nodes of regulation network, these genes are regulated by the upstream hubgenes. Our results corroborate that these genes closely relate to tillering.

Meanwhile, our results showed that some genes related to tillering, N and other genes fall in other modules and these modules are not closely related to certain traits (tiller, N, etc.). We speculate that these genes may have some novel unknown molecular functions, which of course requires further experimental verification.

Obviously, these differences between varieties are the results of common regulation of many genes. Changes of certain pivotal gene expression will lead to changes in other genes. The difference in these expression patterns led to the difference in the response of N rate between *indica* and *japonica* rice.

In addition, consistent with previous reports on AS of pre-mRNA [39–43], the major category of AS events revealed in our current study is intron retention (RI), irrespective of rice varieties. Despite the trending differences in the percentage of AS events between the two varieties, we have not found significant distribution differences among the categories event-wise. However, as we did not compare the components of AS genes between the groups, it would be interesting to further analyze if there are differences in this aspects of AS events of certain genes between varieties or among N rates [41–43].

4. Materials and Methods

4.1. Plant Materials

Two representative rice varieties Nipponbare (NPB, *Oryza sativa* L. subsp. *japonica*) and Yangdao6 (YD6, *Oryza sativa* L. subsp. *indica*) were employed in this study. NPB and 9311 (a sister line of YD6) are widely being used as reference sequencing varieties, in molecular genetic analysis and practical rice production. NPB generally produces more tillers under lower N rate; whereas YD6 bears fewer tillers, yet produces more tillers under higher N rate.

4.2. Growth Conditions, N Rate Treatment and Measurement

The experiments were conducted at the Yangzhou University, Jiangsu Province of China. Plants were grown in plastic pots (29 cm in diameter, 30 cm in height), which were filled with a mixture of soil and vermiculite at 5:1 (*v/v*) ratio. The soil type is sandy loam, containing 1.02 g·kg^{-1} total N, 22.73 mg·kg^{-1} available phosphorus, 49.24 mg·kg^{-1} available potassium, and 13.98 g·kg^{-1} total organic matter. The soil osmotic conductivity was 0.11 ms·cm^{-1} and pH was at 7.66.

Total N fertilizer (urea) enrichment rate was set up at 0 (LN, CK), 9 (MN) and 18 (HN) g N·m^{-2}. Basal N fertilizer (50% of total rate, in the form of urea, dissolved into water first then further diluted and applied; the same practice was followed for all other top-dressings) was premixed into the top 10 cm of soil at 3 days prior to seeding. Top dressing was at 2nd and 4th leaf emerging stage, each at 25% of the respective N rate. Twelve sprouting seeds were planted in each pot at an even spacing. After germination, the pot was irrigated with a shallow layer (1–2 cm) of water. Four biological replicates were included, with 16 pots in each treatment.

Enumeration of tiller was conducted at 4th, 6th and 8th leaf emerging stage. Plants were sampled for dry matter measurement, and subsequently for soluble sugar, starch and N content determination. Total soluble sugar and starch detection were by anthrone reagent following Brooks et al. [44]. N determination was by Kjeldhal method with Kjeltec 8400 Analyzer Unit (Foss Analytical AB, Hoganas, Sweden) following the manufacturer's recommendation.

4.3. Samples Preparation for RNA Extraction, cDNA Library Construction and Sequencing

To investigate the transcriptome changes in response to N enrichment, tissues for RNA isolation were collected when 4th leaf started spreading out. After removing leaves and the out layers of sheaths with a surgical blade, only tissues surrounding the shoot apical meristem (SAM, about 5mm in length) were collected. SAM isolation operation was carried out on the ice, and the SAM tissues were wrapped in aluminum and snap frozen in liquid N$_2$ before transferring to a deep freezer at −80 °C till use.

A total of 12 tissue samples were collected for RNA-Seq, representing two varieties (NPB and YD6) and three N levels (LN, MN and HN), with two biological replicates in each combination. Another set of 12 samples was collected for validation purpose. Total RNA was extracted using the RNAiso Plus Total RNA extraction reagent (Cat#9109, TAKARA, Kusatsu, Japan). Qualified total RNA was further purified by RNeasy micro kit (Cat#74004, QIAGEN, Duesseldorf, Germany). The rRNA was removed using Ribo-Zero rRNA Removal Kits (CAT# MRZMB126, Epicentre, Illumina, San Diego, CA, USA). RNA integrity was evaluated using the Agilent 2100 Bioanalyzer (Agilent, Santa Clara, CA, USA). Samples with RNA Integrity Number (RIN) ≥ 7 were subjected to the subsequent sequencing reactions. The libraries were constructed by using TruSeq Stranded mRNA LT Sample Prep Kit (CAT#15032612, Illumina, San Diego, CA, USA), and sequenced on the Illumina sequencing platform HiSeq 2500, consequentially 100–150 bp paired-end reads were generated.

4.4. RNA-Seq Data Processing and Gene Expression Calculation

Raw reads were first filtered by Fastx program to remove disqualified, short or ribosome RNA reads, resulting in clean reads. Clean reads % was calculated as (clean reads/raw reads) %; mRNA % was as ((clean reads-rRNA reads)/clean reads) %; Subsequent genome mapping was by spliced mapping algorithm in Tophat program [45] to a reference genome, *Oryza sativa japonica* NPB version 7.0 from ftp://ftp.plantbiology.msu.edu, and generated BAM files.

To estimate the abundance of gene expression, reads number of uniquely mapped genes were normalized to Fragments Per Kilobase of exon model per Million mapped reads (FPKM) by Cufflinks program [46]. FPKM was calculated as (total exon fragments in reads)/((mapped reads in millions) × exon length in Kilo base pair)), where total exon fragments was the number of reads mapped to exons, mapped reads was the number of reads mapped to the reference genome, and exon length was total base pair number of exons in Kilo base pair.

To generate a list of differentially expressed genes (DEG) between treatments, FPKM was used to calculate the fold change and false discovery rate (FDR, an adjusted *p*-value, i.e., *q*-value) by Cuffdiff program [18]. The threshold of DEG was set as *q*-value≤ 0.05 and fold change ≥ 2.

4.5. Novel Gene, Alternative Splicing and Enrichment Analysis

When using Cuffcompare algorithm from Cufflinks software in mapping, reads that could not be mapped to the reference genome and the transcript FPKM ≥ 10 in any single sample were deemed as novel genes. For alternative splicing (AS) screening, the BAM files were processed by Cufflinks program to generate assembled transcripts followed by Cuffmerge algorithm to the final transcriptome assembly, which was subjected to Astalavista (version 3.1) algorithm to reveal AS [19,47,48]. Functional enrichment analysis of gene ontology (GO) and KEGG (Kyoto Encyclopedia of Genes and Genomes) were by String 10.5 [49].

4.6. Quantitative Real Time RT-PCR Validation

To validate the RNA-Seq data, twenty-one DEG at different expression level were selected to confirm their expression in the corresponding samples by quantitative real-time reverse transcription polymerase chain reaction (qRT-PCR) with a BioRad CFX-96 system (BioRad, Hercules, CA, USA), following the method in [50]. Briefly, one µg of total RNA from the same batch of RNA for high throughput RNA-Seq was used for the first strand of cDNA synthesis using iScript (Cat#1708891, BioRad, Hercules, CA, USA) according to the supplier's protocol. The PCR reaction was conducted in a total volume of 12 µL reaction mix, with one µL of cDNA template, 400 nM forward primer, 400 nM reverse primer and six µL of SsoFast EvaGreen Supermix (Cat. #1725200, Bio-Rad, Hercules, CA, USA), and three technical replicates. Two tubulin genes (LOC_Os11g14220 and LOC_Os03g51600) were deemed as the reference genes. Gene-specific primer sequences for qRT-PCR are listed in Table S2. Quantification was determined by BioRad CFX manager software (V3.1). Fold change relative to control level was determined by the $2^{-\Delta\Delta Ct}$ method [50]. PCR amplifications of each sample were used in triplicate.

4.7. Co-Expression Network Analysis for Construction of Modules

WGCNA (v1.29) package in R was performed to construct the gene co-expression regulation network [20]. Detailed analysis procedures and methods were followed in accordance with Zhang et al. [51]. Among the 37,824 genes, 21,700 genes with an averaged FPKM from three replicates > 1 were used for the WGCNA unsigned co-expression network analysis. Through testing the independence and the average connectivity degree of different modules with different power value, the appropriate power value in this study was determined as seven. The modules were obtained by the automatic network construction function with default parameters in the WGCNA software package. The correlation between the modules and traits were calculated by the Pearson method using *blockwiseModules* function. The top ten genes with maximum intra-modular connectivity were considered as "highly connected gene" (hubgene) [52]. The top 20 genes including candidate hub gene network was visualized by the Cytoscape (version 3.7.2) [53].

4.8. Statistical Analysis of Genes Expression Data in a Specific Pathway

Multivariate analysis procedure of the general linear model (GLM) method from the IBM SPSS software (version 22, IBM, Armonk, NY, USA) was used in the ANOVA comparison of selected gene expression differences. Correlation analysis was made by the correlate procedure in the SPSS software.

5. Conclusions

This is a pioneer study to reveal the transcriptomic changes of SAM tissue in response to N rate between *indica* and *japonica* rice subspecies, especially for cultivars widely used in the production. N rate influenced tiller number in a different pattern between varieties, with NPB being more sensitive to N enrichment, and YD6 being more tolerant to N rate. Tiller number was positively related to N content in leaf, culm and root tissue, but negatively related to the soluble carbohydrate, irrespective of variety. Higher N enrichment brought more drastic transcription change than moderate N rate; however, varietal background dominated the differences. For the reported 65 tiller genes, less than half of them showed decent expression in SAM at tiller starting phase; among them only nineteen being significantly influenced by N rate, and two genes showing significant interaction between the N rate and variety. GO analysis revealed that the majority of these common DEGs are involved in general stress responses, stimulus responses, and hormonal signaling process. WGCNA network identified specific modules that are associated with the phenotypic traits and candidate hub genes for each module. Several genes associated with tillering and N content fall on certain most relevant modules. These results help us understand the complexity regulatory mechanisms involved in *indica* and *japonica* rice response to N rate.

Supplementary Materials: Supplementary materials can be found at http://www.mdpi.com/1422-0067/20/23/5922/s1.

Author Contributions: Conceptualization, X.Z., J.Z., J.H. and Y.Y.; Data curation, N.H. and L.M.; Formal analysis, X.Z. and N.H.; Investigation, L.M., M.L., Y.G. and C.C.; Methodology, J.J.; Project administration, A.L. and Y.W.; Resources, Y.W., J.H. and Y.Y.; Software, Z.Y.; Validation, S.Y. and G.D.; Visualization, Y.Z.; Writing–original draft, X.Z.; Writing–review & editing, Y.Y.

Funding: This work was funded by the National Natural Science Foundation of China (31571608), Natural Science Foundation of Jiangsu Province (BK20151311), Major Basic Research Project of the Natural Science Foundation of the Jiangsu Higher Education Institutions (15KJA210003), the Open Funds of the Key Laboratory of Plant Functional Genomics of the Ministry of Education (ML201903), Jiangsu Agriculture Science and Technology Independent Innovation Fund (CX(19)3060), and the Social Development Project of Yangzhou (YZ2018067).

Acknowledgments: We gratefully thank Chenwu Xu (College of Agriculture, Yangzhou University, Yangzhou, China) for meaningful help.

Conflicts of Interest: The authors declare no conflict of interest.

Abbreviations

AS	Alternative splicing
DEGs	Differentially expressed genes
FPKM	Fragments per kilo-base per million reads
GO	Gene ontology
KEGG	Kyoto Encyclopedia of Genes and Genomes
N	Nitrogen
NAM	No apical meristem
qRT-PCR	Quantitative real-time polymerase chain reaction
RNA-Seq	RNA-sequencing
SAM	Shoot apical meristem
WSC	Water soluble carbohydrate

References

1. Ding, Y.; Huang, P.; Lin, Q. Relationship between emergence of tiller and nitrogen concentration of leaf blade or leaf sheath on specific node of rice. *J. Nanjing Agric. Univ.* **1995**, *18*, 14–18.
2. Jin, Z.; Zhang, Y.; Pan, D.; Tong, L.; Li, D.; Li, M.; Wang, H.; Han, Y.; Zhang, Z.; Agriculture, S.O. Correlation analysis of plant carbon-nitrogen content and tiller-related gene expression with rice tiller formation. *J. Northeast Agric. Univ.* **2017**, *48*, 10–18.
3. Mathieu, S.; Bernhard, M.; Bernard, N.; Gilles, P.; André, M. Long-term fate of nitrate fertilizer in agricultural soils. *Proc. Natl. Acad. Sci. USA* **2013**, *110*, 18185–18189.
4. Yang, X.; Nian, J.; Xie, Q.; Feng, J.; Zhang, F.; Jing, H.; Zhang, J.; Dong, G.; Liang, Y.; Peng, J. Rice ferredoxin-dependent glutamate synthase regulates nitrogen-carbon metabolomes and is genetically differentiated between japonica and indica subspecies. *Mol. Plant* **2016**, *9*, 1520–1534. [CrossRef]

5. Chu, G.; Chen, S.; Xu, C.; Wang, D.; Zhang, X. Agronomic and physiological performance of indica/japonica hybrid rice cultivar under low nitrogen conditions. *Field Crop. Res.* **2019**, *243*, 107625. [CrossRef]

6. Pinson, S.R.M.; Wang, Y.; Tabien, R.E. Mapping and validation of quantitative trait loci associated with tiller production in rice. *Crop Sci.* **2015**, *55*, 1537. [CrossRef]

7. Minakuchi, K.; Kameoka, H.; Yasuno, N.; Umehara, M.; Luo, L.; Kobayashi, K.; Hanada, A.; Ueno, K.; Asami, T.; Yamaguchi, S. FINE CULM1 (FC1) works downstream of strigolactones to inhibit the outgrowth of axillary buds in rice. *Plant Cell Physiol.* **2010**, *51*, 1127–1135. [CrossRef]

8. Choi, M.S.; Woo, M.O.; Koh, E.B.; Lee, J.; Ham, T.H.; Seo, H.S.; Koh, H.J. Teosinte Branched 1 modulates tillering in rice plants. *Plant Cell Rep.* **2012**, *31*, 57–65. [CrossRef]

9. Li, X.; Qian, Q.; Fu, Z.; Wang, Y.; Xiong, G.; Zeng, D.; Wang, X.; Liu, X.; Teng, S.; Hiroshi, F.; et al. Control of tillering in rice. *Nature* **2003**, *422*, 618–621. [CrossRef]

10. Lin, Q.; Wang, D.; Dong, H.; Gu, S.; Cheng, Z.; Gong, J.; Qin, R.; Jiang, L.; Li, G.; Wang, J.L. Rice APC/CTE controls tillering by mediating the degradation of MONOCULM 1. *Nat. Commun.* **2012**, *3*, 752. [CrossRef]

11. Xu, C.; Wang, Y.; Yu, Y.; Duan, J.; Liao, Z.; Xiong, G.; Meng, X.; Liu, G.; Qian, Q.; Li, J. Degradation of MONOCULM 1 by APC/CTAD1 regulates rice tillering. *Nat. Commun.* **2012**, *3*, 750. [CrossRef] [PubMed]

12. Keishi, K.; Masahiko, M.; Shin, U.; Yuzuki, S.; Ikuyo, F.; Hironobu, O.; Ko, S.; Junko, K. LAX and SPA: Major regulators of shoot branching in rice. *Proc. Natl. Acad. Sci. USA* **2003**, *100*, 11765–11770.

13. Liang, W.; Shang, F.; Lin, Q.; Lou, C.; Zhang, J. Tillering and panicle branching genes in rice. *Gene* **2014**, *537*, 1–5. [CrossRef] [PubMed]

14. Wang, B.; Smith, S.M.; Li, J. Genetic regulation of shoot architecture. *Annu. Rev. Plant Biol.* **2018**, *69*, 437–468. [CrossRef] [PubMed]

15. Wang, Z.; Gerstein, M.; Snyder, M. RNA-Seq: A revolutionary tool for transcriptomics. *Nat. Rev. Genet.* **2009**, *10*, 57–63. [CrossRef] [PubMed]

16. Deng, Y.; Pan, Y.; Luo, M. Detection and correction of assembly errors of rice Nipponbare reference sequence. *Plant Biol.* **2014**, *16*, 643–650. [CrossRef] [PubMed]

17. Streets, A.M.; Huang, Y. How deep is enough in single-cell RNA-seq? *Nat. Biotechnol.* **2014**, *32*, 1005–1006. [CrossRef]

18. Cole, T.; Williams, B.A.; Geo, P.; Ali, M.; Gordon, K.; Van Baren, M.J.; Salzberg, S.L.; Wold, B.J.; Lior, P. Transcript assembly and quantification by RNA-Seq reveals unannotated transcripts and isoform switching during cell differentiation. *Nat. Biotechnol.* **2010**, *28*, 511–515.

19. Sammeth, M.; Foissac, S.; Guigo, R. A general definition and nomenclature for alternative splicing events. *PLoS Comput. Biol.* **2008**, *4*, e1000147. [CrossRef]

20. Langfelder, P.; Horvath, S. WGCNA: An R package for weighted correlation network analysis. *BMC Bioinform.* **2008**, *9*, 559. [CrossRef]

21. Cheng, X.; Peng, J.; Ma, J.; Tang, Y.; Chen, R.; Mysore, K.S.; Wen, J. NO APICAL MERISTEM (MtNAM) regulates floral organ identity and lateral organ separation in Medicago truncatula. *New Phytol.* **2012**, *195*, 71–84. [CrossRef]

22. Krahmer, J.; Ganpudi, A.; Abbas, A.; Romanowski, A.; Halliday, K.J. Phytochrome, carbon sensing, metabolism, and plant growth plasticity. *Plant Physiol.* **2017**, *176*, 1039–1048. [CrossRef] [PubMed]

23. Chen, X.; Yao, Q.; Gao, X.; Jiang, C.; Harberd, N.P.; Fu, X. Shoot-to-root mobile transcription factor HY5 coordinates plant carbon and nitrogen acquisition. *Curr. Biol.* **2016**, *26*, 640–646. [CrossRef]

24. Brooks, J.R.; Griffin, V.K.; Kattan, M.W. A modified method for total carbohydrate analysis of glucose syrups, maltodextrins, and other starch hydrolysis products. *Cereal. Chem.* **1986**, *63*, 465–466.

25. Zhou, W.; Yang, Z.; Wang, T.; Fu, Y.; Chen, Y.; Hu, B.; Yamagishi, J.; Ren, W. Environmental compensation effect and synergistic mechanism of optimized nitrogen management increasing nitrogen use efficiency in indica hybrid rice. *Front. Plant Sci.* **2019**, *10*, 245. [CrossRef] [PubMed]

26. Xu, J.; Zha, M.; Li, Y.; Ding, Y.; Chen, L.; Ding, C.; Wang, S. The interaction between nitrogen availability and auxin, cytokinin, and strigolactone in the control of shoot branching in rice (Oryza sativa L.). *Plant Cell Rep.* **2015**, *34*, 1647–1662. [CrossRef]

27. Lin, C.M.; Koh, S.; Stacey, G.; Yu, S.; Lin, T.; Tsay, T. Cloning and functional characterization of a constitutively expressed nitrate transporter gene, OsNRT1, from rice. *Plant Physiol.* **2000**, *122*, 379–388. [CrossRef]

28. Tang, Z.; Fan, X.; Li, Q.; Feng, H.; Miller, A.; Shen, Q.; Xu, G. Knockdown of a rice stelar nitrate transporter alters long-distance translocation but not root influx. *Plant Physiol.* **2012**, *160*, 2052–2063. [CrossRef]

29. Fan, X.; Tang, Z.; Tan, Y.; Zhang, Y.; Luo, B.; Yang, M.; Lian, X.; Shen, Q.; Miller, A.J.; Xu, G. Overexpression of a pH-sensitive nitrate transporter in rice increases crop yields. *Proc. Natl. Acad. Sci. USA* **2016**, *113*, 7118–7123. [CrossRef]

30. Ishikawa, S.; Ishimaru, Y.; Igura, M.; Kuramata, M.; Abe, T.; Senoura, T.; Hase, Y.; Arao, T.; Nishizawa, N.; Nakanishi, H. Ion-beam irradiation, gene identification, and marker-assisted breeding in the development of low-cadmium rice. *Proc. Natl. Acad. Sci. USA* **2018**, *115*, E4950–E4951. [CrossRef]

31. He, F.; Karve, A.; Maslov, S.; Babst, B. Large-scale public transcriptomic data mining reveals a tight connection between the transport of nitrogen and other transport processes in Arabidopsis. *Front. Plant Sci.* **2016**, *7*, 1207. [CrossRef] [PubMed]

32. Goel, P.; Sharma, N.K.; Bhuria, M.; Sharma, V.; Chauhan, R.; Pathania, S.; Swarnkar, M.K.; Chawla, V.; Acharya, V.; Shankar, R. Transcriptome and co-expression network analyses identify key genes regulating nitrogen use efficiency in Brassica juncea L. *Sci. Rep.* **2018**, *8*, 7451. [CrossRef] [PubMed]

33. Liu, W.; Chao, W.; Fu, Y.; Hu, G.; Si, H.; Li, Z.; Luan, W.; He, Z.; Sun, Z. Identification and characterization of HTD2: a novel gene negatively regulating tiller bud outgrowth in rice. *Planta* **2009**, *230*, 649–658. [CrossRef] [PubMed]

34. Gao, Z.; Qian, Q.; Liu, X.; Yan, M.; Feng, Q.; Dong, G.; Liu, J.; Han, B. Dwarf 88, a novel putative esterase gene affecting architecture of rice plant. *Plant Mol. Biol.* **2009**, *71*, 265–276. [CrossRef] [PubMed]

35. Arite, T.; Iwata, H.; Ohshima, K.; Maekawa, M.; Nakajima, M.M.; Sakakibara, H.; Kyozuka, J. DWARF10, an RMS1/MAX4/DAD1 ortholog, controls lateral bud outgrowth in rice. *Plant J.* **2010**, *51*, 1019–1029. [CrossRef]

36. Zhang, S.; Li, G.; Fang, J.; Chen, W.; Jiang, H.; Zou, J.; Liu, X.; Zhao, X.; Li, X.; Chu, C. The interactions among DWARF10, auxin and cytokinin underlie lateral bud outgrowth in rice. *J. Integr. Plant Biol.* **2010**, *52*, 626–638. [CrossRef]

37. Zhang, Y.; Yan, Y.; Wang, L.; Yang, K.; Xiao, N.; Liu, Y.; Fu, Y.; Sun, Z.; Fang, R.; Chen, X. A novel rice gene, NRR responds to macronutrient deficiency and regulates root growth. *Mol. Plant* **2012**, *5*, 63–72. [CrossRef]

38. Morita, R.; Inoue, K.; Ikeda, K.I.; Hatanaka, T.; Misoo, S.; Fukayama, H. Starch content in leaf sheath controlled by CO_2 responsive CCT protein is a potential determinant of photosynthetic capacity in rice. *Plant Cell Physiol.* **2016**, *57*, 2334–2341. [CrossRef]

39. Nishida, S.; Kakei, Y.; Shimada, Y.; Fujiwara, T. Genome-wide analysis of specific alterations in transcript structure and accumulation caused by nutrient deficiencies in Arabidopsis thaliana. *Plant J.* **2017**, *91*, 741–753. [CrossRef]

40. Wang, Z.; Zhang, H.; Gong, W. Genome-wide identification and comparative analysis of alternative splicing across four legume species. *Planta.* **2019**, *249*, 1133–1142. [CrossRef]

41. Li, W.; Lin, W.; Ray, P.; Lan, P.; Schmidt, W. Genome-wide detection of condition-sensitive alternative splicing in Arabidopsis roots. *Plant Physiol.* **2013**, *162*, 1750–1763. [CrossRef] [PubMed]

42. Calixto, C.; Guo, W.; James, A.; Tzioutziou, N.; Entizne, J.; Panter, P.; Knight, H.; Nimmo, H.; Zhang, R.; Brown, J. Rapid and dynamic alternative splicing impacts the Arabidopsis cold response transcriptome. *Plant Cell.* **2018**, *30*, 1424–1444. [CrossRef] [PubMed]

43. Zhu, G.; Li, W.; Zhang, F.; Guo, W. RNA-seq analysis reveals alternative splicing under salt stress in cotton, Gossypium davidsonii. *BMC Genom.* **2018**, *19*, 73. [CrossRef] [PubMed]

44. Clegg, K.M. The application of the anthrone reagent to the estimation of starch in cereals. *J. Sci. Food Agric.* **1956**, *7*, 40–44. [CrossRef]

45. Trapnell, C.; Pachter LSalzberg, S.L. TopHat: Discovering splice junctions with RNA-Seq. *Bioinformatics* **2009**, *25*, 1105–1111. [CrossRef]

46. Mortazavi, A.; Williams, B.A.; Mccue, K.; Schaeffer, L.; Wold, B.J. Mapping and quantifying mammalian transcriptomes by RNA-Seq. *Nat. Methods* **2008**, *5*, 621–628. [CrossRef]

47. Sylvain, F.; Michael, S. ASTALAVISTA: Dynamic and flexible analysis of alternative splicing events in custom gene datasets. *Nucleic Acids Res.* **2007**, *35*, W297–W299.

48. Michael, S. Complete alternative splicing events are bubbles in splicing graphs. *J. Comput. Biol. A J. Comput. Mol. Cell Biol.* **2009**, *16*, 1117–1140.

49. Szklarczyk, D.; Morris, J.H.; Cook, H.; Kuhn, M.; Wyder, S.; Simonovic, M.; Santos, A.; Doncheva, N.T.; Roth, A.; Bork, P. The STRING database in 2017: Quality-controlled protein–protein association networks, made broadly accessible. *Nucleic Acids Res.* **2017**, *45*, D362–D368. [CrossRef]

50. Livak, K.J.; Schmittgen, T.D. Analysis of relative gene expression data using real-time quantitative PCR and the 2(-Delta Delta C(T)) Method. *Methods* **2001**, *25*, 402–408. [CrossRef]
51. Zhang, X.; Huang, N.; Mo, L.; Lv, M.; Gao, Y.; Wang, J.; Liu, C.; Yin, S.; Zhou, J.; Xiao, N.; et al. Global transcriptome and co-expression network analysis reveal contrasting response of japonica and indica rice cultivar to γ radiation. *Int. J. Mol. Sci.* **2019**, *20*, 4358. [CrossRef] [PubMed]
52. Wisniewski, N.; Cadeiras, M.; Bondar, G.; Cheng, R.K.; Shahzad, K.; Onat, D.; Latif, F.; Korin, Y.; Reed, E.; Fakhro, R. Weighted gene coexpression network analysis (WGCNA) modeling of multiorgan dysfunction syndrome after mechanical circulatory support therapy. *J. Heart Lung Transpl.* **2013**, *32*, S223. [CrossRef]
53. Shannon, P.; Markiel, A.; Ozier, O.; Baliga, N.S.; Wang, J.T.; Ramage, D.; Amin, N.; Schwikowski, B.; Ideker, T. Cytoscape: A software environment for integrated models of biomolecular interaction networks. *Genome Res.* **2003**, *13*, 2498–2504. [CrossRef] [PubMed]

International Journal of
Molecular Sciences

Article

Targeted Mutagenesis of the Rice *FW 2.2*-Like Gene Family Using the CRISPR/Cas9 System Reveals OsFWL4 as a Regulator of Tiller Number and Plant Yield in Rice

Qingsong Gao [1,†], Gang Li [2,†], Hui Sun [1], Ming Xu [1], Huanhuan Wang [1], Jianhui Ji [1], Di Wang [2], Caiyong Yuan [2,*] and Xiangxiang Zhao [1,*]

[1] Jiangsu Collaborative Innovation Center of Regional Modern Agriculture & Environmental Protection/Jiangsu Key Laboratory for Eco-Agricultural Biotechnology around Hongze Lake, Huaiyin Normal University, Huai'an 223300, China

[2] Huaiyin Institute of Agricultural Sciences of Xuhuai Region in Jiangsu, Huai'an 223001, China

*   Correspondence: hysdycy@163.com (C.Y.); xxzhao2013@163.com (X.Z.); Tel.: +86-517-83659907 (C.Y.); +86-517-83525885 (X.Z.)

†   These authors contributed equally to this work.

Received: 30 November 2019; Accepted: 22 January 2020; Published: 26 January 2020

Abstract: The *FW2.2*-like (*FWL*) genes encode cysteine-rich proteins with a placenta-specific 8 domain. They play roles in cell division and organ size control, response to rhizobium infection, and metal ion homeostasis in plants. Here, we target eight rice *FWL* genes using the CRISPR/Cas9 system delivered by *Agrobacterium*-mediated transformation. We successfully generate transgenic T_0 lines for 15 of the 16 targets. The targeted mutations are detected in the T_0 lines of all 15 targets and the average mutation rate is found to be 81.6%. Transfer DNA (T-DNA) truncation is a major reason for the failure of mutagenesis in T_0 plants. T-DNA segregation analysis reveals that the T-DNA inserts in transgenic plants can be easily eliminated in the T_1 generation. Of the 30 putative off-target sites examined, unintended mutations are detected in 13 sites. Phenotypic analysis reveals that tiller number and plant yield of *OsFWL4* gene mutants are significantly greater than those of the wild type. Flag leaves of *OsFWL4* gene mutants are wider than those of the wild type. The increase in leaf width of the mutants is caused by an increase in cell number. Additionally, grain length of *OsFWL1* gene mutants is higher than that of the wild type. Our results suggest that transgene-free rice plants with targeted mutations can be produced in the T_1 generation using the *Agrobacterium*-mediated CRISPR/Cas9 system and that the *OsFWL4* gene is a negative regulator of tiller number and plant yield.

Keywords: *FW2.2*-like gene; tiller number; grain yield; rice; CRISPR/Cas9; genome editing; off-target effect

---

1. Introduction

fw 2.2 is a major quantitative trait locus that regulates fruit size and weight in tomato [1,2]. The underlying gene *FW2.2* regulates fruit size by negatively regulating cell division [2,3]. Homolog identification and sequence analysis have revealed that FW2.2 belongs to a large eukaryotic family of cysteine-rich proteins containing a featured placenta-specific 8 domain [4,5]. *FW2.2*-like (*FWL*) genes have been characterized in various plant species and are reported to play important roles in diverse biological processes, such as cell number and organ size control [4,6–9], nodulation [5,10], and metal ion homeostasis [11–20]. The rice *FWL* gene family contains eight members [9]. Among them, the *OsFWL3* gene is reported to negatively affect grain length and weight by regulating cell division in the glume [9]. However, the *OsFWL4* gene has been reported to affect cadmium (Cd) resistance upon expression in yeast [15]. RNA interference-mediated knockdown of *OsFWL4* has been found to reduce translocation of Cd from the roots to shoots in rice seedlings. More recently, a change in the expression of *OsFWL1* and *OsFWL2* induced by their overexpression or RNA interference has been found to affect Cd tolerance and accumulation in rice [20]. Interestingly, the *OsFWL5/PCR1* gene, which affects Cd and Zn tolerance when expressed in yeast cells, has been reported to regulate metal ion homeostasis and grain size and weight in rice [13,14]. However, whether other Cd-responsive rice *FWL* genes also play a role in plant and organ development in rice remains unknown.

Generating mutants with intended mutations is crucial for functional analysis of plant genes. The clustered regularly interspaced short palindromic repeats (CRISPR)/CRISPR-associated protein 9 (Cas9) system is a powerful tool for genome editing in various organisms, including plants. This system induces DNA double-strand breaks at given genomic sites, which are subsequently repaired by either non-homologous end joining or homologous recombination pathways in the cells [21]. Non-homologous end joining is error-prone and can act throughout the cell cycle. It is therefore commonly utilized to disrupt genes by creating random insertions or deletions (indels) at target sites [22]. In the presence of a homologous DNA template, a double-strand break can be repaired by homologous recombination, leading to target gene replacement or insertion. The precise cleavage of the target DNA using the CRISPR/Cas9 system requires two components, namely, Cas9 nuclease harboring HNH and RuvC endonuclease domains for cleaving and an engineered single-guide RNA (sgRNA) for directing Cas9 to the target site [23]. A prerequisite for binding and cleavage of the target DNA is the presence of a trinucleotide protospacer adjacent motif (PAM) immediately after the target DNA [24]. Sequence specificity can be achieved by changing a 20-nucleotide "guide sequence" in the sgRNA. As this system does not require protein engineering, the nuclease can be easily reprogrammed. With the development of highly efficient CRISPR/Cas9 systems, stable homozygous mutants can be obtained within a single generation in many plants [25–28].

Agrobacterium-mediated transfer DNA (T-DNA) transformation is commonly used for delivering CRISPR/Cas9 DNA into rice cells. In this study, two target sites are designed for each of the eight rice *FWL* genes for gene editing using the *Agrobacterium*-mediated CRISPR/Cas9 system. We generate transgenic T_0 lines from 15 out of 16 constructs and detect targeted mutations in all T_0 lines. Gene editing efficiency, T-DNA segregation patterns, and off-target effects are analyzed. The phenotypes of homozygous and transgene-free mutants with no detected off-target mutations of the *OsFWL1* and *OsFWL4* genes are then examined. Our results suggest that OsFWL4 is a negative regulator of tiller number and plant yield in rice and that OsFWL1 plays a role in modulating rice grain length.

2. Results

2.1. Generation of Rice FWL Gene Mutants Using CRISPR/Cas9

Two target sites were designed in the coding region of each of the eight rice *FWL* genes for CRISPR/Cas9 gene editing (Table 1). The GC content in these target sites was in the range 45–75%. The synthesized oligos were inserted into the CRISPR/Cas9 binary vector (Figure S1). Subsequently, the 16 constructed vectors were transformed into the *Japonica* rice variety Zhonghua 11 using the *Agrobacterium*-mediated method.

Of the 16 vectors, we successfully generated transgenic T_0 lines for 15 vectors (Table 2). We detected targeted mutations in all those T_0 lines. The mutation rates varied from 26.7% to 100%, and the average mutation rate was 81.6% (Table 2), suggesting that the CRISPR/Cas9 system constructed in this study is efficient in rice gene editing. Bi-allelic mutants were detected in T_0 plants from each vector, with detection percentages varying from 20.0% to 87.5% (Table 2). Homozygous mutants were detected in T_0 plants from 13 vectors, with the highest detection percentage being 64.3%. By contrast, heterozygotes and chimeras were detected only in T_0 plants from three vectors (Table 2). The percentage of heterozygotes and chimeras in all T_0 plants was only 1.3% and 4.0%, respectively. Detailed sequencing results of all T_0 mutants are shown in Table S1.

Table 1. Target sites of rice *FWL* genes for clustered regularly interspaced short palindromic repeats (CRISPR)/CRISPR-associated protein 9 (Cas9)-mediated gene editing.

Gene	Locus	Target Name	Target Sequence (5′–3′) [1]	GC Content (%)
OsFWL1	LOC_Os02g52550	Osfwl1a	CTGAAGGACTTACAGTTTCC GGG	45
		Osfwl1b	TGGGCAGGTCGCTGACATCG TGG	65
OsFWL2	LOC_Os02g36940	Osfwl2a	GCGCTGGTGATGCTCCTCAC GGG	65
		Osfwl2b	CATCTTGGCGCGGTAGAAGC AGG	60
OsFWL3	LOC_Os02g36950	Osfwl3a	ATCGCGGAGATCGTCGACCG GGG	65
		Osfwl3b	GTGGACGAGGCAGTCGGGGC AGG	75
OsFWL4	LOC_Os03g61440	Osfwl4a	ATTGAAGCAGGCGAAGAGTC CGG	50
		Osfwl4b	CGCAGCATGGGTCCTCGGGG AGG	75
OsFWL5	LOC_Os10g02300	Osfwl5a	ATCGCAGAAATCGTCGACAG GGG	50
		Osfwl5b	CTCACGGTGCATCTGTGCGA TGG	60
OsFWL6	LOC_Os03g61470	Osfwl6a	TCGACGTCGTGCGGCACCGG CGG	75
		Osfwl6b	GGCAAGATGCGCACTCAGTA CGG	55
OsFWL7	LOC_Os03g61500	Osfwl7a	CCCGTGCATCACGTTCGGGC GGG	70
		Osfwl7b	CATCTTGCCCCGGTAGACGC AGG	65
OsFWL8	LOC_Os03g61480	Osfwl8a	GGGTCGACGTCGTTCGGCAC CGG	70
		Osfwl8b	GTTGAGGTGCCATCCGAGCT TGG	60

[1] The protospacer adjacent motif (PAM) sequences are shown in green.

Table 2. Identification of targeted mutations in T_0 plants.

Target	No. of T_0 Plants Obtained	No. of Plants with Mutations	Zygosity				Combined Percentage of Homozygous and Bi-Allelic Mutants (%)
			Homozygous	Bi-Allelic	Heterozygous	Chimeric	
Osfwl1a	15	7 (46.7%)	–	7 (46.7%)	–	–	46.7
Osfwl1b	14	14 (100.0%)	9 (64.3%)	5 (35.7%)	–	–	100.0
Osfwl2a	16	15 (93.8%)	1 (6.3%)	14 (87.5%)	–	–	93.8
Osfwl2b	14	11 (78.6%)	3 (21.4%)	8 (57.1%)	–	–	78.6
Osfwl3a	7	6 (85.7%)	–	4 (57.1%)	–	2 (28.6%)	57.1
Osfwl3b	25	22 (88.0%)	2 (8.0%)	13 (52.0%)	1 (4.0%)	6 (24.0%)	60.0
Osfwl4a	15	11 (73.3%)	3 (20.0%)	8 (53.3%)	–	–	73.3
Osfwl4b	15	10 (66.7%)	1 (6.7%)	8 (53.3%)	–	1 (6.7%)	60.0
Osfwl5a	14	14 (100.0%)	2 (14.3%)	12 (85.7%)	–	–	100.0
Osfwl5b	14	14 (100.0%)	7 (50.0%)	7 (50.0%)	–	–	100.0
Osfwl6a	15	4 (26.7%)	1 (6.7%)	3 (20.0%)	–	–	26.7
Osfwl6b	16	15 (93.8%)	5 (31.3%)	10 (62.5%)	–	–	93.8
Osfwl7a	14	13 (92.9%)	2 (14.3%)	10 (71.4%)	1 (7.1%)	–	85.7
Osfwl7b	15	12 (80.0%)	5 (33.3%)	6 (40.0%)	1 (6.7%)	–	73.3
Osfwl8b	14	14 (100.0%)	3 (21.4%)	11 (78.6%)	–	–	100.0
Total	223	182 (81.6%)	44 (19.7%)	126 (56.5%)	3 (1.3%)	9 (4.0%)	76.2

Sequencing analyses revealed that most mutations were short indels; 62.3% of indels were 1 bp changes (Figure 1A,B). A majority of the 1 bp insertions (83.2%) were either A or T, which is consistent with previous reports [27,29].

Of the 223 T_0 plants, 41 plants did not contain mutations. To test whether failed editing of these plants was caused by a lack of the CRISPR/Cas9 construct, the presence of *hygromycin phosphotransferase* (*HPT*), sgRNA, and *Cas9* transgenes in these 41 plants was examined. Two plants did not contain *HPT*, sgRNA, and *Cas9* sequences (Table S2), which suggests that these plants escaped hygromycin selection. Twenty-five plants did not contain sgRNA and/or *Cas9* sequences (Table S2) which suggests that incompleteness of the sgRNA/*Cas9* expression cassette led to failed mutagenesis in these plants. Interestingly, when unmutated T_0 plants without the complete sgRNA/*Cas9* construct were excluded, all targets except Osfwl1a, Osfwl4a, and Osfwl6a had a mutation rate of 100% (Table 2 and Table S2). The score of sgRNA activity in all targets predicted using the sgRNA Scorer 2.0 varied from −0.64 to 1.09 (Table S3), indicating moderate efficiency of the sgRNAs [30].

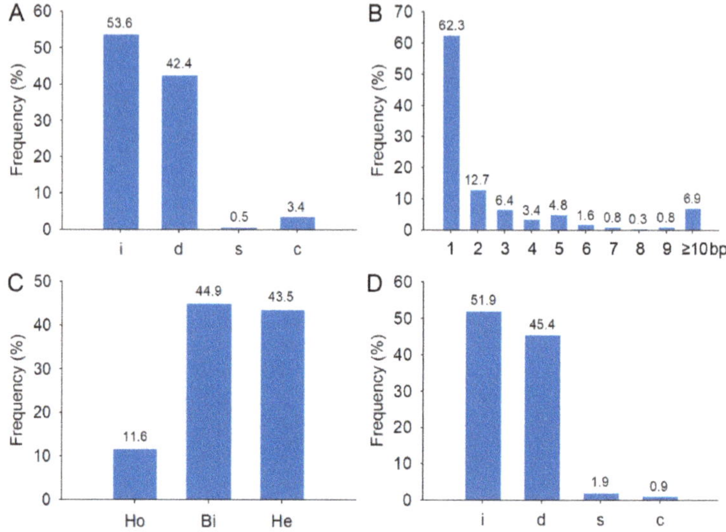

Figure 1. Characterization of on-target and off-target mutations. (**A**) Frequencies of different types of on-target mutations. (**B**) Frequencies of different lengths of on-target mutations. (**C**) Zygosity of off-target mutations. (**D**) Types of off-target mutations. Legend: i, insertion; d, deletion; s, substitution; c, combined mutation; Ho, homozygote; Bi, bi-allele; He, heterozygote.

The inheritance patterns of targeted mutations in later generations were also examined. Mutations of most homozygous T_0 plants were stably transmitted to the T_1 generation (Table S4). However, unexpected genotypes were detected in the T_1 generation of four of the five bi-allelic T_0 plants. Additionally, a large proportion of the progeny of a chimeric T_0 plant were chimeras (Tables S4 and S5). The transmission of mutations of several randomly selected T_1 lines that did not contain transgenes ('transgene-free'; see Section 2.2) in the T_2 generation was also examined. The genotypes of all these lines were faithfully transmitted to T_2 plants (Table S6).

2.2. Segregation of T-DNA in the T_1 Generation

The presence of CRISPR/Cas9 DNA and marker genes in gene-edited plants may cause adverse effects, such as an increased risk of off-target changes, and may trigger regulation concerns when these plants are used in crop breeding [22,31,32]. To test whether the T-DNA fragment carrying the CRISPR/Cas9 construct could be segregated out in the progeny of T_0 mutants, the presence of *HPT*, sgRNA, and *Cas9* transgenes in T_1 plants derived from one of the homozygous or bi-allelic T_0 mutants of each target (except Osfwl3a) was examined (Figure S2). For target Osfwl3a, the progeny of a chimeric T_0 plant (Osfwl3a#4) were used. The genotype of the T_0 mutant for each T_1 line used is shown in Table S1. Transgene-free plants were obtained in several randomly selected T_1 progeny for all lines (Figure S2), suggesting that the number of T-DNA insertion loci was low in T_0 plants.

In most T_1 lines (13 out of 15) analyzed, consistent segregation patterns were observed for all the *HPT*, sgRNA, and *Cas9* transgenes (Figure S2). However, inconsistent segregation patterns were detected in two lines (Osfwl3a#4 and Osfwl4b#6). In line Osfwl3a#4, *HPT* and *Cas9* sequences were not detected in some plants, although the sgRNA sequence was present (Figure S2). Interestingly, the eight T_1 plants of this line that lacked the complete sgRNA/*Cas9* expression cassette were all homozygotes, whereas the other 12 plants containing the complete sgRNA/*Cas9* expression cassette were all chimeras (Figures S2 and S3; Tables S4 and S5). In line Osfwl4b#6, sgRNA and *Cas9* sequences were not detected in all examined T_1 plants, but the *HPT* sequence was present in some of the plants (Figure S2). Next, we examined the T_0 plant of this line for the presence of a T-DNA fragment and found that it also lacked sgRNA and *Cas9* sequences. This indicates that mutations of this line were generated by transient CRISPR/Cas9 expression.

2.3. Phenotypic Analysis of Rice FWL Gene Mutants

Phenotypes of allelic mutant lines that were homozygous, transgene-free, and with no detected off-target effects (see Section 2.4) of the *OsFWL1* and *OsFWL4* genes were analyzed. For *OsFWL1*, T_2 plants of lines Osfwl1a#4 and Osfwl1b#11 (Table S6) were selected for Osfwl1a and Osfwl1b targets, respectively. For *OsFWL4*, T_3 plants of lines Osfwl4a#7 and Osfwl4b#6 (Table S6) were analyzed for Osfwl4a and Osfwl4b targets, respectively.

The number of tillers per *osfwl4a* and *osfwl4b* mutant plants was 45.9% and 41.1% greater, respectively, than that of the wild type (WT; Figure 2A,B). The number of grains per panicle of mutants was not significantly changed (Figure 2C). Although 1000-grain weight of mutants was slightly reduced (Figure 2D), the grain yield per plant was increased by 25.6–35.8% (Figure 2E).

Additionally, the flag leaf width of *osfwl4a* and *osfwl4b* mutants was 7.7% and 6.3% greater, respectively, than that of the WT (Figure 3A,C). However, there was no marked difference in flag leaf length (Figure 3B). Analysis of leaf epidermal cell size revealed no significant difference in cell length and width between WT and mutants (Figure S4). This suggests that the increase in leaf width of mutants was caused by an increase in cell number but not in cell size. In addition, the plant height of mutants was slightly reduced compared with that of the WT (Figure 2A and Figure S5).

The expression profile of the *OsFWL4* gene during the life cycle of rice was examined using qRT-PCR. *OsFWL4* was mainly expressed in the developing endosperm and the stem at the heading stage (Figure 4); it was also expressed in the leaf, root, and panicle. To gain insights into the molecular function of *OsFWL4*, a gene co-expression analysis was performed using the Genevestigator program [33] with the mRNA-Seq datasets. Many positively correlated genes of *OsFWL4* were involved in cell signal transduction, disease resistance, and heavy metal resistance (Table S7). Interestingly, some negatively correlated genes of *OsFWL4* were found to encode the F-box domain- and BTB (bric-a-brac, tramtrack and broad complex) domain-containing proteins, which may play a role in protein ubiquitination [34] (Table S8).

Grain length of *osfwl1a* and *osfwl1b* mutants was 4.2% and 5.5% greater, respectively, than that of the WT (Figure 5A,B). However, there was no difference in grain width between the mutants and WT plants (Figure 5C). Grain thickness of mutants was slightly lower than that of the WT (Figure 5D). Finally, there was no change in the 1000-grain weight of mutants (Figure 5E). Additionally, plant height, leaf size, and grain yield per plant of mutants were not considerably different from those of the WT (Figure S6).

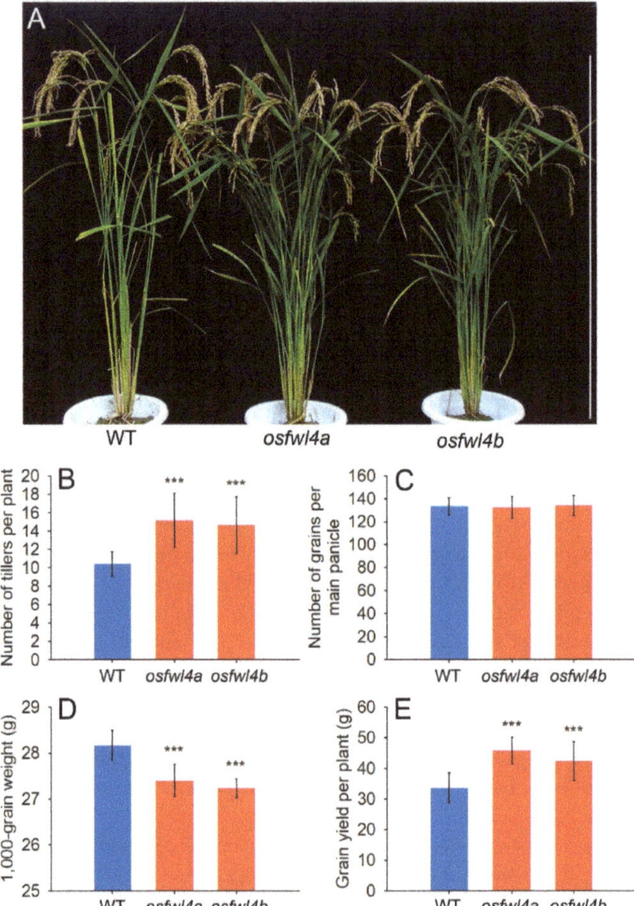

Figure 2. Analysis of yield traits of wild type (WT) and *OsFWL4* gene mutants. (**A**) WT and mutant plants, bar = 1 m. (**B**) Number of tillers per plant of the WT and mutants, $n = 20$. (**C**) Number of grains per main panicle of the WT and mutants, $n = 10$. (**D**) 1000-grain weight of the WT and mutants, $n = 10$. (**E**) Grain yield per plant of the WT and mutants, $n = 12$–15. Error bars are standard deviations. ***$p < 0.001$.

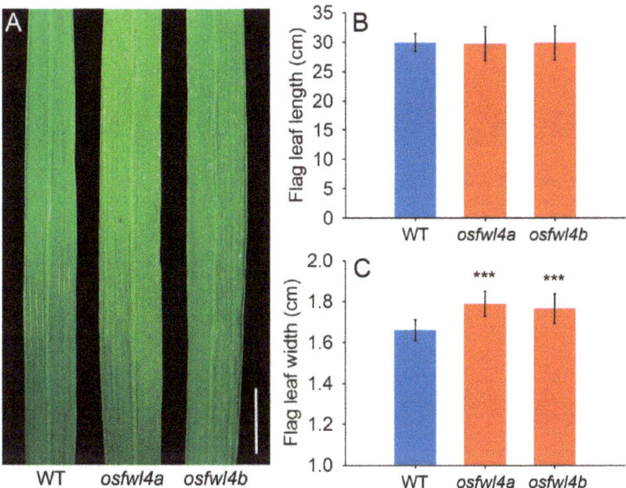

Figure 3. Phenotypes of flag leaves of WT and *OsFWL4* gene mutants. (**A**) Flag leaves of the WT and mutants, bar = 2 cm. (**B**) Flag leaf length of the WT and mutants. (**C**) Flag leaf width of the WT and mutants. The values in (**B**) and (**C**) are means of 20 plants. Error bars are standard deviations. ***$p < 0.001$.

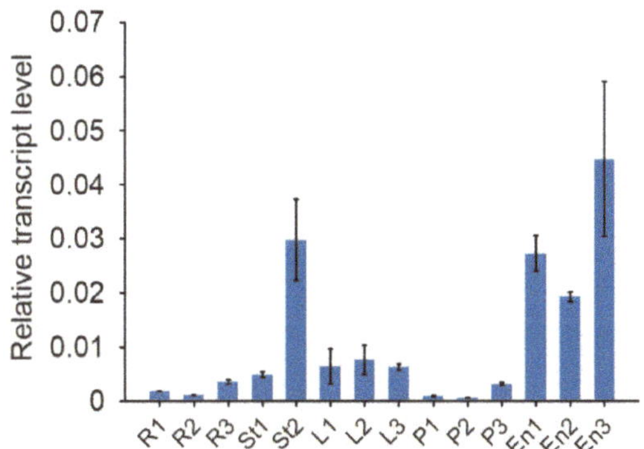

Figure 4. qRT-PCR results of the *OsFWL4* gene in 14 tissue samples of *japonica* rice Zhonghua 11. The rice *Actin1* gene was used as the internal control. Legend: R1–R3, roots in the seedling, tillering, and heading stages, respectively; St1 and St2, stems in the jointing and heading stages, respectively; L1–L3, leaves in the seedling, tillering, and heading stages, respectively; P1–P3, panicles 5, 15, and 20 cm in length, respectively; En1–En3, endosperms 5, 14, and 21 days after pollination. Error bars are standard deviations of three technical repeats.

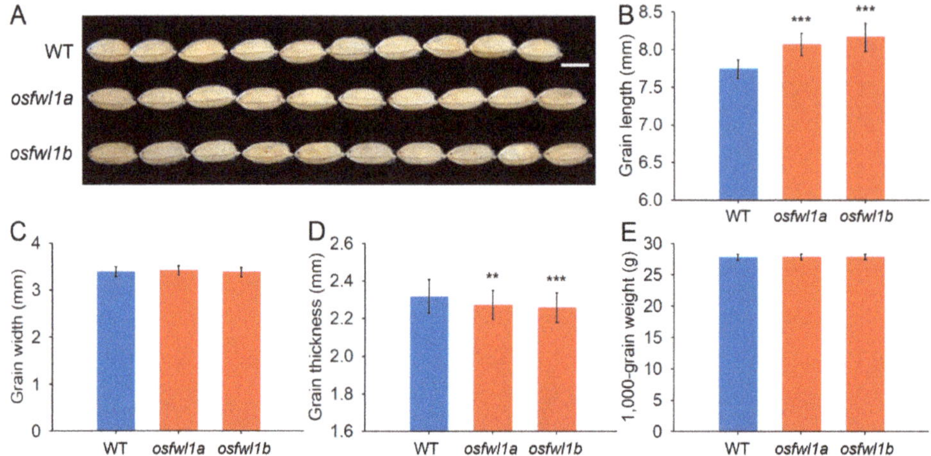

Figure 5. Analysis of grain shape of WT and *OsFWL1* gene mutants. (**A**) Grains of the WT and mutants, bar = 5 mm. (**B**) Grain length of the WT and mutants, $n = 50$. (**C**) Grain width of the WT and mutants, $n = 50$. (**D**) Grain thickness of the WT and mutants, $n = 50$. (**E**) 1000-grain weight of the WT and mutants, $n = 10$. Error bars are standard deviations. **$p < 0.01$; ***$p < 0.001$.

2.4. Analysis of Off-Target Effects

To investigate the potential off-target events in our experiments, the two most probable off-target sites were selected for each of the 15 targets. The potential editing events in these sites were examined in all T_0 plants and several randomly selected T_1 lines. When the same off-target sequence occurred at different genomic loci, only one locus was examined. Of the 30 putative off-target sites, we detected mutations in 13 sites (Table 3). All the five loci that had a single-base mismatch with sgRNA exhibited off-target effects. Mutations were detected in four of the five (80.0%) and three of the 10 (30.0%) loci that had two and three mismatches, respectively. Additionally, one locus (Osfwl2aOFF-2) that had four dispersed mismatches was cleaved in transgenic plants. No mutations were detected in the loci that had five mismatches with sgRNA (Table 3). Most of the off-target sites with mutations were located within the gene region (Table 3).

Table 3. Off-target effect analysis of transgenic plants.

Name of Putative Off-Target Site	Locus	Sequence of Putative Off-Target Site [1]	Region	No. of Mismatching Bases	No. of Plants Tested	No. of Plants with Mutations
Osfwl1aOFF-1	Chr2: 11000661–11000683	CTGAATGACTGACTGTCTCC TGG	LOC_Os02g18850 intron	4	32	0
Osfwl1aOFF-2	Chr2: 1853752–1853774	CTGAAGGACTTGCACATTTC AGG	LOC_Os02g04230 intron	4	32	0
Osfwl1bOFF-1	Chr6: 17267236–17267258	CGGGCAGGACGCCGACATCG CGG	LOC_Os06g29994 CDS	3	32	2
Osfwl1bOFF-2	Chr8: 15832268–15832290	TGTGCAGGTCGATGACATCA TGG	Intergenic	3	32	0
Csfwl2aOFF-1	Chr1: 14111612–14111634	GCGATGGTGATGCTCCTCGC CGG	Intergenic	2	32	4
Csfwl2aOFF-2	Chr6: 463763–463785	ACGCAGGTGAAGCTCCTAAC TGG	LOC_Os06g01800 intron	4	32	7
Csfwl2bOFF-1	Chr4: 24327403–24327425	CATCTTGGGGAGGTAGAAGA AGG	LOC_Os04g40990 CDS	3	30	0
Csfwl2bOFF-2	Chr2: 22320818–22320840	CAGCTTGGAGCGGTAGATGC AGG	LOC_Os02g36950 CDS	3	30	0
Csfwl3aOFF-1	Chr4: 23060613–23060635	ATCGCGGAGATCGTCGACCA GGG	LOC_Os04g38790 CDS	1	27	26
Csfwl3aOFF-2	Chr2: 22312627–22312649	ATCGCGGAGATCATCGACCG GGG	LOC_Os02g36940 CDS	1	27	26
Csfwl3bOFF-1	Chr2: 22312479–22312501	GTGGACGGGGCAGTCGGCGC AGG	LOC_Os02g36940 CDS	2	41	7
Csfwl3bOFF-2	Chr2: 4426323–4426345	GTGGAAGAAGAAGTCGAGGC AGG	LOC_Os02g08330 CDS	4	41	0
Csfwl4aOFF-1	Chr7: 3799011–3799033	TTTGAAGCAGGTGAAGAGTC CGG	LOC_Os07g07580 intron	2	28	23
Csfwl4aOFF-2	Chr9: 2298486–2298508	CATGAGGAAGGCGAGGAGTC CGG	LOC_Os09g04339 CDS	5	28	0
Csfwl4bOFF-1	Chr1: 31161078–31161100	CGCTGCATCTGTCCTCGGGA AGG	Intergenic	4	31	0
Csfwl4bOFF-2	Chr2: 5315090–5315112	AGCAGAAAGGATCCTGGGGG AGG	Intergenic	5	31	0
Csfwl5aOFF-1	Chr6: 7393721–7393743	ATCTCAGAAATAATCGACAG CGG	Intergenic	3	31	0
Csfwl5aOFF-2	Chr4: 19176717–19176739	GTCCCAGGACTCGTCGACAG AGG	LOC_Os04g32020 5' UTR	4	31	0
Csfwl5bOFF-1	Chr4: 34131133–34131155	CGCCCGGTGCATCTGCGCGA TGG	LOC_Os04g57330 5' UTR	3	32	3
Csfwl5bOFF-2	Chr6: 22532424–22532446	ATCACGGTGAGCATGTGCGA TGG	LOC_Os06g38090 intron	5	32	0
Csfwl6aOFF-1	Chr3: 34884082–34884104	TCGACGTCGTGCGGCACCAG CGG	LOC_Os03g61500 CDS	1	33	21
Csfwl6aOFF-2	Chr3: 16449514–16449536	AAGACGTCGAGCGGCACCGG CGG	LOC_Os03g28980 CDS	3	33	21
Csfwl6bOFF-1	Chr3: 34878541–34878563	GGCAAGATGCGCGCACAGTA CGG	LOC_Os03g61490 CDS	2	31	0
Csfwl6bOFF-2	Chr1: 29494129–29494151	AGCTAGACGTGCAATCAGTA CGG	Intergenic	5	31	0
Csfwl7aOFF-1	Chr3: 34870990–34871012	CCCGTGCATCACGTTCGGGA GGG	LOC_Os03g61470 CDS	1	31	17
Csfwl7aOFF-2	Chr10: 21731587–21731609	CCCATGCATCACGTTAGGTC CGG	LOC_Os10g40580 5' UTR	3	31	0
Csfwl7bOFF-1	Chr11: 6506399–6506421	GATCTTGCTCCGGTCGACGC CGG	Intergenic	3	32	0
Csfwl7bOFF-2	Chr2: 22312530–22312552	CATCTTGGCGCGGTAGAAGC AGG	LOC_Os02g36940 CDS	3	32	0
Csfwl8bOFF-1	Chr3: 34883868–34883890	GTTGAGGTCCCATCCGAGCT TGG	LOC_Os03g61500 CDS	1	30	30
Csfwl8bOFF-2	Chr3: 34857174–34857196	GTTGAGGTGCCACCCAAGCT TGG	LOC_Os03g61430 CDS	2	30	13

[1] The PAM sequences are shown in green and the mismatched bases in red.

Among the putative off-target sites with two mismatches, the mismatched nucleotides of Osfwl6bOFF-1 were separated by only one nucleotide in the PAM-proximal region, and no modifications were detected at this site (Table 3). Genotype analysis of the off-target mutations in T_0 plants revealed that 44.9% and 43.5% of the genotypes were bi-alleles and heterozygotes, respectively (Figure 1C). Most off-target mutation events were insertions and deletions as observed in the on-target mutation events (Figure 1A,D).

2.5. Expression Analysis of Cas9 in Transgenic Plants

The expression level of *Cas9* was examined in transgene-positive plants of several randomly selected T_1 lines by qRT-PCR. The *Cas9* mRNA level was approximately 11.6–30.3 fold that of the *OsActin1* gene in different lines (Figure 6).

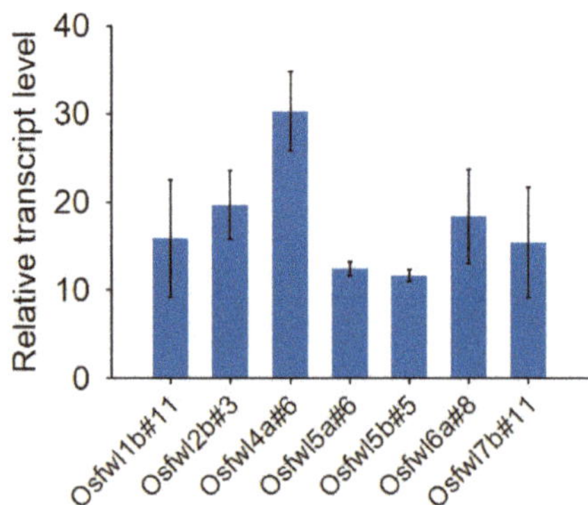

Figure 6. qRT-PCR analysis of *Cas9* expression in the transgene-positive plants of the T_1 lines. The rice *Actin1* gene was used as the internal control. Error bars represent standard deviations of three biological repeats.

3. Discussion

In this study, rice *FWL* family genes were mutated using the *Agrobacterium*-mediated CRISPR/Cas9 system, and the phenotypes of mutants of two genes (*OsFWL1* and *OsFWL4*) were characterized. The results suggest that the *OsFWL4* gene is a negative regulator of tiller number and plant yield in rice and that the *OsFWL1* gene plays a role in modulating rice grain length.

Rice tiller number is an important agronomic trait that largely affects grain yield. The tiller number of the *OsFWL4* gene mutants was increased by up to 45.9% compared with that of the WT (Figure 2A,B). Additionally, flag leaf width of mutants was also increased (Figure 3A,C). Leaf epidermal cell observation revealed that the increase in leaf width of mutants was caused by an increase in cell number but not in cell size (Figure S4). Hence, OsFWL4 may negatively affect cell proliferation during leaf and tiller development. In the mutants, the grain yield per plant was increased by up to 35.8% (Figure 2E), suggesting that the *OsFWL4* gene may be useful in breeding to improve rice yield.

The grain length of the *OsFWL1* gene mutants was significantly higher than that of the WT (Figure 5A,B). Similarly, grain length of the *OsFWL3* gene mutant has also been reported to be increased [9]. However, grain width of the *OsFWL1* gene mutants was not affected and grain thickness was reduced (Figure 5C,D). The decrease in grain thickness might be caused by insufficient grain filling due to enlarged glumes in the mutants. Finally, grain weight of mutants was not changed. Together, these results suggest that rice *FWL* genes play a role in the regulation of organ development in rice.

It has been reported that the *OsFWL4* gene can enhance Cd resistance when expressed in yeast cells and mediate the translocation of Cd from the roots to shoots in rice seedlings [15]. Recently, the *OsFWL1* gene was also reported to mediate Cd homeostasis in rice [20]. Hence, the two rice

FWL genes function in both organ development regulation and Cd homeostasis in rice. Similarly, the *OsFWL5/PCR1* gene has been found to confer Cd resistance and Zn hypersensitivity upon expression in yeast and to modulate grain size and weight and metal ion homeostasis in rice [13,14]. However, how a single *FWL* gene fulfills such diverse roles remains unknown. Interestingly, the OsFWL5/PCR1 protein has been found to be localized as oligomers in the plasma membrane microdomains [13]. Additionally, GmFWL1, an important FWL protein involved in soybean nodulation, has also been demonstrated to be a plasma membrane microdomain-associated protein [5,10]. The plasma membrane microdomains are membrane sub-compartments consisting of special lipids and proteins and are considered signal integration hubs of cells [35]. Hence, the membrane microdomain-associated FWL protein may act in several distinct signaling pathways and thus affect multiple biological processes in plants. Both OsFWL1 and OsFWL4 proteins are located in the plasma membrane, and the OsFWL4 protein is distributed in a punctate manner [9,15]. We speculate that OsFWL4 may also be a microdomain-associated protein. A gene co-expression analysis revealed that *OsFWL4* may be involved in many cell functions (Tables S7 and S8). The *OsFWL1* gene is reported to be co-expressed with the zinc finger and ubiquitination-related protein genes [9].

CRISPR/Cas9 DNA can be delivered into rice cells by *Agrobacterium*-mediated transformation and integrated into the rice genome. Studies have shown that T-DNA truncation frequently occurs in *Agrobacterium*-mediated transformation [36,37]. Detection of CRISPR/Cas9 DNA in unmutated T_0 plants revealed that most (25 out of 39, excluding two plants that escaped hygromycin selection) of them lacked sgRNA and/or *Cas9* transgenes (Table S2). This indicates that the integrity of the sgRNA/*Cas9* expression cassette is an important factor affecting editing efficiency. Truncation of T-DNA can occur at its different ends (left, right, or both ends) and different stages of integration (before or during integration) [36–39]. In rice, truncated T-DNAs were detected in more than 18% of the transformants [40]. Hence, improving the quality of T-DNA integration may aid in further increasing the efficiency of CRISPR/Cas9 gene editing based on *Agrobacterium*-mediated transformation.

T-DNA segregation analysis revealed that transgene-free plants could be obtained in several T_1 plants for all lines examined (Figure S2). This suggests that T-DNA insertions in CRISPR/Cas9 gene-edited plants can be easily eliminated in the T_1 generation. Interestingly, inconsistent segregation of *HPT*, sgRNA, and *Cas9* transgenes was observed in two lines (Osfwl3a#4 and Osfwl4b#6; Figure S2). The absence of sgRNA and *Cas9* transgenes in Osfwl4b#6 T_1 plants was caused by the lack of these sequences in the T_0 plant. In the 20 T_1 plants examined for line Osfwl3a#4, seven plants contained only the sgRNA transgene, 12 plants contained all the three transgenes, and one plant had no transgene (Figures S2 and S3). This inconsistent segregation could be attributed to the presence of two T-DNA insertion sites in this line; one contained the complete T-DNA fragment, whereas the other harbored a truncated T-DNA with only the sgRNA transgene.

The off-target effect is a major concern in the application of CRISPR/Cas9 technology. Several studies have reported that the CRISPR/Cas9 system is highly specific in plants [27,32,41,42]. However, moderate or even high-frequency off-target mutagenesis has also been reported [43–46]. In the present study, potential editing events at 30 putative off-targets of the 15 sgRNAs were examined. We detected mutations in 13 out of the 30 putative off-target sites (Table 3). Analysis of the relationship between mismatch numbers of target-like sequences and off-target activity revealed that all the sequences harboring single mismatches with the sgRNAs and 80.0% of the sequences containing double mismatches were cleaved. These results indicate that at least two mismatches between the sgRNA and potential off-target sequences are required to minimize the off-target effects. Interestingly, an off-target site with up to four mismatches (Osfwl2aOFF-2) was also mutated (Table 3). The first mismatch of this site located at the first base in the 5′ end is usually tolerated by CRISPR/Cas9. Additionally, all four mismatched bases of this site were adenine (Table 3), which led to rN:dT base pairing during sgRNA binding. Generally, the rN:dT mismatches are well tolerated [17,18]. Hence, both the identity and position of mismatched bases might contribute to the cleavage of this site by Cas9. The results suggest that the sgRNAs should be designed carefully to minimize or avoid off-target mutagenesis in plants.

4. Materials and Methods

4.1. Plant Materials and Growth Conditions

The rice variety used for transformation was Zhonghua 11 (*Oryza sativa* L. ssp. *japonica*). The rice plants were grown in experimental fields of the Huaiyin Normal University in Huai'an, China or in Lingshui, China in different growing seasons. Rice plants were also grown in plastic buckets in growth chambers with a 14/10 h light/dark cycle at 30 and 25 °C.

4.2. Construction of the CRISPR/Cas9 Plasmids

Maize ubiquitin promoter was used to drive the expression of the *hSpCas9* gene, which was amplified from the pX260 vector [49]. The construct was inserted into the pCAMBIA1300 vector (Cambia, Canberra, Australia) harboring the hygromycin resistance gene. The *Bsa*I site originally present in the pCAMBIA1300 vector was disrupted by point mutation. Subsequently, a construct containing the OsU6 promoter [50], a negative selection marker gene (*ccdB*) with *Bsa*I sites at both ends, and a fragment encoding the sgRNA scaffold derived from the pX260 vector was cloned into this vector to generate the CRISPR/Cas9 binary vector (Figure S1). Target sequences containing at least one mismatch in the 12-bp PAM proximal region with other genomic sites and relatively high GC content were selected for the rice *FWL* genes. The designed target sequences were synthesized, annealed, and ligated into the *Bsa*I site of the CRISPR/Cas9 binary vector to obtain the CRISPR plasmids for targeted gene editing. The plasmids were propagated in *Escherichia coli* competent cells and subsequently introduced into the *Agrobacterium tumefaciens* strain EHA105 for *Agrobacterium*-mediated transformation of rice [51].

4.3. Detection of On-Target and Off-Target Mutations

The potential off-targets of sgRNAs were predicted using the "offTarget" program in the CRISPR-GE software toolkit [52]. Genomic DNA of rice was extracted using the CTAB (cetyl trimethylammonium bromide) method. The DNA fragments covering the on-target and off-target sites were amplified by PCR using the specific genomic primers. PCR amplifications were performed in a Mastercycler nexus gradient thermal cycler (Eppendorf, Hamburg, Germany). Each reaction contained DNA templates, $1 \times$ PCR buffer, 0.4 mmol L^{-1} dNTPs (deoxynucleotide triphosphates), 0.3 μmol L^{-1} of both forward and reverse primers, and 1 U KOD FX DNA polymerase (Toyobo, Osaka, Japan). Distilled water was added to a final volume of 50 μL. The PCR conditions included an initial incubation at 94 °C for 2 min, followed by 30 cycles of 98 °C for 10 s, 50–55 °C for 30 s, and 68 °C for 0.5–1 min, with a final extension at 68 °C for 5 min. The amplified products were sequenced directly. For some samples, PCR products were cloned and individual clones were sequenced. The superimposed sequencing chromatograms of heterozygous and bi-allelic mutations were decoded using DSDecodeM [53]. The PCR primers used are listed in Table S9.

4.4. Detection of the T-DNA Fragment

The T-DNA fragment in transgenic plants was detected by PCR using three pairs of primers amplifying the *HPT*, sgRNA, and *Cas9* transgenes. Amplifications were carried out in a Mastercycler nexus gradient thermal cycler (Eppendorf, Hamburg, Germany). Each reaction contained DNA templates, $1 \times$ Es Taq MasterMix (Cwbio, Beijing, China), and 0.4 μmol L^{-1} of both forward and reverse primers. Distilled water was added to a final volume of 25 μL. The PCR conditions included an initial incubation at 94 °C for 2 min, followed by 30 cycles of 94 °C for 30 s, 55 °C for 30 s, and 72 °C for 30 s, with a final extension at 72 °C for 2 min. PCR products were separated by electrophoresis on 1.5% agarose gels containing GoldView I nucleic acid dye (Solarbio, Beijing, China). The primers used are listed in Table S9.

4.5. RNA Isolation and qRT-PCR

RNA isolation and qRT-PCR analysis were performed as previously described [54]. Briefly, total RNA was isolated using the TRIzol Total RNA Isolation kit (Sangon Biotech, Shanghai, China) and treated with DNase I (Sangon Biotech, Shanghai, China). Eight hundred nanograms of total RNA was reverse-transcribed using RevertAid Premium Reverse Transcriptase (Thermo Fisher Scientific, Waltham, MA, USA) and diluted ten-fold for PCR amplification. The PCR was performed on a LightCycler480 II instrument (Roche, Basel, Switzerland). Each reaction contained 2 μL of cDNA template, 10 μL of SYBR Green qPCR Master Mix (BBI, Toronto, ON, Canada), and 0.2 μmol L^{-1} gene-specific primers in a final volume of 20 μL. The PCR conditions included an initial incubation at 95 °C for 3 min, followed by 45 cycles of 95 °C for 5 s and 60 °C for 30 s. The specificity of the PCR reactions was determined by melting curve analyses of the products. Relative expression levels were calculated by the $2^{-\Delta\Delta CT}$ method. The rice *Actin1* gene was used as the internal control. The primer sequences are listed in Table S9.

4.6. Leaf Epidermal Cell Observation

Epidermal cells in flag leaves were observed following the method used by Yoshikawa et al. [55].

4.7. Trait Measurement

Plant height, leaf size, and tiller number of WT and mutants were measured in the field at the maturity stage. For tiller number determination, only seed setting tillers were counted. Rice plants were harvested when the grains were fully mature. Grains threshed from each plant were dried, and filled grains were weighed to determine grain yield per plant. Fully filled grains were used for determining grain size and weight. Grain weight was measured based on 100 grains and converted to 1000-grain weight.

5. Conclusions

Collectively, our findings showed that transgene-free rice plants with targeted mutations can be produced in the T$_1$ generation using the *Agrobacterium*-mediated CRISPR/Cas9 system, and that the *OsFWL4* gene plays a role in the regulation of tillering and plant yield in rice. The specific mutants obtained in this study provide valuable materials for functional analysis of rice *FWL* genes.

Supplementary Materials: Supplementary materials can be found at http://www.mdpi.com/1422-0067/21/3/809/s1.

Author Contributions: C.Y. and X.Z. designed the study; Q.G., G.L., H.S., M.X., H.W., J.J., and D.W. performed the experiments; Q.G. and G.L. analyzed the data; and Q.G. and X.Z. wrote the manuscript. All authors have read and agreed to the published version of the manuscript.

Funding: This research was funded by the National Key Research Program of China (2017YFD0100400), the National Natural Science Foundation of China (31601385), the National Science and Technology Major Project (2016ZX08012-002), and the Natural Science Foundation of Jiangsu Province (BK20160429 and BK20180107).

Acknowledgments: We thank Yangwen Qian from the Biogle Genome Editing Center (Changzhou) for assistance with gene transformation. We also thank the anonymous reviewers for their constructive comments and suggestions.

Conflicts of Interest: The authors declare no conflict of interest. The funders had no role in the design of the study; in the collection, analyses, or interpretation of data; in the writing of the manuscript, or in the decision to publish the results.

References

1. Cong, B.; Liu, J.P.; Tanksley, S.D. Natural alleles at a tomato fruit size quantitative trait locus differ by heterochronic regulatory mutations. *Proc. Natl. Acad. Sci. USA* **2002**, *99*, 13606–13611. [CrossRef]
2. Frary, A.; Nesbitt, T.C.; Grandillo, S.; Knaap, E.; Cong, B.; Liu, J.; Meller, J.; Elber, R.; Alpert, K.B.; Tanksley, S.D. *fw2.2*: A quantitative trait locus key to the evolution of tomato fruit size. *Science* **2000**, *289*, 85–88. [CrossRef]

3. Liu, J.; Cong, B.; Tanksley, S.D. Generation and analysis of an artificial gene dosage series in tomato to study the mechanisms by which the cloned quantitative trait locus *fw2.2* controls fruit size. *Plant Physiol.* **2003**, *132*, 292–299. [CrossRef]

4. Guo, M.; Rupe, M.A.; Dieter, J.A.; Zou, J.; Spielbauer, D.; Duncan, K.E.; Howard, R.J.; Hou, Z.; Simmons, C.R. *Cell Number Regulator1* affects plant and organ size in maize: Implications for crop yield enhancement and heterosis. *Plant Cell* **2010**, *22*, 1057–1073. [CrossRef]

5. Libault, M.; Zhang, X.C.; Govindarajulu, M.; Qiu, J.; Ong, Y.T.; Brechenmacher, L.; Berg, R.H.; Hurley-Sommer, A.; Taylor, C.G.; Stacey, G. A member of the highly conserved *FWL* (tomato *FW2.2-like*) gene family is essential for soybean nodule organogenesis. *Plant J.* **2010**, *62*, 852–864. [CrossRef] [PubMed]

6. Dahan, Y.; Rosenfeld, R.; Zadiranov, V.; Irihimovitch, V. A proposed conserved role for an avocado *FW2.2*-like gene as a negative regulator of fruit cell division. *Planta* **2010**, *232*, 663–676. [CrossRef] [PubMed]

7. De Franceschi, P.; Stegmeir, T.; Cabrera, A.; van der Knaap, E.; Rosyara, U.R.; Sebolt, A.M.; Dondini, L.; Dirlewanger, E.; Quero-Garcia, J.; Campoy, J.A.; et al. Cell number regulator genes in *Prunus* provide candidate genes for the control of fruit size in sweet and sour cherry. *Mol. Breeding* **2013**, *32*, 311–326. [CrossRef] [PubMed]

8. Li, Z.; He, C. *Physalis floridana Cell Number Regulator1* encodes a cell membrane-anchored modulator of cell cycle and negatively controls fruit size. *J. Exp. Bot.* **2015**, *66*, 257–270. [CrossRef] [PubMed]

9. Xu, J.; Xiong, W.T.; Cao, B.B.; Yan, T.Z.; Luo, T.; Fan, T.T.; Luo, M.Z. Molecular characterization and functional analysis of "fruit-weight 2.2-like" gene family in rice. *Planta* **2013**, *238*, 643–655. [CrossRef]

10. Qiao, Z.; Brechenmacher, L.; Smith, B.; Strout, G.W.; Mangin, W.; Taylor, C.; Russell, S.D.; Stacey, G.; Libault, M. The Gm*FWL1* (*FW2-2-like*) nodulation gene encodes a plasma membrane microdomain-associated protein. *Plant Cell Environ.* **2017**, *40*, 1442–1455. [CrossRef]

11. Qiao, K.; Tian, Y.; Hu, Z.; Chai, T. Wheat Cell Number Regulator CNR10 enhances the tolerance, translocation, and accumulation of heavy metals in plants. *Environ. Sci. Technol.* **2019**, *53*, 860–867. [CrossRef] [PubMed]

12. Song, W.Y.; Choi, K.S.; Kim, D.Y.; Geisler, M.; Park, J.; Vincenzetti, V.; Schellenberg, M.; Kim, S.H.; Lim, Y.P.; Noh, E.W.; et al. *Arabidopsis* PCR2 is a zinc exporter involved in both zinc extrusion and long-distance zinc transport. *Plant Cell* **2010**, *22*, 2237–2252. [CrossRef] [PubMed]

13. Song, W.Y.; Lee, H.S.; Jin, S.R.; Ko, D.; Martinoia, E.; Lee, Y.; An, G.; Ahn, S.N. Rice PCR1 influences grain weight and Zn accumulation in grains. *Plant Cell Environ.* **2015**, *38*, 2327–2339. [CrossRef] [PubMed]

14. Song, W.Y.; Martinoia, E.; Lee, J.; Kim, D.; Kim, D.Y.; Vogt, E.; Shim, D.; Choi, K.S.; Hwang, I.; Lee, Y. A novel family of cys-rich membrane proteins mediates cadmium resistance in Arabidopsis. *Plant Physiol.* **2004**, *135*, 1027–1039. [CrossRef]

15. Xiong, W.T.; Wang, P.; Yan, T.Z.; Cao, B.B.; Xu, J.; Liu, D.F.; Luo, M.Z. The rice "fruit-weight 2.2-like" gene family member OsFWL4 is involved in the translocation of cadmium from roots to shoots. *Planta* **2018**, *247*, 1247–1260. [CrossRef]

16. Kurusu, T.; Nishikawa, D.; Yamazaki, Y.; Gotoh, M.; Nakano, M.; Hamada, H.; Yamanaka, T.; Iida, K.; Nakagawa, Y.; Saji, H.; et al. Plasma membrane protein OsMCA1 is involved in regulation of hypo-osmotic shock-induced Ca^{2+} influx and modulates generation of reactive oxygen species in cultured rice cells. *BMC Plant Biol.* **2012**, *12*, 11. [CrossRef]

17. Nakagawa, Y.; Katagiri, T.; Shinozaki, K.; Qi, Z.; Tatsumi, H.; Furuichi, T.; Kishigami, A.; Sokabe, M.; Kojima, I.; Sato, S.; et al. *Arabidopsis* plasma membrane protein crucial for Ca^{2+} influx and touch sensing in roots. *Proc. Natl. Acad. Sci. USA* **2007**, *104*, 3639–3644. [CrossRef]

18. Song, W.Y.; Choi, K.S.; Alexis de, A.; Martinoia, E.; Lee, Y. *Brassica juncea* plant cadmium resistance 1 protein (BjPCR1) facilitates the radial transport of calcium in the root. *Proc. Natl. Acad. Sci. USA* **2011**, *108*, 19808–19813. [CrossRef]

19. Yamanaka, T.; Nakagawa, Y.; Mori, K.; Nakano, M.; Imamura, T.; Kataoka, H.; Terashima, A.; Iida, K.; Kojima, I.; Katagiri, T.; et al. MCA1 and MCA2 that mediate Ca^{2+} uptake have distinct and overlapping roles in Arabidopsis. *Plant Physiol.* **2010**, *152*, 1284–1296. [CrossRef]

20. Wang, F.; Tan, H.; Han, J.; Zhang, Y.; He, X.; Ding, Y.; Chen, Z.; Zhu, C. A novel family of PLAC8 motif-containing/PCR genes mediates Cd tolerance and Cd accumulation in rice. *Environ. Sci. Eur.* **2019**, *31*, 82. [CrossRef]

21. Scully, R.; Panday, A.; Elango, R.; Willis, N.A. DNA double-strand break repair-pathway choice in somatic mammalian cells. *Nat. Rev. Mol. Cell Biol.* **2019**, *20*, 698–714. [CrossRef] [PubMed]

22. Chen, K.L.; Wang, Y.P.; Zhang, R.; Zhang, H.W.; Gao, C.X. CRISPR/Cas genome editing and precision plant breeding in agriculture. *Annu. Rev. Plant Biol.* **2019**, *70*, 667–697. [CrossRef] [PubMed]
23. Jinek, M.; Chylinski, K.; Fonfara, I.; Hauer, M.; Doudna, J.A.; Charpentier, E. A programmable dual-RNA-guided DNA endonuclease in adaptive bacterial immunity. *Science* **2012**, *337*, 816–821. [CrossRef] [PubMed]
24. Sternberg, S.H.; Redding, S.; Jinek, M.; Greene, E.C.; Doudna, J.A. DNA interrogation by the CRISPR RNA-guided endonuclease Cas9. *Nature* **2014**, *507*, 62–67. [CrossRef]
25. Feng, C.; Su, H.D.; Bai, H.; Wang, R.; Liu, Y.L.; Guo, X.R.; Liu, C.; Zhang, J.; Yuan, J.; Birchler, J.A.; et al. High-efficiency genome editing using a *dmc1* promoter-controlled CRISPR/Cas9 system in maize. *Plant Biotechnol. J.* **2018**, *16*, 1848–1857. [CrossRef]
26. Wang, Z.P.; Xing, H.L.; Dong, L.; Zhang, H.Y.; Han, C.Y.; Wang, X.C.; Chen, Q.J. Egg cell-specific promoter-controlled CRISPR/Cas9 efficiently generates homozygous mutants for multiple target genes in *Arabidopsis* in a single generation. *Genome Biol.* **2015**, *16*, 144. [CrossRef]
27. Zhang, H.; Zhang, J.; Wei, P.; Zhang, B.; Gou, F.; Feng, Z.; Mao, Y.; Yang, L.; Xu, N.; Zhu, J.K. The CRISPR/Cas9 system produces specific and homozygous targeted gene editing in rice in one generation. *Plant Biotechnol. J.* **2014**, *12*, 797–807. [CrossRef]
28. Zhou, H.; Liu, B.; Weeks, D.P.; Spalding, M.H.; Yang, B. Large chromosomal deletions and heritable small genetic changes induced by CRISPR/Cas9 in rice. *Nucleic Acids Res.* **2014**, *42*, 10903–10914. [CrossRef]
29. Ma, X.; Zhang, Q.; Zhu, Q.; Liu, W.; Chen, Y.; Qiu, R.; Wang, B.; Yang, Z.; Li, H.; Lin, Y.; et al. A robust CRISPR/Cas9 system for convenient, high-efficiency multiplex genome editing in monocot and dicot plants. *Mol. Plant* **2015**, *8*, 1274–1284. [CrossRef]
30. Chari, R.; Yeo, N.C.; Chavez, A.; Church, G.M. sgRNA Scorer 2.0: A species-independent model to predict CRISPR/Cas9 activity. *ACS Synth. Biol.* **2017**, *6*, 902–904. [CrossRef]
31. Mao, Y.; Botella, J.R.; Zhu, J.K. Heritability of targeted gene modifications induced by plant-optimized CRISPR systems. *Cell Mol. Life Sci.* **2017**, *74*, 1075–1093. [CrossRef] [PubMed]
32. Xu, R.F.; Li, H.; Qin, R.Y.; Li, J.; Qiu, C.H.; Yang, Y.C.; Ma, H.; Li, L.; Wei, P.C.; Yang, J.B. Generation of inheritable and "transgene clean" targeted genome-modified rice in later generations using the CRISPR/Cas9 system. *Sci. Rep.* **2015**, *5*, 11491. [CrossRef] [PubMed]
33. Hruz, T.; Laule, O.; Szabo, G.; Wessendorp, F.; Bleuler, S.; Oertle, L.; Widmayer, P.; Gruissem, W.; Zimmermann, P. Genevestigator v3: A reference expression database for the meta-analysis of transcriptomes. *Adv. Bioinform.* **2008**, *2008*, 420747. [CrossRef] [PubMed]
34. Gingerich, D.J.; Hanada, K.; Shiu, S.H.; Vierstra, R.D. Large-scale, lineage-specific expansion of a bric-a-brac/tramtrack/broad complex ubiquitin-ligase gene family in rice. *Plant Cell* **2007**, *19*, 2329–2348. [CrossRef] [PubMed]
35. Yu, M.; Cui, Y.; Zhang, X.; Li, R.; Lin, J. Organization and dynamics of functional plant membrane microdomains. *Cell Mol. Life Sci.* **2019**. [CrossRef] [PubMed]
36. Gheysen, G.; Herman, L.; Breyne, P.; Gielen, J.; Van Montagu, M.; Depicker, A. Cloning and sequence analysis of truncated T-DNA inserts from *Nicotiana tabacum*. *Gene* **1990**, *94*, 155–163. [CrossRef]
37. Herman, L.; Jacobs, A.; Van Montagu, M.; Depicker, A. Plant chromosome/marker gene fusion assay for study of normal and truncated T-DNA integration events. *Mol. Gen. Genet.* **1990**, *224*, 248–256. [CrossRef]
38. Gheysen, G.; Villarroel, R.; Van Montagu, M. Illegitimate recombination in plants: A model for T-DNA integration. *Genes Dev.* **1991**, *5*, 287–297. [CrossRef]
39. Kim, S.R.; Lee, J.; Jun, S.H.; Park, S.; Kang, H.G.; Kwon, S.; An, G. Transgene structures in T-DNA-inserted rice plants. *Plant Mol. Biol.* **2003**, *52*, 761–773. [CrossRef]
40. Yin, Z.; Wang, G.L. Evidence of multiple complex patterns of T-DNA integration into the rice genome. *Theor. Appl. Genet.* **2000**, *100*, 461–470. [CrossRef]
41. Feng, Z.; Mao, Y.; Xu, N.; Zhang, B.; Wei, P.; Yang, D.L.; Wang, Z.; Zhang, Z.; Zheng, R.; Yang, L.; et al. Multigeneration analysis reveals the inheritance, specificity, and patterns of CRISPR/Cas-induced gene modifications in *Arabidopsis*. *Proc. Natl. Acad. Sci. USA* **2014**, *111*, 4632–4637. [CrossRef] [PubMed]
42. Tang, X.; Liu, G.Q.; Zhou, J.P.; Ren, Q.R.; You, Q.; Tian, L.; Xin, X.H.; Zhong, Z.H.; Liu, B.L.; Zheng, X.L.; et al. A large-scale whole-genome sequencing analysis reveals highly specific genome editing by both Cas9 and Cpf1 (Cas12a) nucleases in rice. *Genome Biol.* **2018**, *19*, 84. [CrossRef] [PubMed]

43. Endo, M.; Mikami, M.; Toki, S. Multigene knockout utilizing off-target mutations of the CRISPR/Cas9 system in rice. *Plant Cell Physiol.* **2015**, *56*, 41–47. [CrossRef] [PubMed]

44. Lawrenson, T.; Shorinola, O.; Stacey, N.; Li, C.; Ostergaard, L.; Patron, N.; Uauy, C.; Harwood, W. Induction of targeted, heritable mutations in barley and *Brassica oleracea* using RNA-guided Cas9 nuclease. *Genome Biol.* **2015**, *16*, 258. [CrossRef]

45. Li, M.R.; Li, X.X.; Zhou, Z.J.; Wu, P.Z.; Fang, M.C.; Pan, X.P.; Lin, Q.P.; Luo, W.B.; Wu, G.J.; Li, H.Q. Reassessment of the four yield-related genes *Gn1a*, *DEP1*, *GS3*, and *IPA1* in rice using a CRISPR/Cas9 system. *Front. Plant Sci.* **2016**, *7*, 377. [CrossRef]

46. Zhang, Q.; Xing, H.L.; Wang, Z.P.; Zhang, H.Y.; Yang, F.; Wang, X.C.; Chen, Q.J. Potential high-frequency off-target mutagenesis induced by CRISPR/Cas9 in *Arabidopsis* and its prevention. *Plant Mol. Biol.* **2018**, *96*, 445–456. [CrossRef]

47. Doench, J.G.; Fusi, N.; Sullender, M.; Hegde, M.; Vaimberg, E.W.; Donovan, K.F.; Smith, I.; Tothova, Z.; Wilen, C.; Orchard, R.; et al. Optimized sgRNA design to maximize activity and minimize off-target effects of CRISPR-Cas9. *Nat. Biotechnol.* **2016**, *34*, 184–191. [CrossRef]

48. Hsu, P.D.; Scott, D.A.; Weinstein, J.A.; Ran, F.A.; Konermann, S.; Agarwala, V.; Li, Y.; Fine, E.J.; Wu, X.; Shalem, O.; et al. DNA targeting specificity of RNA-guided Cas9 nucleases. *Nat. Biotechnol.* **2013**, *31*, 827–832. [CrossRef]

49. Cong, L.; Ran, F.A.; Cox, D.; Lin, S.; Barretto, R.; Habib, N.; Hsu, P.D.; Wu, X.; Jiang, W.; Marraffini, L.A.; et al. Multiplex genome engineering using CRISPR/Cas systems. *Science* **2013**, *339*, 819–823. [CrossRef]

50. Feng, Z.; Zhang, B.; Ding, W.; Liu, X.; Yang, D.L.; Wei, P.; Cao, F.; Zhu, S.; Zhang, F.; Mao, Y.; et al. Efficient genome editing in plants using a CRISPR/Cas system. *Cell Res.* **2013**, *23*, 1229–1232. [CrossRef]

51. Nishimura, A.; Aichi, I.; Matsuoka, M. A protocol for *Agrobacterium*-mediated transformation in rice. *Nat. Protoc.* **2006**, *1*, 2796–2802. [CrossRef] [PubMed]

52. Xie, X.; Ma, X.; Zhu, Q.; Zeng, D.; Li, G.; Liu, Y.G. CRISPR-GE: A convenient software toolkit for CRISPR-based genome editing. *Mol. Plant* **2017**, *10*, 1246–1249. [CrossRef] [PubMed]

53. Liu, W.; Xie, X.; Ma, X.; Li, J.; Chen, J.; Liu, Y.G. DSDecode: A web-based tool for decoding of sequencing chromatograms for genotyping of targeted mutations. *Mol. Plant* **2015**, *8*, 1431–1433. [CrossRef] [PubMed]

54. Gao, Q.S.; Xu, S.H.; Zhu, X.Y.; Wang, L.L.; Yang, Z.F.; Zhao, X.X. Genome-wide identification and characterization of the RIO atypical kinase family in plants. *Genes Genom.* **2018**, *40*, 669–683. [CrossRef] [PubMed]

55. Yoshikawa, T.; Eiguchi, M.; Hibara, K.; Ito, J.; Nagato, Y. Rice *slender leaf 1* gene encodes cellulose synthase-like D4 and is specifically expressed in M-phase cells to regulate cell proliferation. *J. Exp. Bot.* **2013**, *64*, 2049–2061. [CrossRef] [PubMed]

International Journal of
Molecular Sciences

Article

Genetic and Global Epigenetic Modification, Which Determines the Phenotype of Transgenic Rice?

Xiaoru Fan [1], Jingguang Chen [1,2], Yufeng Wu [3], CheeHow Teo [4], Guohua Xu [1] and Xiaorong Fan [1,*,†]

[1] State Key Laboratory of Crop Genetics and Germplasm Enhancement, MOA Key Laboratory of Plant Nutrition and Fertilization in Low-Middle Reaches of the Yangtze River, Nanjing Agricultural University, Nanjing 210095, China; 2017203043@njau.edu.cn (X.F.); chenjingguang@caas.cn (J.C.); ghxu@njau.edu.cn (G.X.)

[2] CAAS-IRRI Joint Laboratory for Genomics-Assisted Germplasm Enhancement, Agricultural Genomics Institute in Shenzhen, Chinese Academy of Agricultural Sciences, Shenzhen 518116, China

[3] Bioinformatics Center, Nanjing Agricultural University, Nanjing 210095, China; yfwu@njau.edu.cn

[4] Centre of Research in Biotechnology for Agriculture (CEBAR), University of Malaya, 50603 Kuala Lumpur, Malaysia; cheehow.teo@um.edu.my

[*] Correspondence: xiaorongfan@njau.edu.cn; Tel./Fax: +86-025-84396238

[†] Present Address: College of Resource and Environmental Science, Nanjing Agricultural University, Nanjing 210095, China

Received: 12 January 2020; Accepted: 1 March 2020; Published: 6 March 2020

Abstract: Transgenic technologies have been applied to a wide range of biological research. However, information on the potential epigenetic effects of transgenic technology is still lacking. Here, we show that the transgenic process can simultaneously induce both genetic and epigenetic changes in rice. We analyzed genetic, epigenetic, and phenotypic changes in plants subjected to tissue culture regeneration, using transgenic lines expressing the same coding sequence from two different promoters in transgenic lines of two rice cultivars: Wuyunjing7 (WYJ7) and Nipponbare (NP). We determined the expression of *OsNAR2.1* in two overexpression lines generated from the two cultivars, and in the RNA interference (RNAi) *OsNAR2.1* line in NP. DNA methylation analyses were performed on wild-type cultivars (WYJ7 and NP), regenerated lines (CK, T0 plants), segregation-derived wild-type from *pOsNAR2.1-OsNAR2.1* (SDWT), *pOsNAR2.1-OsNAR2.1*, *pUbi-OsNAR2.1*, and RNAi lines. Interestingly, we observed global methylation decreased in the T_0 regenerated line of WYJ7 (CK-WJY7) and *pOsNAR2.1-OsNAR2.1* lines but increased in *pUbi-OsNAR2.1* and RNAi lines of NP. Furthermore, the methylation pattern in SDWT returned to the WYJ7 level after four generations. Phenotypic changes were detected in all the generated lines except for SDWT. Global methylation was found to decrease by 13% in *pOsNAR2.1-OsNAR2.1* with an increase in plant height of 4.69% compared with WYJ7, and increased by 18% in *pUbi-OsNAR2.1* with an increase of 17.36% in plant height compared with NP. This suggests an absence of a necessary link between global methylation and the phenotype of transgenic plants with *OsNAR2.1* gene over-expression. However, epigenetic changes can influence phenotype during tissue culture, as seen in the massive methylation in CK-WYJ7, T0 regenerated lines, resulting in decreased plant height compared with the wild-type, in the absence of a transformed gene. We conclude that in the transgenic lines the phenotype is mainly determined by the nature and function of the transgene after four generations of transformation, while the global epigenetic modification is dependent on the genetic background. Our research suggests an innovative insight in explaining the reason behind the occurrence of transgenic plants with random and undesirable phenotypes.

Keywords: genetic; epigenetic; global methylation; transgenic; phenotype; *OsNAR2.1*

---

1. Introduction

Transgenic technologies allow gene transfer to completely unrelated organisms and their application in agriculture has increased the global transgenic crop cultivation to 181 million hectares [1]. In addition, as the basis of transgenic technology, tissue culture is also used for the clonal propagation and regeneration of many plants [2,3]. Although cultured material is not expected to show many genetic changes compared with the original material, there are clear examples of tissue culture material showing heritable phenotypic differences [4–6]. Phenotypic changes in tissue culture-derived material have been found to be caused by epigenetic changes [6–8]. For instance, Rhee et al. [7] demonstrated the silencing of epialleles by epigenetic modifications and showed that the pericarp color1 (*p1*) epialleles were capable of functioning in the presence of the correct trans-acting factors in maize. Furthermore, there are many reports on epigenetic changes caused by plant regeneration in rice [9], garlic [10], triticale [11], pineapple [12], torenia, and rye [13,14].

Variation in plant phenotype is determined by both genetic and epigenetic factors [4]. Epigenetics refers to the study of heritable phenotype changes without genetic alteration [15]. DNA methylation is an epigenetic mechanism that involves the transfer of a methyl group to the C5 position of cytosine and contributes to the epigenetic regulation of nuclear gene expression and to genome stability [16]. In plants, DNA methylation occurs in three sequence contexts: CG, CHG, and CHH (H=A, C or T) [16,17]. The modulation of DNA methylation in culture is crucial to regeneration outcomes: successful regenerants of *Oryza sativa* ssp. *japonica* had lower CG methylation levels than failed regenerants [9]. The regeneration process, with or without genetic transformation, affects gene regulation at the transcriptional and post-transcriptional levels and correlates with changes in DNA methylation patterns [18]. Transgenic approaches have been successfully used to produce herbicide and pest-resistant varieties in several crop species. However, there is only sporadic research concerning global DNA methylation changes in transgenic plants. Potential global DNA methylation modification occurring in transgenic plants is still largely unexplored [18–23]. There have been reports of transgenic plants from different genetic backgrounds having random and undesirable phenotypes [24–26]. Transgenic maize with overexpressed *DREB3*, a dehydration-responsive element-binding transcription factor, showed higher yields in some genetic backgrounds but not others [24]. Researchers have suggested that insert transgene position, metabolic imbalances and environmental constraints [24,27] could be the reasons behind undesirable phenotypes of transgenic plants. However, we suspect that epigenetic changes in transgenic lines may be another reason for the occurrence.

Rice (*Oryza sativa* L.) is a major staple food for a large part of the global population. To determine whether genetic and global epigenetic modification influences the phenotype of transgenic rice, we used whole-genome bisulfite sequencing (WGBS) and methylated DNA immunoprecipitation sequencing (MeDIP-seq) methods to determine DNA methylation at the genomic level in various transgenic plants and their controls. The transgenic process includes tissue induction, selection pressure and insertion of the transgene [28]. We used wild-type Wuyunjing7 (WYJ7), its regenerated line (CK), and segregation-derived wild-type from *pOsNAR2.1-OsNAR2.1* (SDWT) to represent different transgenic processing, and wild-type of Nipponbare (NP), its *OsNAR2.1* overexpression line under the ubiquitin promoter (*pUbi-OsNAR2.1*), and the *OsNAR2.1* RNA interference (RNAi) NP line to represent different backgrounds and different transgenic expression. Previous studies have shown that OsNAR2.1 is a partner protein of rice high-affinity nitrate transporters (OsNRT2s) [29–31] and plays a key role in enabling the plant to cope with variable environmental nitrate supplies [29–31]; overexpression of *OsNAR2.1* can lead to an increase in grain yield and higher nitrogen-use efficiency (NUE) in rice cultivation systems [32,33]. We used *OsNAR2.1* transgenic lines as representative lines for the transgenic process.

2. Results

2.1. Transgenic Process Induces Significant Epigenetic Changes in Rice

To investigate global methylation caused by the transgenic process in plants, we used WGBS to sequence samples of four types of rice: Wuyunjing7 wild type (WYJ7), T_0 plants regenerated from callus (CK), wild type plant derived from segregation of *pOsNAR2.1-OsNAR2.1* (SDWT) and *pOsNAR2.1-OsNAR2.1* plants. T_4 generation SDWT plants served as true experimental controls for the *pOsNAR2.1-OsNAR2.1* line to study the transgene insertional effect on the epigenome as it was derived from the T_0 *pOsNAR2.1-OsNAR2.1* heterozygote line (for the selection process see Supplementary Table S1). T_4 generation *pOsNAR2.1-OsNAR2.1* plants with high yield and high NUE phenotypes have been described in Chen *et al.* [33]. We also used MeDIP sequencing on four more samples, including the wild type of Nipponbare (NP), the knockdown plant of *OsNAR2.1* by RNA interference (RNAi) (describe as r1 in Yan *et al.* [29]), and overexpression plant of *OsNAR2.1* by the ubiquitin promoter (*pUbi-OsNAR2.1*). The relationships between different samples are described in Table 1 as: CK = WYJ7 + callus inducing media(CIM) + shoot inducing media(SIM); SDWT = WYJ7 + CIM + SIM + selection pressure (Hygromycin, Hyg) + transformation with the *pOsNAR2.1-OsNAR2.1* construct + segregation of *pOsNAR2.1-OsNAR2.1* heterozygote line; *pOsNAR2.1-OsNAR2.1* = WYJ7 + CIM + SIM + selection pressure (Hyg) + transformation with the *pOsNAR2.1-OsNAR2.1* construct + *pOsNAR2.1-OsNAR2.1* insertion; *pUbi-OsNAR2.1* = NP + CIM + SIM + selection pressure (Hyg) + transformation with the *pUbi-OsNAR2.1* construct + *pUbi-OsNAR2.1* insertion; RNAi = NP + CIM + SIM + selection pressure (Hyg) + RNA interference (RNAi) construct. Table 1 also shows the values for raw reads, uniquely mapped reads, and normalized cytosine methylation (mC) number in each replicate of four samples for WGBS and filtered reads, aligned reads, and peaks counts of three samples for MeDIP. We sequenced the flanking DNA of *pOsNAR2.1-OsNAR2.1* insertion site of in *pOsNAR2.1-OsNAR2.1* plants and found that the insertion of the *pOsNAR2.1-OsNAR2.1* transgene was between LOC_Os02g49950 gene and LOC_Os02g49960 gene in chromosome 2. We also sequenced the flanking DNA of the *pUbi-OsNAR2.1* insertion, which was inserted in chromosome 10, between LOC_Os10g33874 and LOC_Os10g33900 in *pUbi-OsNAR2.1* line (Supplementary Figure S1d,e). Neither of these sites occur in high methylation areas nor in functional genes.

Table 1. Samples description in this study.

Sample	Description	Raw Reads	Uniquely Mapping Reads	Normalized mC Count
WT-WYJ7-1		140,469,848	114,222,086	396,005
WT-WYJ7-2	Wild type Wuyujing 7 cultivar (WYJ7)	140,314,960	114,019,328	380,987
WT-WYJ7-3		137,035,182	111,328,844	406,068
CK1		115,760,650	84,978,096	340,144
CK2	T0 generation plant of WYJ7 +callus inducing media (CIM) + shoot inducing media (SIM)	115,709,516	83,108,206	344,139
CK3		135,865,342	98,592,558	375,679
SDWT1	T4 generation plant of WYJ7+CIM+SIM+ selection pressure (Hyg) + transformation of *pOsNAR2.1-OsNAR2.1* construct + segregation of *pOsNAR2.1-OsNAR2.1* heterozygote line	134,716,330	109,964,522	376,469
SDWT2		127,794,020	104,131,328	393,058
SDWT3		138,032,054	112,763,894	377,411
pOsNAR2.1-OsNAR2.1-1	T4 generation plant of WYJ7+CIM+SIM+selection pressure (Hyg) + transformation of *pOsNAR2.1-OsNAR2.1* construct + one more *pOsNAR2.1-OsNAR2.1* insertion.	144,544,940	79,464,658	354,464
pOsNAR2.1-OsNAR2.1-2		109,740,316	73,616,122	328,232
pOsNAR2.1-OsNAR2.1-3		136,405,336	100,950,580	356,844
		Pass Filtering Reads	Reads Aligned Reads	Peaks count
WT-NP	Wild type Nipponbare cultivar (NP)	13,047,179	12,798,555	37,223
pUbi-OsNAR2.1	T8 generation plant of NP+CIM+SIM+ selection pressure (Hyg)+transformation of *pOsNAR2.1-OsNAR2.1* vector+*pUbi-OsNAR2.1* insertion	20,671,931	20,360,102	44,902
RNAi	T8 generation plant of NP+CIM+SIM+ selection pressure (Hyg)+RNA interference (RNAi) constructs	18,722,855	18,428,653	44,307

Our analysis of the sequencing data showed that the total normalized methylated cytosine (mC) numbers differed significantly among the four samples (Figure 1a), with normalized mC counts of approximately 3.9, 3.5, 3.8, and 3.5 million (M) for WYJ7, CK, SDWT, and *pOsNAR2.1-OsNAR2.1*, respectively. The result suggests that the tissue culture (CK) and the transgenic process (*pOsNAR2.1-OsNAR2.1*) of WYJ7 both lead to global DNA hypomethylation. In contrast to CK and *pOsNAR2.1-OsNAR2.1* plants, the mC level in SDWT plants returned to the WT level (WYJ7) after four selfing generations in the field. Since the CK and SDWT both went through regeneration, the result suggested the loss of the transgene in generation increases the methylation in SDWT. Simultaneously, the *pOsNAR2.1-OsNAR2.1* lines showed a significantly higher percentage of symmetric CG sites and lower percentage of asymmetric CHH sites, and CK lines showed a lower percentage of symmetric CHG sites instead (Figure 1b–d), which indicates that while both *pOsNAR2.1-OsNAR2.1* and CK lines show a decrease in global DNA methylation the sites of mC changes were different. Interestingly, in the other two independent transgenic lines, MeDIP sequencing data showed that peak counts in both *pUbi-NAR2.1* and RNAi lines increased compared with NP (Figure 1e). The results suggest that the tissue culture process leads to a decrease in global methylation, but the transgenic process leads to different global methylation changes in different rice backgrounds.

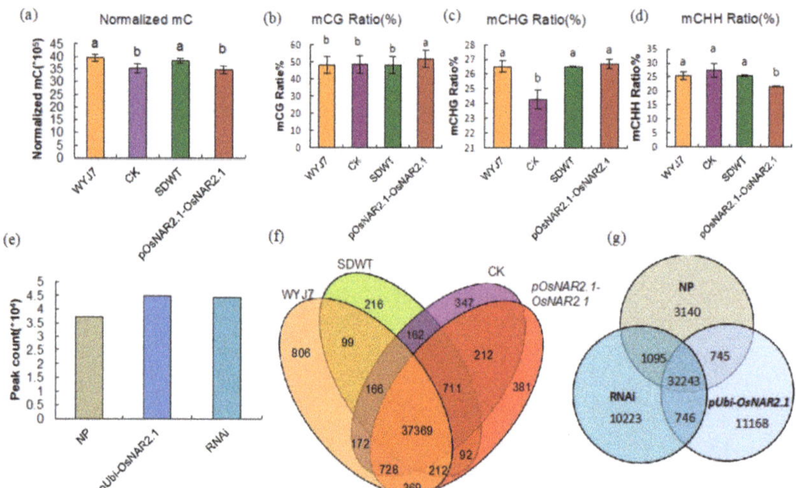

Figure 1. Characteristics of sequencing data (**a**) Normalized DNA cytosine methylation numbers in WYJ7, CK, SDWT and *pOsNAR2.1-OsNAR2.1* samples. (**b-d**) mCG, mCHG and mCHH ratio in WYJ7, CK, SDWT and *pOsNAR2.1-OsNAR2.1* samples. (**e**) Peak count of methylation in NP, *pUbi-OsNAR2.1* and RNAi. (**f**) Venn diagram of unique and shared genes in CpG methylation states for WYJ7, CK, SDWT and *pOsNAR2.1-OsNAR2.1* samples. (**g**) Venn diagram of peaks in NP, RNAi and *pUbi-OsNAR2.1* samples.

2.2. Both Tissue Culture and Transgenic Processes Can Induce Unique Methylation Changes in Genic Regions

The Venn diagram of WGBS sequencing shows the unique and shared CpG methylation areas of 42,042 genes in the WYJ7, CK, SDWT and *pOsNAR2.1-OsNAR2.1* samples (Figure 1f). The results showed that 88.9% of genes in these four samples shared the same CpG methylation area. CK, SDWT, and *pOsNAR2.1-OsNAR2.1* had 347, 216, and 381 genes with unique methylation areas, respectively. If only compared with wild-type WYJ7, CK, SDWT, and *pOsNAR2.1-OsNAR2.1* showed 1432, 1181, and 1396 genes with unique methylation areas. The Venn diagram analysis of the methylation peaks of the MeDIP-seq data sets for NP, RNAi, and *pUbi-OsNAR2.1* samples showed that the majority of the peaks, approximately 32,000, are shared by the three samples (Figure 1g). Both RNAi and *pUbi-OsNAR2.1* samples had 10,223 and 11,168 unique peaks, respectively (Figure 1g). The results indicate that both tissue culture and the transgenic processes could induce unique methylation changes in genic regions although neither knockdown nor overexpression of *OsNAR2.1* in transgenic plants caused the majority of DNA methylation changes.

2.3. Transgenic Process Leads to a Higher Number of Hyper-DMRs Than Hypo-DMRs

We analyzed the differentially methylated regions (DMRs) of wild-type and transgenic lines and found that there were more hypo-DMRs (71%) in CK lines but more hyper-DMRs (78%) in transgenic lines in *pOsNAR2.1-OsNAR2*.1 under WYJ7. Similar results were observed in *pUbi-OsNAR2.1* and RNAi under NP (71% and 73%) and, for the SDWT line, the percentage of the hypo- and hyper- DMRs were very close (49% and 51%) (Figure 2a–c,f,g). In spite of global DNA methylation caused by the transgenic process being background sensitive, the insertion of the transgene was able to induce more hyper-DMRs than hypo-DMRs in both backgrounds. CK and the three transgenic lines contained 5122, 3539, 5253, and 4239 DMRs, respectively, which were considerably more comparable with the 2732 DMRs of SDWT. There were few common DMRs between these hypo- and hyper-DMRs in CK and *pOsNAR2.1-OsNAR2.1* lines, nor in SDWT and *pOsNAR2.1-OsNAR2.1* lines (Figure 2d,e). However, there were nearly 2000 common hyper-DMRs and nearly 500 common hypo-DMR between the *pUbi-OsNAR2.1* and RNAi lines (Figure 2h), caused by insertion of the transgene, rather than by the expression of *OsNAR2.1*.

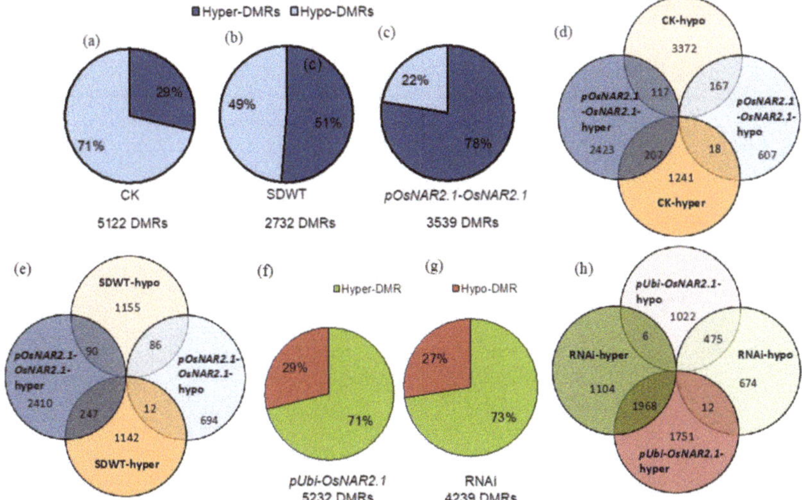

Figure 2. Characteristics of DMRs. (**a-c**) Total number of DMRs and breakdown of hyper- and hypo-DMRs in CK, SDWT and *pOsNAR2.1-OsNAR2.1* samples. (**d**) Venn diagram of unique and shared hyper- and hypo-DMRs in *pOsNAR2.1-OsNAR2.1* and CK samples. (**e**) Venn diagram of unique and shared hyper- and hypo-DMRs in *pOsNAR2.1-OsNAR2.1* and SDWT samples. (**f,g**) Total number of DMRs and breakdown of hyper- and hypo-DMRs in *pUbi-OsNAR2.1* and RNAi samples. (**h**) Venn diagram of unique and shared hyper- and hypo-DMRs in *pUbi-OsNAR2.1* and RNAi samples.

2.4. Tissue Culture Process Causing Random Epigenetic Changes

We analyzed the differentially CpG methylated regions of five comparison groups: WYJ7 and SDWT, WYJ7 and CK, WYJ7 and *pOsNAR2.1-OsNAR2.1*, SDWT and *pOsNAR2.1-OsNAR2.1*, and CK and *pOsNAR2.1-OsNAR2.1*. We found that the mutual DMRs differed among them and calculated the CpG methylation level in these DMRs in different replicates of different samples. The clustered heatmap is shown in Figure 3. The heatmap shows that the DMR clustering pattern is similar between wild-type and lines subjected to transgenic manipulation (SDWT and *pOsNAR2.1-OsNAR2.1*), whereas a more variable pattern is observed for CK replicates. The result shows that different CpG clusters of common DMRs were detected in different CK lines regenerated from the same tissue culture process with the same explant WYJ7 seeds (Figure 3). This suggests that the tissue culture process could randomly alter the epigenetic status. On the other hand, a tissue culture process followed by a transgenic process tends to lead to a more consistent CpG methylation level clustering pattern (Figure 3), suggesting that the influence of the transgenic process on the rice epigenome is stronger than the influence of the tissue culture process.

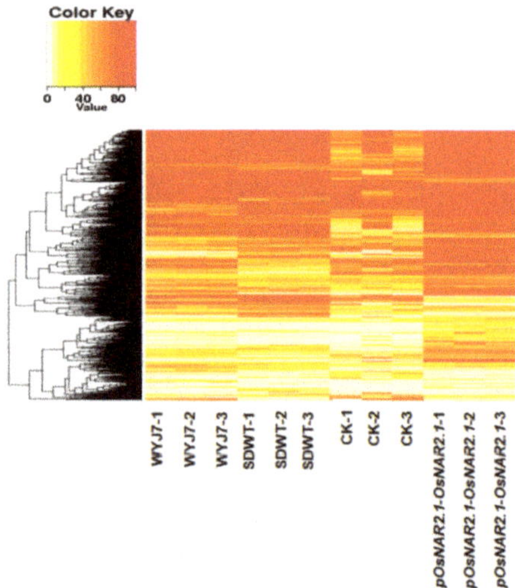

Figure 3. Heatmap for DMRs methylation level cluster. Heatmap representation of hierarchical clustering based on CG methylation levels within DMRs. Rows represent all DMRs identified and columns represent the samples.

2.5. Both Tissue Culture and Transgenic Process Can Change the Phenotype of Plants

We planted all the wild-type and transgenic lines in the field and estimated their gene expression, grain yield, and seed setting rate (Figure 4). For the lines used in the WGBS analysis, the phenotype showed that the plant height was significantly shorter in the CK line and significantly taller in the *pOsNAR2.1-OsNAR2.1* line compared with WYJ7, whereas there was no significant change in the SDWT line (Figure 4a, Supplementary Table S2). The total tiller number was higher in *pOsNAR2.1-OsNAR2.1* line with no significant difference in the CK and SDWT lines (Supplementary Table S2). For the expression of *OsNAR2.1*, while there were no significant differences between the WYJ7, CK and SDWT lines, expression was significantly higher in the *pOsNAR2.1-OsNAR2.1* line (Figure 4b). The grain yield and seed setting rate were significantly lower in the CK line and significantly higher in the *pOsNAR2.1-OsNAR2.1* line compared with WYJ7, with no significant difference in the SDWT line (Figure 4c,d). For the lines used in MeDIP analysis, both the plant height and tiller number (Figure 4e, Supplementary Table S3) were higher in the *pUbi-OsNAR2.1* line and lower in the RNAi line compared with NP. The relative expression of *OsNAR2.1*, grain yield, and seed setting rate were also higher in the *pUbi-OsNAR2.1* line and lower in the RNAi line, compared to NP (Figure 4f–h). As a summary of the global methylation, genetic and phenotype changes status of all materials Table 2 shows that global methylation changes of transgenic lines are dependent on the genetic background. For the CK line, the trend of changes of the phenotypes is same as global methylation, and for all the transgenic lines, the trend for phenotypic change appears the same as the expression of the transgene, but the percentages of the increases in overexpression lines are quite different. We consider that methylation change is one of the reasons causing the difference. These results suggest that both regeneration and the transgenic process can change the phenotype of plants, with or without gene insertion.

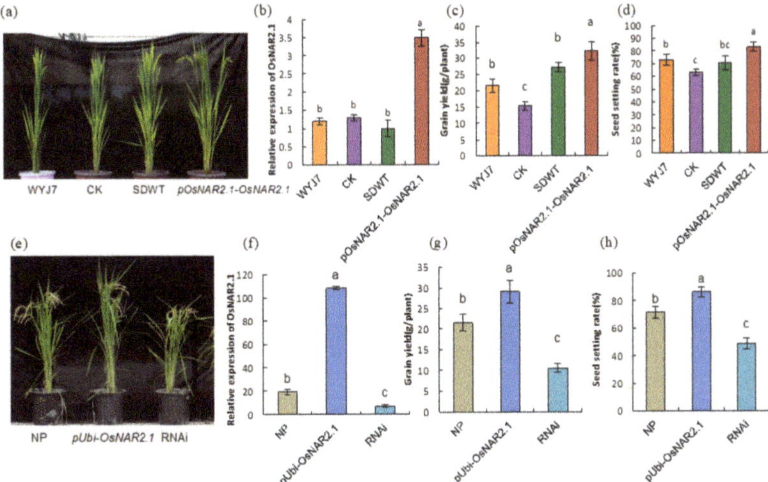

Figure 4. Characteristics of phenotype (**a**) Gross morphology of WYJ7, CK, SDWT and *pOsNAR2.1-OsNAR2.1*. (**b**) Real-time quantitative RT-PCR analysis of *OsNAR2.1* expression in WYJ7, CK, SDWT and *pOsNAR2.1-OsNAR2.1* lines. Error bars: SD ($n = 3$ plants). (**c**) Grain yield for WYJ7, CK, SDWT and *pOsNAR2.1-OsNAR2.1* plants grown in the field. Error bars: SD ($n = 5$ plants). Significant differences between WYJ7 and transgenic lines are indicated by different letters ($P < 0.05$, one-way ANOVA). (**d**) Seed setting rate for WYJ7, CK, SDWT and *pOsNAR2.1-OsNAR2.1* plants grown in the field. Error bars: SD ($n = 5$ plants). Significant differences between WYJ7 and transgenic lines are indicated by different letters ($p < 0.05$, one-way ANOVA). (**e**) Gross morphology of wild-type of NP, *pUbi-OsNAR2.1* and RNAi. (**f**) Real-time quantitative RT-PCR analysis of *OsNAR2.1* expression in NP, *pUbi-OsNAR2.1* and RNAi lines. Error bars: SD ($n = 3$ plants). (**g**) Grain yield for NP, *pUbi-OsNAR2.1* and RNAi plants grown in the field. Error bars: SD ($n = 5$ plants). Significant differences between NP and transgenic lines are indicated by different letters ($p < 0.05$, one-way ANOVA). (**h**) Seed setting rate for NP, *pUbi-OsNAR2.1* and RNAi plants grown in the field. Error bars: SD ($n = 5$ plants). Significant differences between NP and transgenic lines are indicated by different letters ($p < 0.05$, one-way ANOVA).

Table 2. Characteristics of methylation and genetics in samples.

Variety		WYJ7			NP	
		CK	SDWT	*pOsNAR2.1-OsNAR2.1*	*pUbi-OsNAR2.1*	RNAi
	Global methylation	↓	NS	↓	↑	↑
Genetic	Reprograming	+	+	+	+	+
	Transgenic	NS	NS	+	+	+
	Exogenous Gene	NS	NS	1	1	1
	Expression of transgene	NS	NS	↑	↑	NA
Phenotype	Plant height	↓6%	NS	↑5%	↑17%	↓18%
	Yield(g/plant)	↓28%	NS	↑49%	↑35%	↓51%
	Seed setting rate	↓13%	NS	↑14%	↑21%	↓32%

Note: NS means no significant difference. NA means not applicable. + means have, 1 means gene number, arrowhead indicates upregulate and downregulate, red indicates upregulate and green indicates downregulate.

3. Discussion

DNA methylation can provide additional heritable information beyond that of the DNA sequence in plant genomes [2]. Tissue culture is considered a stressful environment and thus trigger epigenetic changes in plants [34]. Culture-induced DNA methylation has been found in different species, including rice, maize, and barley [35–37]. Furthermore, many studies have shown that regeneration under various selection stresses or from various donor tissues induced changes in methylation patterns [38–40]. When uniform callus donor tissue was used in an *Agrobacterium*-mediated transformation procedure in rice [41,42], there was no difference in methylation among donor tissues between experiments. It has previously been reported that tissue culture reduces mC in rice, and this reduction in mC is stable from T_2 to T_6 generation of regenerated plants [18]. Our sequence data confirmed that the tissue culture process leads to a reduction in global DNA methylation in CK and that this reduction was maintained in the *pOsNAR2.1-OsNAR2.1* line at least until the T_4 generation (Figure 1a). Moreover, we showed that the global methylation level returned to a level similar to that of WYJ7 after removal of the transgene by a segregation process in the SDWT line. Since Stroud et al. [18] confirmed that the loss of methylation in regenerated plants is stable across generations, our results suggest that the global methylation status in the SDWT is more unstable and is unable to maintain the mC decrease across the generations. However, the insertion of the transgene stabilized the massive loss of mC in *pOsNAR2.1-OsNAR2.1*. The tissue culture process can cause the loss of methylation in both WYJ7 and NP (Figure 1a) according to both our results and those of Stroud, et al. [20], but *pOsNAR2.1-OsNAR2.1* in WYJ7 showed stabilization of the loss rather than *pUbi-OsNAR2.1* in NP. Therefore, we consider that overexpression of *OsNAR2.1* is not the reason behind the stabilization. We prefer the explanation that the unstabilized methylation of SDWT was caused by the double genetic change, gaining and losing the transgene during generation. Furthermore, the way insertion of the transgene alters global methylation appears to depend on the rice genetic background, since global methylation reduction was observed in the *pOsNAR2.1-OsNAR2.1* WYJ7 line whereas it was increased in the *pUbi-NAR2.1* and RNAi NP lines.

We found global methylation decreased in the WYJ7 *OsNAR2.1*-overexpression line, while increased *OsNAR2.1*-overexpression, and RNAi lines in the NP. We suggest that rice varieties have different sensitivities to DNA methylation in the transgenic process. It has been reported that there are extensive variations in DNA methylation among plant inbred lines, and that DNA methylation can provide unique information in explaining variation of phenotype in maize [43]. Vilperte et al. [5] reported that the methylation status of genes showed significant differences in four different maize backgrounds with the same transgene. In our results, although in both the *pOsNAR2.1-OsNAR2.1* and *pUbi-OsNAR2.1* lines, plant height and yield of per plant were significantly higher than in the wild-type, but the plant height increased 4.69% and 17.36%, and yield of per plant increased 49.07% and 34.97%, respectively, in WYJ7 and NP (Table 1, Supplementary Tables S2 and S3). Even overexpression of the same gene was able to cause different phenotypes in different backgrounds. This result could be caused by several possible factors. The first of these is the original traits of the two wild types: WYJ7 is photo-sensitive late-maturing japonica rice and NP is photo-insensitive early-maturing japonica rice [44,45]. The two wild-types are, therefore, affected differently by the circadian clock. The circadian clock regulates NO_3^- uptake and usage, and thus the expression of *OsNAR2.1* [30,46]. A second factor is the promoter of the transgene. It is well known that different promoters have different effects on transgenes [47]. In our research, the expression of *OsNAR2.1* increased around 2.5 times with the *OsNAR2.1* native promoter and increased four times with the *Ubi* promoters. Wang et al. [48] reported that overexpressed auxin-inducible gene (*ARGOS*) increased plant height in *Arabidopsis* under the 35s promoter but showed no phenotypic change under the *Ubi* promoter. The third factor is background. There have been many reports of the same transgene showing different phenotypes in plants from a variety of genetic backgrounds. It has been reported that, while the *DREB3* transgene was detected in wheats from four different genetic backgrounds, only three lines expressed the transgene, and only two showed phenotypes of higher yield [24]. Knockout of *OsNramp5*, a member of the natural resistance-associated macrophage protein (NRAMP) family, decreased yield in Xidao 1, a japonica rice cultivar, but did not alter yield in indica hybrid rice [26,49]. Even knockout of the same gene in different backgrounds with same genome also resulted in phenotypic variation, as shown by the findings of Yang et al [50]: knockout of *OsNramp5* using CRISPR/Cas9 in two japonica varieties, Nanjing 46 (NJ46) and Huaidao 5 (HD5), resulted in similar plant height, grain number, and seed setting rate, but with increased panicle number in NJ46 but not in HD5. Researchers surmised that damage to the recipient genome caused by insertion fragments, efficiency of the transgene promoter, metabolic imbalances and environmental constraints could be the reasons behind undesirable phenotypes of transgenic plants [24,27,51], while we suspected that the insertion of the transgene causing different methylation change in different backgrounds could be another reason.

The Venn diagram of genes in the methylation region in WYJ7, SDWT, CK, and *pOsNAR2.1-OsNAR2.1* showed that all three had over 1000 unique methylation genes compared with WYJ7 (Figure 1f), and all had 200–400 unique methylation genes compared to each other. The results suggest that, compared with WYJ7, all three samples (CK, SDWT, and *pOsNAR2.1-OsNAR2.1*) have abundant genes with different degrees of methylation, but not all three lines showed phenotypic changes (Figure 4a). In spite of having these differentially methylated genes, SDWT still showed a similar phenotype to WYJ7. These results suggest that the global methylation status change had a stronger influence on plants than changes in the methylation of individual genes.

Our results showed that the regenerated line CK had more hypo-DMRs than hyper-DMRs and that all three transgenic lines had more hyper- than hypo-DMRs, compared to the wild-type. However, in SDWT, the percentage of hyper- and hypo-DMRs are similar, 51% and 49%, respectively. The total number of DMRs in SDWT (2732 DMRs) was much less than that in CK (5122 DMRs) and the three transgenic lines (3539 DMRs in *pOsNAR2.1-OsNAR2.1*, 5232 DMRs in *pUbi-OsNAR2.1* and 4239 DMRs in RNAi) (Figure 2a–c,f,g). Stroud et al. [18] reported that all of their 12 regeneration lines had much more hypo-DMRs than hyper-DMRs, confirming our results. Therefore, it is possible that the reprogramming process could cause an abundance of hypo-DMRs, while insertion of the gene could cause hyper-DMRs instead, regardless of the function of the inserted gene. Furthermore, for the SDWT lines, WT went through both reprogramming and transgene processes but after segregation, it was found that this double genetic change resulted in nearly 3000 DMRs with the numbers of hyper- and hypo-DMRs returning to the baseline level of WYJ7. Since the phenotype of SDWT did not show significant differences compared to WYJ7, this indicated that the balance of hyper- and hypo-DMRs could be an important factor for the stability of the epigenetic status.

We found that the methylation status in the CK line regenerated from the tissue culture process gave rise to a more variable DMR methylation level clustering pattern than in the CK replicates. This is similar to the observations of Hsu et al. [9]. These authors reported that methylomes in different stages of callus showed a high level of variability and the global methylome demonstrated a prominent genetic/cultivar-related impact. Kaeppler et al. [4] showed that epigenetic modifications of genomic DNA are less stable in culture. Interestingly, both the transgenic line (*pOsNAR2.1-OsNAR2.1*) and the wild-type lines derived from the segregation process of *pOsNAR2.1-OsNAR2.1* (SDWT) showed a more consistent DMR methylation level clustering pattern. This suggests that the transgenic process may play a role in stabilizing the global methylation status of tissue cultured plants.

However, it is possible that the CK line has a genetic change in the genome in addition to epigenetic changes. For example, there have been many reports of the occurrence of transposons, such as *Tos17*, in regenerants [34,52]. To clarify the effects of epigenetic changes on the CK phenotype, we have listed the methylation sequence data and agronomic traits of each CK line from three independent calli in Supplementary Table S4. We found that, despite the differences between individuals, there is an overall consistency in the data of CK replicates (Supplementary Table S4). The Table shows that normalized mC number, mCHG ratio of epigenetic and plant height, panicle length, grain number per panicle, and yield of phenotype are all significantly lower in CK. Even if we cannot exclude possible genetic change in CK, the changes, at least, did not influence the phenotypes of the individual CK replicates in our study. It has also been reported that genetic variation during tissue culture is related to the CCGG target, which suggests that genetic and epigenetic changes in regenerants are relevant [34,53,54]. These changes could be the reason for the clustering of variable DMR methylation levels in CK but not causing phenotypic changes.

Previous research has reported that phenotypes are influenced by both genetic and epigenetic mechanisms [20,55,56]. Epigenetic modifications caused by the tissue culture process occur in an apparently random manner and it is thus difficult to predict phenotypic changes resulting from these modifications [4,5,7,57,58]. Here, we summarized global methylation, genetic and phenotype changes of five plant materials from two varieties, as shown in Table 2. The results indicate that, in the transgenic lines (*pOsNAR2.1-OsNAR2.1*, *pUbi-OsNAR2.1*, and RNAi), the phenotype is mainly determined by the nature and function of the transgene, but the global methylation change in transgenic lines is determined by the rice background rather than the function of the transgene, which may cause the difference between the phenotypic changes in the two overexpression lines. However, the CK lines showed fewer phenotypic changes, with the methylation status being significantly decrease from WYJ7 as result of the tissue culture process and showing DMR variation between replicates (Figure 3).

4. Materials and Methods

4.1. Plant Materials for Sequencing

For WGBS sequencing, we used the shoots of wild-type *Oryza sativa* L. ssp. *japonica* cv. Wuyujing7 (WYJ7), T_0 plants regenerated from tissue culture (CK), T_4-generation plants carrying an insertion of the *pOsNAR2.1-OsNAR2.1* construct (*pOsNAR2.1-OsNAR2.1*) as described as Ox1 in Chen et al. [33] (see Supplementary Figure S1 and Table S2 for the transformation construct and plant growth data), and T_4-generation wild-type line lacking the transgene that derived from the segregation process of the T_0 *pOsNAR2.1-OsNAR2.1* heterozygote line (SDWT, see Supplementary Tables S1 and S2). T_1 generation of Ox1-1 (AA) and Ox1-3 (aa) lines shown in Supplementary Table S1 were renamed as *pOsNAR2.1-OsNAR2.1* and SDWT and their T_4 plants were used for further WGBS experiment.

The WYJ7, SDWT, and *pOsNAR2.1-OsNAR2.1* were sterilized for 30 min with 10% (*v/v*) hydrogen peroxide, washed thoroughly with deionized water and then grown in water. CK lines were moved from root medium to water after germination. All samples were collected when all four lines grew to two leaves and one heart period. DNA from each plant was sequenced with three replicates. Before sequencing, we tested for the T-DNA insertion loci of *pOsNAR2.1-OsNAR2.1* (Supplementary Figure S1d). The primers used for TAIL-PCR are listed in Supplementary Table S5.

We used MeDIP-seq to examine three types of plants: wild-type of *Oryza sativa* L. ssp. *japonica* cv. Nipponbare (NP), the T_8-generation plant of knockdown plant of *OsNAR2.1* by RNA interference (RNAi) (describe as r1 in Yan et al. [29]), and the T_8-generation *pUbi-OsNAR2.1* overexpression plant. The transformation constructs and plant growth data are shown in Supplementary Figure S1 and Table S3. We harvested the mixed samples of the first leaf blade, culm and panicles at the anthesis stage for MeDIP-seq (Figure 4e). Details of the growth conditions of the plants in soil are listed in Supplementary Table S3. The level of *OsNAR2.1* expression was reduced in T_8 generation RNAi plants to one half of the WT, whereas was five times higher in the *pUbi-OsNAR2.1* T_8 generation transgenic line (Figure 4f). Other growth characteristics of RNAi and *pUbi-OsNAR2.1* transgenic lines are described in Supplementary Table S3. We tested for T-DNA insertion in three samples (Supplementary Figure S1e); the primers for TAIL-PCR are listed in Supplementary Table S5.

4.2. Growth Conditions of Plants in the Field

All materials were grown in plots at the Baguazhou base of Nanjing Agricultural University in Nanjing, Jiangsu. Nanjing is located in a subtropical monsoon climate zone. The pH of the soil is 6.5, and chemical properties included 0.91 g/kg total N content; 18.91 mg/kg available phosphorus (P) content; 185.67 mg/kg exchangeable potassium (K) and 11.56 g/kg organic matter. We applied

Ca(H$_2$PO$_4$)$_2$, 30 kg P/ha and KCl, 60 kg K/ha as basal applications to the plots three days before transplanting. We use urea as N fertilizer, and applied 40% before transplanting, 30% at tilling, and 40% before the heading stage. Plots size was 2 × 2.5 m and the seedlings planted in a 10 × 10 array [32,59–62]. We randomly chose five seedlings from each plot, avoiding those on the edges. The agronomic characters of plant height, total tiller number per plant, panicle length, seed setting rate per plant, grain weight per panicle, grain number per panicle and yield per plant were measured at the maturity stage. Plant height indicated the height of the highest panicle. One panicle from each plant was chosen for calculating the panicle length, grain weight per panicle and grain number per panicle [33]. The agronomic traits of the samples are listed in Supplementary Tables S2 and S3.

4.3. Gene Expression Analysis

Total RNA was extracted from leaf tissue using TRIzol reagent (Vazyme Biotech Co, Ltd., People's Republic of China). DNase I-treated total RNAs were subjected to reverse transcription (RT) with HiScript II Q Select RT SuperMix for qPCR (+gDNA wiper) kit (Vazyme Biotech Co, Ltd., People's Republic of China) according to the manufacturer's instructions. Quantitative assays were performed using AceQTM qPCR SYBR Green Master Mix (Vazyme Biotech Co, Ltd., China). Relative expression level is normalized to the amount of *OsActin* (LOC_Os03g50885) in the same sample and presented as $2^{-\Delta CT}$. All primers used for RT-qPCR are listed in Supplementary Table S6.

4.4. T-DNA Insertion Loci Analysis

Leaf tissues harvested from the plants at 10 days were grown in water. Genomic DNA isolation was performed using the CTAB extraction procedure [63]. T-DNA insertion loci of two overexpression transgenic lines were determined by TAIL-PCR following the procedures previously described [64]. The primers are listed in Supplementary Table S5.

4.5. WGBS and Data Analyses

Genomic DNA was extracted from 10 day-old seedlings and shipped to the Anoroad Genome Company (Beijing, China) for WGBS. WGBS library preparation and sequencing was performed by the Anoroad Genome Company (Beijing, China).

After downloading, the raw reads were filtered and trimmed to obtain clean reads and the available data were compared with the reference genome of *Oryza_sativa Japonica* IRGSP-1.0 to obtain the alignment results using Bismark (v0.9.0) [65]. The uniquely mapped reads will be used to call the methylated cytosines (mC) in highly enriched regions. For each cytosine site, the methylation level (%) was calculated by: 100 × (reads-supported methylation)/(total reads depth for the site). For the methylated region, the methylation level (%) was calculated by: 100 × methylation level of all cytosine sites in the region / the total number of cytosine sites in the region. The conversion ratio of Lambda DNA was around 99.5% to 99.6% in all samples. Valid coverage of methylated cytosine is the percentage of methylated C among all C on the reference genome, is associated with methylation level. C sites with read depths less than 5 are eliminated firstly. Then, for each C site, valid coverage is calculated by for all C or C within each pattern (CG, CHG and CHH). Coverage is the count of C sites with more than 5 depth divided by total number of C site by its pattern. Normalized mC number (Figure 1, Table 1) was calculated by: total mC number/valid coverage of methylated cytosine. Relative data are listed in Supplementary Table S7.

The differential analysis of methylation was performed at the region-base level, differentially methylated regions (DMRs). Aberrant DNA methylation was compared with the control group, an increased methylation pattern was defined as hyper-methylation, whereas a decreased methylation pattern was defined as hypo-methylation. DMR detection was performed using a methylKit and eDMR package [66] to estimate the boundary of CpG islands with a mixed bimodal normal distribution assumption of the distance between CpGs.

4.6. MeDIP-Seq and Data Analysis

MeDIP-seq for this study was performed by KangChen Bio-tech, Shanghai, China. Sequencing library preparation and data analysis were as performed described by Ding et al. [67]. The available data were aligned to the reference genome of *Oryza_sativa. Japonica* IRGSP-1.0 using BOWTIE software (V2.1.0) [68]. MeDIP peaks were identified by MACS2 [69], and statistically significant MeDIP-enriched regions (peaks) were identified by comparison to a Poisson background model (cut-off q-value = 10^{-2}). DMRs in MeDIP dataset was identified by diffReps [70] ($p < 0.0001$).

4.7. Statistical Analysis

The data of two groups were analyzed by paired sample Student's test (*t*-test) and the data of more than two groups were analyzed by Tukey's test of one-way analysis of variance (ANOVA). Statistically significant differences at $P < 0.05$ between samples were indicated by different letters on the histograms or after mean values. All statistical evaluations were conducted using the IBM SPSS Statistics version 23 software (SPSS Inc., Chicago, IL, USA).

5. Conclusions

We found that both tissue culture and transgenic processes cause global methylation changes. The tissue culture process generally leads to a reduction in global methylation whereas the transgenic process causes global methylation changes that are dependent on the background. Transgenes could cause the global methylation decrease in WYJ7 and increase in NP. In addition, tissue culture caused abundant hypo-DMRs while the transgenic process caused hyper-DMRs. Epigenetic changes such as large amounts of methylation can influence phenotype during the tissue culture process. This happened in CK, resulting in small-sized plants compared to the wild-type WYJ7, even with no transformed gene. In the transgenic lines, while the phenotype is mainly determined by the nature and function of the transgene, but the global epigenetic modification in transgenic lines occurred and resulting in different patterns of changes in plants of different genetic backgrounds. Our results indicate a potential reason behind the occurrence of transgenic plants with random and undesirable phenotypes.

Supplementary Materials: Supplementary materials can be found at http://www.mdpi.com/1422-0067/21/5/1819/s1.

Author Contributions: Conceptualization, X.F. (Xiaorong Fan); Data curation, X.F. (Xiaoru Fan) and J.C.; Formal analysis, X.F. (Xiaoru Fan); Funding acquisition, X.F. (Xiaorong Fan); Investigation, X.F. (Xiaoru Fan); Methodology, Y.W.; Project administration, X.F. (Xiaorong Fan); Resources, X.F. (Xiaorong Fan); Supervision, G.X. and X.F. (Xiaorong Fan); Writing – original draft, X.F. (Xiaoru Fan) and X.F. (Xiaorong Fan); Writing – review & editing, X.F. (Xiaoru Fan), J.C., Y.W., C.T. and X.F. (Xiaorong Fan). All authors have read and agreed to the published version of the manuscript.

Funding: This work is supported by China National Key Program for Research and Development (2016YFD0100700), National Natural Science Foundation (Grant 31372122), the Transgenic Project (Grant 2016ZX08001003-008), the 111 Project (Grant 12009), Innovation project of basic scientific research in central Universities of Ministry of Education (KYYJ201604, KJJQ201701), Innovative Research Team Development Plan of Ministry of Education of China (IRT_17R56; KYT201802), Jiangsu Science and Technology Development Program (Grant BE2019375-1).

Conflicts of Interest: The authors declare no conflict of interest.

Availability of Data and Materials: The WGBS and MeDIP dataset created for this publication are available in the NCBI GEO repository with accession number GSE94319.

References

1. Clive, J. *Global Status of Commercialized Biotech/GM Crops*; International Service for the Acquisition of Agri-biotech Applications (ISAAA): Ithaca, NY, USA, 2014.
2. Han, Z.; Crisp, P.A.; Stelpflug, S.; Kaeppler, S.M.; Li, Q.; Springer, N.M. Heritable Epigenomic Changes to the Maize Methylome Resulting from Tissue Culture. *Genetics* **2018**, *209*, 983–995. [CrossRef]

3. Krikorian, A.D.; Berquam, D.L. Plant cell and tissue cultures: The role of Haberlandt. *Bot. Rev.* **1969**, *35*, 58–88. [CrossRef]

4. Kaeppler, S.M.; Kaeppler, H.F.; Rhee, Y. Epigenetic aspects of somaclonal variation in plants. *Plant Mol. Biol.* **2000**, *43*, 179–188. [CrossRef] [PubMed]

5. Miguel, C.; Marum, L. An epigenetic view of plant cells cultured in vitro: Somaclonal variation and beyond. *J. Exp. Bot.* **2011**, *62*, 3713–3725. [CrossRef] [PubMed]

6. Neelakandan, A.K.; Wang, K. Recent progress in the understanding of tissue culture-induced genome level changes in plants and potential applications. *Plant Cell Rep.* **2012**, *31*, 597–620. [CrossRef] [PubMed]

7. Rhee, Y.; Sekhon, R.S.; Chopra, S.; Kaeppler, S. Tissue Culture-Induced Novel Epialleles of a Myb Transcription Factor Encoded by pericarp color1 in Maize. *Genetics* **2010**, *186*, 843–855. [CrossRef]

8. Ong-Abdullah, M.; Ordway, J.M.; Jiang, N.; Ooi, S.E.; Kok, S.Y.; Sarpan, N.; Azimi, N.; Hashim, A.T.; Ishak, Z.; Rosli, S.K.; et al. Loss of Karma transposon methylation underlies the mantled somaclonal variant of oil palm. *Nature* **2015**, *525*, 533–537. [CrossRef]

9. Hsu, F.; Gohain, M.; Allishe, A.; Huang, Y.; Liao, J.; Kuang, L.; Chen, P. Dynamics of the Methylome and Transcriptome during the Regeneration of Rice. *Epigenomes* **2018**, *2*. [CrossRef]

10. Gimenez, M.D.; Yanez-Santos, A.M.; Paz, R.C.; Quiroga, M.P.; Marfil, C.F.; Conci, V.C.; Garcia-Lampasona, S.C. Assessment of genetic and epigenetic changes in virus-free garlic (Allium sativum L.) plants obtained by meristem culture followed by in vitro propagation. *Plant Cell Rep.* **2016**, *35*, 129–141. [CrossRef]

11. Machczynska, J.; Zimny, J.; Bednarek, P.T. Tissue culture-induced genetic and epigenetic variation in triticale (x Triticosecale spp. Wittmack ex A. Camus 1927) regenerants. *Plant Mol. Biol.* **2015**, *89*, 279–292. [CrossRef]

12. Lin, W.; Zhang, X.; Li, Y.; Liu, S.; Sun, W.; Zhang, X.; Wu, Q. Whole-Genome Bisulfite Sequencing Reveals a Role for DNA Methylation in Variants from Callus Culture of Pineapple (Ananas comosus L.). *Genes* **2019**. [CrossRef] [PubMed]

13. Aydin, M.; Arslan, E.; Taspinar, M.S.; Karadayi, G.; Agar, G. Analyses of somaclonal variation in endosperm-supported mature embryo culture of rye (Secale cereale L.). *Biotechnol. Biotechnol. Equip.* **2016**, *30*, 1082–1089. [CrossRef]

14. Sun, S.L.; Zhong, J.Q.; Li, S.H.; Wang, X.J. Tissue culture-induced somaclonal variation of decreased pollen viability in torenia (Torenia fournieri Lind.). *Bot. Stud.* **2013**, *54*, 36. [CrossRef] [PubMed]

15. Chang, Y.N.; Zhu, C.; Jiang, J.; Zhang, H.; Zhu, J.K.; Duan, C.G. Epigenetic regulation in plant abiotic stress responses. *J. Integr. Plant Biol.* **2019**. [CrossRef]

16. Zhang, H.; Lang, Z.; Zhu, J.K. Dynamics and function of DNA methylation in plants. *Nat. Rev. Mol. Cell. Biol.* **2018**, *19*, 489–506. [CrossRef]

17. Chen, X.; Brigitte, S.; Menz, J.; Ludewig, U. Plasticity of DNA methylation and gene expression under zinc deficiency in Arabidopsis roots. *Plant Cell Physiol.* **2018**, *59*, 1790–1802. [CrossRef]

18. Stroud, H.; Ding, B.; Simon, S.A.; Feng, S.; Bellizzi, M.; Pellegrini, M.; Wang, G.L.; Meyers, B.C.; Jacobsen, S.E. Plants regenerated from tissue culture contain stable epigenome changes in rice. *eLife* **2013**, *2*. [CrossRef]

19. Rajeevkumar, S.; Anunanthini, P.; Sathishkumar, R. Epigenetic silencing in transgenic plants. *Front Plant Sci.* **2015**, *6*, 693. [CrossRef]

20. Vilperte, V.; Tenfen, S.Z.A.; Wikmark, O.G.; Nodari, R.O. Levels of DNA methylation and transcript accumulation in leaves of transgenic maize varieties. *Environ. Sci. Eur.* **2016**, *28*, 29. [CrossRef]

21. Matzke, M.A.; Mette, M.F.; Matzke, A.J.M. Transgene silencing by the host genome defense: Implications for the evolution of epigenetic control mechanisms in plants and vertebrates. *Plant Mol. Biol.* **2000**, *43*, 401–415. [CrossRef]

22. Huettel, B.; Kanno, T.; Daxinger, L.; Aufsatz, W.; Matzke, A.J.M.; Matzke, M. Endogenous targets of RNA-directed DNA methylation and Pol IV in Arabidopsi. *EMBO J.* **2006**, *25*, 2828–2836. [CrossRef] [PubMed]

23. Tsuchiya, T.; Eulgem, T. An alternative polyadenylation mechanism coopted to the Arabidopsis RPP7 gene through intronic retrotransposon domestication. *Proc. Natl. Acad. Sci. USA* **2013**, *110*, 3235–3543. [CrossRef] [PubMed]

24. Shavrukov, Y.; Baho, M.; Lopato, S.; Langridge, P. The TaDREB3 transgene transferred by conventional crossings to different genetic backgrounds of breadwheat improves drought tolerance. *Plant Biotech. J.* **2016**, *14*, 313–322. [CrossRef] [PubMed]

25. Liu, S.; Jiang, J.; Liu, Y.; Meng, J.; Xue, S.; Tan, Y.; Li, Y.; Shu, Q.; Huang, J. Characterization and evaluation of OsLCT1 and OsNramp5 mutants generated through CRISPR/Cas9-mediated mutagenesis for breeding low Cd rice. *Rice Sci.* **2019**, *26*, 88–97. [CrossRef] [PubMed]

26. Mao, B.; Li, Y.; Lv, Q.; Zhang, L.; Chen, C.; He, H.; Wang, W.; Zeng, X.; Shao, Y. Knockout of OsNramp5 using the CRISPR/Cas9 system produces low Cd-accumulating indica rice without compromising yield. *Sci. Rep.* **2017**, *7*, 1–12.

27. Thomsen, H.C.; Eriksson, D.; Moller, I.S.; Schjoreeing, J.K. Cytosolic glutamine synthetase: A target for improvement of crop nitrogen use efficiency? *Trends Plant Sci.* **2014**, *19*, 656–663. [CrossRef]

28. Kyozuka, J.; Shimamoto, K. Transgenic rice plants: Tools for studies in gene regulation and crop improvement. *Control Plant Gene Expr.* **1992**, 173–193.

29. Yan, M.; Fan, X.; Feng, H.; Miller, A.; Shen, Q.; Xu, G. Rice OsNAR2.1 interacts with OsNRT2.1, OsNRT2.2 and OsNRT2.3a nitrate transporters to provide uptake over high and low concentration ranges. *Plant Cell Environ.* **2011**, *34*, 1360–1372. [CrossRef]

30. Feng, H.; Yan, M.; Fan, X.; Li, B.; Shen, Q.; Miller, A.J.; Xu, G. Spatial expression and regulation of rice high-affinity nitrate transporters by nitrogen and carbon status. *J. Exp. Bot.* **2011**, *62*, 2319–2332. [CrossRef]

31. Liu, X.; Huang, D.; Tao, J.; Miller, A.J.; Fan, X.; Xu, G. Identification and functional assay of the interaction motifs in the partner protein OsNAR2.1 of the two-component system for high-affinity nitrate transport. *New Phytol.* **2014**, *204*, 74–80. [CrossRef]

32. Chen, J.; Zhang, Y.; Tan, Y.; Zhang, M.; Zhu, L.; Xu, G.; Fan, X. Agronomic nitrogen-use efficiency of rice can be increased by driving OsNRT2.1 expression with the OsNAR2.1 promoter. *Plant Biotechnol. J.* **2016**, *14*, 1705–1715. [CrossRef] [PubMed]

33. Chen, J.; Fan, X.; Qian, K.; Zhang, Y.; Song, M.; Liu, Y.; Xu, G.; Fan, X. pOsNAR2.1:OsNAR2.1 expression enhances nitrogen uptake efficiency and grain yield in transgenic rice plants. *Plant Biotechnol. J.* **2017**, *15*, 1273–1283. [CrossRef] [PubMed]

34. Bednarek, P.; Orlowska, R. Plant tissue culture environment as a switch-key of (epi)genetic changes. *Plant Cell Tissue Organ* **2020**, *140*, 245–257. [CrossRef]

35. Brown, P.T.H. DNA methylation in plants and its role in tissue culture. *Genome* **1989**, *31*, 717–729. [CrossRef]

36. Muller, E.; Brown, P.T.H.; Hartke, S.; Lorz, H. DNA variation in tissue-culture-derived rice plants. *Theor. Appl. Genet.* **1990**, *80*, 673–679. [CrossRef] [PubMed]

37. Zheng, K.L.; Castiglione, S.; Biasini, M.G.; Biroli, A.; Morandi, C.; Sala, F. Nuclear DNA amplification in cultured cells of Oryza sativa L. *Theor. Appl. Genet.* **1987**, *74*, 65–70. [CrossRef]

38. Dominguez, A.; Fagoaga, C.; Navarro, L.; Moreno, P.; Pena, L. Regeneration of transgenic citrus plants under non selective conditions results in high-frequency recovery of plants with silenced transgenes. *Mol. Genet. Genom.* **2002**, *267*, 544–556. [CrossRef]

39. Wang, Q.M.; Wang, Y.Z.; Sun, L.; Gao, F.; Sun, W.; He, J.; Gao, X.; Wang, L. Direct and indirect organogenesis of Clivia miniata and assessment of DNA methylation changes in various regenerated plantlets. *Plant Cell Rep.* **2012**, *31*, 1283–1296. [CrossRef]

40. Orlowska, R.; Machczynska, J.; Oleszczuk, S.; Zimny, J.; Bednarek, P.T. DNA methylation changes and TE activity induced in tissue cultures of barley (Hordeum vulgare L.). *J. Biol. Res. Thessal.* **2016**, *23*, 19. [CrossRef]

41. Hiei, Y.; Ohta, S.; Komari, T.; Kumashiro, T. Efficient Transformation of Rice (Oryza-Sativa L) Mediated by Agrobacterium and Sequence-Analysis of the Boundaries of the T-DNA. *Plant J.* **1994**, *6*, 271–282. [CrossRef]

42. Yin, Z.; Wang, G. Evidence of multiple complex patterns of T-DNA integration into the rice genome. *Theor. Appl. Genet.* **2000**, *100*, 461–470. [CrossRef]

43. Xu, J.; Chen, G.; Hermanson, P.J. Population-level analysis reveals the widespread occurrence and phenotypic consequence of DNA methylation variation not tagged by genetic variation in maize. *Genome Biol.* **2019**, *20*, 243. [CrossRef] [PubMed]

44. Niu, Z.; Yang, Z.; Li, J.; Hu, M. Characteristics and high-yield cultivation techniques of Wuyunjing 7. *Chin. Rice* **1998**, *4*, 9–10.

45. Jiang, Y. High-yielding rice cultivation technology of Nipponbare. *Shandong Agric. Sci.* **1981**, *2*, 28–29.

46. Xu, G.; Fan, X.; Miller, A.J. Plant nitrogen assimilation and use efficiency. *Annu. Rev. Plant Biol.* **2012**, *63*, 153–182. [CrossRef]

47. Qu, L.Q.; Takaiwa, F. Evaluation of tissue specificity and expression strength of rice seed component gene promoters in transgenic rice. *Plant Biotechnol. J.* **2004**, *2*, 113–125. [CrossRef]

48. Wang, B.; Sang, Y.; Song, J.; Gao, X.Q.; Zhang, X. Expression of a rice OsARGOS gene in Arabidopsis promotes cell division and expansion and increases organ size. *J. Genet. Genom.* **2009**, *36*, 31–40. [CrossRef]

49. Lu, C.; Zhang, L.; Tang, Z.; Huang, X.Y.; Ma, J.F.; Zhao, F.J. Producing cadmium-free Indica rice by overexpressing OsHMA3. *Environ. Int.* **2019**, *126*, 619–626. [CrossRef]

50. Yang, C.; Zhang, Y.; Huang, C. Reduction in cadmium accumulation in japonica rice grains by CRISPR/Cas9-mediated editing of OsNRAMP5. *J. Integr. Agric.* **2019**, *18*, 688–697. [CrossRef]

51. Que, Q.; Wang, H.Y.; English, J.J.; Jorgensen, R.A. The frequency and degree of cosuppression by sense chalcone synthase transgenes are dependent on transgene promoter strength and are reduced by premature nonsense codons in the transgene coding sequence. *Plant Cell* **1997**, *9*, 1357–1368. [CrossRef]

52. Hirochika, H.; Sugimoto, K.; Otsuki, Y.; Tsugawa, H.; Kanda, M. Retrotransposons of rice involved in mutations induced by tissue culture. *Proc. Natl. Acad. Sci. USA* **1993**, *93*, 7783–7788. [CrossRef] [PubMed]

53. Smulders, M.J.M.; de Klerk, G.J. Epigenetics in plant tissue culture. *Plant Growth Regul.* **2011**, *63*, 137–146. [CrossRef]

54. Vázquez, A.M. Insight into somaclonal variation. *Plant Biosyst.* **2001**, *135*, 57–62. [CrossRef]

55. Wang, Q.; Wang, L. An evolutionary view of plant tissue culture: Somaclonal variation and selection. *Plant Cell Rep.* **2012**, *31*, 1535–1547. [CrossRef]

56. Rodriguez Enriquez, J.; Dickinson, H.G.; Grant Downton, R.T. MicroRNA misregulation: An overlooked factor generating somaclonal variation? *Trends Plant Sci.* **2011**, *16*, 242–248. [CrossRef]

57. Krishna, H.; Alizadeh, M.; Singh, D.; Singh, U.; Chauhan, N.; Eftekhari, M.; Sadh, R.K. Somaclonal variations and their applications in horticultural crops improvement. *3 Biotech* **2016**, *6*, 54. [CrossRef]

58. Krizova, K.; Fojtova, M.; Depicker, A.; Kovarik, A. Cell Culture-Induced Gradual and Frequent Epigenetic Reprogramming of Invertedly Repeated Tobacco Transgene Epialleles. *Plant Physiol.* **2009**, *149*, 1493–1504. [CrossRef]

59. Mubeen, K.; Iqbal, A.; Hussain, M.; Zahoor, F.; Siddiqui, M.H.; Mohsin, A.U.; Bakht, H.F.S.G.; Hanif, M. Impact of Nitrogen and Phosphorus on the Growth, Yield and Quality of Maize (Zea Mays L.) Fodder in Pakistan. *Philipp. J. Crop Sci.* **2013**, *38*, 43–46.

60. Ookawa, T.; Hobo, T.; Yano, M.; Murata, K.; Ando, T.; Miura, H.; Asano, K.; Ochiai, Y.; Ikeda, M.; Nishitani, R.; et al. New approach for rice improvement using a pleiotropic QTL gene for lodging resistance and yield. *Nat. Commun.* **2010**, *1*, 132. [CrossRef]

61. Pan, S.; Rasul, F.; Li, W.; Tian, H.; Mo, Z.W.; Duan, M.; Tang, X. Roles of plant growth regulators on yield, grain qualities and antioxidant enzyme activities in super hybrid rice (Oryza sativa L.). *Rice* **2013**, *6*, 9. [CrossRef]

62. Srikanth, B.; Rao, I.S.; Surekha, K.; Subrahmanyam, D.; Voleti, S.R.; Neeraja, C.N. Enhanced expression of OsSPL14 gene and its association with yield components in rice (Oryza sativa) under low nitrogen conditions. *Gene* **2016**, *576*, 441–450. [CrossRef] [PubMed]

63. Saghai, M. Ribosomal DNA spacer-length polymorphisms in barley: Mendelian inheritance, chromosomal location, and population dynamics. *Proc. Natl. Acad. Sci. USA* **1984**, *81*. [CrossRef]

64. Singer, T.; Burke, E. High-throughput TAIL-PCR as a tool to identify DNA flanking insertions. *Methods Mol. Bio.* **2003**, *236*, 241–272.

65. Krueger, F.; Andrews, S.R. Bismark: A flexible aligner and methylation caller for Bisulfite-Seq applications. *Bioinformatics* **2011**, *27*, 1571–1572. [CrossRef] [PubMed]

66. Akalin, A.; Kormaksson, M.; Li, S.; Garrett-Bakelman, F.E.; Figueroa, M.E.; Melnick, A.; Mason, C.E. methylKit: A comprehensive R package for the analysis of genome-wide DNA methylation profiles. *Genome Biol.* **2012**, *13*. [CrossRef] [PubMed]

67. Ding, X.; Zheng, D.Y.; Fan, C.N.; Liu, Z.Q.; Dong, H.; Lu, Y.Y.; Qi, K.M. Genome-wide screen of DNA methylation identifies novel markers in childhood obesity. *Gene* **2015**, *566*, 74–83. [CrossRef]

68. Langmead, B.; Salzberg, S.L. Fast gapped-read alignment with Bowtie 2. *Nat. Methods* **2012**, *9*, 357–359. [CrossRef]

69. Zhang, Y.; Liu, T.; Meyer, C.A.; Eeckhoute, J.; Johnson, D.S.; Bernstein, B.E.; Nussbaum, C.; Myers, R.M.; Brown, M.; Li, W.; et al. Model-based Analysis of ChIP-Seq (MACS). *Genome Biol.* **2008**, *9*. [CrossRef]
70. Shen, L.; Shao, N.; Liu, X.; Maze, I.; Feng, J.; Nestler, E. diffReps: Detecting Differential Chromatin Modification Sites from ChIP-seq Data with Biological Replicates. *PLoS ONE* **2013**, *8*. [CrossRef]

International Journal of
Molecular Sciences

Article

A RING-Type E3 Ubiquitin Ligase, *OsGW2*, Controls Chlorophyll Content and Dark-Induced Senescence in Rice

Kyu-Chan Shim [1], Sun Ha Kim [1], Yun-A Jeon [1], Hyun-Sook Lee [1], Cheryl Adeva [1], Ju-Won Kang [2], Hyun-Jung Kim [3], Thomas H Tai [4,5] and Sang-Nag Ahn [1,*]

[1] Department of Agronomy, Chungnam National University, Daejeon 34134, Korea;
    zktnrl@naver.com (K.-C.S.); sunha82@cnu.ac.kr (S.H.K.); jya0911@cnu.ac.kr (Y.-A.J.);
    leehs0107@gmail.com (H.-S.L.); ccadeva_758@yahoo.com (C.A.)
[2] Department of Southern Area Crop Science, National Institute of Crop Science, RDA, Miryang 50424, Korea;
    kangjw81@korea.kr
[3] LG Chemical, Ltd., Seoul 07796, Korea; hk269@cornell.edu
[4] USDA-ARS Crops Pathology and Genetics Research Unit, Davis, CA 95616, USA; thomas.tai@usda.gov
[5] Department of Plant Sciences, University of California, Davis, CA 95616, USA
* Correspondence: ahnsn@cnu.ac.kr; Tel.: +82-42-821-5728

Received: 22 January 2020; Accepted: 27 February 2020; Published: 2 March 2020

Abstract: Leaf senescence is the final stage of plant development. Many internal and external factors affect the senescence process in rice (*Oryza sativa* L.). In this study, we identified *qCC2*, a major quantitative trait locus (QTL) for chlorophyll content using a population derived from an interspecific cross between *O. sativa* (*cv.* Hwaseong) and *Oryza grandiglumis*. The *O. grandiglumis* allele at *qCC2* increased chlorophyll content and delayed senescence. *GW2* encoding E3 ubiquitin ligase in the *qCC2* region was selected as a candidate for *qCC2*. To determine if *GW2* is allelic to *qCC2*, a *gw2*-knockout mutant (*gw2-ko*) was examined using a dark-induced senescence assay. *gw2-ko* showed delayed leaf senescence in the dark with down-regulated expression of senescence-associated genes (SAGs) and chlorophyll degradation genes (CDGs). The association of the *GW2* genotype with the delayed senescence phenotype was confirmed in an F_2 population. RNA-seq analysis was conducted to investigate 30-day-old leaf transcriptome dynamics in Hwaseong and a backcross inbred line—CR2002—under dark treatment. This resulted in the identification of genes involved in phytohormone signaling and associated with senescence. These results suggested that transcriptional regulation was associated with delayed senescence in CR2002, and RING-type E3 ubiquitin ligase GW2 was a positive regulator of leaf senescence in rice.

Keywords: leaf senescence; rice; quantitative trait loci; transcriptome analysis

1. Introduction

Chlorophyll (Chl) is a photosynthetic pigment that is an essential component of the plant photosystem. It changes solar energy to chemical energy. Because of its photosynthetic ability, increasing Chl content in crops may be an effective way to increase grain yield and biomass production [1]. A number of studies have demonstrated that Chl content is controlled by quantitative trait locus (QTL) in various genetic backgrounds in rice [2–6].

Senescence or biological aging is the final stage of plant development [7]. During senescence, leaf color turns from green to yellow because of chlorophyll degradation [8]. Leaf yellowing is frequently used as a senescence indicator. Stay-green (non-yellowing) mutants maintain leaf greenness after the grain-ripening stage [9]. Stay-green mutants exhibit delayed leaf senescence and have been found in various plant species [8,10]. In rice, several stay-green genes have been cloned and characterized, including *Stay-Green Rice* (*SGR*), *NON-YELLOW COLORING 1* (*NYC1*), *NYC1-LIKE* (*NOL*), *Oryza sativa NAC-like, activated by Apetala3/Pistillata* (*OsNAP*) [10–14]. *SGR*, a highly conserved senescence-associated gene, encodes a novel chloroplast protein, and the expression of *SGR* is up-regulated in both natural and dark-induced senescence. This gene interacts with the light-harvesting chlorophyll-binding protein (LHCP) [10,14]. *NYC1* and *NOL* encode short-chain dehydrogenase/reductase (SDR) [11,12]. These two genes are co-localized in the thylakoid membrane and function as a chlorophyll *b* reductase. *OsNAP* contains a typical NAC structure at the N terminus [13]. This gene plays a role in regulating leaf senescence and acts as a key component, linking ABA signaling in rice. These genes have been used as ideal markers for the onset of the senescence process.

Ubiquitination is the post-translational modification of protein substrates [15]. The ubiquitin-proteasome pathway is known to play an important role in plant seed and organ size determination, such as *DA1*, *DA2*, *EOD1/BB*, *SAMBA*, and *SOD/UBP15* genes in *Arabidopsis* [16–20]. A major function of E3 ubiquitin ligase is to regulate polyubiquitination and to degrade their target substrate proteins. E3 ubiquitin ligase *BigBrother* (*BB*) controls final organ size and seed size acting in parallel with the *DA1* gene in *Arabidopsis,* and these two genes also positively control leaf senescence [17,19,20]. Delayed senescence has been confirmed by measuring the expression of *AHK3*, *CRF6*, and *ARF2*. The negative regulators of senescence—*AHK3* and *CRF6*—are expressed higher in *da1-1_bb/eod1-2* leaves, while the auxin repressor *ARF2* is expressed at a lower level than wild type [19]. *DA2*, an ortholog of *OsGW2*, controls seed and organ size by interacting with *DA1* [18].

OsGW2, a QTL on chromosome 2, controls grain width and weight in rice and encodes RING-type E3 ubiquitin ligase [21]. This gene might function in the degradation by the ubiquitin-proteasome pathway, and loss-of-function in *OsGW2* increases cell numbers and grain size [21]. The *OsGW2* mutant also shows increased transcript levels of *OsPCR1* during the developing grain stage, leading to an increase in Zn concentration in the seed [22]. Yeast two-hybrid and in vitro pull-down assays have shown that OsGW2 directly interacts with expansin-like 1 (EXPLA1), chitinase 14 (CHT14), and phosphoglycerate kinase (PGK) [23,24]. Transcription activation activity has been found in the GW2-C terminus (205 to 260) [24]. Together, the findings that OsGW2 interacts with various proteins and that the mutation of *OsGW2* has a pleiotropic effect in rice raises the possibility that *GW2* is also involved in regulating leaf senescence.

The objective of this study was to identify QTL controlling chlorophyll content and leaf senescence and to characterize candidate genes associated with this trait. *qCC2*, a major QTL for chlorophyll content, was mapped using an introgression line, CR2002, developed from an interspecific cross between *O. sativa cv.* Hwaseong and the wild species *O. grandiglumis*. CR2002 harboring *qCC2* from *O. grandiglumis* showed delayed leaf yellowing and a higher *Fv/Fm* value than Hwaseong. Endogenous expression levels of senescence-associated genes (SAGs) and chlorophyll degradation genes (CDGs) in CR2002 were lower than in Hwaseong. To determine if *GW2* is allelic to *qCC2* and thereby associated with delayed senescence, a *gw2*-knockout mutant (hereafter termed *gw2-ko*) was examined using dark-induced senescence (DIS) assay. *gw2-ko* showed delayed leaf senescence under dark conditions with down-regulated expression of SAGs and CDGs. Delayed senescence was confirmed by segregation analysis in the F_2 generation. Comparative RNA-seq analysis was conducted to identify differentially-expressed transcripts, using 30-day-old leaves from Hwaseong and CR2002 that were subjected to dark treatment. This enabled the identification of which genotype-specific expressions were enriched in CR2002 and reduced in Hwaseong and vice versa, including genes involved in phytohormone biosynthesis and signaling, NAC transcription factors, and senescence-associated genes. Collectively, the results of this study suggested that *OsGW2* controlled chlorophyll content and was a positive regulator of leaf senescence in rice.

2. Results

2.1. QTL Analysis for Chlorophyll Content

CR2002 exhibited a stay-green phenotype and a higher SPAD value (a parameter of leaf greenness) than Hwaseong (Figure 1b). CR2002, which had four *O. grandiglumis* introgressions, also showed differences in several agronomic traits, including grain weight (Table 1, Figure 1a, and Figure S1). To identify loci associated with chlorophyll content and the stay-green phenotype, QTL analysis was conducted using F_3 and F_4 populations derived from a cross between Hwaseong and CR2002. Chlorophyll contents were measured at the heading stage and one month after heading. A total of three significant QTLs located on chromosomes 1, 2, and 6 were detected in the F_3 and F_4 populations (Table 2 and Figure S2). A QTL for chlorophyll content (*qCC2*) was located on chromosome 2 between RM12813 and RM12983. This QTL was repeatedly detected in F_3 and F_4 generations at both stages, heading (Chlorophyll I) and one month after heading (Chlorophyll II). *qCC2* explained 24.6% of the phenotypic variation, and the *O. grandiglumis* allele contributed to the increased chlorophyll content, indicating that *qCC2* was a major QTL responsible for leaf greenness at heading and ripening stage.

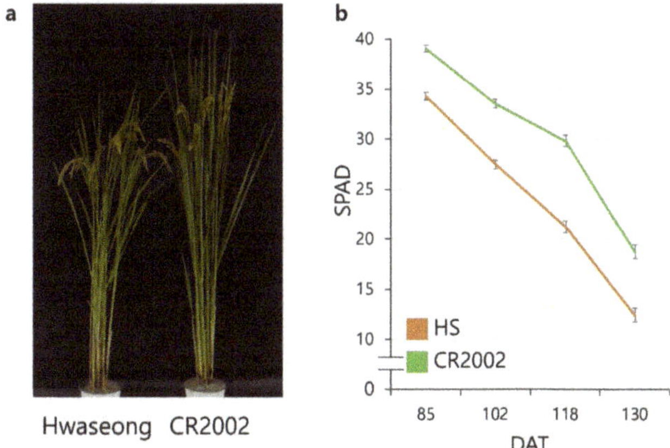

Figure 1. CR2002 showed a higher SPAD value and delayed senescence than Hwaseong in the field condition. Comparison of (**a**) plant morphology and (**b**) SPAD value (*n* = 30) of two parents. Error bars indicate the standard error. DAT: days after transplanting.

Table 1. Comparison of agronomic traits and chlorophyll content between Hwaseong and CR2002.

Trait	Hwaseong	CR2002	*p*-Value
Plant height (cm)	97 ± 4.00 *	101 ± 3.84	0.000
Stem diameter (mm)	4.47 ± 0.24	5.26 ± 0.48	0.000
Panicle length (cm)	21 ± 1.28	19 ± 1.04	0.000
First internode diameter (mm)	1.72 ± 0.12	1.99 ± 0.14	0.000
Grain length (mm)	7.14 ± 0.45	7.46 ± 0.17	0.002
Grain width (mm)	3.29 ± 0.13	3.83 ± 0.10	0.000
Grain thickness (mm)	2.20 ± 0.06	2.56 ± 0.10	0.000
1000-grain weight (g)	25.8 ± 0.68	32.9 ± 1.04	0.000
Chlorophyll content I (mg/m^2)	440 ± 16.38	457 ± 37.57	0.048
Chlorophyll content II (mg/m^2)	181 ± 50.57	216 ± 50.86	0.038

* Data are presented as mean ± standard deviation. The student's *t*-test was conducted for *p*-value. Chlorophyll content I and II were measured at the heading stage and 1 month after heading, respectively.

Table 2. QTLs (quantitative trait loci) for chlorophyll content based on one-way ANOVA in the F_3 and F_4.

Trait [1]	Gen.	QTL	Chr.	Marker	*p*-Value	R^2 (%)	H/H [2]	G/G
Chlorophyll content I	F_3	*qCC1*	1	RM11302-RM11315	0.002	3.48	442	452
	F_4	*qCC2*	2	RM7288-RM12983	0.029	10.76	546	570
Chlorophyll content II	F_3	*qCC2*	2	RM12813	0.018	10.68	173	189
	F_4	*qCC2*	2	RM12813-RM7288	0.001	24.63	497	545
	F_3	*qCC6*	6	RM584	0.018	10.62	454	434

[1] Chlorophyll content I and II were measured at heading stage and heading after 1 month, respectively. [2] H/H and G/G indicate the mean chlorophyll contents of homozygous for Hwaseong and *O. grandiglumis* genotypes, respectively.

The *qCC2* region was confirmed and fine-mapped by substitution mapping (Figure S3). SPAD values were measured four times, from 95 to 125 days after transplanting (DAT), for four substitution lines and two parental lines. CR7036 and CR7039 showed significantly higher SPAD values at 125 DAT than CR7038 and CR7058, suggesting that *qCC2* was located in the marker interval between RM3390 and RM7288 (about 1.4 Mbp).

2.2. Dark-Induced Senescence in Hwaseong and CR2002

To determine whether *qCC2* is associated with senescence under dark condition, detached leaves of Hwaseong and CR2002 were incubated on 3 mM MES buffer at 27 °C under complete darkness (Figure 2). Hwaseong leaves turned yellow four days after incubation (DAI), while CR2002 remained green (Figure 2a). Although the chlorophyll content of CR2002 was higher than Hwaseong at 0 DAI, a larger difference was observed at 4 DAI between CR2002 and Hwaseong (Figure 2b). The efficiency of photosystem II (*Fv/Fm* ratio) of CR2002 was significantly higher than Hwaseong, indicating CR2002 showed delayed senescence in the DIS assay (Figure 2c).

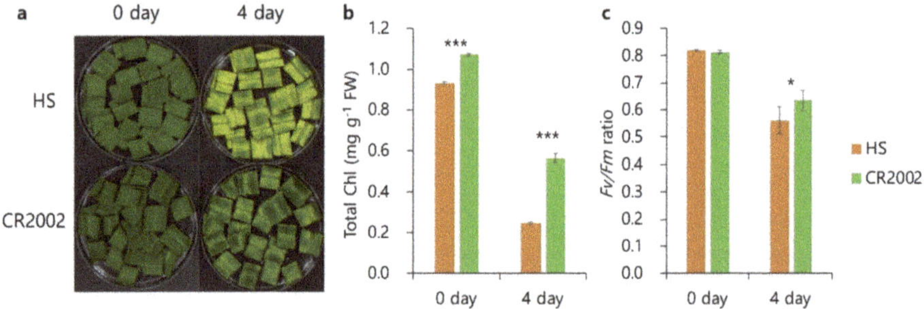

Figure 2. CR2002 showed delayed senescence under dark-induced senescence (DIS) conditions. Hwaseong and CR2002 were grown in a paddy field, and fully expanded flag leaves at the heading stage were used for DIS. (**a**) Detached leaves were incubated in 3 mM MES buffer (pH 5.8) at 27 °C under dark conditions. (**b,c**) Total chlorophyll contents (*n*= 6) and *Fv/Fm* ratio (*n*= 5) were compared between Hwaseong and CR2002. Error bars indicate a standard error, and more than three samples were used for each experiment. * and *** indicate significant difference at $p < 0.05$ and $p < 0.001$ based on Student's *t*-test, respectively.

To further examine the different senescence of Hwaseong and CR2002, transcript levels of SAGs (*OsNAP*, *Osh36*, and *Osl57*) and CDGs (*SGR*, *RCCR1*, *PAO*, *NYC1*, and *NOL*) were measured by qRT-PCR using detached leaf samples (Figure 3). During DIS, transcription levels of the SAGs and CDGs showed increases of 4.9-fold to 325-fold in Hwaseong, whereas increases of 4.7-fold to 200-fold were observed in CR2002. Transcript levels of SAGs and CDGs in CR2002 were significantly lower than Hwaseong at 4 DAI, with the exception of *Osl57*, indicating that the lower expression of SAGs and CDGs in CR2002 might result in the delayed senescence exhibited by CR2002 in the DIS assay.

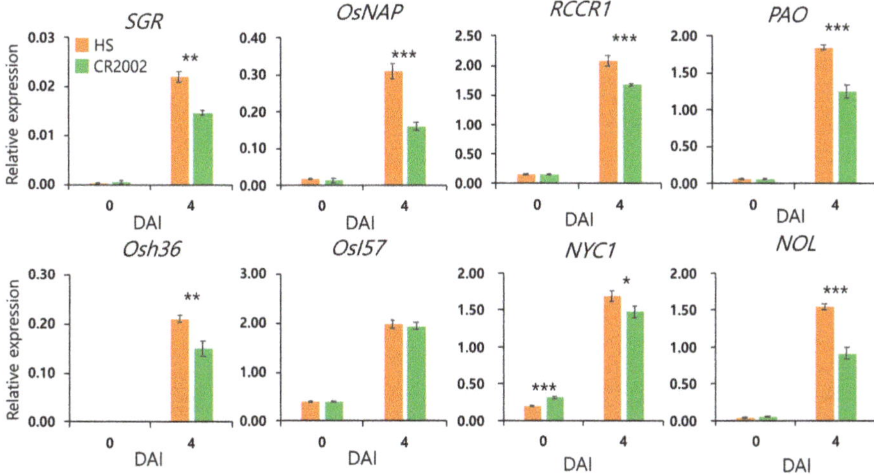

Figure 3. Expression of senescence-associated genes and chlorophyll degradation genes in Hwaseong and CR2002. qRT-PCR was conducted to determine the transcript level of genes. *OsUBQ5* was used for normalization. Error bars indicate the standard deviation of three replications. *, **, and *** indicate significant difference at $p < 0.05$, $p < 0.01$, and $p < 0.001$ based on Student's *t*-test, respectively. DAI: days after incubation.

2.3. Characterization of gw2-ko

The *O. grandiglumis* segment harboring *qCC2* is approximately 1.4 Mb in size. In this region, *GW2* was selected as a candidate gene based on its function as a RING-type E3 ubiquitin ligase. A 1-bp deletion at position 316 in the coding region of the *O. grandiglumis GW2* allele, which causes a premature stop and leads to an increased grain width and thickness, was also observed in this study (Figure S4) [21]. In *Arabidopsis*, ubiquitin-activated peptidase DA1 regulates leaf senescence together with E3 ubiquitin ligase BB [18]. In addition, *DA1* controls seed and final organ size with interacting partner *DA2*, which is orthologous to *OsGW2*. Due to the previously identified pleiotropic effect of *GW2*, the *GW2* gene was selected as a candidate gene for the *qCC2* QTL.

To determine if *GW2* is associated with delayed senescence, a T-DNA insertional mutant (PFG_1B-10017.R) was selected from a T-DNA insertional mutant library [25]. The T-DNA insertion was confirmed by two sets of PCR primers, and homozygous T-DNA lines were selected by genomic PCR (Figure 4a,b). The *OsGW2* knock-out mutant (*gw2-ko*) showed increased grain size, and the transcription of *OsGW2* in wild type (WT Dongjin) and *gw2-ko* was evaluated using semi-quantitative RT-PCR (Figure 4c,d). The *OsGW2* transcript level was examined during DIS in WT. The expression level was down-regulated at 2 DAI, whereas gene expression was up-regulated after 4 DAI, and a similar level of gene expression with 0 DAI was observed (Figure 4e). Consistent expression patterns of *OsGW2* and *OgGW2 (O. grandiglumis GW2)* were found in Hwaseong and CR2002, but CR2002 showed lower transcript levels than Hwaseong at 0 and 4 DAI (Figure S5).

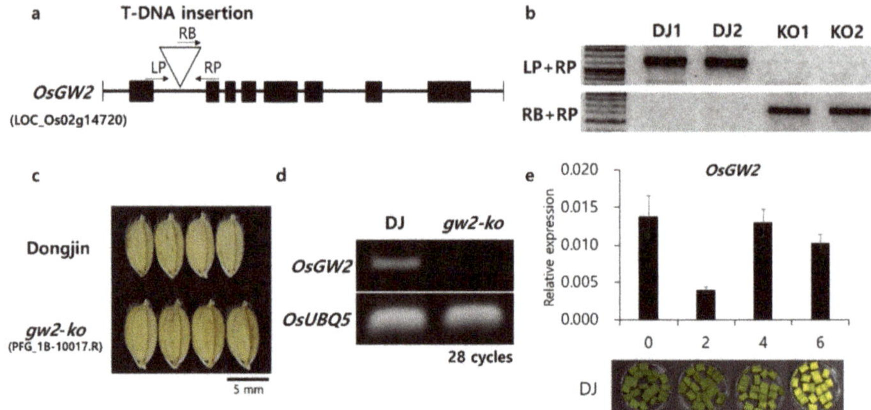

Figure 4. *OsGW2* T-DNA insertion knock-out mutant (*gw2-ko*) in wild-type Dongjin genetic background. (**a**) Schematic diagram showing the location of T-DNA insertion of the PFG_1B-10017.R mutant. (**b**) T-DNA insertion was confirmed using two sets of primers indicated in (**a**), and lines homozygous for the T-DNA insertion were used for this experiment. (**c**) Comparison of grain shape between wild type and *gw2-ko*. (**d**) Confirmation of *OsGW2* knock-out by semi-quantitative RT-PCR. *OsUBQ5* was used as a loading control. (**e**) Gene expression pattern of *OsGW2* during the dark-induced senescence. The transcript level of genes was determined by qRT-PCR. *OsUBQ5* was used for normalization.

The *gw2-ko* mutant showed significantly higher SPAD values than Dongjin from 85 to 130 DAT under the field conditions (Figure 5a,b). To characterize the phenotypic difference between Dongjin and *gw2-ko* during DIS, fully expanded detached leaves were incubated in 3 mM MES buffer (pH 5.8) at 27 °C in the dark (Figure 5c). Leaves of Dongjin turned yellow at 6 DAI, while *gw2-ko* remained green. Total chlorophyll contents of Dongjin and *gw2-ko* at 6 DAI were also significantly different, despite the total chlorophyll contents of *gw2-ko* at 0 DAI being slightly higher than Dongjin (Figure 5d). The *Fv/Fm* ratio remained higher in *gw2-ko* than Dongjin at 5 DAI (Figure 5e). These results demonstrated that leaf senescence was delayed in *gw2-ko*.

An F_2 population (n = 107) derived from a cross between Dongjin and *gw2-ko* was evaluated for DIS to determine if the T-DNA insertion in *OsGW2* was responsible for delayed senescence. After five days of dark incubation, plants homozygous for *gw2-ko* genotype (KO/KO group) showed significantly higher total chlorophyll contents than those homozygous for Dongjin genotype (DJ/DJ group) (Figure 6a,b). This segregation analysis suggested that T-DNA insertion in *OsGW2* caused the delayed senescence exhibited by *gw2-ko*.

Endogenous expression levels of SAGs and CDGs were examined (Figure 7). Among CDGs, *SGR*, *NYC1*, and *NOL* were significantly down-regulated in *gw2-ko* at 6 DAI, but *PAO* and *RCCR1* showed no significant difference between Dongjin and *gw2-ko*. SAGs (*OsNAP*, *Osh36*, and *OsI57*) were also down-regulated in *gw2-ko*. The loss-of-function of *GW2* led to the down-regulation of SAGs and CDGs during DIS.

The physiological phenotypes and endogenous transcript levels during DIS indicated that *gw2-ko* showed delayed leaf senescence compared to wild type and that *GW2* positively regulated leaf senescence in rice.

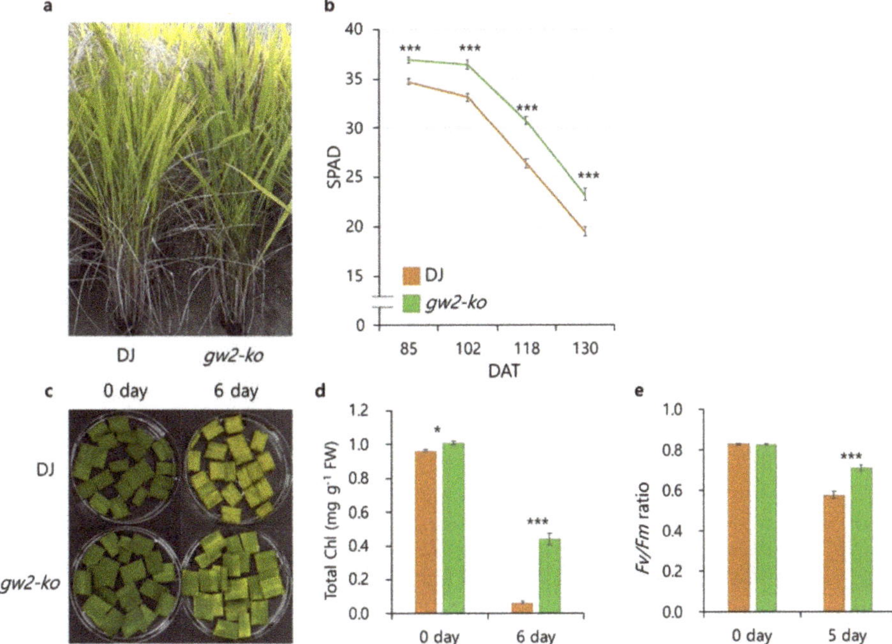

Figure 5. *gw2-ko* mutant showed delayed senescence. Comparison of (**a**) plant morphology (at 138 days after transplanting) and (**b**) SPAD value ($n = 30$) of wild-type Dongjin and *gw2-ko*. (**c**) Wild type and *gw2-ko* were grown in a paddy field, and fully expanded flag leaves at the heading stage were used for DIS. (**d**,**e**) Total chlorophyll contents ($n = 6$) and *Fv/Fm* ratio ($n = 10$) were compared between wild type and *gw2-ko*. Error bars indicate a standard error, and more than six samples were used for each experiment. * and *** indicate significant difference at $p < 0.05$ and $p < 0.001$ based on Student's *t*-test, respectively.

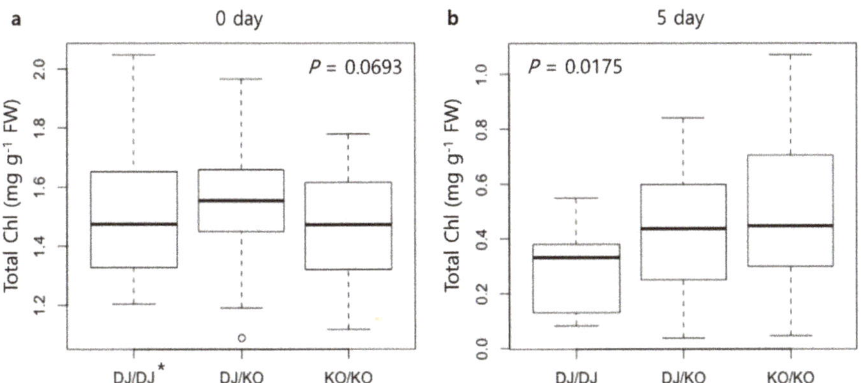

Figure 6. Boxplot of chlorophyll contents at (**a**) 0 day after incubation and (**b**) 5 days after incubation under the dark-induced senescence conditions in the F_2 population derived from a cross between Dongjin and *gw2-ko*. Genotype was determined with two sets of primers, as indicated in Figure 4a—LP+RP and RB+RP. One-way ANOVA was conducted to determine the significant difference between genotypes. *DJ/DJ: homozygous for Dongjin, DJ/KO: heterozygous, KO/KO: homozygous for *gw2-ko*.

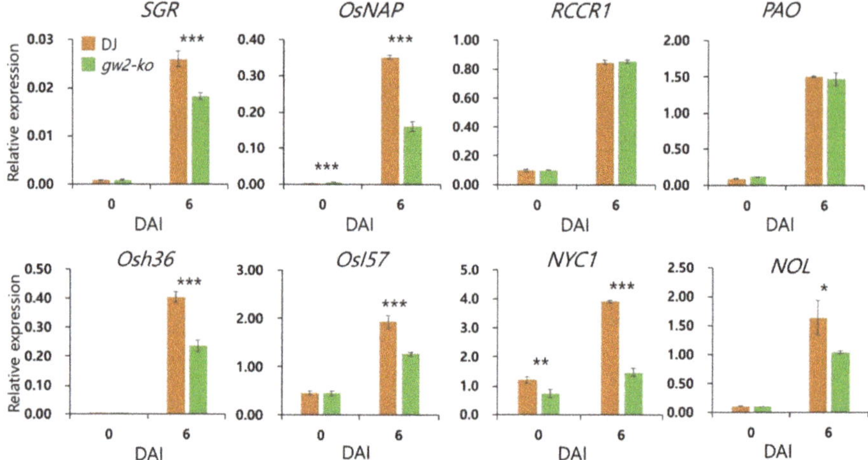

Figure 7. Expression of senescence-associated genes and chlorophyll degradation genes compared in Dongjin and *gw2-ko*. qRT-PCR was conducted to determine the transcript level of genes. *OsUBQ5* was used for normalization. Error bars indicate the standard deviation of three replications. *, **, and *** indicate significant difference at $p < 0.05$, $p < 0.01$, and $p < 0.001$ based on Student's *t*-test, respectively. DAI: days after incubation.

2.4. Transcriptome Analysis using Hwaseong and CR2002

To gain a better understanding of the mechanism for delayed senescence under dark-treatment conditions, the transcriptome of leaves from Hwaseong and CR2002 plants before (control) and after dark-treatment for 3 days were investigated by RNA-seq. A total of 264 million reads were generated using the Illumina Hiseq 4000 platform, and adapter and low-quality read trimming were conducted (Table S2). A total of 262 million high-quality reads were mapped against the reference genome sequence, and an average mapping rate of 66% was observed (Table S2). A multi-dimensional scaling plot showed that three replicates were clustered, indicating samples were consistently generated (Figure S6).

Differentially expressed genes (DEGs) were identified in Hwaseong and CR2002 under dark treatment compared with control conditions. DEGs were determined based on the criteria of a greater than 2-fold change with significance at Benjamini-Hochberg false discovery rate adjusted $p < 0.05$. A total of 10,339 DEGs was detected in the comparison of 0 day and 3 days of dark treatment in Hwaseong, and 9885 DEGs were found in CR2002 genotype during the dark treatment (Figure 8a). A Venn diagram of the DEGs indicated that Hwaseong and CR2002 shared 7901 of them (Table S5). Among these 7901 genes, only 29 genes showed different regulation patterns between Hwaseong and CR2002. For example, Os09g0511600 was up-regulated in Hwaseong by 5.13 log₂FC (fold-change), while down-regulated in CR2002 by -8.26 log₂FC. In addition, a total of 2438 and 1984 genes were specifically identified as DEGs in Hwaseong and CR2002, respectively (Tables S3 and S4). For Hwaseong, 1240 and 1198 genes were specifically up- and down-regulated during the dark treatment, respectively (Figure 8a and Table S3). A total of 838 up-regulated genes and 1146 down-regulated genes were specifically identified during the dark treatment in CR2002 (Figure 8a and Table S4).

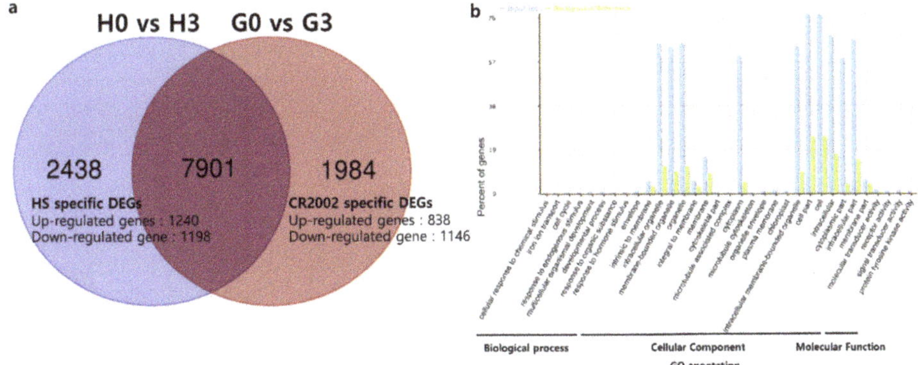

Figure 8. Summary of the numbers of differentially expressed genes (DEGs) upon incubation of leaves of two lines (Hwaseong and CR2002) in complete darkness. (**a**) A Venn diagram, showing the number of genes shared and distinct to each genotype. (**b**) Significantly enriched gene ontology (GO) terms for CR2002-specific DEGs. Blue and green bars indicate the input list of CR2002 and background/reference, respectively. H0, H3, G0, and G3 indicate Hwaseong day 0, Hwaseong day 3, CR2002 day 0, and CR2002 day 3 in dark condition, respectively.

To know functional information of the DEGs specifically detected in Hwaseong and CR2002, gene ontology (GO) analysis was performed with agriGO v2.0. For the 1984 CR2002-specific DEGs, a total of 33 GO terms was significantly enriched (Figure 8b). Among 33 GO terms, biological process, cellular component, and molecular function ontologies included 8, 21, and 4 GO terms, respectively. For Hwaseong-specific DEGs, a total of 33 GO terms were also significantly overrepresented. Of these GO terms, 6, 23, and 4 were included in the biological process, cellular component, and molecular function ontologies, respectively (Figure S7). Hwaseong-specific DEGs included intracellular signaling cascade, organelle membrane, cytosol, cytoskeleton, Golgi membrane, endomembrane system, and Golgi apparatus GO terms, while iron ion transport, cytoplasmic part, envelope, cellular response to chemical stimulus, cell cycle, microtubule-associated complex, and organelle envelope were included only in CR2002-specific DEGs.

To understand the mechanism involved in delayed senescence in CR2002, the SAGs, CDGs, and genes involved in phytohormone biosynthesis and signaling were examined from the DEGs. A total of 109 genes were identified (Table S6). Among these genes, we focused on those specifically detected in Hwaseong or CR2002 (Table 3). During the dark treatment, *OsNIT1*, *OsDWF1*, and *OsRR3* were up-regulated in CR2002, and *OsERF5*, *OsABI1*, *OsJAZ1*, and *OsGLU* were down-regulated. Phytohormone signaling and biosynthesis genes were mainly found in the CR2002 DEGs. When these seven DEGs were subjected to RT-qPCR using the same G0/G3 samples as for RNA-Seq, similar expression patterns were observed, suggesting the involvement of these DEGs in delayed senescence in CR2002 (Figure S8). For Hwaseong, 18 genes were identified from the DEGs, and 10 and 8 genes were down- and up-regulated, respectively. Hwaseong-specific DEGs included phytohormone biosynthesis and signaling genes, NAC transcription factors, and senescence-associated genes.

Table 3. List of differentially expressed genes (DEGs) associated with senescence, chlorophyll degradation, and phytohormone specifically detected in CR2002 and Hwaseong.

Gene	CR2002		Symbol	Description
	log_2FC	Adj. p-Value		
Os07g0410300	−3.676	0.037	OsERF5	Conserved hypothetical protein.
Os09g0532400	−2.653	0.031	OsABI1	Signal transduction response regulator, receiver region domain containing protein.
Os10g0392400	−1.577	0.044	OsJAZ1	Tify domain containing protein.
Os07g0658400	−1.013	0.012	OsGLU	Ferredoxin-dependent glutamate synthase, Leaf senescence, and nitrogen remobilization.
Os02g0635200	1.081	0.000	OsNIT1	Similar to Nitrilase 2.
Os10g0397400	1.360	0.003	OsDWF1	Dim/dwf1 protein, Cell elongation protein DIMINUTO/Dwarf1, Brassinosteroid (BR) biosynthesis.
Os02g0830200	2.592	0.003	OsRR3	A-type response regulator, Cytokinin signaling.

Gene	Hwaseong		Symbol	Description
	log_2FC	Adj. p-value		
Os11g0143300	−2.966	0.011	OsRR9	A-type response regulator, Cytokinin signaling.
Os12g0139400	−2.671	0.006	OsRR10	A-type response regulator, Cytokinin signaling.
Os11g0143200	−2.423	0.047	OsCPD1	Similar to Cytochrome P450 90A1.
Os03g0856700	−2.126	0.027	OsGA20ox1	Gibberellin 20 oxidase 1.
Os04g0673300	−1.981	0.002	OsRR6	A-type response regulator, Cytokinin signaling.
Os02g0164900	−1.524	0.002	OsARF6	Similar to Auxin response factor 3.
Os01g0208600	−1.462	0.021	OsSCAR1	SCAR-like protein 2, Component of the suppressor of cAMP receptor/Wiskott-Aldrich syndrome protein family verprolin-homologous (SCAR/WAVE) complex, Actin organization, Panicle development, Regulation of water loss.
Os01g0718300	−1.328	0.005	OsBRI1	Brassinosteroid LRR receptor kinase, Similar to Brassinosteroid-insensitive 1.
Os01g0723100	−1.323	0.001		Senescence-associated family protein.
Os03g0265100	−1.229	0	PLS2	Glycosyl transferase, group 1 domain containing protein.
Os07g0209000	1.096	0	OsDGL	Dolichyl-diphosphooligosaccharide-protein glycosyltransferase 48 kDa subunit precursor, N-glycosylation.
Os02g0324700	1.184	0.001		Similar to senescence-associated protein.
Os01g0927600	1.192	0	OsARF2	Similar to Auxin response factor 2 (ARF1-binding protein).
Os01g0752500	1.321	0.006	OsERF2	APETELA2/ethylene response factor (AP2/ERF) type transcription factor, Negative regulation of disease resistance, Negative regulation of salt tolerance.
Os03g0327800	1.872	0.013	OsNAP	NAC Family transcriptional activator, Abiotic stress response, Positive regulator of leaf senescence.
Os10g0477600	1.942	0.024	ONAC120	Similar to NAM / CUC2-like protein.
Os11g0126900	4.026	0.048	ONAC122	NAC-domain protein, Drought tolerance.
Os04g0578000	5.520	0.028	OsACS2	ACC synthase, Ethylene biosynthesis.

3. Discussion

3.1. qCC2 is Associated with Chlorophyll Content and Leaf Senescence

Chlorophyll content, one of the most important traits due to its role in photosynthesis, is regulated by polygenic loci. Several studies have identified a number of QTLs across various genetic resources and environments [3–6]. A number of studies reported QTLs on the short arm of chromosome 2. Four QTLs have been mapped for stay-green traits on chromosome 2 between RM145 and RM322 using 190 doubled haploid lines from a cross between 'Zhenshan 97' and 'Wuyujing 2' [2]. Jiang et al. (2012) reported QTL for chlorophyll content from heading to the maturity stage (*qCHM2*) located on the short arm of chromosome 2 between RM145 and RM29 [5]. In this study, three QTLs for chlorophyll content were detected using progeny from an interspecific cross between *O. sativa* and *O. grandiglumis*. *qCC2* was located on chromosome 2 between RM3390 and RM7288, which were approximately 1.4 Mbp apart. This QTL was detected in F_3 and F_4 generations and at two different developmental stages, respectively, and confirmed by substitution mapping (Figure S3). *qCC2* shared a similar location with QTLs reported in previous studies [2,5]. Although allelism tests are needed to clarify the relationships among these QTLs, these studies indicated the presence of genes associated with chlorophyll content and leaf senescence on the short arm region of chromosome 2.

When Hwaseong and CR2002 were tested for dark-induced senescence, CR2002 showed higher chlorophyll content with delayed leaf senescence than Hwaseong (Figure 2a,b). *Fv/Fm* values were also significantly different at 4 days after incubation (Figure 2c). Endogenous expression levels of SAGs and CDGs were severely down-regulated in CR2002 than Hwaseong, and these results indicated that *qCC2* was associated with leaf senescence in CR2002.

3.2. GW2 Positively Regulates Leaf Senescence

GW2 encodes a RING-type E3 ubiquitin ligase and resides in the *qCC2* region (1.4 Mbp). The *O. grandiglumis GW2* allele in CR2002 has the 1-bp deletion in 4th exon, which causes a premature stop and truncated protein (Figure S4) [21]. Many studies have suggested ubiquitination as a candidate pathway for the regulation of senescence. *SAUL1* encodes E3 ubiquitin ligase, which is required for suppression of premature senescence in *Arabidopsis* [26]. Also, *spl11* encodes E3 ubiquitin ligase, which regulates cell death in rice [27]. E3 ubiquitin ligase *Big Brother* (*BB*) gene and ubiquitin receptor *DA1* gene negatively regulate leaf size and promote senescence [19]. Mutations in *DA1* and *BB* enhance leaf growth, an effect that is synergistically increased in the double mutant. A *da1-1/eod1-2* double mutant especially exhibits a longer lifespan than wild type or the single mutants [19]. In addition, DA1 physically interacts with DA2, an ortholog of *GW2*, and *DA2* acts synergistically with *DA1* to regulate seed size in *Arabidopsis* [18]. Based on these previous results, *GW2* was chosen as a candidate gene for the delayed senescence phenotype in CR2002. The transcript level of *GW2* was down-regulated at 2 DAI and showed a similar expression level at 0 DAI and at 4 and 6 DAI in the dark (Figure 4e and Figure S5). This expression pattern was similar to the *NYC4* gene, which functions in the degradation of chlorophyll and chlorophyll-protein complexes during DIS [28]. The effect of *gw2-ko* was examined in DIS (Figure 5c). *gw2-ko* showed delayed senescence and down-regulated expression of SAGs and CDGs. Compared to Hwaseong and Dongjin, most of the SAGs and CDGs genes were consistently down-regulated in CR2002 and *gw2-ko* during DIS. However, transcript levels of *RCCR1* (*red chlorophyll catabolite reductase 1*), *PAO* (*Pheophorbide a oxygenase*), and *OsI57* (*putative 3-ketoacyl-CoA thiolase*) did not show significant differences between Dongjin and *gw2-ko* or Hwaseong and CR2002 in DIS. These results suggested that *RCCR1* and *PAO* might function down-stream in the chlorophyll degradation pathway [8]. Because the transcript levels of *RCCR1* and *PAO* in *gw2-ko* were not significantly different from wild-type Dongjin, *GW2* might affect genes up-stream in the chlorophyll degradation pathway.

3.3. Differentially Expressed Genes Associated with Leaf Senescence

Comparative RNA-seq analysis was conducted to identify differentially-expressed transcripts, using 30-day-old leaves from Hwaseong and CR2002 to gain a better understanding of the mechanism involved in delayed senescence. A total of 10,339 and 9885 DEGs were identified in Hwaseong and CR2002 under dark treatment compared with the control, respectively (Figure 8a). To examine the difference between Hwaseong and CR2002, Hwaseong- and CR2002-specific DEGs were extracted. In CR2002-specific DEGs, 33 GO terms were enriched, and these included iron ion transport, response to hormone stimulus, cell cycle, response to endogenous stimulus, and chloroplast. Transporter genes are closely associated with leaf senescence process. During senescence, remobilization of nutrients from senescing cells to developing tissues is mediated by transporters, and transporter genes show notable changes in expression [13,29]. Chloroplast GO term contains 15 CR2002-specific DEGs, including Os07g0462000 (*OsSG1*), Os07g0558500 (*NYC4*), Os02g0152400 (*RbcS1*) [28,30]. Os07g0462000 encodes glutamate-cysteine ligase and possibly controls chlorophyll content and stay-green phenotype [30]. Os07g0558500 (*NYC4*) is an ortholog of *THF1* in *Arabidopsis*. The *AtTHF1* is known as a multi-function protein involved in acclimation to high light, sugar sensing, and disease resistance, while *nyc4-1* mutant shows the stay-green phenotype in rice [28]. Os02g0152400 encodes the rubisco small subunit 1, which reflects involvement in the regulation of rubisco catalytic activity.

Transcription factors (TFs) play an important role in regulating leaf senescence. NAC (NAM, ATAF1, and CUC2) transcription factor family is one of the large TF families, and many NAC TFs are up-regulated during leaf senescence [31]. In addition, several NAC TFs regulating leaf senescence have been characterized in various plant species [13,32–36]. Transcript levels of NAC transcription factors (*OsNAP*, *ONAC120*, and *ONAC122*) were specifically up-regulated in Hwaseong (Table 3). *OsNAP* is a positive regulator of leaf senescence, and its orthologs *TtNAM-B1* and *AtNAP* have shown similar function in wheat and *Arabidopsis* [13,32,33]. *OsNAC122* (or *OsNAC10*, Os11g0126900) over-expression and root-specific expression transgenic have shown significantly improved drought, high salinity, and low-temperature tolerance in rice [37].

Phytohormone signaling and biosynthesis-related genes were identified in Hwaseong- and CR2002-specific DEGs. During dark treatment, ethylene (*OsERF5*), ABA (*OsABI1*), and jasmonic acid (*OsJAZ1*) signaling genes were significantly down-regulated, while auxin synthesis (*OsNIT2*), brassinosteroid biosynthesis (*OsDWF1*), and cytokinin signaling (*OsRR3*) genes were up-regulated in CR2002 (Table 3). In contrast, ethylene biosynthesis and signaling (*OsACS2* and *OsERF2*) and auxin signaling (*OsARF2*) genes were up-regulated in Hwaseong, whereas cytokinin signaling (*OsRR6*, *OsRR9*, and *OsRR10*), brassinosteroid signaling (*OsBRI1*), and *gibberellin 20-oxidase* (*OsGA20ox1*) were down-regulated (Table 3). These results indicated that delayed leaf senescence in CR2002 (*gw2-ko*) was associated with phytohormone signaling or biosynthesis pathways in rice. The rice stay-green mutant *oself3.1* has shown down-regulated transcript levels of ABA-, ethylene-, JA-associated genes in microarray analysis [7]. The results from the present study were consistent with the report that *OsELF3.1* promoted leaf senescence by modulating signaling pathways [7]. To further investigate this relationship with phytohormones, gene expression of *GW2* in response to plant hormones was examined using the RiceXPro database. *GW2* did not show large changes in response to plant hormones (data not shown) [38]. In durum wheat, a *GW2* knock-down mutant has shown enhanced *cytokinin dehydrogenase 1* (*CKX1*) transcript level and down-regulation of *cytokinin dehydrogenase 2* (*CKX2*) and *gibberellin oxidase 3* (*GA3-ox*) [39]. In addition, a wheat near-isogenic line (NIL) harboring the *TaGW2-6A* allelic variant has shown an increase in not only grain size but also endogenous cytokinin content (Z+ZR) [40]. Compared to the control (Chinese Spring), the expression level of cytokinin biosynthesis genes (*TaIPT2*, *TaIPT3*, *TaIPT5*, and *TaIPT8*) are found to be up-regulated, while cytokinin degradation genes (*TaCKX1*, *TaCKX2*, and *TaCKX6*) are found to be down-regulated in the NIL during the endosperm development [40]. These results supported the possibility that delayed senescence in CR2002 and *gw2-ko* was mediated by phytohormone-related pathways.

3.4. GW2 as an E3 Ubiquitin Ligase

The ubiquitination process is a post-translational modification. E3 ligases function in the regulation of polyubiquitination and are involved in target protein degradation. Several studies have reported interacting partners of GW2. Target proteins of GW2 have been revealed through yeast two-hybrid screening [24]. Yeast two-hybrid and pull-down assays have shown that expansin-like 1 (EXPLA1) directly interacts with *GW2*. In addition, in vitro assays have identified the ubiquitination of EXPLA1 by *GW2* at lysine 279 (K279). Proteomic analysis has shown that chitinase 14 (CHT14) and phosphoglycerate kinase (PGK) directly interact with GW2 [23]. However, genes that are directly associated with leaf senescence have not been reported. Identification of GW2 substrate protein(s) responsible for leaf senescence would be key to understanding the post-translational molecular mechanism of leaf senescence in rice.

4. Materials and Methods

4.1. Plant Materials

CR2002 was derived from a cross between *O. sativa* Korean elite line 'Hwaseong' as a recurrent parent and wild species *O. grandiglumis* (2n = 48, CCDD, Acc. No. 101154) [41]. Four *O. grandiglumis* chromosome segments were found in CR2002 on chromosomes 1, 2, 6, and 10. CR2002 was backcrossed with Hwaseong to produce an F_2 population. A total of 705 F_2 plants were genotyped using SSR markers from the introgression regions. Fifty-eight F_3 plants with different combinations of *O. grandiglumis* homozygous segments were selected and advanced to the F_3. Among the 58 F_3 lines, 17 that have different recombination break-points within the *qCC2* region were selected and advanced to the F_4 generation. QTL analysis was performed using two replications of the 58 F_3 lines and 17 F_4 lines, which were grown in the Chungnam National University paddy field during 2015 and 2016, respectively. For substitution mapping, four F_4 recombinant lines in the *qCC2* region were investigated in 2017. To determine whether *OsGW2* is associated with leaf senescence, a *gw2*-knockout (*gw2-ko*) mutant (PFG_1B-10017.R) in the *japonica* rice 'Dongjin' background was used for evaluating leaf senescence [42,43]. An F_2 population (*n* = 107) derived from a cross between Dongjin and *gw2-ko* was generated and used for DIS assay to confirm whether the T-DNA insertion in *OsGW2* locus was responsible for delayed senescence.

4.2. Phenotypic Evaluation

A total of 10 traits, including chlorophyll content and three-grain morphology traits (Table 1), were evaluated in the parental lines (*n*= 10) using the methods, as described in Yoon et al. (2006) [41].

To evaluate chlorophyll content, three methods were used. First, fluorescence ratio F_{735}/F_{700} was measured in F_3 and F_4 generation plants, and the relative chlorophyll content (mg/m^2) was evaluated using CCM-300 (OPTI-SCIENCE, Hudson, NH, USA). This data was used for QTL analysis. Second, SPAD values were measured to compare parental lines (Hwaseong, CR2002, Dongjin, and *gw2-ko*) using a SPAD-502Plus (KONICA MINOLTA, Tokyo, Japan). Third, total chlorophyll was extracted from detached leaves and used for dark-induced senescence using pre-chilled 80% acetone [7]. The absorbance of supernatants was measured at 645 and 663 nm using a UV/VIS spectrophotometer (Hanson Tech., Seoul, Korea) [44]. For the dark-induced senescence (DIS) assay, fully expanded flag leaves were detached and placed in 3 mM MES buffer (pH 5.8) and incubated at 27 °C under complete darkness [45]. *Fv/Fm* ratio was measured using a Hansatech PEA MK2 (Hansatech, Norfolk, UK). The middle part of leaf samples from DIS was adapted in the dark for 30 min to complete the oxidation of Q_A, and then the *Fv/Fm* value was measured. A student's *t*-test was performed for trait comparisons.

4.3. DNA Extraction and QTL Analysis

DNA was extracted from leaf tissues using the extraction method, as described in Causse et al. (1994) [46]. Genotyping was conducted in 58 F_3 and 17 F_4 lines using 11 polymorphic

SSR markers between CR2002 and Hwaseong. PCR was conducted, as described in Shim et al. (2019) [47]. PCR products were separated on 3% metaphor agar stained with StaySafe Nucleic Acid Gel Stain (RBC, Taiwan) or 4% polyacrylamide denaturing gel stained with Silver Staining Kit (Bioneer, Korea), respectively.

QTLs were determined by single-marker analysis (SMA). SMA was conducted to establish the effect of each marker on each trait. In SMA, a QTL was declared when the phenotype was associated with SSR genotype at $p < 0.05$ by one-way analysis of variance (ANOVA). The observed phenotypic variation was estimated using the coefficient of determination (R^2). For Tukey's test, Minitab19 software was used. Student's *t*-test was conducted using Microsoft Excel. Boxplots were drawn using R version 3.5.3.

4.4. RNA Isolation and Quantitative Real-Time PCR

Total RNA of detached leaves from DIS was isolated using NucleoSpin RNA (Macherey Nagel, Deuren, Germany), according to the manufacturer's instructions. Following reverse-transcription into the first-strand cDNA with SmartGene Mixed cDNA synthesis kit (SJ Bioscience, Daejeon, Korea), real-time PCR was performed using a CFX Connect Real-Time System (Bio-Rad, Hercules, CA, USA). Each reaction contained 10 µL of 2× SYBR Green Master Mix reagent, 2 µL of diluted cDNA samples, 1 µL of 10 pmol gene-specific primers in a final volume of 20 µL. The real-time PCR protocol was: 15 min at 95 °C to denature and activate an enzyme, followed by 40 cycles at 95 °C for 20 s (denaturation), 60 °C for 40 s (annealing), 72 °C for 30 s (extension). qRT-PCR was conducted according to the $2^{-\Delta\Delta Ct}$ method [48]. Rice *UBIQUITIN5* (*OsUBQ5*) was used for normalization, and relative expression levels were compared by Student's *t*-test. Primers used in this study are listed in Table S1.

4.5. RNA-Seq Analysis

The 30-day-old Hwaseong and CR2002 plants were dark-treated to induce senescence for three days. From plants each before (day 0) and 3 days after dark-treatment (day 3), the 5th leaves were collected with three replicates, and total RNA was extracted, as described above. A total of 12 transcriptome libraries were constructed using TruSeq RNA Sample Prep Kit v2 (Illumina, San Diego, CA, USA), according to the manufacturer's instructions. The library construction was conducted by GnC Bio Inc. (Daejeon, Korea). The constructed libraries were sequenced on a HiSeq 4000 instrument with 150 bp of paired-end reads. The sequence data generated from this study have been deposited at the National Center for Biotechnology Information Sequence Read Archive (SRA) under accession number PRJNA602373. Sequencing results were checked using FastQC (version 0.11.7). Raw sequences were processed to obtain clean reads by removing adapters and low-quality reads, using SCYTHE (https://github.com/vsbuffalo/scythe) and SICKLE (https://github.com/najoshi/sickle), respectively. High-quality reads were mapped on rice reference genome IRGSP-1.0, and transcript abundances were quantified using Salmon version 0.8.2 [49]. Differentially expressed genes (DEGs) were determined with more than 2-fold change with significance at Benjamini-Hochberg false discovery rate adjusted $p < 0.05$ using Limma-Voom package [50,51]. GO enrichment was carried out using agriGO v2.0 singular enrichment analysis [52].

5. Conclusions

A mapping population derived from an interspecific cross between *O. sativa* and *O. grandiglumis* facilitated the discovery of *qCC2*, a major QTL for chlorophyll content. The *O. grandiglumis* allele at *qCC2* increased chlorophyll content and delayed senescence. Building on several previous studies, *GW2*, which encodes E3 ubiquitin ligase, was selected as the most probable candidate gene for this phenotype, and this was confirmed through genetic analysis of a knockout mutant. The *gw2-ko* mutant showed delayed leaf senescence under dark conditions with down-regulation of SAGs and CDGs, and the delayed senescence trait completely segregated with the *gw2-ko* gene. On the basis of this work, the RING-type ubiquitin ligase *GW2* appeared to function as a positive regulator of leaf senescence in rice. This study also represented the first investigation of global gene expression in rice leaves subjected to complete darkness. The results of this whole-genome transcriptome analysis of Hwaseong and CR2002 highlighted the processes and genes that are likely to play a role in the mechanisms underlying leaf senescence in rice (e.g., carbohydrate metabolism, pyrophosphate-dependent energy conservation, and ethylene signaling). The QTL analysis and the genotype-specific regulation suggested that senescence was regulated at the transcriptional level, although the possibility of post-translational regulation by GW2 requires further investigation.

Supplementary Materials: Supplementary materials can be found at http://www.mdpi.com/1422-0067/21/5/1704/s1.

Author Contributions: K.-C.S. and S.H.K. performed experiments and analyzed data. K.-C.S. and S.-N.A. conceived the study, designed, and supervised the study. Y.-A.J., H.-S.L., J.-W.K., and C.A. investigated the traits. H.-J.K. and T.H.T. assisted in analyzing the RNA-seq data. K.-C.S., T.H.T., and S.-N.A. wrote and edited the manuscript. All authors have read and agreed to the published version of the manuscript.

Funding: This work was carried out with the support of "Cooperative Research Program for Agriculture Science and Technology Development (Project No. PJ01321401)", Rural Development Administration and the Basic Science Research Program through the National Research Foundation (NRF) of Korea funded by the Ministry of Education (NRF-2017R1A2B2007554).

Acknowledgments: The *gw2*-knockout (*gw2-ko*) mutant seeds were provided by G. An, Kyung Hee University. We thank Cheol Seong Jang and Ju Hee Kim, Kangwon National University, for their critical comments on this work and Young-Sook Kim for her technical assistance.

Conflicts of Interest: The authors declare no conflict of interest. The funders had no role in the design of the study; in the collection, analyses, or interpretation of data; in the writing of the manuscript, or in the decision to publish the results.

Abbreviations

DIS	Dark-induced senescence
SNP	Single nucleotide polymorphism
InDel	Insertion and deletion
qRT-PCR	Quantitative real-time PCR
SSR	Simple sequence repeat
CDG	Chlorophyll degradation gens
QTL	Quantitative trait loci
SAG	Senescence-associated gene
WT	Wild type

References

1. Wang, F.H.; Wang, G.X.; Li, X.Y.; Huang, J.L.; Zheng, J.K. Heredity, physiology and mapping of a chlorophyll content gene of rice (*Oryza sativa* L.). *J. Plant Physiol.* **2008**, *165*, 324–330. [CrossRef] [PubMed]

2. Jiang, G.H.; He, Y.Q.; Xu, C.G.; Li, X.H.; Zhang, Q. The genetic basis of stay-green in rice analyzed in a population of doubled haploid lines derived from an *indica* by *japonica* cross. *Theor. Appl. Genet.* **2004**, *108*, 688–698. [CrossRef] [PubMed]

3. Teng, S.; Qian, Q.; Zeng, D.L.; Kunihiro, Y.; Fujimoto, K.; Huang, D.N.; Zhu, L.H. QTL analysis of leaf photosynthetic rate and related physiological traits in rice (*Oryza sativa* L.). *Euphytica* **2004**, *135*, 1–7. [CrossRef]

4. Takai, T.; Kondo, M.; Yano, M.; Yamamoto, T. A Quantitative Trait Locus for Chlorophyll Content and its Association with Leaf Photosynthesis in Rice. *Rice* **2010**, *3*, 172–180. [CrossRef]

5. Jiang, S.K.; Zhang, X.J.; Zhang, F.M.; Xu, Z.J.; Chen, W.F.; Li, Y.H. Identification and Fine Mapping of *qCTH4*, a Quantitative Trait Loci Controlling the Chlorophyll Content from Tillering to Heading in Rice (*Oryza sativa* L.). *J. Hered.* **2012**, *103*, 720–726. [CrossRef]

6. Abdelkhalik, A.F.; Shishido, R.; Nomura, K.; Ikehashi, H. QTL-based analysis of leaf senescence in an *indica/japonica* hybrid in rice (*Oryza sativa* L.) *Theor. Appl. Genet.* **2005**, *110*, 1226–1235. [CrossRef]

7. Sakuraba, Y.; Han, S.H.; Yang, H.J.; Piao, W.L.; Paek, N.C. Mutation of *Rice Early Flowering3.1* (*OsELF3.1*) delays leaf senescence in rice. *Plant Mol. Biol.* **2016**, *92*, 223–234. [CrossRef]

8. Hortensteiner, S. Stay-green regulates chlorophyll and chlorophyll-binding protein degradation during senescence. *Trends Plant Sci.* **2009**, *14*, 155–162. [CrossRef]

9. Thomas, H.; Smart, C.M. Crops That Stay Green. *Ann. Appl. Biol.* **1993**, *123*, 193–219. [CrossRef]

10. Park, S.Y.; Yu, J.W.; Park, J.S.; Li, J.; Yoo, S.C.; Lee, N.Y.; Lee, S.K.; Jeong, S.W.; Seo, H.S.; Koh, H.J.; et al. The senescence-induced staygreen protein regulates chlorophyll degradation. *Plant Cell* **2007**, *19*, 1649–1664. [CrossRef]

11. Kusaba, M.; Ito, H.; Morita, R.; Iida, S.; Sato, Y.; Fujimoto, M.; Kawasaki, S.; Tanaka, R.; Hirochika, H.; Nishimura, M.; et al. Rice NON-YELLOW COLORING1 is involved in light-harvesting complex II and grana degradation during leaf senescence. *Plant Cell* **2007**, *19*, 1362–1375. [CrossRef] [PubMed]

12. Sato, Y.; Morita, R.; Katsuma, S.; Nishimura, M.; Tanaka, A.; Kusaba, M. Two short-chain dehydrogenase/reductases, NON-YELLOW COLORING 1 and NYC1-LIKE, are required for chlorophyll *b* and light-harvesting complex II degradation during senescence in rice. *Plant J.* **2009**, *57*, 120–131. [CrossRef] [PubMed]

13. Liang, C.Z.; Wang, Y.Q.; Zhu, Y.N.; Tang, J.Y.; Hu, B.; Liu, L.C.; Ou, S.J.; Wu, H.K.; Sun, X.H.; Chu, J.F.; et al. OsNAP connects abscisic acid and leaf senescence by fine-tuning abscisic acid biosynthesis and directly targeting senescence-associated genes in rice. *Proc. Natl. Acad. Sci. USA* **2014**, *111*, 10013–10018. [CrossRef]

14. Jiang, H.W.; Li, M.R.; Liang, N.B.; Yan, H.B.; Wei, Y.L.; Xu, X.; Liu, J.F.; Xu, Z.; Chen, F.; Wu, G.J. Molecular cloning and function analysis of the stay green gene in rice. *Plant J.* **2007**, *52*, 197–209. [CrossRef] [PubMed]

15. Ardley, H.C.; Robinson, P.A. E3 ubiquitin ligases. *Essays Biochem.* **2005**, *41*, 15–30. [CrossRef]

16. Li, N.; Li, Y.H. Signaling pathways of seed size control in plants. *Curr. Opin. Plant Biol.* **2016**, *33*, 23–32. [CrossRef]

17. Li, Y.H.; Zheng, L.Y.; Corke, F.; Smith, C.; Bevan, M.W. Control of final seed and organ size by the *DA1* gene family in *Arabidopsis thaliana*. *Genes Dev.* **2008**, *22*, 1331–1336. [CrossRef]

18. Xia, T.; Li, N.; Dumenil, J.; Li, J.; Kamenski, A.; Bevan, M.W.; Gao, F.; Li, Y.H. The Ubiquitin Receptor DA1 Interacts with the E3 Ubiquitin Ligase DA2 to Regulate Seed and Organ Size in *Arabidopsis*. *Plant Cell* **2013**, *25*, 3347–3359. [CrossRef]

19. Vanhaeren, H.; Nam, Y.J.; De Milde, L.; Chae, E.; Storme, V.; Weigel, D.; Gonzalez, N.; Inze, D. Forever Young: The Role of Ubiquitin Receptor DA1 and E3 Ligase BIG BROTHER in Controlling Leaf Growth and Development. *Plant Physiol.* **2017**, *173*, 1269–1282. [CrossRef]

20. Disch, S.; Anastasiou, E.; Sharma, V.K.; Laux, T.; Fletcher, J.C.; Lenhard, M. The E3 ubiquitin ligase BIG BROTHER controls *Arabidopsis* organ size in a dosage-dependent manner. *Curr. Biol.* **2006**, *16*, 272–279. [CrossRef]

21. Song, X.J.; Huang, W.; Shi, M.; Zhu, M.Z.; Lin, H.X. A QTL for rice grain width and weight encodes a previously unknown RING-type E3 ubiquitin ligase. *Nat. Genet.* **2007**, *39*, 623–630. [CrossRef]

22. Song, W.Y.; Lee, H.S.; Jin, S.R.; Ko, D.; Martinoia, E.; Lee, Y.; An, G.; Ahn, S.N. Rice PCR1 influences grain weight and Zn accumulation in grains. *Plant Cell Environ.* **2015**, *38*, 2327–2339. [CrossRef]

23. Lee, K.H.; Park, S.W.; Kim, Y.J.; Koo, Y.J.; Song, J.T.; Seo, H.S. Grain width 2 (GW2) and its interacting proteins regulate seed development in rice (*Oryza sativa* L.). *Bot. Stud.* **2018**, *59*, 1–7. [CrossRef]

24. Choi, B.S.; Kim, Y.J.; Markkandan, K.; Koo, Y.J.; Song, J.T.; Seo, H.S. GW2 Functions as an E3 Ubiquitin Ligase for Rice Expansin-Like 1. *Int. J. Mol. Sci.* **2018**, *19*. [CrossRef]

25. An, S.Y.; Park, S.; Jeong, D.H.; Lee, D.Y.; Kang, H.G.; Yu, J.H.; Hur, J.; Kim, S.R.; Kim, Y.H.; Lee, M.; et al. Generation and analysis of end sequence database for T-DNA tagging lines in rice. *Plant Physiol.* **2003**, *133*, 2040–2047. [CrossRef]

26. Raab, S.; Drechsel, G.; Zarepour, M.; Hartung, W.; Koshiba, T.; Bittner, F.; Hoth, S. Identification of a novel E3 ubiquitin ligase that is required for suppression of premature senescence in Arabidopsis. *Plant J.* **2009**, *59*, 39–51. [CrossRef]

27. Zeng, L.R.; Qu, S.H.; Bordeos, A.; Yang, C.W.; Baraoidan, M.; Yan, H.Y.; Xie, Q.; Nahm, B.H.; Leung, H.; Wang, G.L. *Spotted leaf11*, a negative regulator of plant cell death and defense, encodes a U-box/armadillo repeat protein endowed with E3 ubiquitin ligase activity. *Plant Cell* **2004**, *16*, 2795–2808. [CrossRef]

28. Yamatani, H.; Sato, Y.; Masuda, Y.; Kato, Y.; Morita, R.; Fukunaga, K.; Nagamura, Y.; Nishimura, M.; Sakamoto, W.; Tanaka, A.; et al. *NYC4*, the rice ortholog of Arabidopsis *THF1*, is involved in the degradation of chlorophyll protein complexes during leaf senescence. *Plant J.* **2013**, *74*, 652–662. [CrossRef]

29. Lee, S.; Jeong, H.; Lee, S.; Lee, J.; Kim, S.J.; Park, J.W.; Woo, H.R.; Lim, P.O.; An, G.; Nam, H.G.; et al. Molecular bases for differential aging programs between flag and second leaves during grain-filling in rice. *Sci. Rep.* **2017**, *7*, 1–16. [CrossRef]

30. Zhao, Y.; Qiang, C.G.; Wang, X.Q.; Chen, Y.F.; Deng, J.Q.; Jiang, C.H.; Sun, X.M.; Chen, H.Y.; Li, J.; Piao, W.L.; et al. New alleles for chlorophyll content and stay-green traits revealed by a genome wide association study in rice (*Oryza sativa*). *Sci. Rep.* **2019**, *9*, 1–11. [CrossRef]

31. Gregersen, P.L.; Holm, P.B. Transcriptome analysis of senescence in the flag leaf of wheat (*Triticum aestivum* L.). *Plant Biotechnol. J.* **2007**, *5*, 192–206. [CrossRef]

32. Uauy, C.; Distelfeld, A.; Fahima, T.; Blechl, A.; Dubcovsky, J. A NAC gene regulating senescence improves grain protein, zinc, and iron content in wheat. *Science* **2006**, *314*, 1298–1301. [CrossRef]

33. Guo, Y.F.; Gan, S.S. AtNAP, a NAC family transcription factor, has an important role in leaf senescence. *Plant J.* **2006**, *46*, 601–612. [CrossRef]

34. Zhou, Y.; Huang, W.F.; Liu, L.; Chen, T.Y.; Zhou, F.; Lin, Y.J. Identification and functional characterization of a rice NAC gene involved in the regulation of leaf senescence. *BMC Plant Biol.* **2013**, *13*. [CrossRef]

35. Mao, C.J.; Lu, S.C.; Lv, B.; Zhang, B.; Shen, J.B.; He, J.M.; Luo, L.Q.; Xi, D.D.; Chen, X.; Ming, F. A Rice NAC Transcription Factor Promotes Leaf Senescence via ABA Biosynthesis. *Plant Physiol.* **2017**, *174*, 1747–1763. [CrossRef]

36. Sakuraba, Y.; Piao, W.; Lim, J.H.; Han, S.H.; Kim, Y.S.; An, G.; Paek, N.C. Rice ONAC106 Inhibits Leaf Senescence and Increases Salt Tolerance and Tiller Angle. *Plant Cell Physiol.* **2015**, *56*, 2325–2339. [CrossRef]

37. Jeong, J.S.; Kim, Y.S.; Baek, K.H.; Jung, H.; Ha, S.H.; Do Choi, Y.; Kim, M.; Reuzeau, C.; Kim, J.K. Root-Specific Expression of *OsNAC10* Improves Drought Tolerance and Grain Yield in Rice under Field Drought Conditions. *Plant Physiol.* **2010**, *153*, 185–197. [CrossRef]

38. Sato, Y.; Takehisa, H.; Kamatsuki, K.; Minami, H.; Namiki, N.; Ikawa, H.; Ohyanagi, H.; Sugimoto, K.; Antonio, B.A.; Nagamura, Y. RiceXPro Version 3.0: Expanding the informatics resource for rice transcriptome. *Nucleic Acids Res.* **2013**, *41*, D1206–D1213. [CrossRef]

39. Sestili, F.; Pagliarello, R.; Zega, A.; Saletti, R.; Pucci, A.; Botticella, E.; Masci, S.; Tundo, S.; Moscetti, I.; Foti, S.; et al. Enhancing grain size in durum wheat using RNAi to knockdown *GW2* genes. *Theor. Appl. Genet.* **2019**, *132*, 419–429. [CrossRef]

40. Geng, J.; Li, L.Q.; Lv, Q.; Zhao, Y.; Liu, Y.; Zhang, L.; Li, X.J. *TaGW2-6A* allelic variation contributes to grain size possibly by regulating the expression of cytokinins and starch-related genes in wheat. *Planta* **2017**, *246*, 1153–1163. [CrossRef]

41. Yoon, D.B.; Kang, K.H.; Kim, H.J.; Ju, H.G.; Kwon, S.J.; Suh, J.P.; Jeong, O.Y.; Ahn, S.N. Mapping quantitative trait loci for yield components and morphological traits in an advanced backcross population between *Oryza grandiglumis* and the *O. sativa japonica* cultivar Hwaseongbyeo. *Theor. Appl. Genet.* **2006**, *112*, 1052–1062. [CrossRef]

42. Jeon, J.S.; Lee, S.; Jung, K.H.; Jun, S.H.; Jeong, D.H.; Lee, J.; Kim, C.; Jang, S.; Lee, S.; Yang, K.; et al. T-DNA insertional mutagenesis for functional genomics in rice. *Plant J.* **2000**, *22*, 561–570. [CrossRef]

43. Jeong, D.H.; An, S.; Park, S.; Kang, H.G.; Park, G.G.; Kim, S.R.; Sim, J.; Kim, Y.O.; Kim, M.K.; Kim, S.R.; et al. Generation of a flanking sequence-tag database for activation-tagging lines in japonica rice. *Plant J.* **2006**, *45*, 123–132. [CrossRef]

44. Wu, G.; Lee, C.-H. Chlorophyll a/b Ratio as a Criterion for the Reliability of Absorbance Values Measured for the Determination of Chlorophyll Concentration. *J. Life Sci.* **2019**, *29*, 509–513.

45. Oh, M.H.; Kim, Y.J.; Lee, C.W. Leaf senescence in a stay-green mutant of *Arabidopsis thaliana*: Disassembly process of photosystem I and II during dark-incubation. *J. Biochem. Mol. Biol.* **2000**, *33*, 256–262.

46. Causse, M.A.; Fulton, T.M.; Cho, Y.G.; Ahn, S.N.; Chunwongse, J.; Wu, K.S.; Xiao, J.H.; Yu, Z.H.; Ronald, P.C.; Harrington, S.E.; et al. Saturated Molecular Map of the Rice Genome Based on an Interspecific Backcross Population. *Genetics* **1994**, *138*, 1251–1274.

47. Shim, K.C.; Kim, S.; Le, A.Q.; Lee, H.S.; Adeva, C.; Jeon, Y.A.; Luong, N.H.; Kim, W.J.; Akhtamov, M.; Ahn, S.N. Fine Mapping of a Low-Temperature Germinability QTL *qLTG1* Using Introgression Lines Derived from *Oryza rufipogon*. *Plant Breed. Biotechnol.* **2019**, *7*, 141–150. [CrossRef]

48. Livak, K.J.; Schmittgen, T.D. Analysis of relative gene expression data using real-time quantitative PCR and the 2(T)(-Delta Delta C) method. *Methods* **2001**, *25*, 402–408. [CrossRef]

49. Patro, R.; Duggal, G.; Love, M.I.; Irizarry, R.A.; Kingsford, C. Salmon provides fast and bias-aware quantification of transcript expression. *Nat. Methods* **2017**, *14*, 417. [CrossRef]
50. Law, C.W.; Chen, Y.S.; Shi, W.; Smyth, G.K. voom: Precision weights unlock linear model analysis tools for RNA-seq read counts. *Genome Biol.* **2014**, *15*, R29. [CrossRef]
51. Ritchie, M.E.; Phipson, B.; Wu, D.; Hu, Y.F.; Law, C.W.; Shi, W.; Smyth, G.K. limma powers differential expression analyses for RNA-sequencing and microarray studies. *Nucleic Acids Res.* **2015**, *43*, e47. [CrossRef]
52. Tian, T.; Liu, Y.; Yan, H.Y.; You, Q.; Yi, X.; Du, Z.; Xu, W.Y.; Su, Z. agriGO v2.0: A GO analysis toolkit for the agricultural community, 2017 update. *Nucleic Acids Res.* **2017**, *45*, W122–W129. [CrossRef]

International Journal of
Molecular Sciences

Article

OsCAF1, a CRM Domain Containing Protein, Influences Chloroplast Development

Qiang Zhang [1,†], Lan Shen [1,†], Zhongwei Wang [2], Guanglian Hu [1], Deyong Ren [1], Jiang Hu [1], Li Zhu [1], Zhenyu Gao [1], Guangheng Zhang [1], Longbiao Guo [1], Dali Zeng [1] and Qian Qian [1,*]

[1]   State Key Laboratory of Rice Biology/China National Rice Research Institute, Chinese Academy of Agricultural Sciences, Hangzhou 310006, China
[2]   Biotechnology Research Center, Chongqing Academy of Agricultural Sciences, Chongqing 401329, China
*   Correspondence: qianqian188@hotmail.com; Tel.: +86-571-63371418
†   These authors contributed equally to this study.

Received: 6 July 2019; Accepted: 3 September 2019; Published: 6 September 2019

Abstract: The chloroplast RNA splicing and ribosome maturation (CRM) domain proteins are involved in the splicing of chloroplast gene introns. Numerous CRM domain proteins have been reported to play key roles in chloroplast development in several plant species. However, the functions of CRM domain proteins in chloroplast development in rice remain poorly understood. In the study, we generated *oscaf1* albino mutants, which eventually died at the seedling stage, through the editing of *OsCAF1* with two CRM domains using CRISPR/Cas9 technology. The mesophyll cells in *oscaf1* mutant had decreased chloroplast numbers and damaged chloroplast structures. OsCAF1 was located in the chloroplast, and transcripts revealed high levels in green tissues. In addition, the OsCAF1 promoted the splicing of group IIA and group IIB introns, unlike orthologous proteins of AtCAF1 and ZmCAF1, which only affected the splicing of subgroup IIB introns. We also observed that the C-terminal of OsCAF1 interacts with OsCRS2, and OsCAF1–OsCRS2 complex may participate in the splicing of group IIA and group IIB introns in rice chloroplasts. OsCAF1 regulates chloroplast development by influencing the splicing of group II introns.

Keywords: chloroplast RNA splicing and ribosome maturation (CRM) domain; intron splicing; chloroplast development; rice

1. Introduction

Chloroplasts are important organelles in plants. A series of metabolic processes occur in chloroplasts, including photosynthesis and anabolism of compounds such as tetrapyrroles, terpenoids, lipids, amino acids, and hormones [1]. Chloroplasts are considered semi-autonomous organelles because their development is not only influenced by their own genetic material but also by fine regulation by nuclear-encoded genes [2,3]. Studies have shown that by plastid-encoded polymerases (PEPs) and nucleus-encoded polymerases (NEPs) influence the development of chloroplasts [4–6]. Chloroplast genes encode approximately 100 proteins, some of which have one or more introns that cannot be self-spliced, and the primary RNA transcription of such chloroplast genes require splicing by ribozymes potentially via chemical steps similar to spliceosome-mediated splicing in the nucleus [7–9]. In plants, based on the primary sequences, predicted structures, and splicing mechanisms, the introns of the chloroplast are mainly classified into two categories, including group I and group II [10]. Group II introns are mainly divided further into two subgroups, including subgroup IIA introns and subgroup IIB introns [11]. *Arabidopsis thaliana*, maize, and rice chloroplast genomes all have only 1 group I intron, and 20, 17, and 17 group II introns, respectively [8]. In plant chloroplasts, such introns have lost the capacity to self-splice in vivo and require nuclear gene-encoded proteins as co-factors to participate in splicing [8,12].

Numerous studies have shown that nuclear-encoded pentatricopeptide repeat (PPR) proteins participate in chloroplast RNA editing and splicing, which are critical for chloroplast development and function [8,13]. Currently, the large PPR protein families, including AtOTP51, ZmPPR5, OsPPR6, and OsPPR1, are considered to participate in different chloroplast group II intron splicing activities [14–18]. Previous studies have shown that the disruption of the normal functions of such PPR proteins could lead to abnormal chloroplast development, which would lead to albino seedlings or death of plants [14–18]. In addition to the PPR proteins mentioned above, chloroplast RNA splicing and ribosome maturation (CRM) domain proteins participate in the splicing of chloroplast gene introns [8].

In plants, the splicing of chloroplast introns requires splicing factors encoded by nuclear genes and the abnormal splicing of chloroplast intron will affect the development of chloroplast [14–16]. Several proteins with CRM domains have been identified as splicing co-factors, including CFM2, CFM3, CRS1, CAF1, and CAF2 [8]. Both ZmCFM2 and AtCFM2 have four CRM domains, and they participate in the splicing of *ndhA* and *ycf3-1* subgroup IIB and group I introns. In *A. thaliana*, AtCFM2 potentially promotes the splicing of *clpP* introns [19]. CFM3 and CRS1 contain three CRM domains, and AtCRS1 and ZmCRS1 have been associated with the splicing of the *atpF* intron, which belongs to subgroup IIA introns [20,21]. In addition, CFM3 has been reported to be dual-localized in chloroplasts and mitochondria, and CFM3a is required for the splicing of group II introns, including *ndhB*, *rpl16*, *rps16*, *petD*, and *petB* introns in chloroplasts [19]. The CAF1 and CAF2 contain two CRM domains, which are required for the splicing of group IIB introns, including *ndhA*, *ndhB*, *petB*, *petD*, *rpl16*, *rps16*, *trnG*, and *ycf3-1* in maize and *A. thaliana* [11,21]. In *A. thaliana*, AtCAF2 potentially promotes the splicing of *rpoC1* and *ClpP* introns that are absent in the maize chloroplast genome [21]. In addition, CAF1 and CAF2 could interact with CRS2, forming the CRS2–CAF1 and CRS2–CAF2 complexes, which participate in the splicing of chloroplast group II introns and the regulation of chloroplast development in maize [22].

Previous studies have shown that there are 14 proteins that contain one or more CRM domains in rice [8]. To date, researchers have only studied the functions of OsCRS1 and OsCFM3, which contain three CRM domains. The studies revealed that *oscrs1* mutants exhibited albino leaf phenotypes in rice [23]. In addition, OsCRS1 has been reported to participate in the splicing of *ndhA*, *ndhB*, *petD*, *ycf3-1*, and *trnL* introns and it could influence the expression of PEP-dependent genes such as *psaA* and *psbA* [23]. Furthermore, *oscfm3* T-DNA insertion mutants exhibited albino seedling phenotypes; however, the OsCFM3 participates in the intron splicing of *ndhB*, *petD*, *rps16*, and *rpl16* in rice [19]. The functions of other CRM domain proteins have hardly been reported in rice. Previous studies have shown that homologous *CAF1* plays important roles in chloroplast development in *A. thaliana* and *Zea mays* L. In rice, the function of *OsCAF1* remains unknown and investigating its function would enhance our understanding of mechanisms of chloroplast development.

In this study, we investigated the function of OsCAF1 in rice chloroplast development, and *oscaf1* mutants were obtained using the CRISPR-Cas9 system. The results of our study revealed that *oscaf1* mutants exhibited albino seedlings phenotype along with a damaged chloroplast structure. In rice, chloroplast development is regulated by the expression of PEP-dependent genes, which could be affected by *OsCAF1*. In addition, we observed that the leaves of *oscaf1* mutants accumulated higher hydrogen peroxide (H_2O_2) contents. In addition, OsCAF1 could interact with the OsCRS2 via C-terminal, and then form an OsCAF1–OsCRS2 complex that regulates the splicing of chloroplast gene introns. Chloroplast RNA splicing analysis showed that OsCAF1 influenced the splicing of both chloroplast subgroup IIA and subgroup IIB introns, which is different from the orthologous proteins of AtCAF1 and ZmCAF1, which both influence the splicing of chloroplast subgroup IIB introns. The results of the present study reveal the functions of OsCAF1 in the regulation of rice chloroplast development.

2. Results

2.1. Knockout of OsCAF1 Produced the Albino Seedling Phenotype

According to a previous study, the ortholog of *ZmCAF1* in rice was *Os01g0495900*, which was named *OsCAF1*, and the function of OsCAF1 has not been determined to date [8]. We observed that the cDNA of *OsCAF1* had a 2106-bp nucleotide and encoded 701 amino acids (aa). To determine the function of OsCAF1 in rice chloroplast development, *oscaf1* mutants were generated using the CRISPR/Cas9 system (Figure 1A,B), and 26 transgenic plants were obtained in the T_0 generation. Sequencing and phenotypic analysis revealed that *oscaf1* homozygous mutants exhibited the albino phenotype in the rice seeding stage (Figure 1C,D), and then died after the three leaves stage. In both of #3 and #4 mutants, a premature stop codon was created by a frame shift in the coding region of *OsCAF1*. However, the *OsCAF1/oscaf1* heterozygous exhibited green leaves (Figure 1C,D), which exhibited normal growth and phenotypes under chamber conditions. Because of *oscaf1/oscaf1* homozygous lethal mutants, the heterozygous lines were used to generate the T_1 generation. The different homozygous mutations in the allele's of *OsCAF1* in T_1 generation were obtained from #1 and #2 plants and also exhibited albino seedling phenotypes. These results indicated that *OsCAF1* mutations caused the albino seedling phenotype. To study the function of OsCAF1, *oscaf1* mutants from T_1 generation of #1 plant were selected for further analysis.

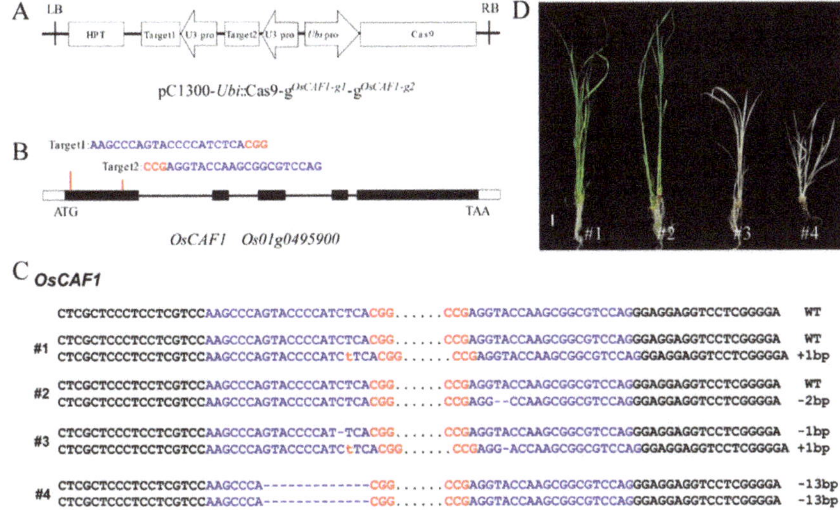

Figure 1. Knockout of *OsCAF1* produces the albino phenotype in seedlings. (**A**) Diagram of CRISPR/ Cas9 system for editing OsCAF1. (**B**) Schematic diagram of the targeted site in OsCAF1. (**C**) Mutation types of four positive plants in T_0 generation. The targeted sequences are in blue letters. The protospacer adjacent motif (PAM) sequences are in red letters. The number of nucleotides at both sides of the target sequences is 18 bp. (**D**) Phenotypes of four positive plants at seedling stage. Scale bar, 1 cm.

2.2. The oscaf1 Mutant Exhibited Defects in Chloroplast Development

The *oscaf1* mutant exhibited an albino seedling phenotype and eventually died after the three leaves stage (Figure 2A). Compared with the wild type (WT), photosynthetic pigment concentrations showed that chlorophyll a (Chla), chlorophyll b (Chlb), and carotenoid (Car) contents decreased significantly in *oscaf1* mutants (Figure 2B). According to the results of chlorophyll fluorescence measurements, the Fv/Tm values of *oscaf1* mutants were low (Figure 2C). The results suggested that *oscaf1* mutants had impaired chloroplast development and photosynthesis. Subsequently, we analyzed the expression levels of chloroplast development and photosynthesis-related genes in WT and *oscaf1* mutants using qRT-PCR. The results showed that compared with the WT, the transcript levels of *HZMA1*, *PORA*, *CAB1*, *psaA*, *rbcL*, *CHLI*, *CHLH*, *ATPa*, *psbA*, *TaX2*, *ATPB*, and *ATPE* in *oscaf1* mutants decreased, while the relative expression levels of *YGLI*, *HSA1*, *HSA2*, *RPOB*, *RPOC1*, and *RPOC2* increased (Figure 2D). Particularly, in *oscaf1* mutants, the expression levels of *HZMA1*, *PORA*, *CAB1*, *psaA*, *rbcL*, *CHLH*, and *psbA* decreased significantly, and the expression levels of *YGLI*, *HSA1*, *RPOB*, *RPOC1*, and *RPOC2* increased markedly in *oscaf1* mutants (Figure 2D). Overall, the results suggested that OsCAF1 influences chloroplast development and photosynthesis in rice.

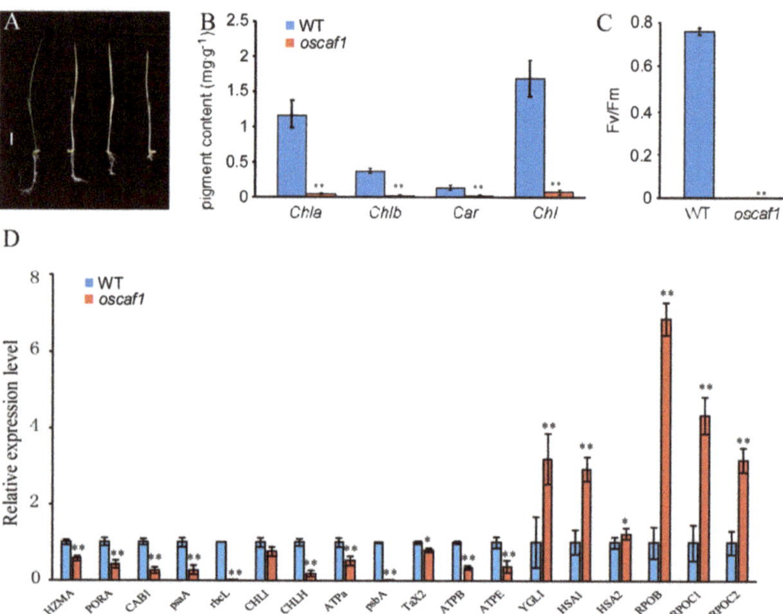

Figure 2. *oscaf1* showed defects in chloroplast development and photosynthesis. (**A**) Phenotype observations of WT (left) and *oscaf1* mutant (right). Scale bar, 1 cm. (**B**) Pigment content of WT and *oscaf1* at seedling stage. (**C**) Photochemical efficiency of PSII in the light (Fv/Fm) of WT and *oscaf1* mutants at seedling stage. (**D**) Relative expression levels of chloroplast development and photosynthesis genes in WT and *oscaf1* mutants. The data represent mean ± SE from three independent biological duplicate, * and ** indicated $p < 0.05$ and $p < 0.01$, respectively.

To investigate chloroplast development in *oscaf1* mutants further, we observed the ultrastructure of chloroplasts at the three leaves stages of WT and *oscaf1* mutants using transmission electron microscopy (TEM). We observed that the chloroplast number in the mesophyll cells of the *oscaf1* mutant was less than in WT (Figure 3B). In addition, the chloroplasts in WT mesophyll cells were well developed, with normal-looking and distinctly stacked grana and thylakoids (Figure 3A,C). Conversely, the *oscaf1* mutant had abnormal chloroplast architecture along with abnormally structured thylakoid and abnormally stacked grana (Figure 3B,D). The results suggested that OsCAF1 plays a key role in early chloroplast development in rice.

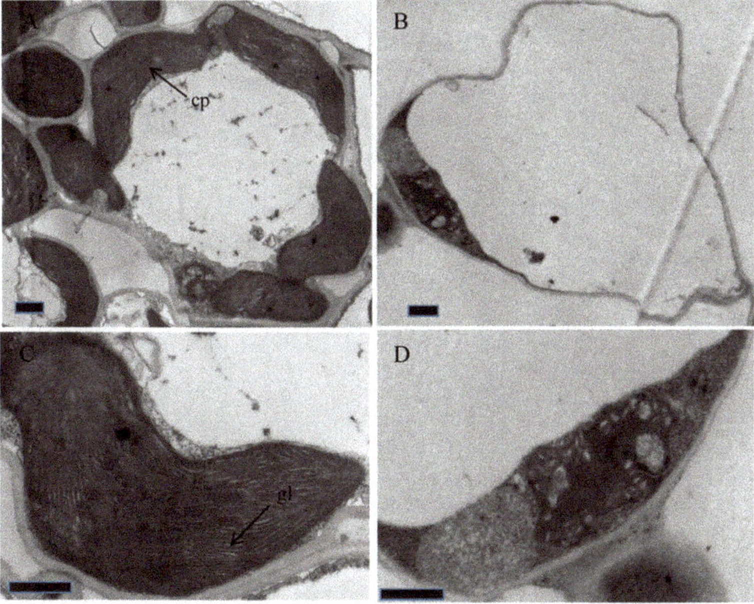

Figure 3. Chloroplast ultrastructure observation of WT (**A**,**C**) and *oscaf1* mutants (**B**,**D**) mesophyll cell at the three leaves stage. cp, chloroplast, gl, grana lamella. Bars = 2.0 μm.

2.3. Increased Reactive Oxygen Species (ROS) Contents and Reduced ROS-scavenging Gene Expression Levels in oscaf1 Mutant

In plants, the overproduction of ROS might cause cell death in leaves [24]. The *oscaf1* mutants died after the three leaves stage; therefore, we tested the ROS levels in leaves of the WTs and *oscaf1* mutants. The *oscaf1* mutant leaves displayed a more intense brown color following 3,3-diaminobenzidine (DAB) staining, indicating that the levels of H_2O_2 in *oscaf1* mutant leaves were higher than in WT leaves (Figure 4A). In addition, following nitroblue tetrazolium (NBT) staining, *oscaf1* mutant leaves had larger blue leaf areas than WT leaves. The results suggested that oscaf1 mutant leaves generate more O^{2-} than WT (Figure 4B). The results of the experiments indicated that oscaf1 leaves accumulated high ROS levels. Moreover, we measured the H_2O_2 contents in WT and *oscaf1* mutant leaves. The results indicated that the H_2O_2 content in the mutant was significantly increased compared to that in WT (Figure 4C). This result was consistent with DAB staining. Meanwhile, the ROS-scavenging gene expression levels were investigated in WT and *oscaf1* mutants. According to the qRT-PCR results, the expression levels of *APX1*, *APX2*, *SODA1*, *SODB*, *CatA*, and *CatC* were significantly decreased in *oscaf1* mutants compared to WT (Figure 4B). The ROS-scavenging genes reduced expression in *oscaf1* might have had feedback from lack of OsCAF1. Overall, a decrease in ROS-scavenging gene expression levels would impair the ROS detoxification system, eventually leading to the accumulation of ROS, such as H_2O_2 in *oscaf1* mutant leaves.

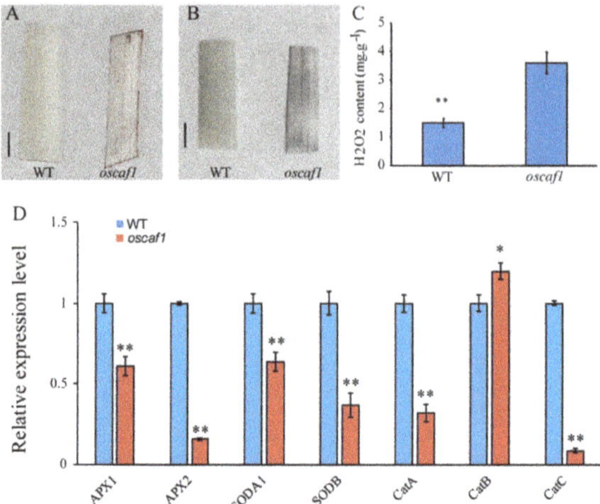

Figure 4. DAB, NBT staining, H$_2$O$_2$ contents and gene expression level analysis in WT and *oscaf1* mutants. (**A**) DAB staining of leaves from WT (left) and oscaf1 mutants (right). Scale bar, 1 cm. (**B**) NBT staining of leaves from WT (left) and oscaf1 mutants (right). Scale bar, 1 cm. (**C**) Measurement of H$_2$O$_2$ content in WT and *oscaf1* mutant (**D**) Relative expression levels of genes involved in ROS-scavenging. The data represent mean ± SE from three independent biological duplicate, * and ** indicated $p < 0.05$ and $p < 0.01$, respectively.

2.4. OsCAF1 Involved in Splicing of Multiple Chloroplast Group II Introns

Previous studies have shown that *ZmCAF1* and *AtCAF1* participate in intron splicing in the chloroplast [21]. To test whether OsCAF1 was involved in the RNA splicing of chloroplast genes, we amplified the chloroplast genes that contained at least one intron in WT and *oscaf1* mutants using the RT-PCR method. We observed that the introns of chloroplast *atpF*, *rpl2*, and *rps12* could not be spliced, whereas the intron splicing efficiency of *ndhA*, *ndhB*, and *ycf3* decreased in group II introns in *oscaf1* mutants (Figure 5). In *oscaf1*, the splicing efficiency of group I introns (*trnL*) was normal. The results indicated that OsCAF1 might participate in the splicing of multiple chloroplast group II introns.

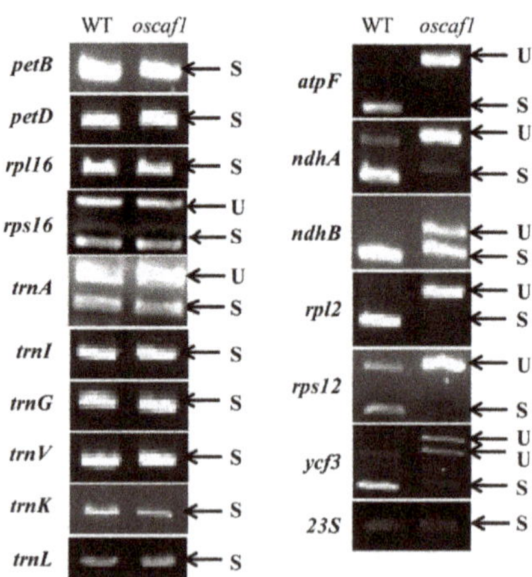

Figure 5. Splicing analysis of chloroplast gene transcripts in WT and *oscaf1* mutants. U, unspliced transcripts; S, spliced transcripts.

2.5. Subcellular Localization and Expression Patterns of OsCAF1

OsCAF1 was localized in the chloroplasts, which was predicted using the TargetP website (http://www.cbs.dtu.dk/services/TargetP/). To determine the actual subcellular localization of OsCAF1, we fused the OsCAF1 full-length cDNA with GFP driven by the 2× CaMV 35S promoter and transiently transfected in rice protoplasts. The results showed that the green fluorescent signals of OsCAF1–GFP co-localized with chloroplast autofluorescence in transformed rice protoplasts (Figure 6A). The assays indicated that OsCAF1 was localized in the rice chloroplast.

The qRT-PCR assays were used to investigate the expression patterns of OsCAF1 in different tissues at the three leaves stage. According to our results, OsCAF1 was highly expressed in green tissue, particularly in the leaves, suggesting that the OsCAF1 mainly functions in green tissue (Figure 6B).

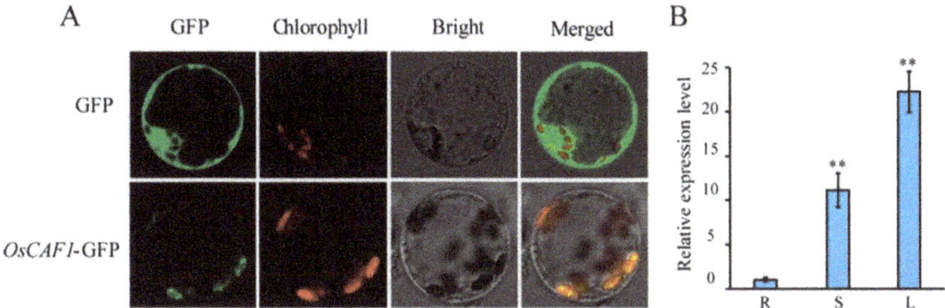

Figure 6. Subcellular localization and expression patterns of OsCAF1. (**A**) Subcellular localization of OsCAF1 in rice protoplasts. (**B**) Expression analysis of OsCAF1 in different tissues. The tissues include roots (R), stems (S), and leaves (L). The values in the figures represent the means ± SE (n = 3). ** indicate $p < 0.01$.

2.6. OsCAF1 through C-terminal Interacts with OsCRS2

In maize, CRS2 could interact with CAF1 and form a CRS2–CAF1 complex in chloroplasts [22]. Therefore, we investigated whether OsCAF1 interacted with OsCRS2 and formed an OsCRS2–OsCAF1 complex in rice. We obtained rice *OsCRS2 (Os01g0132800)* by homologous alignment with *ZmCRS2* and observed that *OsCRS2* was also expressed highly in green tissue (Supplementary Figure S1A). The results of the yeast two-hybrid (Y2H) experiment revealed that OsCAF1 interacted with OsCRS2 and formed an OsCRS2–OsCAF1 complex in rice (Figure 7A). To determine which OsCAF1 section was essential for interactions with OsCRS2, we first generated three truncated forms of OsCAF1, including OsCAF1-N, OsCAF1-M, which contains two CRM domains, and OsCAF1-C (Figure 7B). The results showed that only OsCAF1-C could interact with OsCRS2, while OsCAF1-N and OsCAF1-M failed to interact with OsCRS2 (Figure 7C). The result indicated that the C-terminal of OsCAF1 was essential for interactions with OsCRS2. To further explore the function of the of OsCAF1 C-terminal, we analyzed the alignment of the C-terminals of OsCAF1 residues 387–701 aa in *Oryza sativa* with different species, including *Z. mays, A. thaliana, Panicum hallii, Brachypodium distachyon, Sorghum bicolor, Setaria italica, Triticum aestivum, Elaeis guineensis,* and *Phoenix dactylifera*. The sequence alignment results showed that the OsCAF1 C-terminal residues of 585–598 aa and 685–701 aa were highly conserved (Supplementary Figure S2). To test the functional regions of the OsCAF1 in C-terminal, we divided the C-terminal containing OsCAF1 residues C1 (387–584 aa), C2 (585–598 aa), C3 (599–684 aa), and C4 (685–701 aa) (Figure 7B) for further study. The results of the Y2H experiment showed that OsCAF1-C1 could bind OsCRS2 (Figure 7C). The results indicated that the C1-terminal of OsCAF1 was required for interactions with OsCRS2 in rice; however, the two highly conserved regions of the C-terminal of OsCAF1 could not interact with OsCRS2.

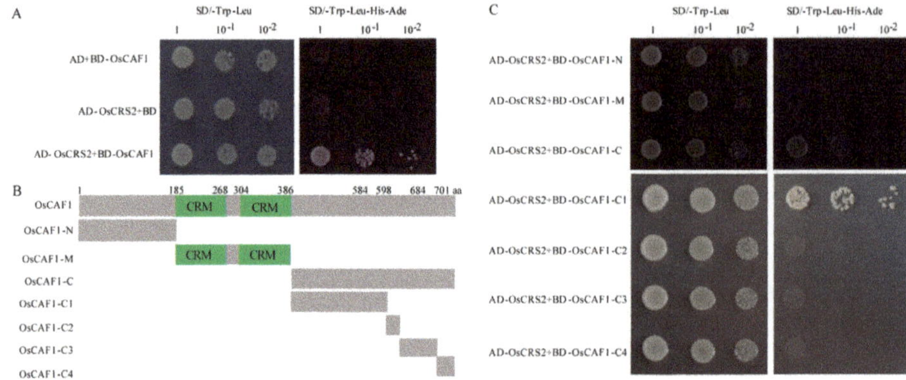

Figure 7. C-terminal of OsCAF1 interacts with OsCRS2. (**A**) OsCAF1 interacts with OsCRS2. The full-length CDS sequences of OsCAF1 and OsCRS2 were cloned into PGBKT7 (BD) and PGADT7 (AD) vector, respectively, to generate BD-OsCAF1 and AD-OsCRS2 constructs. The successfully cloned two constructs were then co-transformed into yeast AH109 cells. AD+ BD-OsCAF1 and BD+AD-OsCRS2 constructs were used as control. (**B**) The protein structure of OsCAF1. Diagram of OsCAF1 protein structure and different structural domains of OsCAF1, OsCAF1-N (1-184), OsCAF1-M (185-386), OsCAF1-C (387-701), OsCAF1-C1 (387-584), OsCAF1-C2 (585-598), OsCAF1-C3 (599-684), and OsCAF1-C4 (685-701). (**C**) Yeast two-hybrid results of the interaction of OsCRS2 and OsCAF1. Truncated versions of OsCAF1 were cloning into BD vector and to generate BD-OsCAF1-N, BD-OsCAF1-M, BD-OsCAF1-C, BD-OsCAF1-C1, BD-OsCAF1-C2, BD-OsCAF1-C3 and BD-OsCAF1-C4 vectors. These vectors were co-transferred to AH109 yeast cells with AD-OsCRS2, respectively. Yeast cells were selected on synthetic dextrose medium lacking Leu and Trp (SD/-Trp-Leu) and the medium lacking Leu, Trp, His, and Ade (SD/-Trp-Leu-His-Ade) with different dilution series (1, 10^{-1}, and 10^{-2}).

The *oscaf1* mutant exhibited albinism and survived for only approximately three weeks. The chloroplast numbers in *oscaf1* mutants decreased, accompanied by abnormal chloroplast development (Supplementary Figure S3) and damaged thylakoid membranes compared with the WT, which led to a decrease in chlorophyll contents in *oscaf1* mutants. Similar phenotypes of CRM domain proteins have been reported in rice, such as *oscfm3* and *oscrs1* [19,23]. OsCAF1 might play an important role in chloroplast development. Among the altered chloroplast and photosynthesis-related gene expressions in *oscaf1* mutants, PEP-dependent genes, including *PsaA*, *PsbA*, and *RbcL*, and encoding photosynthesis-associated proteins significantly decreased in *oscaf1* mutants, although NEP-dependent genes, *RpoC1* and *RpoC2*, markedly increased in *oscaf1* mutants. In addition, the expression levels of *AtpB* and *AtpE*, transcribed by both NEP and PEP, significantly decreased in *oscaf1* mutants. Therefore, OsCAF1 might regulate chloroplast development by influencing PEP activities.

The ROS contents such as H_2O_2 increased in the three leaves stage leaves of *oscaf1* mutants, compared with the WT. In addition, the expression levels of ROS-scavenging related genes decreased in *oscaf1* mutant leaves. Considering the results of a previous study [19], the functional obstruction of OsCAF1 could lead to abnormal chloroplast development, ROS over-accumulation in leaves, and eventually the death of rice seedlings after the three leaves stage.

Numerous leaf-color associated mutants have been identified and cloned in different plants [25–27]. Abnormal splicing of chloroplast gene introns could cause chlorophyll deficiency, which then leads to visible losses in green color in leaves or death of plants [25–27]. In rice, *wsl4* exhibit bleached appearances before the four leaves stage, and after the five leaves stage, all leaves turned to a normal color similar to WT under normal growth environments. The bleached appearance phenomenon in *wsl4* was caused by defective intron splicing of chloroplast genes, including *atpF*, *rpl2*, *ndhA*, and *rps12* [26]. The results of such studies indicate that the abnormal splicing of the four introns does not cause lethal phenotypes at seedling stages. In the present study, the splicing of *ndhB* and *ycf3* introns was also affected. The ndhB protein was an NADPH dehydrogenase and bound to the thylakoid membrane, influenced NDH activity, and regulated cyclic electron transport around photosystem I (PS I) and respiratory electron transport [28]. In tobacco, the Δ*ndhB* mutant exhibited abnormal intron splicing of only chloroplast *ndhB*, and the Δ*ndhB* mutant could grow normally under normal conditions [28]. In addition, both *crr2-1* and *crr2-2*, with impaired intron splicing of *ndhB* in *Arabidopsis*, could grow normally under normal conditions [29]. In the present study, NDH activity or electron transfer efficiency of PS I may be affected in *oscaf1* mutants; however, the impairing of the *ndhB* intron was not the key factor influencing seedling death after the three leaves stage. Another intron splicing impaired gene, *ycf3*, the chloroplast-encoded Ycf3 protein, was responsible for the stability of the PS I protein complex [16]. Previous studies have shown that numerous mutants, such as *otp51*, *osppr6*, and *tha8*, have albino seedling lethal phenotypes similar to the one in the *oscaf1* mutant, and all the mutants have similar genetic defects including reduced intron splicing efficiency of *ycf3*, similar to the results in the present study [14,16,27]. Therefore, the Ycf3 protein could be required for early chloroplast development. In *oscaf1* mutants, the reduced splicing efficiency of *ycf3* was responsible for mutant deaths after the three leaves stages.

In *Arabidopsis*, AtCAF1 is required for the splicing of group IIB introns, including *ndhA*, *petD*, *rpl16*, *rps16*, *trnG*, *ycf3*, *rpoC1*, and *ClpP* [21]. ZmCAF1 in maize is also required for the splicing of group IIB introns, almost similar to *A. thaliana*, excluding *rpoC1* and *ClpP* [11]. Previous results have shown that CAF1 is not required for group IIA intron splicing in maize and *A. thaliana* [11,21]. Notably, our results indicated that the efficiency of intron splicing of group IIA introns, including *atpF*, *rpl2*, and *rps12*, and group IIB introns, including *ndhB*, *ycf3*, and *ndhA*, were affected. Particularly, the group IIA introns were not spliced in *oscaf1* mutants completely. The results indicated that OsCAF1 could play a key role in the regulation of intron splicing in group IIA and group IIB introns in rice, which was different from maize and *A. thaliana*. Previous study showed that OsCRS1 not only participated in the splicing of *atpF* introns but also promoted the splicing of other introns, including *trnL*, *rpl2*, *ndhA*, *ndhB*, *petD*, and *ycf3* [22]. Therefore, the functions of OsCRS1 and OsCAF1 could partially overlap but are not redundant in rice. In addition, the intron splicing of subgroup IIB introns, including *petD*, *rpl16*, *rps16*, and *trG* does not require OsCAF1 in rice, while the splicing of the subgroup IIB introns requires ZmCAF1 and AtCAF1 in maize and *A. thaliana* [11,21]. Moreover, in rice, there are three additional proteins containing two CRM domains, and the functions of the proteins may be different from OsCAF1. Intron splicing of subgroup IIA and subgroup IIB introns was regulated by OsCAF1, which, in turn, influenced chloroplast development in rice.

Nuclear-encoded protein ZmCRS2 could interact with ZmCAF1 to form the CRS2–CAF1 complex, which regulates the splicing of group IIB introns in maize chloroplasts [22]. In addition, our study demonstrated that OsCAF1 could interact with OsCRS2 through the C-terminal region, suggesting that the interaction mechanism of CAF1 and CRS2 might be conserved in maize and rice. The homologous alignment analysis results of the C-terminal region of CAF1 among various plants suggested high conservation at C-terminal positions 585–598 aa and 685–701 aa. However, the results of our study showed that the C1 region of OsCAF1, 387–584 aa, was essential for interactions with OsCRS2 rather than the conserved regions of the C2 and C4 regions. The C2 and C4 conserved regions of OsCAF1 could interact with other proteins, which requires further investigations. In addition, the *OsCRS2* expression levels decreased significantly in *oscaf1* (Supplementary Figure S2B), suggesting the mutation of OsCAF1 also influenced *OsCRS2* expression. Based on the above findings, the C1 region of OsCAF1 interacts with OsCRS2, and the generated complex, OsCRS2–OsCAF1, could participate in the splicing of subgroup IIA and subgroup IIB introns in rice chloroplast.

OsCAF1, which has two CRM domains, is an RNA intron-splicing factor in rice. *OsCAF1* influences the expression of chloroplast-associated genes, affects the accumulation of H_2O_2 in rice leaves, and plays a key role in early chloroplast development in rice. We demonstrated that OsCAF1 is potentially involved in the splicing of both group IIA and group IIB introns of chloroplast transcripts in rice.

In addition, the C1-terminal region of OsCAF1 could interact with OsCRS2 and the OsCRS2–OsCAF1 complex could participate in the splicing of group II introns.

3. Materials and Methods

3.1. Plant Materials and Growth Conditions

The *oscaf1* mutants were obtained using the CRISPR-Cas9 gene editing system in the *japonica* rice variety, Nipponbare, which was used as the WT. For chlorophyll content measurement and genetic material and RNA extraction, the WT and *oscaf1* plants were grown in growth chambers under a 16 h light and 8 h of a dark cycle at a constant temperature of 30 °C.

3.2. Knockout of OsCAF1 Using the CRISPR/Cas9 System

The rice genome editing strategy was based on the CRISPR/Cas9 system according to a previous study [30]. Two gRNA target sequences (AAGCCCAGTACCCCATCTCACGG, CCGAGGTACCA AGCGGCGTCCAG) were designed from exon 1 to construct the intermediate vector (Figure 1B) namely SK-gRNA- g$^{OsCAF1-g1}$ and SK-gRNA- g$^{OsCAF1-g2}$. The binary expression vector transferred into *Agrobacterium tumefaciens* strain EHA105 was constructed using the isocaudamer ligation method; the intermediate vectors were digested with *Kpn* I/*Xho* I, *Sal* I/*Bgl* II, and then assembled into the pC1300-Cas9 binary vector (digested with *Kpn* I/*BamH* I). The schematic for the development of the plant expression vector is illustrated in Figure 1A,B. The intermediate vector SK-gRNA and the expression vector pC1300-Cas9 were provided by Kejian Wang and were kept in our laboratory. All of the primers are listed in Supplementary Table S1.

3.3. Transmission Electron Microscopy (TEM)

The WT and *oscaf1* seedlings were grown in growth chambers, as above. The transmission electron microscopy (TEM) analysis was performed according to a previous study [23]. Briefly, the third leaves from WT and *oscaf1* were collected and cut into approximately 0.5 cm^2 pieces. Leaf samples were fixed using 2.5% glutaraldehyde and 1% OsO$_4$, dehydrated using an ethanol series, and embedded in resin. The samples were stained again with uranyl acetate and alkaline lead citrate and finally observed under a HitachiH-7500 TEM (Tokyo, Japan).

3.4. RNA Extraction and Quantitative Real-time PCR (qRT-PCR) Assay

Total rice RNA from leaves of *oscaf1* and WT plants was extracted using an RNA extraction Kit (TaKaRa, Japan). First-strand cDNA was synthesized using a ReverTra Ace qPCR RT Kit (TOYOBO, Japan). The qRT-PCR was conducted using a SYBR green real-time PCR master mix (TOYOBO, Japan) on a Bio-Rad CFX96 system according to the manufacturer's instructions. The qRT-PCR procedure was as follows: 5 min at 95 °C followed by 40 cycles of 95 °C for 10 s and 58 °C for 1 min. *OsActin1* were used as internal controls. The fluorescence data were analysis by Lin-RegPCR program [31,32] to calculate primer efficiency and obtain Ct values. The genes relative expression levels were calculated according to previous study [33]. All qRT-PCR primers are listed in Supplementary Table S1. For date statistical significance was analyzed using ANOVA with Tukey post hoc pairwise comparisons. * and ** indicate $p < 0.05$ and $p < 0.01$, respectively.

3.5. Chloroplast RNA Splicing Analysis

The cDNA of WT and *oscaf1* plants seeding leaves were obtained according to the procedures above. The RT-PCR procedure was as follows: 95 °C for 5 min, followed by 32 cycles of 95 °C for 30 s, 60 °C for 30 s, 72 °C for 1 min, and a final elongation step at 72 °C for 8 min. The corresponding RT-PCR primers were used for chloroplast RNA splicing analysis according to previous studies [18,26].

3.6. Chlorophyll Concentration and F_v/F_m Measurement

Total chlorophyll content in the plants was determined spectrophotometrically. In brief, we obtained 0.1 g leaf samples at the three leaves stage, cut them into pieces, and immersed them into 15 mL 95% ethanol for 48 h in darkness at 4 °C. The chlorophyll concentrations were measured using a UV-1800PC (Mapada, China) spectrophotometer at 665 nm, 649 nm, and 470 nm. According to the method of Lichtenthaler [34], we calculated Chla, Chlb, and Car content pigment concentrations.

In three leaves stage, WT and *oscaf1* mutants were treated with darkness for 30 min, and Fv/Fm was measured using a handheld fluorometer according to the manufacturer's instructions. There were 5 WT and mutant plants used for analysis, respectively. The values in the figures represent the means ± SE. Statistical significance was analyzed using Student's *t*-test * and ** indicate $p < 0.05$ and $p < 0.01$, respectively.

3.7. Subcellular Localization of OsCAF1

We cloned the coding sequence of *OsCAF1*, and the PCR product was fused to the N-terminus of the green fluorescent protein (GFP) in the pAN580 vector. Both the OsCAF1–GFP fusion construct and the empty GFP vector were transformed into rice protoplasts as previously described [35]. Protoplasts with GFP signals were observed under a laser scanning confocal microscope (Zeiss LSM700, Germany). The PCR amplification primers are listed in Supplementary Table S1.

3.8. Yeast Two-hybrid Analysis

The coding sequences of *OsCAF1* and *OsCRS2* (*Os01g0132800*) were amplified using gene-specific primers. Full-length *OsCAF1* and partial-length *OsCAF1* were cloned into pGBKT7 (BD) to generate the BD-OsCAF1, BD-OsCAF1-N, BD-OsCAF1-M, BD-OsCAF1-C, BD-OsCAF1-C1, BD-OsCAF1-C2, BD-OsCAF1-C3, and BD-OsCAF1-C4 plasmids. In addition, full-length OsCRS2 was cloned into pGADT7 (AD) to generate the AD-OsCRS2 plasmid. The pairwise plasmid was co-transformed into the AH109 yeast strain and growth on SD-T/L in a 28 °C incubator. Next, the co-transformed yeast strains were transferred to SD-T/L/H/A for interaction analysis.

3.9. Measurement of H_2O_2 Content

The H_2O_2 was extracted using 3-amino-1,2,4-triazole, and according to the method of Brennan and Frenkel [36]. In brief, we obtained 0.1-g leaf samples at the three leaves stage. The samples were homogenized in 3 mL of 3-amino-1,2,4-triazole (10 mM) and centrifugation for 20 min under 6000× g. Next, 1 mL (0.1% titanium tetrachloride to 20% H_2SO_4) was added to 1.5 mL supernatant. Centrifuge the reaction solution to remove insoluble materials and measured absorbance at 410 nm against a blank. The standard curve was used to calculate the content of H_2O_2.

Supplementary Materials: Supplementary materials can be found at http://www.mdpi.com/1422-0067/20/18/4386/s1.

Author Contributions: Formal analysis, Q.Z., Z.W., G.H., J.H., Z.G. and L.G.; Funding acquisition, L.S., D.R., L.Z., G.Z., D.Z., Project administration Q.Q.; Writing—original draft, Q.Z. and L.S.

Funding: This research was supported by the National Natural Science Foundation of China (31801333), the 'Collaborative Innovation Project' of the Chinese Academy of Agricultural Sciences, the Zhejiang Provincial 'Ten Thousand Talent Program' Project (2018R52025), and the Central Public-interest Scientific Institution Basal Research Fund of China National Rice Research Institute (2017RG001-4).

Conflicts of Interest: The authors declare no conflict of interest.

References

1. Li, Y.; Zhang, J.; Li, L.; Gao, L.; Xu, J.; Yang, M. Structural and comparative analysis of the complete chloroplast genome of *pyrus hopeiensis*-"wild plants with a tiny population"-and three other *pyrus* species. *Int. J. Mol. Sci.* **2018**, *19*, 3262. [CrossRef]

2. Bobik, K.; Burch-Smith, T.M. Chloroplast signaling within, between and beyond cells. *Front. Plant Sci.* **2015**, *6*, 781. [CrossRef]

3. Zerges, W. Translation in chloroplasts. *Biochimie* **2000**, *82*, 583–601. [CrossRef]

4. Hedtke, B.; Börner, T.; Weihe, A. Mitochondrial and chloroplast phage-type RNA polymerases in *Arabidopsis*. *Science* **1997**, *277*, 809–811. [CrossRef]

5. Pogson, B.J.; Albrecht, V. Genetic dissection of chloroplast biogenesis and development: An overview. *Plant Physiol.* **2011**, *155*, 1545–1551. [CrossRef]

6. Yua, Q.B.; Ma, Q.; Kong, M.M.; Zhao, T.T.; Zhang, X.L.; Zhou, Q.; Huang, C.; Chong, K.; Yang, Z.N. AtECB1/MRL7, a thioredoxin-like fold protein with disulfide reductase activity, regulates chloroplast gene expression and chloroplast biogenesis in *Arabidopsis thaliana*. *Mol. Plant* **2014**, *7*, 206–217. [CrossRef]

7. Nakai, M. New perspectives on chloroplast protein import. *Plant Cell Physiol.* **2018**, *59*, 1111–1119. [CrossRef]

8. de Longevialle, A.F.; Small, I.D.; Lurin, C. Nuclearly encoded splicing factors implicated in RNA splicing in higher plant organelles. *Mol. Plant* **2010**, *3*, 691–705. [CrossRef]

9. Börner, T.; Aleynikova, A.Y.; Zubo, Y.O.; Kusnetsov, V.V. Chloroplast RNA polymerases: role in chloroplast biogenesis. *Biochim. Biophys. Acta* **2015**, *1847*, 761–769. [CrossRef]

10. Saldanha, R.; Mohr, G.; Belfort, M.; Lambowitz, A.M. Group I and group II introns. *FASEB J.* **1993**, *7*, 15–24. [CrossRef]

11. Ostheimer, G.J.; Williams-Carrier, R.; Belcher, S.; Osborne, E.; Gierke, J.; Barkan, A. Group II intron splicing factors derived by diversification of an ancient RNA-binding domain. *EMBO J.* **2003**, *22*, 3919–3929. [CrossRef]

12. Bonen, L.; Vogel, J. The ins and outs of group II introns. *Trends Genet.* **2001**, *17*, 322–331. [CrossRef]

13. Barkan, A.; Small, I. Pentatricopeptide repeat proteins in plants. *Annu. Rev. Plant. Biol.* **2014**, *65*, 415–442. [CrossRef]

14. de Longevialle, A.F.; Hendrickson, L.; Taylor, N.L.; Delannoy, E.; Lurin, C.; Badger, M.; Millar, A.H.; Small, I. The pentatricopeptide repeat gene *OTP51* with two LAGLIDADG motifs is required for the cis-splicing of plastid *ycf3* intron 2 in *Arabidopsis thaliana*. *Plant J.* **2008**, *56*, 157–168. [CrossRef]

15. Beick, S.; Schmitz-Linneweber, C.; Williams-Carrier, R.; Jensen, B.; Barkan, A. The pentatricopeptide repeat protein PPR5 stabilizes a specific tRNA precursor in maize chloroplasts. *Mol. Cell Biol.* **2008**, *28*, 5337–5347. [CrossRef]

16. Tang, J.; Zhang, W.; Wen, K.; Chen, G.; Sun, J.; Tian, Y.; Tang, W.; Yu, J.; An, H.; Wu, T.; et al. OsPPR6, a pentatricopeptide repeat protein involved in editing and splicing chloroplast RNA, is required for chloroplast biogenesis in rice. *Plant Mol. Biol.* **2017**, *95*, 345–357. [CrossRef]

17. Gothandam, K.M.; Kim, E.S.; Cho, H.; Chung, Y.Y. OsPPR1, a pentatricopeptide repeat protein of rice is essential for the chloroplast biogenesis. *Plant Mol. Biol.* **2005**, *58*, 421–433. [CrossRef]

18. Tan, J.; Tan, Z.; Wu, F.; Sheng, P.; Heng, Y.; Wang, X.; Ren, Y.; Wang, J.; Guo, X.; Zhang, X.; et al. A novel chloroplast-localized pentatricopeptide repeat protein involved in splicing affects chloroplast development and abiotic stress response in rice. *Mol. Plant* **2014**, *7*, 1329–1349. [CrossRef]

19. Asakura, Y.; Bayraktar, O.A.; Barkan, A. Two CRM protein subfamilies cooperate in the splicing of group IIB introns in chloroplasts. *RNA* **2008**, *14*, 2319–2332. [CrossRef]

20. Keren, I.; Klipcan, L.; Bezawork-Geleta, A.; Kolton, M.; Shaya, F.; Ostersetzer-Biran, O. Characterization of the molecular basis of group II intron RNA recognition by CRS1-CRM domains. *J. Biol. Chem.* **2008**, *283*, 23333–23342. [CrossRef]

21. Asakura, Y.; Barkan, A. *Arabidopsis* orthologs of maize chloroplast splicing factors promote splicing of orthologous and species-specific group II introns. *Plant Physiol.* **2006**, *142*, 1656–1663. [CrossRef]

22. Ostheimer, G.J.; Rojas, M.; Hadjivassiliou, H.; Barkan, A. Formation of the CRS2-CAF2 group II intron splicing complex is mediated by a 22-amino acid motif in the COOH-terminal region of CAF2. *J. Biol. Chem.* **2006**, *281*, 4732–4738. [CrossRef]

23. Liu, C.; Zhu, H.; Xing, Y.; Tan, J.; Chen, X.; Zhang, J.; Peng, H.; Xie, Q.; Zhang, Z. Albino leaf 2 is involved in the splicing of chloroplast group I and II introns in rice. *J. Exp. Bot.* **2016**, *67*, 5339–5347. [CrossRef]

24. Noctor, G.; Reichheld, J.P.; Foyer, C.H. ROS-related redox regulation and signaling in plants. *Semin. Cell Dev. Biol.* **2018**, *80*, 3–12. [CrossRef]

25. Huang, W.; Zhu, Y.; Wu, W.; Li, X.; Zhang, D.; Yin, P.; Huang, J. The Pentatricopeptide repeat protein SOT5/EMB2279 is required for plastid *rpl2* and *trnK* intron splicing. *Plant Physiol.* **2018**, *177*, 684–697. [CrossRef]

26. Wang, Y.; Ren, Y.; Zhou, K.; Liu, L.; Wang, J.; Xu, Y.; Zhang, H.; Zhang, L.; Feng, Z.; Wang, L.; et al. White stripe leaf4 encodes a novel P-type PPR protein required for chloroplast biogenesis during early leaf development. *Front. Plant Sci.* **2017**, *8*, 1116. [CrossRef]
27. Khrouchtchova, A.; Monde, R.A.; Barkan, A. A short PPR protein required for the splicing of specific group II introns in angiosperm chloroplasts. *RNA* **2012**, *18*, 1197–1209. [CrossRef]
28. Endo, T.; Shikanai, T.; Takabayashi, A.; Asada, K.; Sato, F. The role of chloroplastic NAD(P)H dehydrogenase in photoprotection. *FEBS Lett.* **1999**, *457*, 5–8. [CrossRef]
29. Hashimoto, M.; Endo, T.; Peltier, G.; Tasaka, M.; Shikanai, T. A nucleus-encoded factor, CRR2, is essential for the expression of chloroplast *ndhB* in *Arabidopsis*. *Plant J.* **2003**, *36*, 541–549. [CrossRef]
30. Shen, L.; Wang, C.; Fu, Y.; Wang, J.; Liu, Q.; Zhang, X.; Yan, C.; Qian, Q.; Wang, K. QTL editing confers opposing yield performance in different rice varieties. *J. Integr. Plant Biol.* **2018**, *60*, 89–93. [CrossRef]
31. Ramakers, C.; Ruijter, J.M.; Deprez, R.H.; Moorman, A.F. Assumption-free analysis of quantitative real-time polymerase chain reaction (PCR) date. *Neurosci. Lett.* **2003**, *339*, 62–66. [CrossRef]
32. Ruijter, J.M.; Ramaker, C.; Hoogaars, W.M.; Karlen, Y.; Bakker, O.; van den Hoff, M.J.; Moorman, A.F. Amplification efficiency: linking baseline and bias in the analysis of quantitative PCR data. *Nucleic Acids Res.* **2009**, *37*, e45. [CrossRef]
33. Pfaffl, M.W.; Horgan, G.M.; Dempfle, L. Relative expression software tool (REST) for group-wise comparison and statistical analysis of relative expression results in real-time PCR. *Nucleic Acids Res.* **2002**, *30*, e36. [CrossRef]
34. Lichtenthaler, H.K. Chlorophylls and carotenoids: pigments of photosynthetic biomembranes. *Methods Enzymol.* **1987**, *148*, 350–382.
35. Zhang, Y.; Su, J.; Duan, S.; Ao, Y.; Dai, J.; Liu, J.; Wang, P.; Li, Y.; Liu, B.; Feng, D.; et al. A highly efficient rice green tissue protoplast system for transient gene expression and studying light/chloroplast-related processes. *Plant Methods* **2011**, *7*, 30. [CrossRef]
36. Brennan, T.; Frenkel, C. Involvement of hydrogen peroxide in the regulation of senescence in pear. *Plant Physiol.* **1977**, *59*, 411–416. [CrossRef]

International Journal of
Molecular Sciences

Article

Effect of Low Temperature on Chlorophyll Biosynthesis and Chloroplast Biogenesis of Rice Seedlings during Greening

Yuqing Zhao [1,†], Qiaohong Han [1,†], Chunbang Ding [1], Yan Huang [1], Jinqiu Liao [1], Tao Chen [1], Shiling Feng [1], Lijun Zhou [1], Zhongwei Zhang [2], Yanger Chen [1], Shu Yuan [2] and Ming Yuan [1,*]

[1]   College of Life Science, Sichuan Agricultural University, Ya'an 625014, China;
      yuqing@stu.sicau.edu.cn (Y.Z.); xiaoyangyang26@126.com (Q.H.); dcb@sicau.edu.cn (C.D.);
      shirley11hy@163.com (Y.H.); liaojinqiu630@sicau.edu.cn (J.L.); chentao293@163.com (T.C.);
      fengshilin@outlook.com (S.F.); zhoulijun@sicau.edu.cn (L.Z.); anty9826@163.com (Y.C.)
[2]   College of Resources, Sichuan Agricultural University, Chengdu 611130, China;
      zzwzhang@sicau.edu.cn (Z.Z.); roundtree@sohu.com (S.Y.)
*   Correspondence: yuanming@sicau.edu.cn
†   These authors contribute equally to this work.

Received: 14 December 2019; Accepted: 17 February 2020; Published: 19 February 2020

Abstract: Rice (*Oryza sativa* L.) frequently suffers in late spring from severe damage due to cold spells, which causes the block of chlorophyll biosynthesis during early rice seedling greening. However, the inhibitory mechanism by which this occurs is still unclear. To explore the responsive mechanism of rice seedlings to low temperatures during greening, the effects of chilling stress on chlorophyll biosynthesis and plastid development were studied in rice seedlings. Chlorophyll biosynthesis was obviously inhibited and chlorophyll accumulation declined under low temperatures during greening. The decrease in chlorophyll synthesis was due to the inhibited synthesis of δ-aminolevulinic acid (ALA) and the suppression of conversion from protochlorophyllide (Pchlide) into chlorophylls (Chls). Meanwhile, the activities of glutamate-1-semialdehyde transaminase (GSA-AT), Mg-chelatase, and protochlorophyllide oxidoreductase (POR) were downregulated under low temperatures. Further investigations showed that chloroplasts at 18 °C had loose granum lamellae, while the thylakoid and lamellar structures of grana could hardly develop at 12 °C after 48 h of greening. Additionally, photosystem II (PSII) and photosystem I (PSI) proteins obviously declined in the stressed seedlings, to the point that the PSII and PSI proteins could hardly be detected after 48 h of greening at 12 °C. Furthermore, the accumulation of reactive oxygen species (ROS) and malondialdehyde (MDA) and cell death were all induced by low temperature. Chilling stress had no effect on the development of epidermis cells, but the stomata were smaller under chilling stress than those at 28 °C. Taken together, our study promotes more comprehensive understanding in that chilling could inhibit chlorophyll biosynthesis and cause oxidative damages during greening.

Keywords: *Oryza sativa* L.; chilling stress; chlorophyll biosynthesis; chloroplast biogenesis; epidermal characteristics

---

1. Introduction

Seedlings, growing in the darkness before emerging from the soil, undergo etiolation with long hypocotyls and closed cotyledons which contain undeveloped plastids called etioplasts that have no chlorophyll [1,2]. The greening process initiates when exposed to light as the seedlings come out of soil, and this de-etiolation process includes the reduction of hypocotyl elongation rate, cotyledon opening, chlorophyll synthesis, and chloroplast biogenesis, and subsequently, seedlings absorb light energy and transition to autotrophy [3,4].

Chlorophyll (Chl) has unique and essential roles in harvesting and transducing light energy in antenna systems, and charge separation and electron transport in reaction centers [5,6]. Chlorophyll level is an important index used to evaluate photosynthetic capacity. Chlorophyll content decreases under cold stress, which might be because low temperature suppresses chlorophyll biosynthesis, probably by inhibiting the activities of chlorophyll biosynthetic enzymes [7]. A previous study showed that the optimum temperature of divinyl reductase (DVR) activity was 30 °C, and low activity was observed at 10 °C and no activity was found at 50 °C [7].

Chls share a common biosynthetic pathway with other tetrapyrroles, including siroheme, heme, and phytochromobilin in plants, algae, and many bacteria [6,8,9]. Since tetrapyrrole intermediates are all photosensitizers that are easily activated by light, leading to highly toxic levels of reactive oxygen species (ROS) and photooxidative damage and growth retardation, Chl biosynthesis must be finely controlled to ensure healthy growth during the greening process [9,10]. Chl biosynthesis is a very complex process that is executed via a series of coordinated reactions catalyzed by numerous enzymes [11]. The process of Chl biosynthesis can be divided into three distinct phases. The first phase involves the synthesis of protoporphyrin IX (Proto IX) from glutamate [12]. The first committed precursor is δ-aminolevulinic acid (ALA), and its synthesis is a key control point in Chl biosynthesis. This step is catalyzed by glutamyl-tRNA reductase (GluTR), which is encoded by *HEMA* gene [8]. δ-Aminolevulinic acid dehydratase (ALAD) catalyzes aggregation of two ALA molecules into porphobilinogen (PBG). Four PBGs polymerize and further cyclize to yield uroporphyrinogen III (urogen III), when then decarboxylates to form coproporphyrinogen III (coprogen III). Coprogen III occurs via oxidative decarboxylation and oxygen-dependent aromatization reaction to form protoporphyrin IX (Proto IX) [9]. The second phase includes the synthesis of chlorophyll *a* (Chl*a*) from Proto IX. Proto IX is the branch point of the chlorophyll and heme biosynthetic pathways [6]. The insertion of Mg^{2+} into Proto IX forms Mg-protoporphyrin IX (Mg-proto IX), which is catalyzed by Mg-chelatase, but the insertion of Fe^{2+} into Proto IX catalyzed by ferrochelatase starts the heme branch [9]. Meanwhile, the photoreduction of Pchlide to chlorophyllide (Chlide), catalyzed by protochlorophyllide oxidoreductase (POR), is another important step. POR is the key enzyme in the light-dependent greening of higher plants [13]. Two POR proteins (PORA and PORB) and three POR proteins (PORA, PORB, and PORC) have been found in rice and *Arabidopsis*, respectively [14,15]. DVR catalyzes 3,8-divinyl-chlide *a* to yield chlorophyllide a (Chlide a), which further forms Chl *a* catalyzed by chlorophyll synthase (CHLG). The third phase is the interconversion of Chl *a* and chlorophyll *b* (Chl *b*) catalyzed by chlorophyllide a oxygenase (CAO) which is known as the chlorophyll cycle [16].

Chloroplasts are responsible for photosynthesis and the production of hormones and metabolites [17,18], which develop from proplastids that are present in the immature cells of plant meristems. Chloroplast biogenesis involves a coordinated function of plastid- and nuclear-encoded genes [19]. The process of chloroplast biogenesis accompanies the biosynthesis of chlorophylls, chloroplast proteins, carotenoids, and lipids, and the assembly of photosynthetic protein complexes including the light-harvesting complex (LHC), photosystem I (PSI), photosystem II (PSII), cytochrome f/b_6 (cytf/b_6), and adenosine triphosphate (ATP) synthase [1,17]. Many factors have direct impacts on chloroplast biogenesis, such as light, water, salt, leaf age, etc. [1,20].

Low temperature is one of the most severe weather events that destructively affect crop growth, quality, and yield [21,22]. Cold or chilling stress impairs the chloroplast microstructure, photosynthetic metabolism, and energy production [23], directly leading to inhibited photosynthesis, which is a severe threat to crop production. Additionally, the efficiency of photosynthetic electron transport in plants is significantly decreased under chilling stress, resulting in a burst of ROS that directly causes cellular oxidative damages and increases membrane rigidity [24–27]. Low temperature also disrupts the carbon reduction cycle and the control of stomatal conductance [28].

Rice (*Oryza sativa* L.) is a kind of thermophilic cereal crop that feeds almost half of the world's population. Originating from tropical and subtropical areas, rice is highly sensitive to low temperatures, particularly during the greening process of the early seedling growth. In China, cold spells in late spring cause serious suppression of rice seed germination and young seedling growth, especially in the middle–lower Yangtze River region and southern China. In the past 30 years, 30%–50% rice seedlings suffered from chilling or cold stress, resulting in a reduction of 3–5 megatons every year [29].

Chlorophyll biosynthesis is seriously blocked by low temperatures during greening. However, the inhibitory mechanism by which this occurs is still unclear. Understanding how rice responds to low temperatures will provide valuable information and genetic resources for improving cold-stress tolerance. Therefore, the object of this study was to explore the response mechanism of chlorophyll biosynthesis and chloroplast biogenesis during the greening process under chilling stress and to provide valuable data for crop production.

2. Results

2.1. Effect of Chilling Stress on Plant Growth

To investigate the effect of chilling stress on the growth of rice seedlings during greening, 6 day etiolated seedlings were treated with 28 °C, 18 °C, and 12 °C for 48 h in light (120 μmol m^{-2} s^{-1}). After 48 h of greening, leaves were green and fully expanded at 28 °C (Figure 1A,B) and the leaves at 18 °C looked yellow-green and were fully expanded, while the leaves at 12 °C were yellow and incompletely expanded (Figure 1A,B). Compared with plants growing at 28 °C, chill-treated rice seedlings exhibited significantly lower shoot and root lengths after 48 h illumination (Figure 1C). These results showed that chilling stress significantly inhibited the greening process and the growth of rice seedlings.

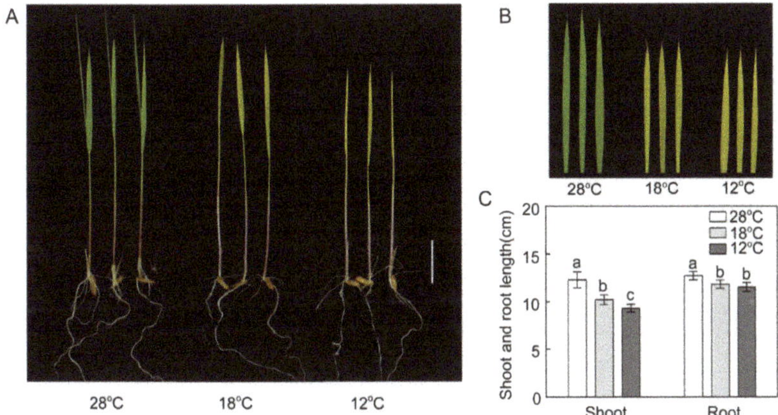

Figure 1. Effect of chilling stress on rice seedling growth (**A,B**) and shoot/root length (**C**) after 48 h of greening. Six day old etiolated seedlings were treated with 18 °C or 12 °C chilling stress. Date represent means ± SD of 10 replicate samples. Bars with different letters above the columns of figures indicate significant differences according to Duncan's multiple range test at $p < 0$. Bar = 2 cm.

2.2. Effect of Chilling Stress on Dry Weight, Protein, Chlorophyll, and Carotenoid Content

The dry weight (DW) of shoots was increased by 1.8%, 17.3%, and 56.5%, respectively, at 28 °C after 0.5 h, 12 h, and 48 h of light exposure (Figure 2A). However, the accumulation of dry matter was significantly inhibited under chilling stress. Shoot DWs were 10.3% and 20.7% lower after 12 h and 48 h at 18 °C compared with those at 28 °C. More severely, DWs at 12 °C were 13.2% and 34.1% lower after 12 h and 48 h of light exposure compared with those at 28 °C, and showed no significant change compared with the seedlings before the light exposure.

At 28 °C, protein contents of seedlings rose to 16.38, 22.74, and 37.90 mg·g^{-1} fresh weight (FW) after 0.5 h, 12 h, and 48 h of greening, respectively (Figure 2B). Low temperature reduced protein accumulation of rice seedlings during the greening process. There were only 22.41 mg·g^{-1} FW at 18 °C and 17.31 mg·g^{-1} FW at 12 °C after 48 h of light exposure.

Chlorophyll during the greening process was significantly increased at 28 °C. However, chlorophyll levels were 77.1% lower at 18 °C and 97.5% lower at 12 °C than at 28 °C (Figure 2C). Carotenoids at 28 °C increased to 0.21 mg·g^{-1} FW after 48 h of greening, while chill-stressed rice seedlings accumulated much lower levels of carotenoids (Figure 2D).

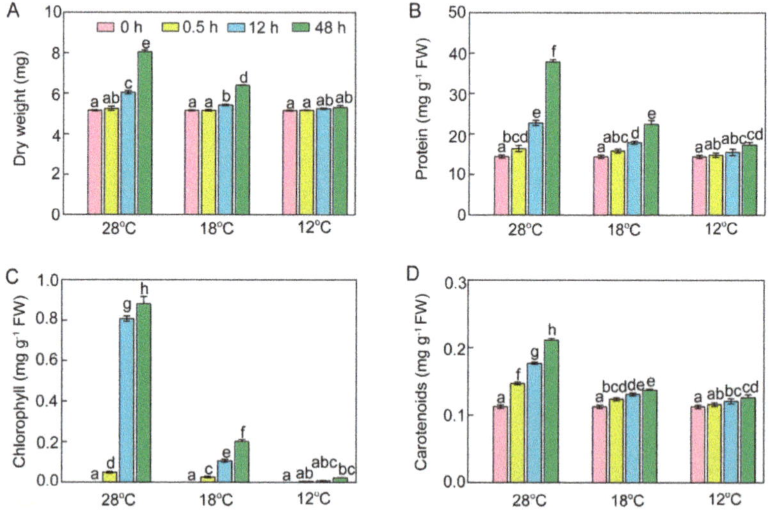

Figure 2. Dry weight (DW, **A**), soluble protein (**B**), chlorophyll (**C**), and carotenoids (**D**) of control (28 °C) and chill-stressed (18 °C and 12 °C) rice seedlings after 0 h, 0.5 h, 12 h, and 48 h greening. Six day old etiolated seedlings were treated with 18 °C or 12 °C cold stress. Seedlings were harvested at 0 h, 0.5 h, 12 h, and 48 h of greening and their DW, protein, chlorophyll, and carotenoid contents were measured. The error bars represent standard deviations of three independent biological replicates. Different letters indicate significantly different at $p < 0.05$ according to Duncan's multiple range tests.

2.3. Effect of Chilling Stress on the Accumulation of Chlorophyll Intermediates in Rice Seedlings during Greening

To explore how chilling stress inhibits chlorophyll biosynthesis in rice seedlings during greening, the accumulation of chlorophyll intermediates was measured.

ALA is the first intermediate of chlorophyll biosynthesis. As shown in Figure 3A, light promoted ALA accumulation. After 48 h of light exposure, ALA contents were 19.4% lower at 18 °C and 46.8% lower at 12 °C compared with those at 28 °C. These results indicate that chilling stress inhibited the synthesis of ALA during greening.

PBG content at 28 °C was increased by 14.8%, 104.9%, and 155.2% after 0.5 h, 12 h, and 48 h of light exposure, respectively (Figure 3B), and there was no significant difference of PBG content between chill-treated rice seedlings and the control seedlings after 12 h and 48 h of greening. Urogen III and coprogen III levels were significantly increased at 28 °C during greening (Figure 3C,D). Under low temperatures, the contents of urogen III and coprogen III during greening were higher than those at 28 °C. Proto IX level was gradually decreased after exposure to light, and the rate of decline was accelerated under low temperatures (Figure 3E).

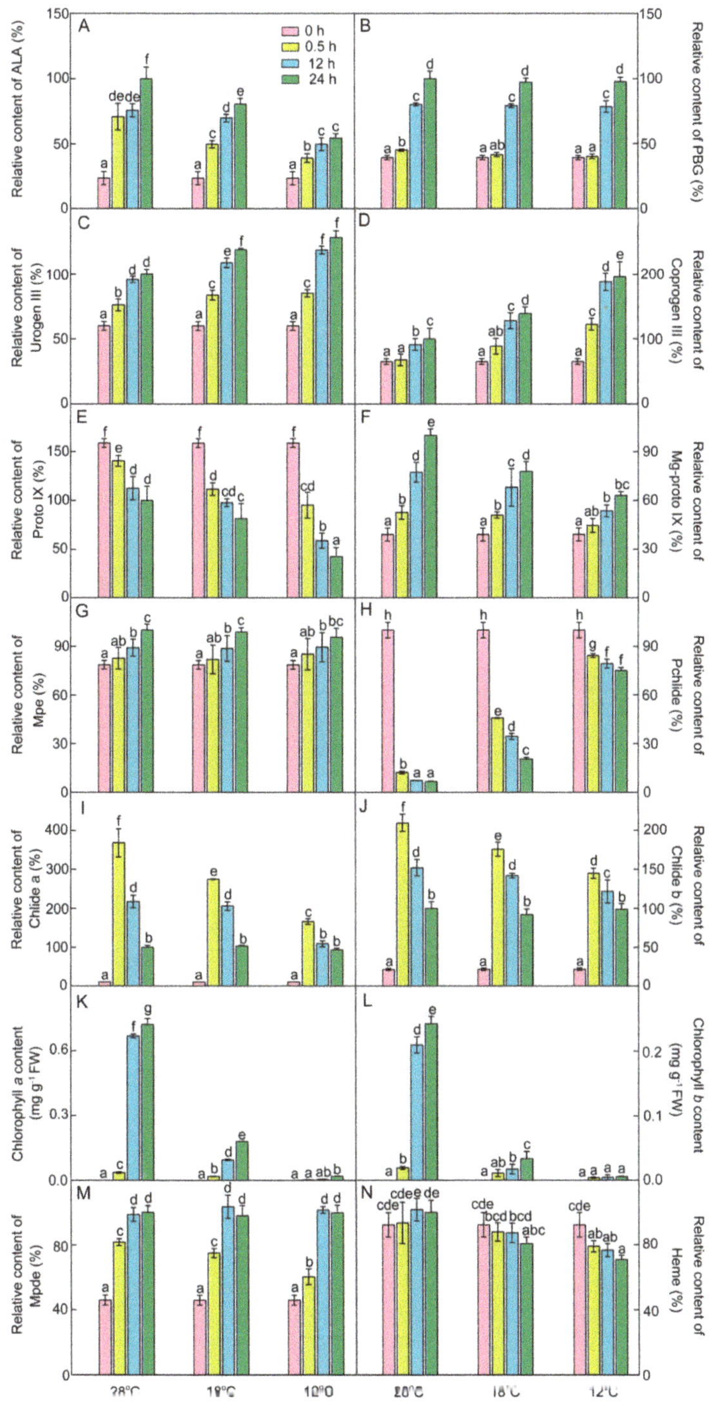

Figure 3. Chlorophyll biosynthesis intermediates during the greening period. δ-Amino levulinic acid (ALA, **A**), porphobilinogen (PBG, **B**), uroporphyrinogen III (urogen III, **C**), coproporphyrinogen III (coprogen III, **D**), protoporphyrin IX (Proto IX, **E**), Mg-protoporphyrin IX (Mg-proto IX, **F**), Mg-protoporphyrin monomethyl ester (Mpe, **G**), protochlorophyllide (Pchlide, **H**), chlorophyllide *a* (Chlide *a*, **I**), chlorophyllide *b* (Chlide *b*, **J**), chlorophyll *a* (Chl *a*, **K**), chlorophyll *b* (Chl *b*, **L**), Mg-protoporphyrin IX diester (Mpde, **M**), and heme (**N**) content of control (28 °C) and cold-stressed (18 °C and 12 °C) rice seedlings after 0 h, 0.5 h, 12 h, and 48 h greening. Six day old etiolated seedlings were treated with 18 °C or 12 °C cold stress. Seedlings were harvested at 0 h, 0.5 h, 12 h, and 48 h of greening and their chlorophyll biosynthesis intermediates contents were measured. The relative content of intermediates at 28 °C after 48 h of light exposure was defined as 100%, except Pchlide; the relative content of Pchlide at 0 h was defined as 100% due to its massive accumulation in the dark. The error bars represent standard deviations of three independent biological replicates. Different letters indicate significantly different at $p < 0.05$ according to Duncan's multiple range tests.

Mg-proto IX is the first intermediate of the Mg branch in the chlorophyll biosynthesis pathway. As shown in Figure 3F, Mg-proto IX increased by 34.5%, 97.5%, and 155.3% after 0.5 h, 12 h, and 48 h of light exposure at 28 °C, respectively. However, the level of Mg-proto IX was significantly decreased during greening under the cold condition. These results demonstrated that chilling stress also inhibited the synthesis of Mg-proto IX during greening. Mg-Proto monomethyl ester (Mpe) content was increased during greening, and was barely influenced by the low temperature (Figure 3G).

In the dark, massive amounts of Pchlide were accumulated, because the protochlorophyllide oxidoreductase (POR) in angiosperms is strictly light-dependent. The etiolated rice seedlings accumulated plenty of Pchlide (Figure 3H). The content of Pchlide was drastically decreased at 28 °C after 0.5 h of light exposure, while Chlide a and Chlide b contents were increased rapidly and then decreased during greening (Figure 3I,J). Compared with 28 °C, the levels of Pchlide were higher but the Chlide a and Chlide b contents were lower in chill-treated groups after 0.5 h and 12 h of light exposure. Obviously, the cold treatment lowered the conversion efficiency of Pchlide to Chlide, especially at 12 °C.

The contents of Chl *a* and Chl *b* were almost undetectable in etiolated seedlings (Figure 3K,L), which were significantly increased at 28 °C during greening. Under low temperatures, the synthesis of Chl *a* and Chl *b* was significantly suppressed during greening, especially at 12 °C. The Chl *a* and Chl *b* contents at 12 °C showed little difference from those in the dark. These results suggested that chilling stress greatly inhibited the synthesis of Chl *a* and Chl *b* during greening.

Mpe can also convert to Mg-protoporphyrin IX diester (Mpde) and then further form Chlide a ester. Mpde contents at both 28 °C and low temperatures were increased during greening (Figure 3M). Heme is the product of the Fe branch, which usually acts as a cofactor in respiration and photosynthesis. Heme contents at different time points had no significant difference during greening (Figure 3N). However, the contents of heme after 48 h of light exposure were decreased by 19.3% and 29.2% at 18 °C and 12 °C, respectively. These results also indicate that chilling stress inhibited the pathway of Fe^{2+} branch during greening.

In summary, the inhibition of chlorophyll biosynthesis under chilling stress may be attributed to inhibited synthesis of ALA and hampered conversion from Pchlide into Chls.

2.4. Effect of Chilling Stress on Enzyme Activities in Chlorophyll Biosynthesis

To further investigate the inhibitory mechanism of chlorophyll biosynthesis, we next examined some key enzymes involved in chlorophyll biosynthesis. Enzymatic activity of glutamate-1-semialdehyde transaminase (GSA-AT), which catalyzes glutamate-1-semialdehyde to ALA, was significantly increased at 28 °C and 18 °C (Figure 4A) during greening. However, low temperature decreased GSA-AT activity, and the activity of GSA-AT at 12 °C had no significant difference from that recorded in the dark (Figure 4A). ALA dehydratase (ALAD) activity was slightly increased after light exposure, and chilling stress had no significant effect on ALAD activity during greening (Figure 4B). Mg-chelatase is a key enzyme which initiates the Mg branch of the chlorophyll biosynthesis pathway. Mg-chelatase activity was increased during greening, and chilling stress had an inhibitory effect on the activity of Mg-chelatase (Figure 4C). POR is a light-dependent enzyme in angiosperms that converts Pchlide to Chlide. POR activity went down gradually during greening (Figure 4D) and chill-treated seedlings showed lower POR activities, especially at 12 °C. In short, the lower ALA content might be attributable to the fact that cold stress inhibited GSA-AT activity, and the inhibition of conversion from Pchlide into Chlide might have been due to the low POR activity under chilling stress.

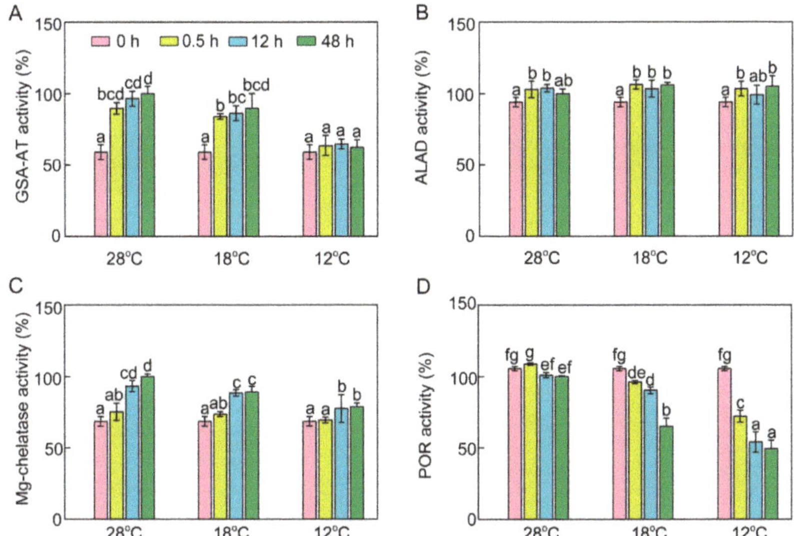

Figure 4. Effects of chilling stress (18 °C and 12 °C) on activities of enzymes involved in chlorophyll biosynthesis. Glutamate-1-semialdehyde transaminase (GSA-AT, **A**), ALA dehydratase (ALAD, **B**), Mg-chelatase (**C**), protochlorophyllide oxidoreductase (POR, **D**) activities of control (28 °C) and chill-stressed (18 °C and 12 °C) rice seedlings after 0 h, 0.5 h, 12 h, and 48 h greening. Six day old etiolated seedlings were treated with 18 °C or 12 °C chilling stress. Seedlings were harvested at 0 h, 0.5 h, 12 h, and 48 h of greening and their activities of enzymes involved in chlorophyll biosynthesis were measured. The activities of enzymes at 28 °C after 48 h of light exposure were defined as 100%. Values are means ± SD from three independent biological replicates. Different letters indicate significant differences according to Duncan's multiple range tests at $p < 0.05$. Each data point is the average of three replicates. The error bars represent SD.

2.5. Effect of Chilling Stress on Transcriptional Expression of Chlorophyll Biosynthetic Genes

To decipher the effects of chilling stress on transcription levels of chlorophyll biosynthetic genes, we analyzed the relative expression of four key chlorophyll biosynthetic genes by RT-qPCR. At 28 °C, the expression of *HEMA* was increased initially, and then decreased after 12 h of light exposure (Figure 5A). This phenomenon might have been due to avoidance of the oxidative stress caused by excessive tetrapyrrole intermediate accumulation. However, the chilling stressed rice seedlings accumulated more transcript of *HEMA* after 12 h of light exposure. At 28 °C, the RNA of *CHLH* after 0.5 h of light exposure was slightly higher than that in etiolated seedlings, but the expression of *CHLH* was much higher at 18 °C (Figure 5B). The expression of most chlorophyll biosynthetic genes was inhibited in the dark, but *PORA* transcripts accumulated in the etiolated seedlings but were degraded rapidly upon illumination (Figure 5C). In contrast, *PORB* mRNA level did not fluctuate at 28 °C after illumination. Interestingly, the mRNA level of *PORB* was very high at 18 °C and significantly lowered at 12 °C (Figure 5D). The expression of *DVR* increased at 28 °C and 18 °C after 12 h of light irradiation, but at 12 °C it had no significant difference from that in the dark (Figure 5E). When the temperature was below 18 °C, the expressions of *CHLH*, *PORB*, and *DVR* were severely repressed. Taken together, the repressed expression of chlorophyll biosynthetic genes might be responsible for the inhibition of chlorophyll biosynthesis under low temperatures.

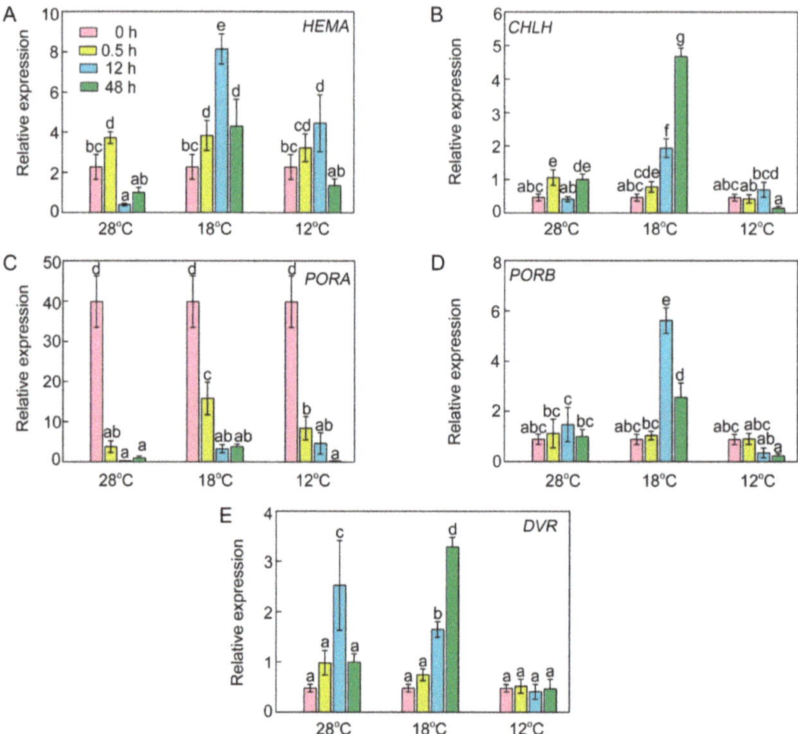

Figure 5. Effect of chilling stress on relative expression of chlorophyll biosynthetic genes *HEMA* (**A**), *CHLH* (**B**), *PORA* (**C**), *PORB* (**D**), and *DVR* (**E**). The expression levels of genes at 28 °C after 48 h of light exposure were set to 1. *OsACTIN1* was used as an internal standard. Values are means ± SD from three independent biological replicates. Different letters indicate significant differences according to Duncan's multiple range tests (*p < 0.05*).

2.6. *Effect of Chilling Stress on Plastid Proteins during Greening*

To determine whether chilling stress affected plastid protein biosynthesis, immunoblotting analysis was performed (Figure 6). SDS-PAGE showed that the proteins with molecular weight from 20 kDa to 35 kDa under low temperatures were much lower than those at 28 °C after 12 h and 48 h of light irradiation (Figure 6C). In etiolated seedlings, PSI (Lhca1, Lhca2, Lhca3, Lhca4, and PsaD) and PSII (D1, D2, CP43, Lchb1, Lchb2, Lchb3, Lchb4, Lchb5, and Lchb6) proteins were undetected, and large amounts of PSI and PSII proteins were rapidly synthesized at 28 °C during greening (Figure 6A,B). Chilling stress inhibited the accumulation of PSI and PSII proteins during greening (Figure 6A,B, Supplementary Figures S1 and S2), especially at 12 °C, where we hardly detected PSI and PSII proteins.

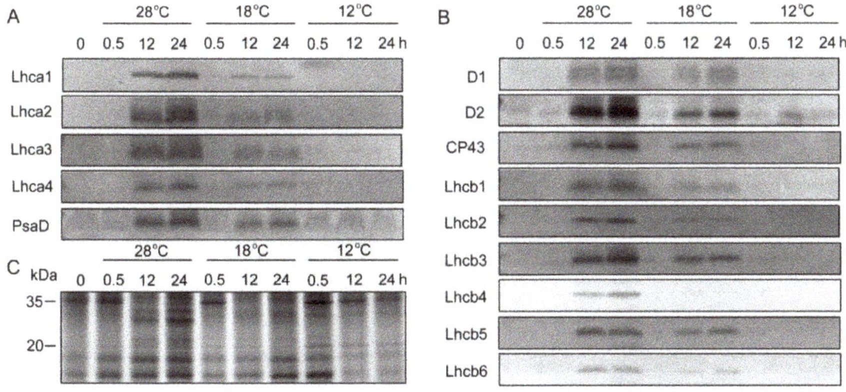

Figure 6. Immunoblot analysis of thylakoid proteins in control and chill-stressed rice seedlings. Six day old etiolated seedlings were treated with 18 °C or 12 °C chilling stress. Thylakoid proteins were isolated from control and chill-stressed seedlings after 0 h, 0.5 h, 12 h, and 48 h of greening. Immunoblot analyses were performed with antibodies specific for representative photosystem I (PSI) (**A**) and photosystem II (PSII) (**B**). The SDS–PAGE of 20 ug plastid protein stained by Coomassie blue (**C**).

2.7. Effect of Chilling Stress on Chloroplast Biogenesis during Greening

Chloroplast biogenesis normally depends on a stable supply and correct stoichiometry of chlorophyll and photosynthetic proteins during greening. To further investigate the effect of low temperature on plastid development, plastid morphology was analyzed via transmission electron microscopy (TEM). The results showed that the proplastid developed into the etioplast that contains the prolamellar bodies (PLBs) in etiolated seedlings (Figure 7A). When seedlings were illuminated (120 μmol m^{-2} s^{-1}) for 48 h at 28 °C, PLBs disappeared and thylakoids formed and grana thylakoids stacked regularly (Figure 7B). Grana stacking was inhibited and thylakoids were much looser at 18 °C (Figure 7C), and no functional thylakoid structure was formed after 48 h of greening at 12 °C (Figure 7D). Taken together, these results suggest that chilling stress inhibited the biogenesis of chloroplast, which might have been due to the lack of chlorophylls and photosynthetic proteins.

Chlorophyll fluorescence is an important indicator of the work status of chloroplasts. Compared with the seedlings grown under low temperature, the seedlings at 28 °C showed higher quantitative values of maximum PSII yield (Fv/Fm) and lower non-photochemical quenching (NPQ) (Figure 8). Minimal fluorescence yield (F$_0$) showed no big fluctuations between 28 °C, 18 °C, and 12 °C treatments, but maximal fluorescence yield (Fm) was significantly lower in the chill-stressed seedlings.

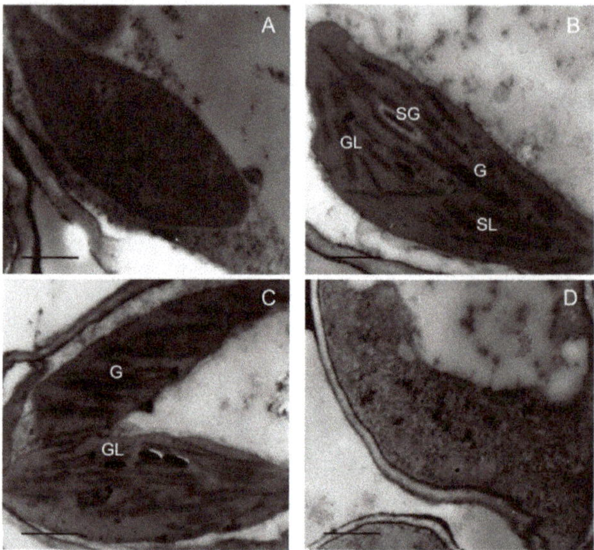

Figure 7. Effect of chilling stress (18 °C and 12 °C) on chloroplast biogenesis of rice seedlings. Plastid ultrastructure of etiolated seedlings (**A**); chloroplast ultrastructure after 48 h of greening under normal temperature (28 °C) condition (**B**); chloroplast ultrastructure after 48 h of greening at 18 °C (**C**); chloroplast ultrastructure after 48 h of greening at 12 °C (**D**). G: granum; SL: stroma lamellae; GL: grana lamellae; SG: starch grain. Bar = 1 μm.

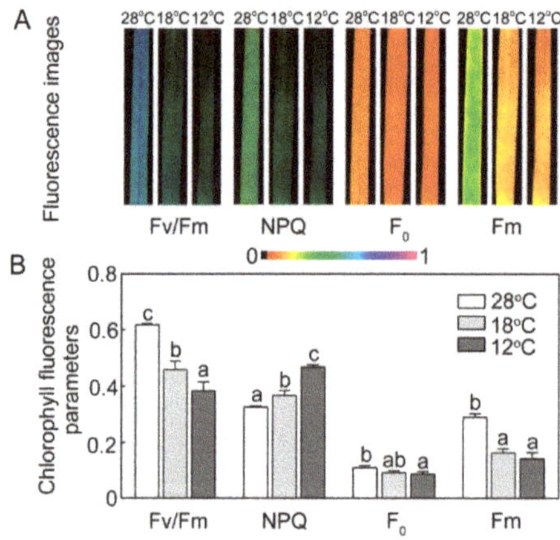

Figure 8. Effect of chilling stress on chlorophyll fluorescence parameters of rice seedlings. The chlorophyll fluorescence images (**A**) and chlorophyll fluorescence parameters (Fv/Fm, NPQ, F_0, Fm) (**B**) after 48 h of greening. Six day old etiolated seedlings were treated with 18 °C or 12 °C chilling stress. Values are means ± SD from three independent biological replicates. Different letters indicate significant differences according to Duncan's multiple range tests ($p < 0.05$).

2.8. Effect of Chilling Stress on ROS Accumulation, Oxidation, and Electrolyte Leakage during Greening

As crucial indexes of oxidative damages under chilling stress, the contents of hydrogen peroxide (H_2O_2), superoxide anion radical ($O_2{}^{\cdot-}$), and malondialdehyde (MDA) were determined. Histochemical detection and quantification analysis showed that H_2O_2 increased slightly, but the superoxide anion radical ($O_2{}^{\cdot-}$) levels remained almost stable at 28 °C during greening. An ROS burst occurred in chill-treated leaves, especially at 12 °C (Figure 9A–D), indicating that chilling stress could induce ROS accumulation. MDA content and electrolyte leakage (EL) were quantified to examine the lipid peroxidation and the damage to cellular membranes. MDA had a slight increase and EL had no remarkable change at 28 °C during greening, but both MDA and EL significantly increased under low temperatures during greening, especially at 12 °C (Figure 9E,F). These results indicate that chilling stress induced ROS accumulation and caused lipid peroxidation and finally destroyed the integrity of membranes during greening.

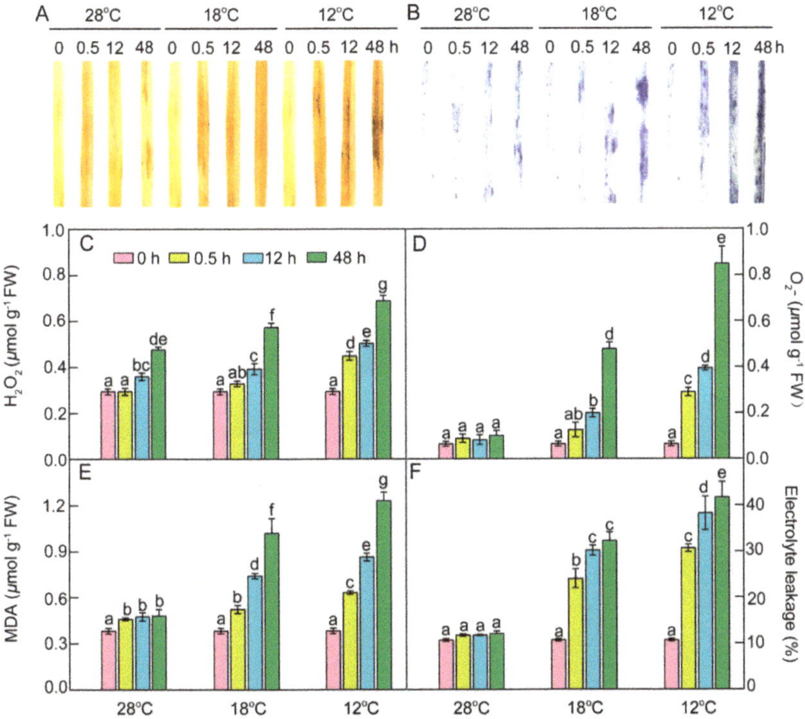

Figure 9. Effects of chill-stressed (18 °C and 12 °C) on H_2O_2 (**A,C**), $O_2^{\cdot-}$ (**B,D**), MDA content (**E**), and EL (**F**) of rice seedlings. Histochemical detection (**A,B**), content of H_2O_2 (**C**), $O_2^{\cdot-}$ (**D**), MDA (**E**), and EL (**F**) of control (28 °C) and chill-stressed (18 °C and 12 °C) rice seedlings after 0 h, 0.5 h, 12 h, and 48 h greening. Six day old etiolated seedlings were treated with 18 °C or 12 °C chilling stress. Seedlings were harvested at 0 h, 0.5 h, 12 h, and 48 h of greening and their ROS levels were measured. Values are means ± SD from three independent biological replicates. Different letters indicate significant differences according to Duncan's multiple range tests ($p < 0.05$).

2.9. Effect of Chilling Stress on Cell Death during Greening

We further examined the effect of low temperature on cell death of leaves using Trypan-blue staining (Figure 10). Few cells could be stained at 28 °C, but the number of dead cells significantly increased under low temperatures during greening, especially at 12 °C. These results indicate that the low temperature aggravated cell death during greening.

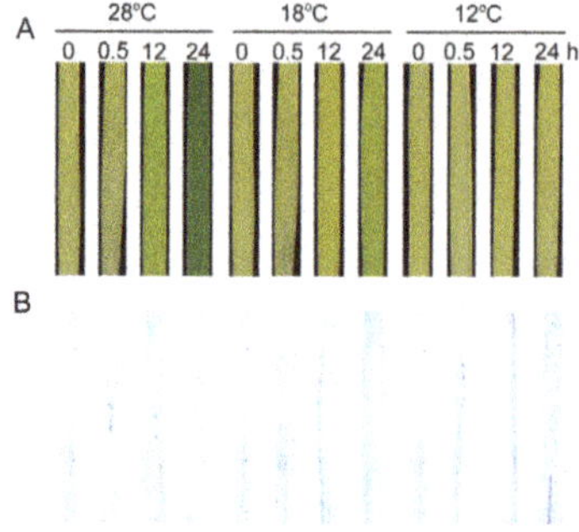

Figure 10. Effect of chilling stress on cell death of rice seedlings. Six day old etiolated seedlings were treated with 18 °C or 12 °C chilling stress. Phenotypes of rice leaves at different time points (**A**); Trypan-blue staining (**B**) of control (28 °C) and chill-stressed (18 °C and 12 °C) leaves after 0 h, 0.5 h, 12 h, and 48 h greening.

2.10. Effect of Chilling Stress on Epidermis of Rice Leaves during Greening

To investigate whether low temperature affected the epidermal characteristics, we observed the epidermis cells after 48 h of greening. The upper and lower epidermis layers of rice are mainly composed of stomata apparatus and epidermis cells. In addition, there were some trichomes. The shapes and sizes of upper and lower epidermis cells showed no significant difference between the seedlings at 28 °C and low temperatures (Figure 11), while stomata under low temperature were smaller than those at 28 °C (Table 1). Meanwhile, chilling stress had no significant effect on the number of trichomes (Table 1).

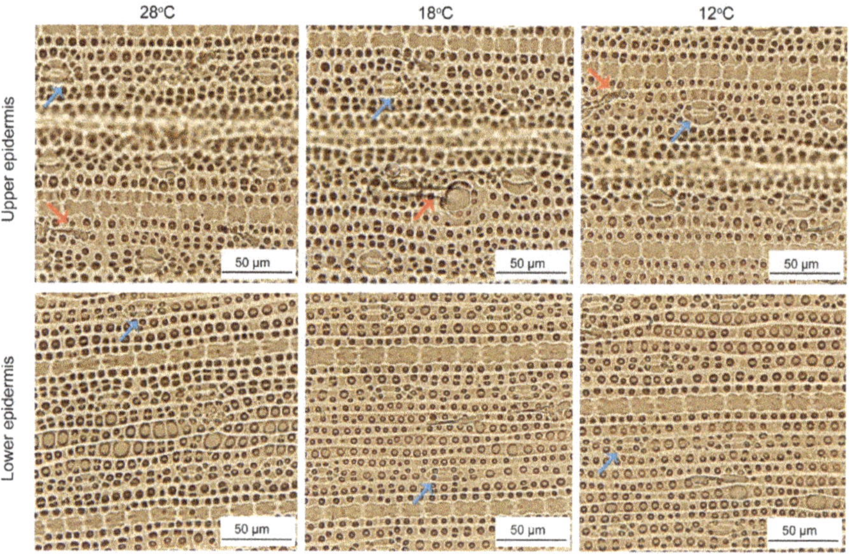

Figure 11. The epidermis cells development of control and chill-stressed rice leaves. Six day old etiolated seedlings were treated with 18 °C or 12 °C chilling stress. Seedlings were harvested at 48 h of greening and their epidermal cells' characteristics were observed. Red arrows and blue arrows represent trichomes and stoma, respectively.

Table 1. Effect of chilling stress on stomatal characteristics and trichomes number of rice seedlings.

Type of Leaf		Stomata Length (µm)	Stomata Width (µm)	Stomata Size (µm²)	Stomata Density (No·mm⁻²)	Trichomes (Per Unit leaf)
28 °C	Upper epidermis	27.50 ± 3.28a	18.40 ± 1.23a	384.32 ± 24.75a	195.51 ± 14.97bc	49.61 ± 10.63a
	Lower epidermis	21.92 ± 1.73c	15.60 ± 1.34b	307.98 ± 20.86b	210.94 ± 25.69a	48.51 ± 6.91a
18 °C	Upper epidermis	23.43 ± 1.44b	15.32 ± 1.74b	303.59 ± 20.76bc	193.67 ± 16.56bc	52.18 ± 10.08a
	Lower epidermis	21.80 ± 1.52c	13.66 ± 1.08c	294.31 ± 16.00c	206.53 ± 30.24ab	51.45 ± 10.11a
12 °C	Upper epidermis	23.19 ± 1.62b	13.57 ± 0.89c	299.30 ± 17.94bc	184.12 ± 22.68c	51.08 ± 9.38a
	Lower epidermis	22.62 ± 1.62c	13.63 ± 0.91c	270.20 ± 26.74d	195.88 ± 14.94bc	51.45 ± 5.33a

The values are expressed as mean ± SD (n = 50); different letters represent significant difference ($p < 0.05$).

3. Discussion

Chlorophyll biosynthesis is affected by various biotic and abiotic factors. Previous studies have reported that water and salt stresses lead to the severe inhibition of chlorophyll biosynthesis during de-etiolation [1,30]. Temperature is one of the major environmental factors that can inhibit chlorophyll biosynthesis and chloroplast biogenesis, and thus affect photosynthesis [31–33]. Our previous study also showed that cold stress dramatically decreases the net photosynthetic rate, stomatal conductance, intercellular CO_2 concentration, and water use efficiency in rice seedlings [34]. In this study, our results demonstrated that chlorophyll was significantly increased with the period of light exposure during greening at 28 °C (Figure 2C). However, chlorophyll biosynthesis was obviously inhibited in chill-stressed rice seedlings. Our study further found that ALA synthesis was significantly inhibited in chill-stressed seedlings (Figure 3A). That is to say, the early step of chlorophyll biosynthesis was inhibited by the low temperature, which ultimately led to a significant reduction of chlorophyll and heme contents. Similar changes have also been observed in water- and salt-stressed rice/wheat seedlings during early seedling development [1,30]. Meanwhile, GSA-AT activity was reduced by low temperature during de-etiolation (Figure 4A), suggesting that decreased ALA synthesis in chill-stressed rice seedlings might have been due to the downregulated GSA-AT activity.

In the whole chlorophyll synthesis process, Mg-proto IX and Chlide levels were reduced significantly in stressed seedlings (Figure 3F,I,J), and Mg-chelatase and POR activities were decreased synchronously in chill-stressed seedlings (Figure 4C,D). Declines in Mg-chelatase and POR activities were also observed in water-stressed rice [1]. In plants, Mg-chelatase is composed of three non-identical subunits that are encoded by *CHLI*, *CHLD*, and *CHLH* [35,36]. Previous studies have demonstrated that the mutations of *Chl1* and *Chl9* genes which encode CHLD and CHLI could reduce Mg-chelatase activity, and thus inhibit chlorophyll synthesis [37]. Additionally, ATP is required for the catalytic activity of Mg-chelatase, while ferrochelatase is inhibited by ATP. More Pchlide is allocated to the Mg branch when ATP levels are higher in the light; conversely, the Mg branch is inhibited in the dark [8]. In addition, we found that *CHLH* expression level was much higher at 18 °C than that at 28 °C, but its expression was the lowest at 12 °C. *CHLH* expression in etiolated seedlings was suppressed with lower histone acetylation levels by PIF3, but increased rapidly during greening with higher acetylation levels of histones [38]. Chilling stress might repress histone acetylation levels of *CHLH* at 12 °C. Numerous studies have indicated that post-translational regulation plays an essential role in chlorophyll biosynthesis [39–41]. The caseinolytic protease activity counteracts the binding of GluTR-binding protein to assure an appropriate content of GluTR and an adequate ALA synthesis at the post-translational level [42]. This might be why the increased transcriptional levels of *DVR* and *CHLH* under chilling stress were not directly proportional to their corresponding products (Figures 2 and 5). Previous studies have also shown that in water- and chill-stressed rice/cucumber seedlings, Mg-chelatase activity and its gene/protein expression were downregulated [1,43]. The light-dependent POR is a plastid (pro)thylakoid-membrane-associated protein, which binds to NADPH and Pchlide to form a ternary complex in etioplasts [44]. There are three POR isoenzymes in *Arabidopsis thaliana*. *PORA* transcripts accumulate in etiolated seedlings, but the expression of *PORA* is strongly downregulated when exposed to light. The *PORB* transcript can be detected throughout the growth and development of plants, while the expression of *PORC* is induced by light and is predominantly present in fully matured green tissues [8,45]. However, there are only two isoenzymes in rice, namely PORA and PORB [1]. In the present study, POR activity was significantly inhibited in stressed seedlings, especially at 12 °C (Figure 4D). *PORA* transcription decreased dramatically when etiolated rice seedlings were exposed to light at 28 °C, but chill-treated rice seedlings had a relatively high *PORA* mRNA level (Figure 5). Similarly, POR activity and *PORB* transcript abundance were downregulated in water- and chill-stressed rice/cucumber seedlings [1,43]. However, we found that a slightly lowered temperature (18 °C) could significantly increase the expression of *PORB*. As another important tetrapyrrole, heme content showed a 30% drop in seedlings at 12 °C compared to that at 28 °C after 24 h of greening (Figure 3A), which was far less than the declined proportion of chlorophylls, indicating that Chl biosynthesis is more sensitive to low temperatures than heme synthesis. Meanwhile, this result indicated that more tetrapyrrole metabolic intermediates were allocated to the heme branch than to the chlorophyll branch under chilling stress. Thus, the inhibition of chlorophyll biosynthesis under chilling stress might be attributable to the blocked synthesis of ALA and the inhibition of conversion from Pchlide into Chls.

It has been found that the development of thylakoid is inhibited when chlorophyll biosynthesis is reduced [37]. Chlorophyll content was decreased and thylakoid membrane was not stacked in *porB porC* double mutants [46]. These results indicated that the inhibition of chlorophyll biosynthesis affected the biogenesis of chloroplasts, which led to the reduction of granum lamellae and thylakoid membrane proteins. Thus, a stable supply and correct stoichiometry of chlorophyll are necessary for chloroplast biogenesis. In the present study, chloroplast biogenesis was significantly affected by chilling stress (Figure 7C,D). The grana lamellae were disorganized at 18 °C and no grana lamellar structure was formed at 12 °C (Figure 7D). The synthesis of thylakoid membrane protein is of importance to the assembly of the photosystem during greening. Western blotting results showed that thylakoid protein synthesis was obviously inhibited under chilling stress (Figure 6). PSII is considered a primary target of photodamage, and the D1 protein is the most vulnerable component in the PSII reaction center under stress conditions [47]. In this study, the content of D1 was obviously lower under low temperatures, and was almost undetectable at 12 °C. Moreover, low temperatures greatly inhibited the content of the peripheral antenna proteins of PSII, including Lhcb1, Lhcb2, Lhcb3, Lhcb4 (CP29), Lhcb5 (CP26), and Lhcb6 (CP24) and peripheral antenna proteins of PSI, including Lhca1, Lhca2, Lhca3, Lhca4, and PsaD (Figure 6), especially at 12 °C.

Chlorophyll fluorescence analysis has been proven to be a powerful method for obtaining the functional status of PSII [48]. *OsAsr1* rice seedlings have a high value of Fv/Fm, which is correlated with an enhanced cold tolerance [33], suggesting that Fv/Fm could be an indicator of cold tolerance. The decline of Fv/Fm in stressed seedlings may be due to the partial inactivation of PSII reaction centers [49]. In the present study, after 48 h of greening, chill-treated seedlings showed an obvious reduction of Fv/Fm (Figure 8A,B), and the lower Fv/Fm might have been because the low temperature suppressed the assembly and formation of PSII. Meanwhile, the ultrastructural changes of the plastids also indicated that chilling stress affected grana stacking and thylakoid integrity (Figure 6C,D), thereby resulting in a decrease in PSII activity in the stressed plants. Nonphotochemical quenching is a self-protection mechanism in plants. Previous studies have shown that effective heat dissipation in plants can reduce the occurrence of photoinhibition induced by stresses [49]. In this study, the increase of NPQ in stressed seedlings indicated that more excess light energy needed to be dissipated because of the low activity of PSII at low temperatures.

ROS plays double roles under cold stress. On one hand, ROS as a signal can trigger stress-responsive gene expression and the MKK6-MPK3 signaling pathway [26]. On the other hand, excessive ROS in plants can directly induce membrane lipid peroxidation, cell integrity damage, and cell death [50]. PSI and PSII in chloroplasts are the major sources of ROS in plants. Numerous studies have indicated that ROS production increases significantly in plants and the balance of ROS is disturbed under stress conditions [49,51]. Cold signals can be sensed by rice cells through changes in membrane rigidity and osmotic pressure [52]. The membrane rigidity increases under cold stress, resulting in a high electrolyte leakage [53,54]. In accordance with ROS accumulation, the level of lipid peroxidation (MDA) and damage to the cellular membranes were also higher under chilling stress (Figure 9E,F). The increased EL indicated the enhancement of membrane rigidification, which is required for the cold-activated SAMK signaling cascade via cytoskeleton, Ca^{2+} fluxes, and CDPKs [55]. Cold stress initially promotes Ca^{2+} influx into the cytoplasm, which might be controlled by Ca^{2+} channels that are activated by membrane rigidification [56]. Furthermore, the calcium signaling cascade interprets and amplifies the rice-sensing cold signal and subsequently activates the DREB-CRT/DRE pathway, which is important for the cold response [57]. At the same time, our observations showed that leaf cell death was significantly increased under the cold-stress condition during greening (Figure 10). Proline and soluble sugars served as osmoprotectants against oxidative damage, which are also considered indicators to assess the potential cold tolerance of plants [34]. Our previous study indicated that proline and soluble sugar accumulation is enhanced in cold-stressed rice seedlings [34].

Over the past decades, researchers have made extensive efforts to improve cold tolerance in crops, especially in rice. The increasing global food demand, together with rapid population growth and frequent occurrence of chilling forces scientists to speed up and push forward the improvement of rice cold tolerance. ALA synthesis is the rate-limiting step in the whole tetrapyrrole metabolic net, and it is obviously inhibited in chill-treated rice seedlings (Figure 3A). Based on this finding, we propose that application of exogenous ALA or overexpression of *HEMA* or *GSA* gene in plants may overcome inadequate chlorophyll biosynthesis and maintain the structural and functional integrity of chloroplasts, and thus improve cold tolerance. Because tetrapyrrole intermediates are easily activated by light, leading to photooxidative damage, the concentration of exogenous ALA should be applied accurately. Several investigations have recognized that ALA pretreatment enhances plants' tolerance to chilling by increasing the activities of antioxidant enzymes to eliminate excessive ROS and improving chlorophyll fluorescence and photosynthesis [58–60]. Thus, our results suggested the protective role of exogenous ALA and contribute to further illustrate its mechanism. Nevertheless, exogenous ALA increased chlorophyll accumulation in etiolated oilseed rape, but failed to enhance its cold tolerance [61]. Application of exogenous ALA is an exciting field to explore, and might be beneficial for increasing chlorophyll content and improving cold resistance of rice seedlings. Thus, more refined investigations of the effect of exogenous ALA in etiolated seedlings under chilling stress are still required.

Given that transgenic technologies have been developed intensively and almost all chlorophyll biosynthetic genes have been identified in many crops, genetically engineered crops provide a new opportunity to solve the threats from environmental stresses. Thus, genetic modification of chlorophyll biosynthetic genes might be a promising approach for improving plant cold tolerance. A previous study suggested that *CHLG*-over-expressing plants have increased ALA synthetic capacity and increased chelatase activity, indicating that overexpression of *CHLG* could stimulate chlorophyll biosynthesis [62]. Overexpression of *HEMA*, *DVR*, and *CHLG* in plants may be helpful to improve the cold tolerance of rice seedlings during greening by increasing chlorophyll biosynthesis.

4. Materials and Methods

4.1. Plant Material and Growth Conditions

Rice (*Oryza sativa* L.) cultivar DM You 6188 was used as experimental material, and was purchased from Ya'an seed store. Seeds were sterilized with 3% (*m/v*) sodium hypochlorite for 10 min, washed five times with distilled water and soaked in distilled water for 36 h, then placed on moist filter papers with 1/4 strength Hoagland nutrient solution and grown in the dark at 28 °C before chilling treatment. After 6 days, the etiolated seedlings were transferred to vermiculite with 1/4 strength Hoagland nutrient solution in the dark. The seedlings were then exposed to light ($120 \ \mu mol \ m^{-2} \ s^{-1}$) and transferred to 28 °C, 18 °C, and 12 °C, respectively. The first leaf was used to measure physiological and biochemical parameters at 0 h, 0.5 h, 12 h, and 48 h after exposure to light, and all experiments were repeated at least three times.

4.2. Determination of Shoot Dry Weight (DW)

The shoots were collected at 0 h, 0.5 h, 12 h, and 48 h after being exposed to light, washed with tap water and rinsed twice with distilled water, gently wiped with a paper towel, and then oven-dried to a constant weight at 80 °C for DW determination.

4.3. Determination of Chlorophyll, Carotenoids and Protein

Chlorophyll and carotenoids were extracted from 0.1 g fresh rice seedlings with 80% acetone. The absorbance of the extract was recorded at 663, 646, and 470 nm according to Lichtenthaler and Wellburn using a spectrophotometer (UV-1750, Shimadzu, Japan) [63]. Protein content was determined by the Bradford method [64].

4.4. Determination of Chlorophyll Precursors

δ-Aminolevulinic acid (ALA) was measured according to Dei [65]. Briefly, 0.5 g of fresh leaf was ground in 10 mL of 4% trichloracetic acid with an ice bath, then centrifuged at $18,000 \times g$ for 15 min. Next, 500 µL of the supernatant was mixed with 2.35 mL of 1 mol/L sodium acetate and 1.5 mL of acetyl-acetone. The mixture was then heated in boiling water for 10 min. After cooling to 25 °C, 2 mL of the mixture was added to 2 mL of Ehrlich-Hg and reacted in the dark for 15 min. The absorption was recorded at 553 nm. The content of ALA was evaluated from the calibration curve prepared from known concentration of ALA.

Porphobilinogen (PBG) was extracted as described by Bogorad [66] with some modifications. Briefly, 0.5 g of fresh leaf was homogenized with 5 mL of extraction solution (0.6 mol/L Tris, 0.1 mol/L EDTA, pH 8.2) in an ice bath and centrifuged for 10 min at $18,000 \times g$. Next, 2 mL of the mixture was mixed with 2 mL of Ehrlich-Hg and reacted in the dark for 15 min, with absorbance measured at 553 nm.

Uroporphyrinogen III (urogen III) and coproporphyrinogen III (coprogen III) were assessed according to Bogorad [66] and Rebeiz et al. [67], with some modifications. To determine urogen III content, 1.0 g fresh sample was extracted in an ice bath with 10 mL 0.067 mol/L PBS pH 6.8, then centrifuged for 10 min at $18,000 \times g$. Next, 5 mL of the supernatant was mixed 0.25 mL of 1% $Na_2S_2O_3$. The mixture was illuminated by strong light for 20 min, after which the pH was adjusted to 3.5 with 1 mol/L formic acid. The mixture was extracted three times with 5 mL of ether. The water phase was used to measure the absorbance at 405.5 nm. For analysis of coprogen III, the ether phase was extracted with 0.1 mol/L HCl three times. The HCl phase was used to measure the absorbance at 399.5 nm.

Protoporphyrin IX (Proto IX) was measured based on the method of Rebeiz et al. [67]. Fresh leaves (1.0) were homogenized with extracted solution (acetone: 0.1 mol/L $NH_3 \cdot H_2O$, 9:1, *v/v*) in an ice bath, then centrifuged for 10 min at $18,000 \times g$. Next, 5 mL of the supernatant was mixed with 2 mL n-hexane. The acetone phase was used to record the fluorescence emission spectra at 400, 622, 633, and 640 nm.

Similarly, Mg-protoporphyrin IX (Mg-proto IX), Mg-Proto monomethyl ester (Mpe), Mg-protoporphyrin IX diester (Mpde), protochlorophyllide (Pchlide), and chlorophyllide (Chlide) were determined based on their fluorescence emission spectrums [67]. Heme was extracted and quantified as described by Wilks [68].

4.5. Measurement of Chlorophyll Biosynthetic Enzyme Activities

Fresh leaves were collected at 0 h, 0.5 h, 12 h, and 48 h after exposure to light, and immediately grounded with extracting buffers at 4 °C. Glutamate 1-semialdehyde aminotransferase (GSA-AT) activity was analyzed according to Shalygo et al. [62]. ALA dehydratase (ALAD) activity was evaluated as described by Kumar et al. [43]. Mg-chelatase activity was determined based on Yaronskaya et al. [69]. POR activity was measured according to Rebeiz et al. [67].

4.6. Isolation of RNA and Quantitative Real-Time PCR

Total RNA was extracted from rice leaves using a Column Plant RNA_{OUT} V1.0 Kit (Tiandz Inc., Beijing, China) according to the manufacturer's instructions. The first strand of cDNA was synthesized using PrimeScript^{TM} RT reagent Kit with gDNA Eraser (Perfect Real Time) (TaKaRa Bio Inc., Dalian, China), following standard protocol. The quantitative real-time PCR was carried out with the diluted cDNA and SYBR Premix Ex Taq^{TM} II (TaKaRa Bio Inc., Dalian, China) using CFX96 TouchTM Real-Time PCR Detection Systems (Bio-Rad, Chicago, USA), as described previously [70]. The relative expression level of *OsACTIN1* was normalized. The primers used for quantitative real-time PCR are listed in Supplementary Table S1.

4.7. Isolation of Thylakoid Proteins and Western Blotting

Thylakoid membrane proteins were isolated as described by Fristedt et al. [71]. Isolated thylakoid membrane protein was separated by SDS-PAGE (5% acrylamide stacking gel + 15% separation gel + 6 M urea) [72]. Western blotting analysis was performed according to Chen et al. [73]. The primary antibodies (all raised in rabbits) including anti-Arabidopsis D1, D2, CP43, LHCb1, LHCb2, LHCb3, LHCb4, LHCb5, LHCb6, LHCa1, LHCa2, LHCa3, and LHCa4, and horseradish-peroxidase-conjugated secondary antibody were purchased from *Agrisera* (Umea, Sweden). The Western blotting signal was detected by a chemiluminescent detection system (ECL, GE Healthcare, Buckinghamshire, UK). The quantification of immunoblots was done with Quantity One software (Bio-Rad, Hercules, CA, United States).

4.8. Observation of Transmission Electron Microscopy

Transmission electron microscopy (TEM) analysis of leaves was carried out following a previous method [74]. Leaf tissue was fixed with 3% glutaraldehyde in 0.1 M sodium cacodylate buffer (pH 6.9) at 4 °C overnight, after being washed with phosphate buffer three times. Samples were post-fixed with 2.5% osmium tetroxide, then dehydrated in a gradient solution of alcohol–acetone mixture and embedded in Epon Ultrathin cross sections were cut with an ultramicrotome (Ultracut F-701704, Reichert-Jung, Reichert, Austria), which was then stained with uranyl acetate and observed using a transmission electron microscope (TEM H-9500, Itachi, Tokyo, Japan) operating at 75 kV.

4.9. Measurement of Chlorophyll Fluorescence

Chlorophyll fluorescence was imaged using a modulated imaging chlorophyll fluorometer (the Imaging PAM M-Series Chlorophyll Fluorometer System, Heinz-Walz Instruments, Effeltrich, Germany) according to the instructions. The rice leaves were adapted in the dark for 30 min prior to the fluorescence assay, and minimum fluorescence yield (F_0), maximum fluorescence yield (Fm) and nonphotochemical quenching (NPQ), and maximal quantum yield of PSII photochemistry (Fv/Fm) were then determined according to the method of Zhou et al. [75].

4.10. Analysis of Reactive Oxygen Species

Histochemical staining of hydrogen peroxide (H_2O_2) and superoxide anion radicals ($O_2{}^{\cdot-}$) was performed using 3, 3-diaminobenzidine (DAB) and nitro blue tetrazolium (NBT), respectively [76]. Leaf tissue was immersed in NBT (1 mg/mL) solution for 2 h or in DAB (0.5 mg/mL) solution for 12 h in the dark. The stained leaves were decolorized in boiling ethanol (90%, *v/v*) for 2 h. H_2O_2 was quantitated according to Velikova et al. [77] and calculated using a standard curve of H_2O_2 reagent. The quantification of $O_2{}^{\cdot-}$ was determined as described by Nahar et al. [78] and calculated using a standard curve of a $NaNO_2$ reagent.

4.11. Determination of Malondialdehyde (MDA) and Electrolyte Leakage (EL)

The level of membrane lipid peroxidation was estimated by MDA content, which was determined by thiobarbituric acid (TBA) assay [78]. Fresh leaves (0.5 g) were homogenized in 5 mL of 5% (*m/v*) trichloroacetic acid (TCA) and centrifuged at 4 °C for 10 min at 8000× *g*. Next, 2 mL of 0.5% TCA containing 0.67% TBA was added to 2 mL of the supernatant. The mixture was incubated at 95 °C for 30 min and then instantly cooled on ice and centrifuged at 4 °C for 10 min at 5000× *g*. The absorbance of the supernatant was recorded at 450, 532 nm, and 600 nm.

EL of rice leaves was determined using an electrical conductivity meter (DDS-309+, Chengdu, China) following Ning et al. [79]. Relative EL was expressed as the ratio of initial conductivity to the conductivity after the samples were boiled for 15 min to achieve 100% electrolyte leakage.

4.12. Trypan-Blue Staining

Dead cells were visually detected using a Trypan-blue staining method as described by Liang et al. [80] with some modifications. Leaves were stained with lactophenol–Trypan blue solution for 20 min under vacuum conditions. The stained leaves were then decolorized in boiling ethanol (90%, *v/v*) for 2 h. Samples were equilibrated with 70% glycerol for scanning.

4.13. Characteristics of Epidermis Cell and Stomata

The upper and lower epidermis layers of rice leaves after 48 h of greening were observed using light microscopy by transparent nail polish imprint method [81]. The epidermis cells' characteristics and stomatal characteristics, including stomatal length, width, size, and density, were observed via Olympus fluorescence microscopy (DP71) (40 × 10) and measured and counted by an image analysis system (Image-Pro Plus 6.3). Five microscope fields were randomly selected in each slide. Each treatment group was repeated 50 times to measure all the characteristic parameters under the microscope.

4.14. Statistical Analysis

All experiments were repeated at least three times, and mean values are presented with standard deviation (SD). The data analysis was performed using IBM SPSS Statistics. Duncan's multiplication range test was used for comparison among different treatments. The difference was considered to be statistically significant when $p < 0.05$.

5. Conclusions

In summary, our results showed that chilling stress induced ROS accumulation, leading to lipid peroxidation and cell death during greening. In particular, our data highlighted the detailed regulation of photosynthetic pigments under chilling stress during greening, suggesting that the inhibition of chlorophyll biosynthesis may be attributable to the blocked synthesis of ALA and the inhibition of conversion from Pchlide into chlorophylls. In this case, we have proposed that application of ALA or overexpression of *HEMA*, *DVR*, and *CHLG* may be good for increasing chlorophyll biosynthesis and improving cold resistance of rice seedlings during greening.

Supplementary Materials: Supplementary materials can be found at http://www.mdpi.com/1422-0067/21/4/1390/s1. Figure S1: Relative content of PSI proteins in control and chill-stressed rice seedlings. Figure S2: Relative content of PSI proteins in control and chill-stressed rice seedlings. Table S1: The primers used for quantitative real-time PCR.

Author Contributions: M.Y. and S.Y. designed the experiments; Y.Z., Q.H., C.D., Y.H., J.L., and T.C. performed the experiments. S.F., L.Z., Z.Z., and Y.C. analyzed the data and results. All authors have read and agreed to the published version of the manuscript.

Funding: This work was supported by Sichuan Agricultural University.

Conflicts of Interest: The authors declare no conflict of interest.

References

1. Dalal, V.K.; Tripathy, B.C. Modulation of chlorophyll biosynthesis by water stress in rice seedlings during chloroplast biogenesis. *Plant Cell Environ.* **2012**, *35*, 1685–1703. [CrossRef] [PubMed]
2. Zhang, D.; Li, Y.H.; Zhang, X.Y.; Zha, P.; Lin, R.C. The SWI2/SNF2 chromatin-remodeling ATPase BRAHMA regulates chlorophyll biosynthesis in *Arabidopsis*. *Mol. Plant* **2017**, *10*, 155–167. [CrossRef]
3. Le, L.P.; Böddi, B.; Kovacevic, D.; Juneau, P.; Dewez, D.; Popovic, R. Spectroscopic analysis of desiccation-induced alterations of the chlorophyllide transformation pathway in etiolated Barley leaves. *Plant Physiol.* **2001**, *127*, 202–211. [CrossRef]
4. Tang, W.J.; Wang, W.Q.; Chen, D.Q.; Ji, Q.; Jing, Y.J.; Wang, H.Y.; Lin, R.C. Transposase-derived proteins FHY3/FAR1 interact with phytochrome-interacting factor1 to regulate chlorophyll biosynthesis by modulating *HEMB1* during deetiolation in *Arabidopsis*. *Plant Cell* **2012**, *24*, 1984–2000. [CrossRef] [PubMed]
5. Eckhardt, U.; Grimm, B.; Hörtensteiner, S. Recent advances in chlorophyll biosynthesis and breakdown in higher plants. *Plant Mol. Biol.* **2004**, *56*, 1–14. [CrossRef] [PubMed]
6. Masuda, T. Recent overview of the Mg branch of the tetrapyrrole biosynthesis leading to chlorophylls. *Photosynth. Res.* **2008**, *96*, 121–143. [CrossRef]
7. Nagata, N.; Tanaka, R.; Tanaka, A. The major route for chlorophyll synthesis includes [3,8-divinyl]-chlorophyllide a reduction in *Arabidopsis thaliana*. *Plant Cell Physiol.* **2007**, *48*, 1803–1808. [CrossRef]
8. Moulin, M.; Smith, A.G. Regulation of tetrapyrrole biosynthesis in higher plants. *Biochem. Soc. Trans.* **2005**, *33*, 737–742. [CrossRef]
9. Yuan, M.; Zhao, Y.Q.; Zhang, Z.W.; Chen, Y.E.; Ding, C.B.; Yuan, S. Light regulates transcription of chlorophyll biosynthetic genes during chloroplast biogenesis. *Crit. Rev. Plant Sci.* **2017**, *36*, 35–54. [CrossRef]
10. Tanaka, R.; Tanaka, A. Tetrapyrrole biosynthesis in higher plants. *Annu. Rev. Plant Biol.* **2007**, *58*, 321–346. [CrossRef]
11. Tanaka, A.; Tanaka, R. Chlorophyll metabolism. *Curr. Opin. Plant Biol.* **2006**, *9*, 248–255. [CrossRef] [PubMed]
12. Vavilin, D.V.; Vermaas, W.F. Regulation of the tetrapyrrole biosynthetic pathway leading to heme and chlorophyll in plants and cyanobacteria. *Physiol. Plant.* **2010**, *115*, 9–24. [CrossRef] [PubMed]
13. Yuan, M.; Zhang, D.W.; Zhang, Z.W.; Chen, Y.E.; Yuan, S.; Guo, Y.R.; Lin, H.H. Assembly of NADPH: Protochlorophyllide oxidoreductase complex is needed for effective greening of barley seedlings. *J. Plant Physiol.* **2012**, *169*, 1311–1316. [CrossRef] [PubMed]
14. Sakuraba, Y.; Rahman, M.L.; Cho, S.H.; Kim, Y.S.; Koh, H.J.; Yoo, S.C.; Paek, N.C. The rice faded green leaf locus encodes protochlorophyllide oxidoreductase B and is essential for chlorophyll synthesis under high light conditions. *Plant J.* **2013**, *74*, 122–133. [CrossRef] [PubMed]
15. Masuda, T.; Takamiya, K. Novel insights into the enzymology, regulation and physiological functions of light-dependent protochlorophyllide oxidoreductase in angiosperms. *Photosynth. Res.* **2004**, *81*, 1–29. [CrossRef] [PubMed]
16. Rüdiger, W. Biosynthesis of chlorophyll b and the chlorophyll cycle. *Photosynth. Res.* **2002**, *74*, 187–193. [CrossRef]
17. Chen, G.; Bi, Y.R.; Li, N. EGY1 encodes a membrane-associated and ATP-independent metalloprotease that is required for chloroplast development. *Plant J.* **2005**, *41*, 364–375. [CrossRef]
18. Pogson, B.J.; Albrecht, V. Genetic dissection of chloroplast biogenesis and development: An overview. *Plant Physiol.* **2011**, *155*, 1545–1551. [CrossRef]
19. Leister, D. Chloroplast research in the genomic age. *Trends Genet.* **2003**, *19*, 47–56. [CrossRef]
20. Weston, E.; Thorogood, K.; Vinti, G.; López-Juez, E. Light quantity controls leaf-cell and chloroplast development in *Arabidopsis thaliana* wild type and blue-light-perception mutants. *Planta* **2000**, *211*, 807–815. [CrossRef]
21. Taşgın, E.; Atici, O.; Nalbantoğlu, B.; Popova, L.P. Effects of salicylic acid and cold treatments on protein levels and on the activities of antioxidant enzymes in the apoplast of winter wheat leaves. *Phytochemistry* **2006**, *67*, 710–715. [CrossRef] [PubMed]
22. Hu, Z.R.; Fan, J.B.; Xie, Y.; Amombo, E.; Liu, A.; Gitau, M.M.; Khaldun, A.B.M.; Chen, L.; Fu, J.M. Comparative photosynthetic and metabolic analyses reveal mechanism of improved cold stress tolerance in bermudagrass by exogenous melatonin. *Plant Physiol. Biochem.* **2016**, *100*, 94–104. [CrossRef] [PubMed]

23. Bilska, A.; Pawe, S. Closure of plasmodesmata in maize (*Zea mays*) at low temperature: A new mechanism for inhibition of photosynthesis. *Anna. Bot.* **2010**, *106*, 675–686. [CrossRef] [PubMed]

24. Chaves, M.M.; Flexas, J.; Pinheiro, C. Photosynthesis under drought and salt stress: Regulation mechanisms from whole plant to cell. *Ann. Bot.* **2009**, *103*, 551–560. [CrossRef] [PubMed]

25. Lawlor, D.W.; Tezara, W. Causes of decreased photosynthetic rate and metabolic capacity in water-deficient leaf cells: A critical evaluation of mechanisms and integration of processes. *Ann. Bot.* **2009**, *103*, 561–579. [CrossRef]

26. Xie, G.; Kato, H.; Sasaki, K.; Imai, R. A cold-induced thioredoxin h of rice, OsTrx23, negatively regulates kinase activities of OsMPK3 and OsMPK6 in vitro. *FEBS Lett.* **2009**, *583*, 2734–2738. [CrossRef]

27. Nakashima, K.; Tran, L.S.; Nguyen, D.V.; Fujita, M.; Maruyama, K.; Todaka, D. Functional analysis of a NAC-type transcription factor OsNAC6 involved in abiotic and biotic stress responsive gene expression in rice. *Plant J.* **2007**, *51*, 617–630. [CrossRef]

28. Allen, D.J.; Ort, D.R. Impacts of chilling temperatures on photosynthesis in warm-climate plants. *Trends Plant Sci.* **2001**, *6*, 36–42. [CrossRef]

29. Fang, C.X.; Zhang, P.L.; Jian, X.; Chen, W.S.; Lin, H.M.; Li, Y.Z.; Lin, W.X. Overexpression of *Lsi1* in cold-sensitive rice mediates transcriptional regulatory networks and enhances resistance to chilling stress. *Plant Sci.* **2017**, *262*, 115–126. [CrossRef]

30. Abdelkader, A.F.; Aronsson, H.; Sundqvist, C. High salt stress in wheat leaves causes retardation of chlorophyll accumulation due to a limited rate of protochlorophyllide formation. *Physiol. Plant.* **2010**, *130*, 157–166. [CrossRef]

31. Sharma, P.; Sharma, N.; Deswal, R. The molecular biology of the low temperature response in plants. *Bioessays* **2005**, *27*, 1048–1059. [CrossRef] [PubMed]

32. Kanneganti, V.; Gupta, A.K. Overexpression of *OsiSAP8*, a member of stress associated protein (SAP) gene family of rice confers tolerance to salt, drought and cold stress in transgenic tobacco and rice. *Plant Mol. Biol.* **2008**, *66*, 445–462. [CrossRef] [PubMed]

33. Kim, S.J.; Lee, S.C.; Hong, S.K.; An, K.; An, G.; Kim, S.R. Ectopic expression of a cold responsive *OsAsr1* cDNA gives enhanced cold tolerance in transgenic rice plants. *Mol. Cell* **2009**, *27*, 449–458. [CrossRef] [PubMed]

34. Han, Q.H.; Huang, B.; Ding, C.B.; Zhang, Z.W.; Chen, Y.E.; Hu, C.; Zhou, L.J.; Huang, Y.; Liao, J.Q.; Yuan, S.; et al. Effects of melatonin on anti-oxidative systems and photosystem II in cold-stressed rice Seedlings. *Front. Plant Sci.* **2017**, *8*, 785. [CrossRef] [PubMed]

35. Jensen, P.E.; Reid, J.D.; Hunter, C.N. Modification of cysteine residues in the ChlI and ChlH subunits of magnesium chelatase results in enzyme inactivation. *Biochem. J.* **2000**, *352*, 435–441. [CrossRef] [PubMed]

36. Alberti, M.; Burke, D.H.; Hearst, J.E. Structure and sequence of the photosynthesis gene cluster. In *Anoxygenic Photosynthetic Bacteria*; Springer: Dordrecht, The Netherlands, 2004.

37. Zhang, H.T.; Li, J.J.; Yoo, J.H.; Yoo, S.C.; Cho, S.H.; Koh, H.J.; Seo, H.S.; Paek, N.C. Rice Chlorina-1 and Chlorina-9 encode ChlD and ChlI subunits of Mg-chelatase, a key enzyme for chlorophyll synthesis and chloroplast development. *Plant Mol. Biol.* **2006**, *62*, 325–337. [CrossRef]

38. Liu, X.C.; Chen, C.Y.; Wang, K.C.; Luo, M.; Tai, R.; Yuan, L.Y. Phytochrome interacting factor3 associates with the histone deacetylase HDA15 in repression of chlorophyll biosynthesis and photosynthesis in etiolated *Arabidopsis* seedlings. *Plant Cell* **2013**, *25*, 1258–1273. [CrossRef]

39. Czarnecki, O.; Grimm, B. Post-translational control of tetrapyrrole biosynthesis in plants, algae, and cyanobacteria. *J. Exp. Bot.* **2012**, *63*, 1675–1687. [CrossRef]

40. Richter, A.S.; Banse, C.; Grimm, B. The GluTR-binding protein is the heme-binding factor for feedback control of glutamyl-tRNA reductase. *eLife* **2019**, *8*, e46300. [CrossRef]

41. Schmied, J.; Hou, Z.W.; Hedtke, B.; Grimm, B. Controlled partitioning of glutamyl-tRNA reductase in stroma- and membrane-associated fractions affects the synthesis of 5-aminolevulinic acid. *Plant Cell Physiol.* **2018**, *59*, 2204–2231. [CrossRef]

42. Apitz, J.; Nishimura, K.; Schmied, J.; Wolf, A.; Hedtke, B.; van Wijk, K.J.; Grimm, B. Posttranslational control of ALA synthesis includes GluTR degradation by Clp protease and stabilization by GluTR-binding protein. *Plant Physiol.* **2016**, *170*, 2040–2051. [CrossRef] [PubMed]

43. Kumar, T.A.; Charan, T.B. Temperature-stress-induced impairment of chlorophyll biosynthetic reactions in cucumber and wheat. *Plant Physiol.* **1998**, *117*, 851–858. [CrossRef] [PubMed]

44. Beale, S.I. Enzymes of chlorophyll biosynthesis. *Photosynth. Res.* **1999**, *60*, 43–73. [CrossRef]
45. Oosawa, N.; Masuda, T.; Awai, K.; Fusada, N.; Shimada, H.; Ohta, H. Identification and light-induced expression of a novel gene of NADPH-protochlorophyllide oxidoreductase isoform in *Arabidopsis thaliana*. *FEBS Lett.* **2000**, *474*, 133–136. [CrossRef]
46. Frick, G.; Su, Q.; Apel, K.; Armstrong, G.A. An *Arabidopsis porB porC* double mutant lacking light-dependent NADPH: Protochlorophyllide oxidoreductases B and C is highly chlorophyll-deficient and developmentally arrested. *Plant J.* **2010**, *35*, 141–153. [CrossRef]
47. Li, H.; Xu, H.L.; Zhang, P.J.; Gao, M.Q.; Wang, D.; Zhao, H.J. High temperature effects on D1 protein turnover in three wheat varieties with different heat susceptibility. *Plant Growth Regul.* **2016**, *81*, 1–9. [CrossRef]
48. Wang, P.; Sun, X.; Chang, C.; Feng, F.J.; Liang, D.; Cheng, L.L.; Ma, F.W. Delay in leaf senescence of *Malus hupehensis* by long-term melatonin application is associated with its regulation of metabolic status and protein degradation. *J. Pineal Res.* **2013**, *55*, 424–434.
49. Chen, Y.E.; Cui, J.M.; Su, Y.Q.; Yuan, S.; Yuan, M.; Zhang, H.Y. Influence of stripe rust infection on the photosynthetic characteristics and antioxidant system of susceptible and resistant wheat cultivars at the adult plant stage. *Front. Plant Sci.* **2015**, *6*, 779–790. [CrossRef]
50. Pamplona, R. Advanced lipoxidation end-products. *Chem. Biol. Interact.* **2011**, *192*, 14–20. [CrossRef]
51. Shi, H.T.; Jiang, C.; Ye, T.T.; Tan, D.X.; Reiter, R.J.; Zhang, H.; Liu, R.Y.; Chan, Z.L. Comparative physiological, metabolomic, and transcriptomic analyses reveal mechanisms of improved abiotic stress resistance in Bermuda grass [*Cynodon dactylon* (L). Pers.] by exogenous melatonin. *J. Exp. Bot.* **2015**, *66*, 681–694. [CrossRef]
52. Los, D.A.; Murata, N. Membrane fluidity and its roles in the perception of environmental signals. *BBA Biomembr.* **2004**, *1666*, 142–157. [CrossRef] [PubMed]
53. Yun, K.Y.; Park, M.R.; Mohanty, B.; Herath, V.; Xu, F.; Mauleon, R.; Wijaya, E.; Bajic, V.B.; Bruskiewich, R.; de Los Reyes, B.G. Transcriptional regulatory network triggered by oxidative signals configures the early response mechanisms of japonica rice to chilling stress. *BMC Plant Biol.* **2010**, *10*, 16. [CrossRef]
54. Tian, Y.; Zhang, H.W.; Pan, X.W.; Chen, X.L.; Zhang, Z.J.; Lu, X.Y.; Huang, R.F. Overexpression of ethylene response factor *TERF2* confers cold tolerance in rice seedlings. *Transgenic Res.* **2011**, *20*, 857–866. [CrossRef] [PubMed]
55. Sangwan, V.; Örvar, B.L.; Beyerly, J.; Hirt, H.; Dhindsa, R.S. Opposite changes in membrane fluidity mimic cold and heat stress activation of distinct plant MAP kinase pathways. *Plant J.* **2002**, *31*, 629–638. [CrossRef] [PubMed]
56. Chinnusamy, V.; Zhu, J.; Zhu, J.K. Gene regulation during cold acclimation in plants. *Physiol. Plant.* **2006**, *126*, 52–61. [CrossRef]
57. Zhang, Q.; Jiang, N.; Wang, G.L.; Hong, Y.H.; Wang, Z.L. Advances in understanding cold sensing and the cold-responsive network in rice. *Adv. Crop Sci. Tech.* **2013**, *1*, 104.
58. Liu, T.; Xu, J.J.; Zhang, J.; Li, J.M.; Hu, X.H. Exogenous 5 aminolevulinic acid pretreatment ameliorates oxidative stress triggered by low temperature stress of *Solanum lycopersicum*. *Acta Physiol. Plant.* **2018**, *40*, 210. [CrossRef]
59. Balestrasse, K.B.; Tomaro, M.L.; Batlle, A.; Noriega, G.O. The role of 5-aminolevulinic acid in the response to cold stress in soybean plants. *Phytochemistry* **2010**, *71*, 2038–2045. [CrossRef]
60. Guo, X.T.; Li, Y.S.; Yu, X.C. Promotive effects of 5-aminolevulinic acid on photosynthesis and chlorophyll fluorescence of tomato seedlings under suboptimal low temperature and suboptimal photon flux density stress. *Hortic. Sci.* **2012**, *39*, 97–99.
61. Liu, D.; Kong, D.D.; Fu, X.K.; Ali, B.; Xu, L.; Zhou, W.J. Influence of exogenous 5-aminolevulinic acid on chlorophyll synthesis and related gene expression in oilseed rape de-etiolated cotyledons under water-deficit stress. *Photosynthetica* **2016**, *54*, 468–474. [CrossRef]
62. Shalygo, N.V.; Czarnecki, O.; Peter, E.; Grimm, B. Expression of chlorophyll synthase is also involved in feedback-control of chlorophyll biosynthesis. *Plant Mol. Biol.* **2009**, *71*, 425–436. [CrossRef] [PubMed]
63. Lichtenthaler, H.K.; Wellburn, A.R. Determination of total carotenoids and chlorophylls a and b of leaf in different solvents. *Biochem. Soc. Trans.* **1983**, *11*, 591–592. [CrossRef]
64. Bradford, M.M. A rapid and sensitive method for the quantitation of microgram quantities of protein utilizing the principle of protein-dye binding. *Anal. Biochem.* **1976**, *72*, 248–254. [CrossRef]

65. Dei, M. Benzyladenine-induced stimulation of 5-aminolevulinic acid accumulation under various light intensities in levulinic acid-treated cotyledons of etiolated cucumber. *Physiol. Plant.* **1985**, *64*, 153–160. [CrossRef]

66. Bogorad, L.; Colowick, S.P.; Kaplan, N.O. *Methods in Enzymology*; Academic Press: New York, NY, USA, 1962.

67. Rebeiz, C.A.; Mattheis, J.R.; Smith, B.B.; Mattheis, J.R.; Rebeiz, C.C.; Dayton, D.F. Chloroplast biogenesis: Biosynthesis and accumulation of Mg-protochlorophyrin IX monoester and other metalloporphyrins by isolated etioplasts and developing chloroplasts. *Arch. Biochem. Biophys.* **1975**, *167*, 351–365. [CrossRef]

68. Wilks, A. *Analysis of Heme and Hemoproteins*; Humana Press: Totowa, NJ, USA, 2002.

69. Yaronskaya, E.; Ziemann, V.; Walter, G.; Averina, N.; Borner, T.; Grimm, B. Metabolic control of the tetrapyrrole biosynthetic pathway for porphyrin distribution in the barley mutant *albostrians*. *Plant J.* **2010**, *35*, 512–522. [CrossRef]

70. Chen, J.B.; Li, X.Y.; Cheng, C.; Wang, Y.H.; Mao, Q.; Zhu, H.T.; Zeng, R.Z.; Fu, X.L.; Liu, Z.Q.; Zhang, G.Q. Characterization of epistatic interaction of QTLs LH8 and EH3 controlling heading date in rice. *Sci. Rep.* **2014**, *4*, 4263. [CrossRef]

71. Fristedt, R.; Granath, P.; Vener, A.V. A protein phosphorylation threshold for functional stacking of plant photosynthetic membranes. *PLoS ONE* **2010**, *5*, e10963. [CrossRef]

72. Aro, E.M.; Mccaffery, S.; Anderson, J.M. Photoinhibition and D1 protein degradation in peas acclimated to different growth irradiances. *Plant Physiol.* **1993**, *103*, 835–843. [CrossRef]

73. Chen, Y.E.; Yuan, S.; Du, J.B.; Xu, M.Y.; Zhang, Z.W.; Lin, H.H. Phosphorylation of photosynthetic antenna protein CP29 and photosystem II structure changes in monocotyledonous plants under environmental stresses. *Biochemistry* **2009**, *48*, 9757–9763. [CrossRef]

74. Sakamoto, W.; Uno, Y.; Zhang, Q.; Miura, E.; Kato, Y. Arrested differentiation of proplastids into chloroplasts in variegated leaves characterized by plastid ultrastructure and nucleoid morphology. *Plant Cell Physiol.* **2009**, *50*, 2069–2083. [CrossRef] [PubMed]

75. Zhou, S.B.; Liu, K.; Zhang, D.; Li, Q.F.; Zhu, G.P. Photosynthetic performance of *Lycoris radiata* var. radiata to shade treatments. *Photosynthetica* **2010**, *48*, 241–248. [CrossRef]

76. Yang, Y.N.; Qi, M.; Mei, C.S. Endogenous salicylic acid protects rice plants from oxidative damage caused by aging as well as biotic and abiotic stress. *Plant J.* **2010**, *40*, 909–919. [CrossRef] [PubMed]

77. Velikova, V.; Yordanov, I.; Edreva, A. Oxidative stress and some antioxidant systems in acid rain-treated bean plants: Protective role of exogenous polyamines. *Plant Sci.* **2000**, *151*, 59–66. [CrossRef]

78. Nahar, K.; Hasanuzzaman, M.; Alam, M.M.; Fujita, M. Roles of exogenous glutathione in antioxidant defense system and methylglyoxal detoxification during salt stress in mung bean. *Biol. Plant.* **2015**, *59*, 745–756. [CrossRef]

79. Ning, J.F.; Ai, S.Y.; Yang, S.H.; Cui, L.H.; Cheng, Y.; Sun, L.L. Physiological and antioxidant responses of *Basella alba* to NaCl or Na_2SO_4 stress. *Acta Physiol. Plant.* **2015**, *37*, 126. [CrossRef]

80. Liang, C.Z.; Zheng, G.Y.; Li, W.Z.; Wang, Y.Q.; Hu, B.; Wang, H.R. Melatonin delays leaf senescence and enhances salt stress tolerance in rice. *J. Pineal Res.* **2015**, *59*, 91–101. [CrossRef]

81. Berger, D.; Altmann, T. A subtilisin-like serine protease involved in the regulation of stomatal density and distribution in *Arabidopsis thaliana*. *Genes Dev.* **2000**, *14*, 1119–1131.

International Journal of
Molecular Sciences

Article

Genome-Wide Identification of QTLs for Grain Protein Content Based on Genotyping-by-Resequencing and Verification of *qGPC1-1* in Rice

Yi-Bo Wu [1,†], Guan Li [1,†], Yu-Jun Zhu [1], Yi-Chen Cheng [1], Jin-Yu Yang [1], Hui-Zhe Chen [1], Xian-Jun Song [2] and Jie-Zheng Ying [1,*]

[1] State Key Laboratory of Rice Biology and Chinese National Center of Rice Improvement, China National Rice Research Institute, Hangzhou 310006, China; 15978677129@163.com (Y.-B.W.); li@ricescience.org (G.L.); yjzhu2013@163.com (Y.-J.Z.); 13872209530@163.com (Y.-C.C.); yang_15271211512@163.com (J.-Y.Y.); chenhuizhe@163.com (H.-Z.C.)

[2] Key Laboratory of Plant Molecular Physiology, Institute of Botany, the Chinese Academy of Sciences, Beijing 100093, China; songxj@ibcas.ac.cn

* Correspondence: yingjiezheng@caas.cn; Tel.: +86-571-63370364

† These authors contributed equally to this study.

Received: 22 November 2019; Accepted: 7 January 2020; Published: 9 January 2020

Abstract: To clarify the genetic mechanism underlying grain protein content (GPC) and to improve rice grain qualities, the mapping and cloning of quantitative trait loci (QTLs) controlling the natural variation of GPC are very important. Based on genotyping-by-resequencing, a total of 14 QTLs were detected with the Huanghuazhan/Jizi1560 (HHZ/JZ1560) recombinant inbred line (RIL) population in 2016 and 2017. Seven of the fourteen QTLs were repeatedly identified across two years. Using three residual heterozygote-derived populations, a stably inherited QTL named as *qGPC1-1* was validated and delimited to a ~862 kb marker interval JD1006–JD1075 on the short arm of chromosome 1. Comparing the GPC values of the RIL population determined by near infrared reflectance spectroscopy (NIRS) and Kjeldahl nitrogen determination (KND) methods, high correlation coefficients (0.966 and 0.983) were observed in 2016 and 2017. Furthermore, 12 of the 14 QTLs were identically identified with the GPC measured by the two methods. These results indicated that instead of the traditional KND method, the rapid and easy-to-operate NIRS was suitable for analyzing a massive number of samples in mapping and cloning QTLs for GPC. Using the gel-based low-density map consisted of 208 simple sequence repeat (SSR) and insert/deletion (InDel) markers, the same number of QTLs (fourteen) were identified in the same HHZ/JZ1560 RIL population, and three QTLs were repeatedly detected across two years. More stably expressed QTLs were identified based on the genome resequencing, which might be attributed to the high-density map, increasing the detection power of minor QTLs. Our results are helpful in dissecting the genetic basis of GPC and improving rice grain qualities through molecular assisted selection.

Keywords: quantitative trait locus; grain protein content; single nucleotide polymorphism; residual heterozygote; rice (*Oryza sativa*); specific length amplified fragment sequencing; Kjeldahl nitrogen determination; near infrared reflectance spectroscopy

---

1. Introduction

Rice grain quality, including appearance, milling, cooking and eating, as well as nutritional qualities, determines the market value, and is getting more and more concern from rice researchers, producers and consumers [1,2]. Grain protein content (GPC) is not only one key factor determining nutritional quality, but is also closely associated with cooking and eating qualities [3,4]. Generally, the increase of GPC may consequently lead to low eating quality.

Compared to the protein content of other cereal crops such as wheat and barley, GPC in rice is relatively low, with a mean about 8.0% and a range of 4.9% to 19.3% in the *indica* subspecies and 5.9% to 16.5% in the *japonica* subspecies [5]. As a typical quantitative trait, GPC in rice is easily affected by environmental conditions, especially the level of nitrogen fertilizer in the field, which makes it very difficult to manipulate in a traditional breeding program. Therefore, illuminating the genetic basis of GPC makes a lot of sense in constructing a molecular marker-assisted selection system to improve rice grain quality [6,7].

Quantitative trait locus (QTL) analysis is the main strategy for dissecting the genetic mechanism underlying a target quantitative trait. During the past two decades, hundreds of QTLs for GPC in rice were detected throughout the entire 12 chromosomes, using different mapping populations, including the recombinant inbred line (RIL) [3,8–13], double haploid population [14–17], chromosome segment substitution line [4,18,19] and the backcross-inbred population [20]. As GPC is sensible to environmental factors, QTLs controlling the GPC are difficult to be repeatedly identified in different populations, or in the same population under different environments [19]. Till now, only two QTLs, *qPC1* and *qGPC-10*, have been map-based cloned and functionally analyzed. *qPC1* was found on the long arm of chromosome 1, which encodes a putative amino acid transporter OsAAP6 and functions as a positive regulator of GPC in rice [6]. *qGPC-10* located on chromosome 10 encodes a glutelin type-A2 precursor, and is also a positive regulator of GPC [7]. Besides, another stably inherited QTL *qPC-1* that is nonallelic to *qPC1* was validated and delimited to a 41-kb region on the long arm of chromosome 1 [19]. Owing to the detection instability of GPC QTLs, it is important to confirm the genetic effect of the QTLs detected in the primary mapping before their map-based cloning and application in the improvement of rice nutritional quality.

Genotyping and phenotyping of the mapping population are two essential components for QTL analysis. With the development of next-generation sequencing, genotyping-by-sequencing becomes a feasible technique to rapidly identify a huge number of single nucleotide polymorphisms (SNPs) throughout the whole genome. Then, a high-density linkage map can be constructed with saturated SNP markers, while most current maps are low-density, and only contain hundreds of gel-based DNA markers, such as restriction fragment length polymorphism and simple sequence repeat (SSR) markers. The detection power of minor QTLs can be improved, and the confidence interval of the QTL can be reduced in a high-resolution linkage map [21]. Among the most common genotyping-by-sequencing methods, specific length amplified fragment sequencing (SLAF-seq) is acceptable as an efficient and high-resolution technology with a relatively lower sequencing cost [22]. With the availability of rice genome draft, genotyping-by-resequencing (GBR) has been applied in linkage mapping and genome-associated analysis to map QTLs for important agronomic traits. Nevertheless, for GPC, the information of QTL identified by GBR of a bi-parent population is still limited in rice.

GPC is traditionally measured by the Kjeldahl nitrogen determination (KND), which is time-consuming and needs a large amount of chemicals such as strong acid and alkali. Therefore, this KND method is difficult for the measurement of a massive number of samples, which is usually necessary in the map-based cloning of a target QTL. Compared to the KND method, near infrared reflectance spectroscopy (NIRS) is a promising technique that is fast and easy-to-use [23]. A lot of QTLs were identified, and two major QTLs have been successfully isolated through the NIRS method. However, it keeps unknown whether there is difference in the detection power of QTLs for GPC between the two methods.

In this study, we mainly completed the following research objectives. First, we analyzed the correlation between the GPC values measured by the KND and NIRS methods, and validated the feasibility of mapping QTLs for GPC using NIRS instead of KND. Second, we identified QTLs for GPC with a high-resolution genetic map containing 18,194 SNP markers in an RIL population, which was derived from a cross between an *indica* variety Huanghuazhan (HHZ) and a *japonica* accession Jizi1560

(JZ1560). Third, we compared the GPC QTLs identified with the high- and low-density map in the same HHZ/JZ1560 RIL population. Finally, one stably inherited major QTL (*qGPC1-1*) located on the short arm of chromosome 1 was validated using three secondary populations developed from three residual heterozygotes (RH) with the heterozygous genotype at the target interval.

2. Results

2.1. Phenotypic Variation and Correlation Analysis

In 2016 and 2017, the GPC of the HHZ/JZ1560 RIL population and the parents was measured by NIRS and KND, respectively. The descriptive statistics of phenotypic variation are shown in Table 1. Significant phenotypic differences were observed between the parents, and the GPC of JZ1560 was higher than that of HHZ in both years. Frequency distributions of GPC in the brown rice flour of RILs and the parents were plotted (Figure 1). The GPC of the RILs showed similar distribution either measured by the KND or NIRS method in each year. Phenotypic variation was continuously distributed with low skewness and kurtosis, showing a typical pattern of quantitative variation. Using the KND method, the GPC of the RILs showed a wide variation from 7.27% to 15.78% in 2016 and from 7.39% to 15.98% in 2017 (Figure 1). Continuous segregation and significant transgressive segregation at two directions suggested polygenic control underlying this trait.

Table 1. Descriptive statistics of the grain protein content (GPC, %) in the recombinant inbred line (RIL) population and the parents in the two years.

Year	Method	RIL Population ($n = 280$)					Parental Mean ($n = 16$)		p
		Mean ± SD	CV	Range	Skewness	Kurtosis	HHZ	JZ1560	
2016	NIRS	12.44 ± 1.59	0.13	8.35–16.35	0.20	−0.26	10.14 ± 0.02	13.16 ± 0.01	<0.001
	KND	11.68 ± 1.62	0.14	7.27–15.78	0.18	−0.30	9.39 ± 0.11	13.06 ± 0.33	
2017	NIRS	10.67 ± 1.40	0.13	7.29–15.90	0.28	0.06	8.44 ± 0.03	10.92 ± 0.01	<0.001
	KND	10.64 ± 1.39	0.13	7.39–15.98	0.29	0.15	8.54 ± 0.00	10.62 ± 0.36	

NIRS: near infrared reflectance spectroscopy; KND: Kjeldahl nitrogen determination; p: two-tailed p value of student's *t* test.

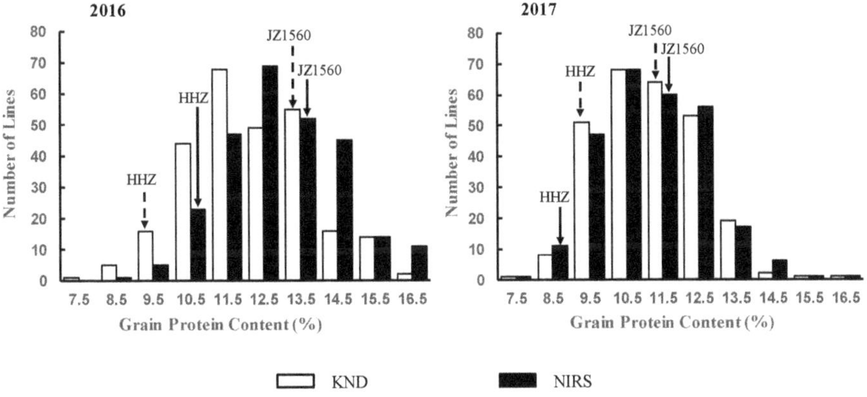

Figure 1. Frequency distributions of grain protein content (GPC) measured by near infrared reflectance spectroscopy (NIRS) and Kjeldahl nitrogen determination (KND) methods in the recombinant inbred line (RIL) population. Parental values are indicated by arrows.

Significant positive correlations between the GPC values measured by the two methods or in different years were found in the RIL population (Table 2). However, correlation coefficients for GPC determined by KND and NIRS methods as 0.966 in 2016 and 0.983 in 2017 were much higher than that of GPC measured with the same method in different years as 0.638 for NIRS and 0.631 for KND. This implied that GPC determined by NIRS was quite consistent with that by the KND method, and GPC was strongly influenced by environmental factors.

Table 2. Correlation coefficients of GPC between two methods and between different years from 280 RILs.

	16NIRS	16KND	17NIRS
16KND	0.966 **		
17NIRS	0.638 **	0.651 **	
17KND	0.614 **	0.631 **	0.983 **

16NIRS, 17NIRS: GPC of 280 RILs in 2016 and 2017 measured by near infrared reflectance spectroscopy (NIRS); 16KND, 17KND: GPC of 280 RILs in 2016 and 2017 measured by Kjeldahl nitrogen determination (KND). ** highly significant correlations at the 0.01 level.

2.2. QTL Analysis of Grain Protein Content in the HHZ/JZ1560 RIL Population

Combining the high-density genetic map containing 18,194 SNP markers with the GPC means of each RIL, 14 QTLs were detected on the whole genome except for chromosomes 6, 9 and 12 with each QTL explaining 0.81%–18.59% of the phenotypic variations (Table 3, Figures 2 and 3). Among the QTLs, 12 were identified in 2016 and nine in 2017. The detailed description of each QTL including peak location, peak LOD value, additive effect and percentage of total phenotypic variations (R^2) are showed in Table 3. Except for *qGPC2*, *qGPC8* and *qGPC10*, the enhancing alleles for GPC were derived from JZ1560 at the remaining 11 loci as the brown rice of JZ1560 contained significantly higher GPC (Table 1).

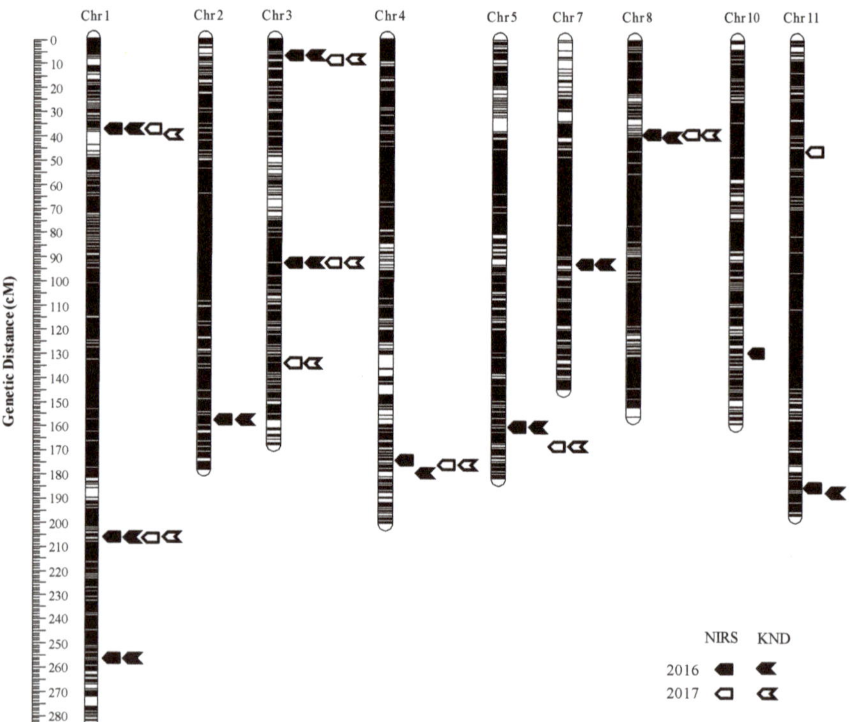

Figure 2. High-density linkage map based on genotyping-by-resequencing showing the most likely positions of QTLs for GPC measured by NIRS and KND methods in the HHZ/JZ1560 RIL population.

Table 3. QTLs for GPC based on genotyping-by-resequencing in the RIL population.

QTL	Year	Method	Marker Interval	Position (cM)	LOD	A	R^2 (%)
qGPC1-1	2016	NIRS	Marker37041–37509	36.46	7.96	0.57	9.14
	2016	KND	Marker37041–37509	36.46	9.72	0.55	10.89
	2017	NIRS	Marker37041–37509	36.46	9.59	0.58	11.85
	2017	KND	Marker37508	38.74	8.72	0.47	11.05
qGPC1-2	2016	NIRS	Marker138502	205.52	2.12	0.45	5.53
	2016	KND	Marker138502–136654	205.52	2.44	0.38	5.20
	2017	NIRS	Marker136541–138500	205.88	3.07	0.41	5.88
	2017	KND	Marker138502–139278	205.52	3.06	0.36	6.41
qGPC1-3	2016	NIRS	Marker208637–210208	255.84	3.00	0.21	1.18
	2016	KND	Marker207139–209159	255.84	2.62	0.15	0.81
qGPC2	2016	NIRS	Marker413084–424309	156.64	3.96	−0.36	3.68
	2016	KND	Marker413084–424409	156.64	3.91	−0.31	3.47
qGPC3-1	2016	NIRS	Marker449522–449025	5.95	11.15	0.78	16.86
	2016	KND	Marker449522–449025	5.95	12.68	0.72	18.59
	2017	NIRS	Marker450509–451398	7.80	6.34	0.46	7.21
	2017	KND	Marker450216–451025	7.61	6.77	0.41	8.40
qGPC3-2	2016	NIRS	Marker567334–573384	91.73	2.41	0.30	2.50
	2016	KND	Marker571649–571070	91.73	2.52	0.26	2.53
	2017	NIRS	Marker571649–572521	91.73	2.31	0.36	4.63
	2017	KND	Marker571649–573384	91.73	2.38	0.30	4.53
qGPC3-3	2017	NIRS	Marker620694–621697	133.44	5.74	0.39	5.44
	2017	KND	Marker620694–621697	133.44	6.11	0.36	6.29
qGPC4	2016	NIRS	Marker779713-795626	173.67	3.90	0.45	5.61
	2016	KND	Marker790607–796001	178.83	4.58	0.39	5.47
	2017	NIRS	Marker792925–793575	175.36	2.93	0.35	4.26
	2017	KND	Marker792925–796001	175.36	3.56	0.31	4.89
qGPC5	2016	NIRS	Marker940773–956686	159.63	4.09	0.35	3.47
	2016	KND	Marker951188–956259	159.63	3.19	0.24	2.12
	2017	NIRS	Marker942375–968984	167.68	5.02	0.41	6.09
	2017	KND	Marker941389–968984	167.68	4.61	0.32	5.21
qGPC7	2016	NIRS	Marker1254363–1255283	92.11	3.46	0.50	6.84
	2016	KND	Marker1254363–1255919	92.11	3.83	0.44	7.15
qGPC8	2016	NIRS	Marker1330696–1330654	38.51	7.81	−0.57	9.23
	2016	KND	Marker1330714–1331214	39.43	7.10	−0.46	7.71
	2017	NIRS	Marker1330696–1330654	38.51	2.12	−0.28	2.70
	2017	KND	Marker1330696	38.33	2.55	−0.25	3.20
qGPC10	2016	NIRS	Marker1714254–1714662	129.12	2.84	−0.43	5.18
qGPC11-1	2017	NIRS	Marker1789482–1788848	45.47	2.08	0.33	3.89
qGPC11-2	2016	NIRS	Marker1901374–1901256	184.63	2.03	0.46	5.84
	2016	KND	Marker1902918–1902895	186.73	2.06	0.36	4.66

NIRS: near infrared reflectance spectroscopy; KND: Kjeldahl nitrogen determination; A: additive effect of replacing a maternal allele with a paternal allele; R^2: proportion of the phenotypic variance explained by the QTL.

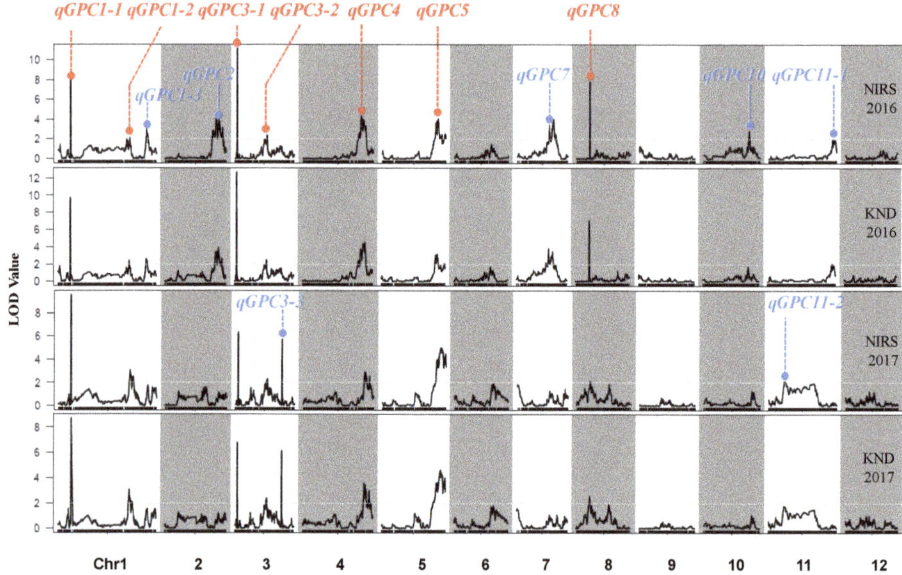

Figure 3. The identified QTLs for GPC measured by the NIRS and KND methods through analyzing the SNP genotypes and corresponding phenotypes of the 280 RILs. Red font indicates that QTLs were detected in both 2016 and 2017, and blue font indicates that QTLs were identified in either year 2016 or 2017.

In order to find the difference in the detection power of QTLs using different measurement methods of GPC, we further compared the QTLs for GPC determined by the NIRS and KND methods. In 2016, we identified 12 QTLs for GPC measured by NIRS and 11 QTLs for GPC measured by KND, which explained 55.07% and 48.25% of the total phenotypic variations, respectively. Eleven QTLs were commonly mapped using the two measurement methods, and one QTL (*qGPC10*) with a small genetic effect was only detected using the NIRS method (Figures 2 and 3, Table 3). Similar results were observed in 2017. Eight common QTLs were identified by both the measurement methods, and only one minor QTL (*qGPC11*) was mapped on chromosome 11 using the NIRS method. This indicated that QTLs for GPC were coincided between the two GPC measurement methods. Seven of the fourteen QTLs, *qGPC1-1*, *qGPC1-2*, *qGPC3-1*, *qGPC3-2*, *qGPC4*, *qGPC5* and *qGPC8*, were repeatedly identified in both years. The remaining seven QTLs were only detected in one year.

QTL analysis was also performed using a low-density genetic map containing 208 SSR and InDel markers and a total of 14 QTLs were detected in the same HHZ/JZ1560 RIL population (Figure 4, Table S1). Compared with the seven stably inherited QTLs identified in the high-density genetic map, only three QTLs including *qGPC1*, *qGPC3-1* and *qGPC5* were mapped at the same region and showed the similar effects for the two years using the low-density genetic map, suggesting that the high-density genetic map increased the detection power of QTLs for GPC.

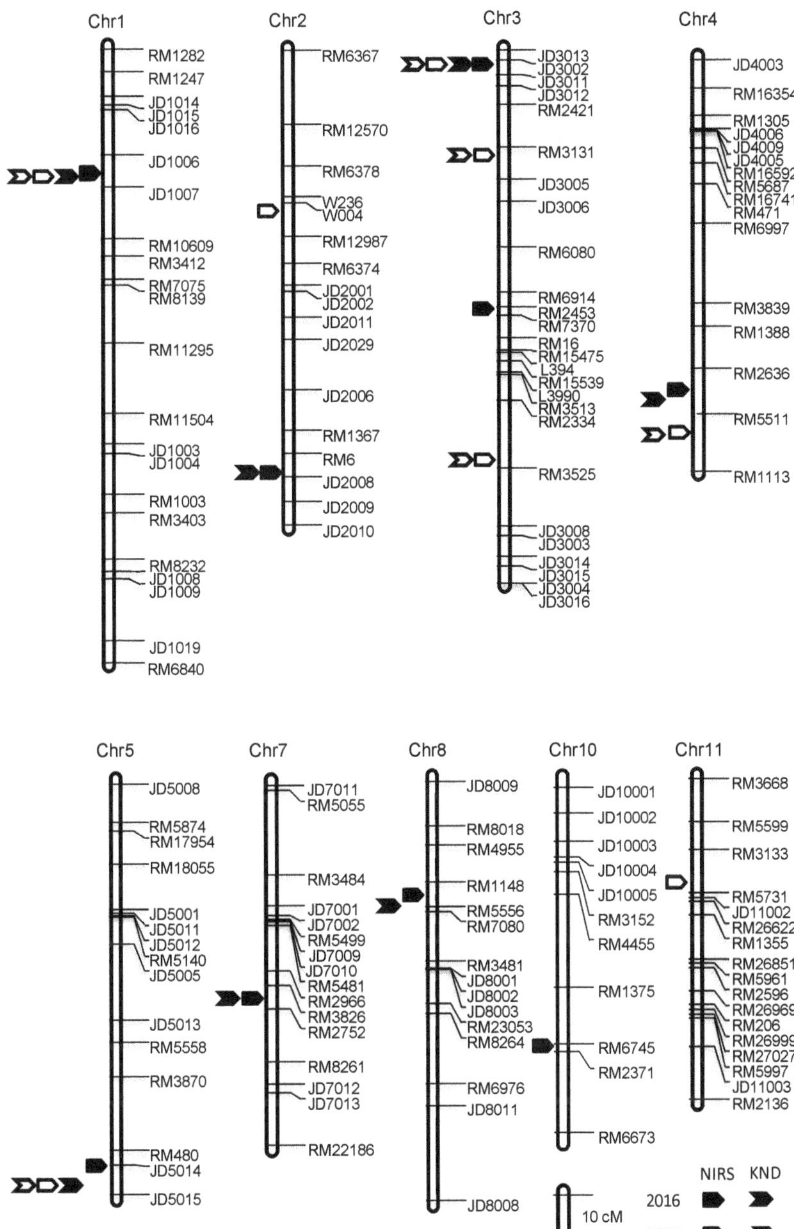

Figure 4. Low-density linkage map containing 208 gel-based SSR and InDel markers showing the most likely positions of QTLs for GPC measured by NIRS and KND methods in the HHZ/JZ1560 RIL population.

The *qGPC1-1* was detected on the short arm of chromosome 1 across two years using the high-density genetic map and accounted for 9.14% to 11.85% of the phenotypic variations. The allele from JZ1560 at this locus increased GPC by 0.47%-0.58%. Corresponding to *qGPC1-1*, *qGPC1* was mapped at the same location with the flanking markers JD1006 and JD1007 using the low-density genetic map in both years

(Figure 4, Table S1). The *qGPC1* identified in this study contributed 11.78% to 13.33% of phenotypic variations with a relatively large additive effect ranging from 0.54% to 0.71% (Table S1).

2.3. Validation and Delimitation of qGPC1-1 Using RH-derived F₂ Populations

To confirm the genetic effect and location of *qGPC1-1*, three RH individuals were selected from one F_8 RIL line with the heterozygous genotype covering the target marker interval of JD1006–JD1007. Three RH-derived F_2 populations, named as WB01, WB02 and WB03, were developed from the three plants with sequential heterozygous segments extending from JD1006 to JD1007, respectively. Based on the sequence differences between the parents HHZ and JZ1560 identified by 30-fold whole genome re-sequencing, additional six InDel markers were developed and used to genotype the three populations (Table S2). GPC was continuously distributed and ranged from 8.33% to 11.90%, 8.27% to 12.19% and 8.32% to 10.47% in WB01, WB02 and WB03 populations, respectively (Figure S1).

Three segmental linkage maps were constructed for WB01, WB02 and WB03, respectively (Figure 5). Combined the genotype and phenotype information, *qGPC1-1* was identified in WB01 and WB02 populations, with the JZ1560 allele always increasing GPC (Table 4; Figure 5). This QTL explained 26.00% and 27.40% of phenotypic variations with the similar additive effects of 0.36% and 0.39% in WB01 and WB02, respectively. No QTL for the GPC was detected in the WB03 population. Therefore, *qGPC1-1* should be located within the common segregating regions of WB01 and WB02, but outside the segregating region of WB03. As shown in Figure 5, *qGPC1-1* was delimited to the interval between JD1006 and JD1075 (~862 kb) with a common segregating region JD1068–JD1075 and one flanking cross-over region JD1006–JD1068.

Table 4. QTLs for the GPC detected in the three residual heterozygote-derived F_2 populations.

Population Name	Segregating Region	Sample	LOD	A	D	R^2 (%)
WB01	JD1006–JD1078	180	7.59	0.36	0.11	26.00
WB02	JD1068–JD1007	137	6.25	0.39	0.04	27.40
WB03	JD1075–JD1007	115	ns	ns	ns	ns

A: additive effect of replacing a maternal allele with a paternal allele; *D*: dominance effect; R^2: proportion of the phenotypic variance explained by the QTL; ns: no significance.

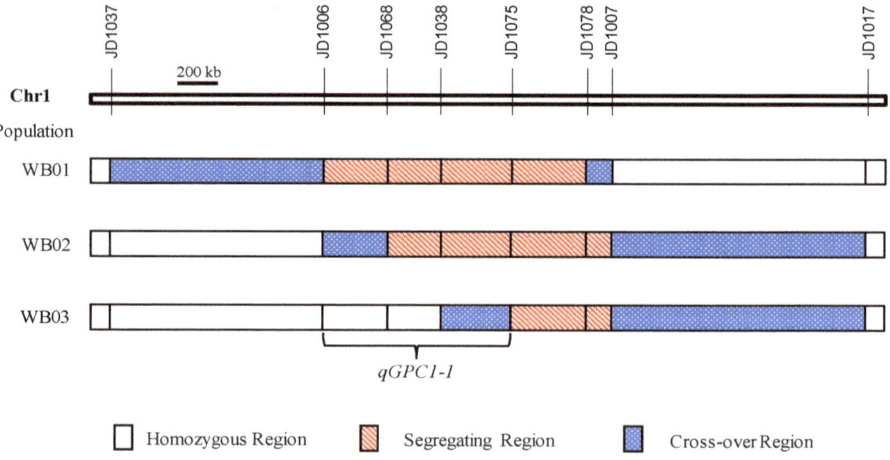

Figure 5. Genotypic compositions of the three residual heterozygote-derived F_2 populations in the segregating regions.

3. Discussion

Elucidating the genetic mechanism of GPC accumulation is very important for regulating rice grain qualities in breeding. In the present study, we characterized the genetic basis of GPC and identified a total of 14 QTLs using the high-resolution map in the HHZ/JZ1560 RIL population. Although GPC is sensitive to environmental conditions and the QTLs for GPC are difficult to be repeatedly identified in different environments, the majority of the 14 QTLs have been reported in the previous studies. On the long arm of chromosome 1, QTLs for GPC have been reported in some studies including *pro1* between RM226 and RM297 [15], *qPC-1* between R888 and R1485 [13], *qPC1* between RM472 and RM104 [6], *qPC-1* between RM1196 and RM302 [19] and *TGP1b* between RM1297 and RM1067 [4]. The *qGPC1-2* and *qGPC1-3* were located in the adjacent chromosome regions with these reported QTLs, and the *indica* variety HHZ allele decreased the GPC. Two minor QTLs, *qGPC2* and *qGPC11-1*, were also mapped at the similar locations with *pro2* between RM6 and RM112 and *pro11* between RM209 and RM229 [15]. On the chromosome 3, we detected three QTLs including *qGPC3-1*, *qGPC3-2* and *qGPC3-3*, and the enhancing alleles were all from the *japonica* rice JZ1560. The *qGPC3-1* with the largest effect in 2016 and *qGPC3-2* were repeatedly detected as *qPC-3.1* in the interval XNpb212–G1318 and *qPC-3.2* in the interval R758–XNpb15, respectively [13]. The *qGPC3-3* was located in the overlapping confidence interval with the QTLs for protein content in several previous reports [13,14,18]. These results indicated that there are multiple genetic factors controlling GPC on chromosome 3. The *qGPC4* was detected within the *qPC-4* region between RG214 and RG620 [12]. We still noted that *qGPC5* was repeatedly identified as *qRPC5* for rice protein content in the interval RG435–RG172a using a doubled haploid population [16]. Although *qGPC7* was only detected in 2016, four QTLs were located in the same or adjacent regions as reported by previous studies [3,4,8,17]. On the chromosome 8, *qGPC8* showed overlapping intervals with *cp8.1*, *qPC-8a* and *TGP8*, which have been identified using different populations in different environments [4,18,24]. The *qPC11-2* with minor effect was mapped near to *qPC11* between RM202 and RM206 [25]. No QTL for GPC has been mapped in the region of *qGPC10* on chromosome 10 before, therefore *qGPC10* might be newly detected in this study. Unlike many previous studies, we did not detect the QTL for GPC near to the *Wx* locus on chromosome 6 [4,8,12,25]. Over all, the locations of QTLs for GPC showed a significant similarity between our studies and the previous findings.

Of the 14 QTLs identified in this study, the *qGPC1-1* was a stably expressed QTL with a relatively large effect and it was repeatedly detected in both years. This QTL was also identified in the similar chromosome region in different populations and environments, implying that *qGPC1-1* plays an important role in controlling GPC [4,11,18]. Based on the primary mapping result, *qGPC1-1* was further validated and delimited in the interval JD1006–JD1075, corresponding to the 6.0–6.8 Mb region on the short arm of chromosome 1 in the Nipponbare genome [26]. The *japonica* rice JZ1560 allele contributed to the increase of GPC in the RH-derived F_2 population (Table 4). Dissecting the genetic mechanism underlying GPC is important for the improvement of rice grain quality, and the main obstacle to date is the absence of key genes/QTLs regulating GPC. Primary mapping leads to a large confident interval and poor repeatability of target QTLs, which makes it difficult to find tightly linked markers for marker-assisted selection. Validation and delimitation of *qGPC1-1* contributed to the facilitation of marker-assisted selection in rice breeding for high nutritional quality. Furthermore, based on these results, fine mapping and map-based cloning of *qGPC1-1* is under way.

Mapping and isolation of QTLs need a high efficiency method to measure GPC. Cloning of QTLs controlling the natural variation of GPC is the important step toward uncovering the regulatory mechanism underlying this quantitative trait. However, map-based cloning of a QTL needs phenotype and genotype information of a massive number of samples, suggesting a rapid and easy operation method for GPC is necessary. Using the NIRS system, the GPC value can be directly measured once the brown rice is grinded into flour.

Compared with the NIRS, the KND method for GPC in rice needs further lengthy operation, which is time-consuming and laborious. More importantly, high correlation between the GPC values determined by the NIRS and KND methods was observed in the HHZ/JZ1560 RIL population across two years. Only two minor QTLs (*qGPC10* and *qGPC11-1*) were detected using the single measurement method of GPC, the remaining twelve QTLs were identical to be identified by both methods in 2016 or 2017 (Figures 2 and 3, Table 3). Comparative analysis between the two methods suggested that NIRS could be a feasible strategy for the mapping and map-based cloning of QTLs for GPC instead of the KND method. In recent years, NIRS has been successfully employed in the isolation and characterization of two major QTLs for GPC, *qPC1* and *qGPC-10* [6,7].

Accompanied with the development of DNA sequence techniques, the sequencing cost decreases continuously and more and more high-density genetic maps have been constructed to detect QTLs for different traits through genotyping-by-sequencing [21,27,28]. In the present study, QTLs for GPC were mapped simultaneously using a high-density genetic map with 18,194 SNP markers identified by GBR and a low-density map with 208 gel-based SSR and InDel markers in the same RIL population. Although the same number of QTLs were identified using the different resolution genetic maps, 7 of 14 and 3 of 14 QTLs were repeatedly detected across two years in the high- and low-density genetic maps, respectively. Identification of more stably expressed QTLs might be attributed to the increased detection power resulted by the saturated SNP markers in the high-density genetic map.

4. Materials and Methods

4.1. Plant Materials and Field Experiments

The HHZ/JZ1560 mapping population consisting of 280 RILs was developed using the single-seed descendent method by Ying et al. [29]. HHZ is an *indica* variety as the female parent with small grains and being widely cultivated in China, while JZ1560 is a *japonica* rice accession with very large grains (Figure S2). Three RH-derived F_2 populations (WB01, WB02 and WB03) consisting 180, 137 and 115 individuals, respectively, were originated from 3 RH plants that were selected from one line of the F_7 HHZ/JZ1560 RIL population. The three populations were used to validate the target QTL, *qGPC1-1*.

All the populations, together with the two parents, were planted with the spacing of 16.7 cm between plants and 26.7 cm between rows during the rice-growing seasons in the experimental field of China National Rice Research Institute, Hangzhou (120.2° E, 30.3° N), China. Eighteen plants per RIL were transplanted and the middle four plants were harvested in bulk at maturity for the measurement of GPC in 2016 and 2017. The three RH-derived populations were tested for one year in 2018. Field management was conducted according to the common practice in rice production. Fertilizer was applied for each cultivation year as follows: 375 kg/ha compound fertilizer (N-P_2O_5-K_2O: 14:16:15) as the basal fertilizer, 52 kg/ha nitrogen at seedling stage, 69 kg/ha nitrogen at 5 days after transplanting and 150 kg/ha potassium at 20 days after transplanting.

4.2. GPC Measurement

GPC of brown rice flour for each RIL was carefully measured by the KND and NIRS methods, respectively. Rice grains were harvested after maturity and stored at room temperature for at least three months before the GPC measurement. The fully filled grains were de-hulled into brown rice by a rice huller (JLGJ-45, Taizhou, Zhejiang, China), then the brown rice was grinded into flour by a grinder (IKA TUBE-MILL, Staufen, Germany). Brown rice flour samples were directly used to determine GPC by NIRS (Foss, Sweden). Each sample was assayed twice and the mean values were used for further analysis. GPC was calculated according to the modified NIRS model constructed by Xie et al. [23].

For the KND method to measure GPC, 0.2 g brown rice flour, 1.0 g catalyst ($CuSO_4$:Na_2SO_4 = 1:10) and 4.0 mL of H_2SO_4 were first added into a 100-mL digestion tube in turn. Then the mixture was immediately heated at 420 °C for 2 h in a digestion stove. After the solution being digested, the mixture was cooled to room temperature. Then, 10 mL ddH_2O was added into the digestion tube. The mixed solution was analyzed using a Kjeltec 8400 Autoanalyzer (Foss, Sweden). GPC of the rice flour was calculated from the total N content by multiplying a conversion factor of 5.95 [30]. The assay for each sample was conducted with two replicates and the means were used for further analysis.

4.3. Genetic Map and DNA Marker Analysis

To compare the detection power of GPC QTL, high-density and low-density linkage maps of the RIL population were both used to map QTL, respectively. The high-density linkage map constructed by one of the GBR method, specific length amplified fragment, was composed of 18,194 SNP markers and spanned 2132.56 cM with an average genetic distance of 0.12 cM. In our previous study, we constructed the SLAF library for each RIL and the products were further sequenced using Illumina HiSeq 2500 system (Illumina, Inc.; San Diego, CA, USA) [29]. Polymorphism loci between the parents were identified for the selection of high-quality SNP markers after filtering out the low-quality raw reads. SNP markers with more than 20% missing data and the segregation distortion were further filtered out. A total of 18,194 high quality SNP markers were used to genotype the 280 RILs. The low-density linkage map constructed by the gel-based method consisted of 121 SSR and 87 InDel markers and spanned 1399.40 cM with an average distance of 7.61 cM.

For the three RH-derived populations, leaf samples of each individual were extracted for genomic DNA through the modified CTAB method [31]. To genotype the three RH-derived populations, a total of six InDel markers (Table S2) in the mapping interval of *qGPC1-1* were designed with Primer3.0 (http://primer3.ut.ee/) based on the 30-fold genome resequencing of the parents, HHZ and JZ1560. According to our previous study [32], the PCR was performed in 10-μL reactions containing 2 × *Taq* MasterMix (CW0682, CWBIO) 5-μL, 0.4 μM of each primer and 50 ng DNA template. The PCR program was set as an initial denaturation at 94 °C for 2 min, then followed by 30 cycles of 30 s at 94 °C, 30 s at 55 °C and 30 s at 72 °C, and finally 2 min at 72 °C. The PCR products were analyzed on 2.5% agarose gels.

4.4. QTL Mapping and Statistical Analysis

Using the high-density genetic map, QTL analysis for GPC was performed by R/qtl software with the method of composite interval mapping (CIM) [33], taking the two years for the RIL population as two environments. A threshold of LOD > 2.0 was used for detecting a putative QTL. Using the low-density genetic map, QTLs controlling GPC were identified by Windows QTL Cartographer 2.5 [34]. Identification of QTLs was performed using CIM. The LOD threshold for claiming a QTL was also fixed as a LOD score of 2.0. QTL analysis in the three RH-derived populations was also conducted by CIM using Windows QTL Cartographer 2.5 [34].

Basic descriptive statistics, including mean, standard deviation, coefficient of variation, range, skewness and kurtosis, and correlation analysis for GPC in the RIL population were performed with Microsoft Excel 2016 for Windows. The *t*-test of the two parents was performed with the SAS program [35].

5. Conclusions

Based on genotyping-by-resequencing, a total of 14 QTLs controlling GPC were identified with an *indica/japonica* (HHZ/JZ1560) RIL population in 2016 and 2017. Of the 14 QTLs, 13 QTLs showed similar chromosome regions with the QTLs for GPC documented in previous studies. The *qGPC10* with a minor effect was newly detected in this study. Seven of the fourteen QTLs were repeatedly identified across two years. The stably inherited *qGPC1-1* with a relatively large effect was validated and delimited to a ~862 kb region flanked by JD1006 and JD1075 on the short arm of chromosome 1, which is helpful for the construction of a marker-assisted selection system to improve rice grain qualities and further map-based cloning of this QTL. Our results indicated that instead of the KND method with lengthy operation, the NIRS with rapid and easy operation was a feasible strategy for measuring a massive collection of samples in the mapping and map-based cloning of GPC QTL. More stably expressed QTLs identified in the genetic map based on genotyping-by-resequencing suggested that high-density map enhanced the detection power of minor QTLs.

Supplementary Materials: Supplementary materials can be found at http://www.mdpi.com/1422-0067/21/2/408/s1.

Author Contributions: Conceptualization, X.-J.S.; Data curation, Y.-B.W., G.L. and Y.-C.C.; Formal analysis, Y.-J.Z.; Investigation, J.-Y.Y.; Methodology, H.-Z.C. and J.-Z.Y.; Supervision, J.-Z.Y.; Writing—original draft, Y.-B.W.; Writing—review & editing, G.L. All authors have read and agreed to the published version of the manuscript.

Funding: This research was funded by Chinese High-yielding Transgenic Program (Grant No. 2016ZX08001-004).

Acknowledgments: We thank L. Xie at China National Rice Research Institute for the measurements of grain protein content through near infrared reflectance spectroscopy.

Conflicts of Interest: The authors declare no conflict of interest.

References

1. Li, Y.; Fan, C.; Xing, Y.; Yun, P.; Luo, L.; Yan, B.; Peng, B.; Xie, W.; Wang, G.; Li, X.; et al. *Chalk5* encodes a vacuolar H($^+$)-translocating pyrophosphatase influencing grain chalkiness in rice. *Nat. Genet.* **2014**, *46*, 398–404. [CrossRef] [PubMed]

2. Zhao, D.S.; Li, Q.F.; Zhang, C.Q.; Zhang, C.; Yang, Q.Q.; Pan, L.X.; Ren, X.Y.; Lu, J.; Gu, M.H.; Liu, Q.Q. *GS9* acts as a transcriptional activator to regulate rice grain shape and appearance quality. *Nat. Commun.* **2018**, *9*, 1240. [CrossRef] [PubMed]

3. Zhong, M.; Wang, L.Q.; Yuan, D.J.; Luo, L.J.; Xu, C.G.; He, Y.Q. Identification of QTL affecting protein and amino acid contents in rice. *Rice Sci.* **2011**, *18*, 187–195. [CrossRef]

4. Kashiwagi, T.; Munakata, J. Identification and characteristics of quantitative trait locus for grain protein content, *TGP12*, in rice (*Oryza sativa* L.). *Euphytica* **2018**, *214*, 165. [CrossRef]

5. Lin, R.; Luo, Y.; Liu, D.; Huang, C. Determination and analysis on principal qualitative characters of rice germplasm. In *Rice Germplasm Resources in China*; Ying, C., Ed.; Agricultural Science and Technology Publisher of China: Beijing, China, 1993; pp. 83–93.

6. Peng, B.; Kong, H.; Li, Y.; Wang, L.; Zhong, M.; Sun, L.; Gao, G.; Zhang, Q.; Luo, L.; Wang, G.; et al. *OsAAP6* functions as an important regulator of grain protein content and nutritional quality in rice. *Nat. Commun.* **2014**, *5*, 4847. [CrossRef]

7. Yang, Y.; Guo, M.; Sun, S.; Zou, Y.; Yin, S.; Liu, Y.; Tang, S.; Gu, M.; Yang, Z.; Yan, C. Natural variation of *OsGluA2* is involved in grain protein content regulation in rice. *Nat. Commun.* **2019**, *10*, 1949. [CrossRef]

8. Tan, Y.F.; Sun, M.; Xing, Y.Z.; Hua, J.P.; Sun, X.L.; Zhang, Q.F.; Corke, H. Mapping quantitative trait loci for milling quality, protein content and color characteristics of rice using a recombinant inbred line population derived from an elite rice hybrid. *Theor. Appl. Genet.* **2001**, *103*, 1037–1045. [CrossRef]

9. Wada, T.; Uchimura, Y.; Ogata, T.; Tsubone, M.; Matsue, Y. Mapping of QTLs for physicochemical properties in *japonica* rice. *Breed. Sci.* **2006**, *56*, 253–260. [CrossRef]

10. Zhang, W.; Bi, J.; Chen, L.; Zheng, L.; Ji, S.; Xia, Y.; Xie, K.; Zhao, Z.; Wang, Y.; Liu, L.; et al. QTL mapping for crude protein and protein fraction contents in rice (*Oryza sativa* L.). *J. Cereal Sci.* **2008**, *48*, 539–547. [CrossRef]

11. Kepiro, J.L.; McClung, A.M.; Chen, M.H.; Yeater, K.M.; Fjellstrom, R.G. Mapping QTLs for milling yield and grain characteristics in a tropical *japonica* long grain cross. *J. Cereal Sci.* **2008**, *48*, 477–485. [CrossRef]

12. Yu, Y.H.; Li, G.; Fan, Y.Y.; Zhang, K.Q.; Min, J.; Zhu, Z.W.; Zhuang, J.Y. Genetic relationship between grain yield and the contents of protein and fat in a recombinant inbred population of rice. *J. Cereal Sci.* **2009**, *50*, 121–125. [CrossRef]

13. Zheng, L.; Zhang, W.; Chen, X.; Ma, J.; Chen, W.; Zhao, Z.; Zhai, H.; Wan, J. Dynamic QTL analysis of rice protein content and protein index using recombinant inbred lines. *J. Plant Biol.* **2011**, *54*, 321–328. [CrossRef]

14. Yoshida, S.; Ikegami, M.; Kuze, J.; Sawada, K.; Hashimoto, Z.; Ishii, T.; Nakamura, C.; Kamijima, O. QTL analysis for plant and grain characters of sake-brewing rice using a doubled haploid population. *Breed. Sci.* **2002**, *52*, 309–317. [CrossRef]

15. Aluko, G.; Martinez, C.; Tohme, J.; Castano, C.; Bergman, C.; Oard, J.H. QTL mapping of grain quality traits from the interspecific cross *Oryza sativa* x *O. glaberrima*. *Theor. Appl. Genet.* **2004**, *109*, 630–639. [CrossRef] [PubMed]

16. Hu, Z.L.; Li, P.; Zhou, M.Q.; Zhang, Z.H.; Wang, L.X.; Zhu, L.H.; Zhu, Y.G. Mapping of quantitative trait loci (QTLs) for rice protein and fat content using doubled haploid lines. *Euphytica* **2004**, *135*, 47–54. [CrossRef]

17. Bruno, E.; Choi, Y.S.; Chung, I.K.; Kim, K.M. QTLs and analysis of the candidate gene for amylose, protein, and moisture content in rice (*Oryza sativa* L.). *3 Biotech* **2017**, *7*, 40. [CrossRef] [PubMed]

18. Liu, X.; Wan, X.; Ma, X.; Wan, J. Dissecting the genetic basis for the effect of rice chalkiness, amylose content, protein content, and rapid viscosity analyzer profile characteristics on the eating quality of cooked rice using the chromosome segment substitution line population across eight environments. *Genome* **2011**, *54*, 64–80.

19. Yang, Y.; Guo, M.; Li, R.; Shen, L.; Wang, W.; Liu, M.; Zhu, Q.; Hu, Z.; He, Q.; Xue, Y.; et al. Identification of quantitative trait loci responsible for rice grain protein content using chromosome segment substitution lines and fine mapping of *qPC-1* in rice (*Oryza sativa* L.). *Mol. Breed.* **2015**, *35*, 130. [CrossRef]

20. Zheng, L.; Zhang, W.; Liu, S.; Chen, L.; Liu, X.; Chen, X.; Ma, J.; Chen, W.; Zhao, Z.; Jiang, L.; et al. Genetic relationship between grain chalkiness, protein content, and paste viscosity properties in a backcross inbred population of rice. *J. Cereal Sci.* **2012**, *56*, 153–160. [CrossRef]

21. Wang, Y.; VandenLangenberg, K.; Wen, C.; Wehner, T.C.; Weng, Y. QTL mapping of downy and powdery mildew resistances in PI 197088 cucumber with genotyping-by-sequencing in RIL population. *Theor. Appl. Genet.* **2018**, *131*, 597–611. [CrossRef]

22. Sun, X.; Liu, D.; Zhang, X.; Li, W.; Liu, H.; Hong, W.; Jiang, C.; Guan, N.; Ma, C.; Zeng, H.; et al. SLAF-seq: An efficient method of large-scale *de novo* SNP discovery and genotyping using high-throughput sequencing. *PLoS ONE* **2013**, *8*, e58700. [CrossRef] [PubMed]

23. Xie, L.H.; Tang, S.Q.; Chen, N.; Luo, J.; Jiao, G.A.; Shao, G.N.; Wei, X.J.; Hu, P.S. Optimisation of near-infrared reflectance model in measuring protein and amylose content of rice flour. *Food Chem.* **2014**, *142*, 92–100. [CrossRef] [PubMed]

24. Li, J.; Xiao, J.; Grandillo, S.; Jiang, L.; Wan, Y.; Deng, Q.; Yuan, L.; McCouch, S.R. QTL detection for rice grain quality traits using an interspecific backcross population derived from cultivated Asian (*O. sativa* L.) and African (*O. glaberrima* S.) rice. *Genome* **2004**, *47*, 697–704. [CrossRef] [PubMed]

25. Yun, Y.T.; Chung, C.T.; Lee, Y.J.; Na, H.J.; Lee, J.C.; Lee, S.G.; Lee, K.W.; Yoon, Y.H.; Kang, J.W.; Lee, H.S.; et al. QTL mapping of grain quality traits using introgression lines carrying *Oryza rufipogon* chromosome segments in *japonica* rice. *Rice (NY)* **2016**, *9*, 62. [CrossRef] [PubMed]

26. International Rice Genome Sequencing Project. The map-based sequence of the rice genome. *Nature* **2005**, *436*, 793–800. [CrossRef] [PubMed]

27. Lu, X.; Xiong, Q.; Cheng, T.; Li, Q.T.; Liu, X.L.; Bi, Y.D.; Li, W.; Zhang, W.K.; Ma, B.; Lai, Y.C.; et al. A *PP2C-1* allele underlying a quantitative trait locus enhances soybean 100-seed weight. *Mol. Plant* **2017**, *10*, 670–684. [CrossRef] [PubMed]

28. Zhu, Z.; Li, X.; Wei, Y.; Guo, S.; Sha, A. Identification of a novel QTL for panicle length from wild rice (*Oryza minuta*) by specific locus amplified fragment sequencing and high density genetic mapping. *Front. Plant Sci.* **2018**, *9*, 1492. [CrossRef]

29. Ying, J.Z.; Ma, M.; Bai, C.; Huang, X.H.; Liu, J.L.; Fan, Y.Y.; Song, X.J. *TGW3*, a major QTL that negatively modulates grain length and weight in rice. *Mol. Plant* **2018**, *11*, 750–753. [CrossRef]

30. Sotelo, A.; Sousa, V.; Montalvo, I.; Hernandez, M.; Hernandez-Arago, L. Chemical composition of different fractions of 12 Mexican varieties of rice obta. *Cereal Chem.* **1990**, *67*, 209–212.

31. Murray, M.G.; Thompson, W.F. Rapid isolation of high molecular weight plant DNA. *Nucleic Acids Res.* **1980**, *8*, 4321–4325. [CrossRef]

32. Zhang, Z.H.; Zhu, Y.J.; Wang, S.L.; Fan, Y.Y.; Zhuang, J.Y. Importance of the interaction between heading date genes *Hd1* and *Ghd7* for controlling yield traits in rice. *Int. J. Mol. Sci.* **2019**, *20*, 516. [CrossRef] [PubMed]

33. Arends, D.; Prins, P.; Jansen, R.C.; Broman, K.W. R/qtl: High-throughput multiple QTL mapping. *Bioinformatics* **2010**, *26*, 2990–2992. [CrossRef] [PubMed]

34. Wang, S.; Basten, C.J.; Zeng, Z.B. *Windows QTL Cartographer 2.5*; Department of Statistics, North Carolina State University: Raleigh, NC, USA, 2006.

35. SAS Institute Inc. *SAS/STAT User's Guide*; SAS Institute: Cary, NC, USA, 1999.

International Journal of
Molecular Sciences

Article

Broad-Spectrum Disease Resistance Conferred by the Overexpression of Rice RLCK BSR1 Results from an Enhanced Immune Response to Multiple MAMPs

Yasukazu Kanda [1,2], Hitoshi Nakagawa [1], Yoko Nishizawa [1], Takashi Kamakura [2] and Masaki Mori [1,2,*]

[1]   Institute of Agrobiological Sciences, NARO (NIAS), Tsukuba 305-8602, Japan; kanday@affrc.go.jp (Y.K.); nakagawa_hitoshi@mac.com (H.N.); ynishi@affrc.go.jp (Y.N.)
[2]   Graduate School of Science and Technology, Tokyo University of Science, Noda 278-8510, Japan; kamakura@rs.noda.tus.ac.jp
*   Correspondence: morimasa@affrc.go.jp; Tel.: +81-29-838-7008

Received: 4 October 2019; Accepted: 4 November 2019; Published: 6 November 2019

Abstract: Plants activate their immune system through intracellular signaling pathways after perceiving microbe-associated molecular patterns (MAMPs). Receptor-like cytoplasmic kinases mediate the intracellular signaling downstream of pattern-recognition receptors. BROAD-SPECTRUM RESISTANCE 1 (BSR1), a rice (*Oryza sativa*) receptor-like cytoplasmic kinase subfamily-VII protein, contributes to chitin-triggered immune responses. It is valuable for agriculture because its overexpression confers strong disease resistance to fungal and bacterial pathogens. However, it remains unclear how overexpressed BSR1 reinforces plant immunity. Here we analyzed immune responses using rice suspension-cultured cells and sliced leaf blades overexpressing BSR1. BSR1 overexpression enhances MAMP-triggered production of hydrogen peroxide (H_2O_2) and transcriptional activation of the defense-related gene in cultured cells and leaf strips. Furthermore, the co-cultivation of leaves with conidia of the blast fungus revealed that BSR1 overexpression allowed host plants to produce detectable oxidative bursts against compatible pathogens. BSR1 was also involved in the immune responses triggered by peptidoglycan and lipopolysaccharide. Thus, we concluded that the hyperactivation of MAMP-triggered immune responses confers BSR1-mediated robust resistance to broad-spectrum pathogens.

Keywords: disease resistance; microbe-associated molecular pattern (MAMP); *Pyricularia oryzae* (formerly *Magnaporthe oryzae*); *Oryza sativa* (rice); receptor-like cytoplasmic kinase (RLCK); reactive oxygen species (ROS)

---

1. Introduction

Plants combat pathogens by activating their innate immunity. Microbe/pathogen-associated molecular patterns (MAMPs/PAMPs), which have molecular structures that are conserved in fungi or bacteria, alert plants to pathogen attacks. Chitin (a backbone of fungal cell walls), peptidoglycan (a component of bacterial cell walls), lipopolysaccharide (LPS; a component of the outer membranes of Gram-negative bacteria), flagellin, and elongation factor-Tu (EF-Tu) are well-known MAMPs [1]. MAMPs are perceived by corresponding pattern-recognition receptors (PRRs) on host cell surfaces [2]. In rice (*Oryza sativa*), CHITIN ELICITOR RECEPTOR KINASE 1 (OsCERK1), a lysine motif (LysM)-receptor-like kinase (RLK), is well-characterized as a protein component of PRRs. OsCERK1 interacts with a receptor like protein CHITIN ELICITOR BINDING PROTEIN (CEBiP) [3] to recognize the chitin oligomer [4–6] and with rice LYSM-CONTAINING PROTEIN 4/6 (OsLYP4/6) to recognize peptidoglycan [7]. OsCERK1 also functions as a receptor/co-receptor for LPS signaling [8]. Thus, the central RLKs, like OsCERK1, function as hub receptors. PRR complexes activate pattern-triggered immunity (PTI) through intracellular signaling pathways. Protein phosphorylation signals originating in PRRs positively regulate early phase responses, such as oxidative burst and the activation of mitogen-activated protein kinase (MAPK) cascades, followed by the transcriptional activation of defense-related genes [9,10]. Oxidative burst is caused by a rapid production of reactive oxygen species (ROS) by plant NADPH oxidases, called RESPIRATORY BURST OXIDASE HOMOLOG (RBOH) proteins [11]. Host-derived ROS play various roles in PTI. They possess an antimicrobial activity that can kill microbes, and they enhance physical barrier production by promoting lignin synthesis and cross-linking of plant cell walls [12–15]. Hydrogen peroxide (H_2O_2), an ROS produced in oxidative bursts, acts as a signaling molecule to induce the transcriptional activation of defense-related genes, biosynthesis of phytoalexin, and programmed cell death [11,16].

Pathogens have adapted to suppress PTI through the secretion of effectors using type III secretion system (TTSS) and structural variations on MAMPs [17]. Molina and Kahmann (2007) reported that the detoxification of host-derived ROS is required for a biotrophic pathogen of maize *Ustilago maydis* to overcome PTI [18]. The deletion of *YAP*, encoding an oxidative stress-responsive transcription factor, in *U. maydis* increases the sensitivity to H_2O_2 and significantly decreases the pathogenicity. Rice blast fungus (*Pyricularia oryzae*), which causes a serious disease in rice, releases a catalase-peroxidase B (CPXB)-dependent ROS-degrading activity near conidia [19,20]. Enzymes that compose the glutathione and thioredoxin antioxidation system in *P. oryzae* are required for virulence as well as resistance to ROS [21,22]. *P. oryzae* mutant strains Δ*des1* and Δ*sir2*, which lack transcriptional regulators for extracellular peroxidases and superoxide dismutase, cannot form susceptible lesions because they induce defense responses, including the accumulation of host-derived ROS and the upregulation of defense-related genes [23,24]. These findings indicate that host-derived ROS is a crucial factor in host–microbe interactions.

Receptor-like cytoplasmic kinases (RLCKs) contribute to cytoplasmic phosphorylation signaling pathways in PTI. RLCKs are characterized as cytoplasmic proteins that contain a RLK-homologous kinase domain but not a transmembrane domain. *Arabidopsis thaliana* and rice have 147 and 379 RLCK-encoding genes, respectively [25,26]. RLCKs are classified into 17 subfamilies based on their sequence features. Several RLCKs belonging to subfamily-VII are involved in PTI [27]. In *A. thaliana*, BOTRYTIS-INDUCED KINASE 1 (BIK1) is phosphorylated by PRRs composed of BRI1-ASSOCIATED KINASE 1 (BAK1) and FLAGELLIN SENSING 2 (FLS2) or EF-Tu RECEPTOR (EFR) depending on the ligand (flagellin or EF-Tu, respectively)-binding [28,29]. Phosphorylated BIK1 further phosphorylates and activates RBOHD, which is responsible for the oxidative bursts in PTI [30,31]. In rice, OsRLCK176, an ortholog of BIK1, interacts with OsCERK1 to mediate chitin- and peptidoglycan-induced defense responses [7]. OsRLCK57, OsRLCK107, and OsRLCK118, which are highly homologous to OsRLCK176, also have similar functions [32]. OsRLCK118 directly phosphorylates OsRBOHB [33]. OsRLCK185 interacts with OsCERK1 and MAPKKK, connecting PRRs and the MAPK cascade [34,35].

BROAD-SPECTRUM RESISTANCE 1 (BSR1; OsRLCK278), a rice RLCK-VII protein, has unique disease control effects when overexpressed. It was identified in a screen for disease resistance in rice Full-length cDNA OvereXpressor (FOX) *Arabidopsis* lines [36,37]. The screening revealed that transgenic *A. thaliana* plants overexpressing BSR1 were highly resistant to *Pseudomonas syringae* pv. *tomato* DC3000 and *Colletotrichum higginsianum*. Furthermore, overexpression of BSR1 in rice conferred strong resistance against four rice pathogens: rice blast fungus, brown spot fungus (*Cochliobolus miyabeanus*), rice leaf blight bacteria (*Xanthomonas oryzae* pv. *oryzae*), and *Burkholderia glumae*, which is the causal agent of bacterial seedling rot and bacterial grain rot [36,38]. To our knowledge, among the many RLCKs, BSR1 is the only one that can enhance disease resistance when overexpressed. However, the mechanism underlying the broad-spectrum disease resistance conferred by the overexpression of BSR1 remains unknown.

The contribution of BSR1 to the innate immunity of wild-type rice has been analyzed. A knockout of *BSR1* caused significant suppression of chitin-induced defense-responses, including oxidative bursts and the transcriptional activation of defense-related genes [39]. BSR1 has an active protein kinase domain that phosphorylates serine/threonine and tyrosine residues [40]. These indicate that BSR1 should mediate the downstream phosphorylation signaling of OsCERK1, because the perception of chitin completely depends on OsCERK1 [4]. The silencing of *BSR1* decreased resistance to not only fungal but also bacterial diseases [40], suggesting that BSR1 is involved in the signaling pathway activated by bacterial MAMPs downstream of OsCERK1.

In this report, we investigated whether BSR1 contributes to defense responses elicited by bacterial MAMPs. The resulting resistance is almost independent of salicylic acid, a plant hormone related to immunity [40]. Therefore, we focused on the early phase of defense responses, like the oxidative bursts. Furthermore, to reveal the mechanisms underlying broad-spectrum disease resistance in the BSR1-overexpressing rice plants, we analyzed the early defense events using suspension-cultured cells and sliced leaf blades overexpressing BSR1.

2. Results

2.1. BSR1 Contributes to Bacterial MAMP-Induced Oxidative Bursts

To assess the contribution of BSR1 to bacterium-derived MAMP-induced defense responses, we evaluated the effects of BSR1 knockout on defense responses using three independent *BSR1*-knockout lines. These lines were generated in our previous study and contain homozygous frameshift mutations in exon 1 of *BSR1* [39]. Suspension-cultured cells were derived from knockout and non-transgenic (wild-type) lines and treated with the bacterium-derived MAMPs peptidoglycan and LPS. After treatment with peptidoglycan, suspension-cultured cells derived from all three *BSR1*-knockout lines produced lower H_2O_2 concentrations than wild-type cells (Figure 1a; Supplementary Materials Table S2a). At 180 min after addition, 59%–71% of H_2O_2 production was lost in knockout cells. The LPS treatment also induced impaired oxidative bursts in *BSR1*-knockout cells (Figure 1b; Supplementary Materials Table S2b). These cells accumulated 38%–45% lower amounts of H_2O_2 compared with wild-type at 60 min after treatment. Knockout mutations in *BSR1* significantly suppressed the oxidative bursts but they were not completely abolished, indicating functional redundancy for BSR1. Thus, BSR1 plays a role in the induction of oxidative bursts in response to peptidoglycan and LPS.

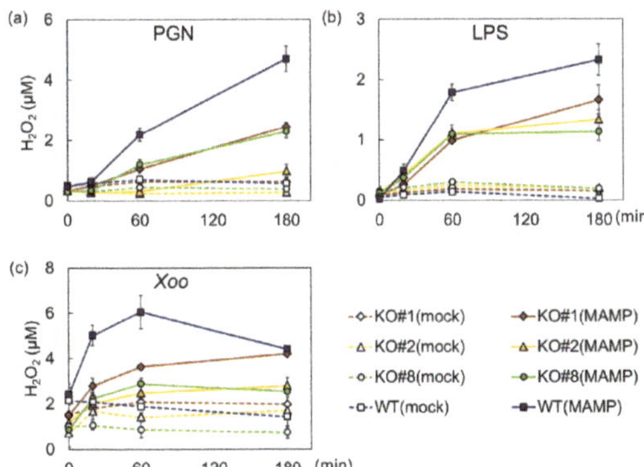

Figure 1. Knockouts of *BSR1* impaired H_2O_2 production in rice cell cultures treated with MAMPs. Suspension-cultured cells were treated with peptidoglycan (**a**), LPS (**b**), or an autoclaved suspension of *Xanthomonas oryzae* pv. *oryzae* (*Xoo*; **c**). H_2O_2 concentrations were measured before treatment and at 20, 60, and 180 min after treatment. Values are presented as the means ± standard deviations of three biological replicates. Experiments were conducted twice with similar results. PGN, peptidoglycan, LPS, lipopolysaccharide; KO, knockout line; KO#1, *bsr1-1*#13-1; KO#2, *bsr1-2*#16-2; KO#8, *bsr1-8*#5-1; WT, wild-type; MAMPs, microbe-associated molecular patterns; BSR1, BROAD-SPECTRUM RESISTANCE 1. The statistical analysis was performed as shown in Supplementary Materials Table S2.

To further provide support for the involvement of BSR1 in oxidative bursts against bacterial infections, autoclaved *X. oryzae* pv. *oryzae* cells were used as the elicitor. Knocking out *BSR1* reduced the production of H_2O_2 by 39%–58% at 60 min after treatment with the autoclaved cells (Figure 1c; Supplementary Materials Table S2c). Thus, BSR1 should contribute to defense responses against not only MAMPs purified from nonpathogenic microbes but also against the cellular components of pathogenic bacteria.

2.2. BSR1 Is Involved in Regulating MAMP-Responsive Genes

In MAMP-treated suspension-cultured cells, the transcriptional activation of defense-related genes was analyzed. Transcript levels of four defense-related genes, diterpenoid phytoalexin (DP) momilactone biosynthetic gene *KAURENE SYNTHASE-LIKE 4 (KSL4)*, DP biosynthetic key transcription factor-encoding gene *DITERPENOID PHYTOALEXIN FACTOR (DPF)* [41], the representative defense marker gene *PROBENAZOLE-INDUCIBLE PROTEIN 1 (PBZ1)*, and flavonoid phytoalexin and lignin biosynthetic gene *PHENYLALANINE AMMONIA-LYASE 1 (PAL1)* were determined. After treatment with peptidoglycan, the inductions of *KSL4*, *DPF*, and *PBZ1* in knockout cells were significantly weaker than in wild-type, although significant changes in *PAL1* transcript level were not detected (Figure 2a). Knocking out *BSR1* resulted in a decrease in *PBZ1* transcript levels under mock-treatment conditions (Figure 2). Our liquid cultivation conditions slightly induced *PBZ1* transcriptional activation, which was mediated by BSR1.

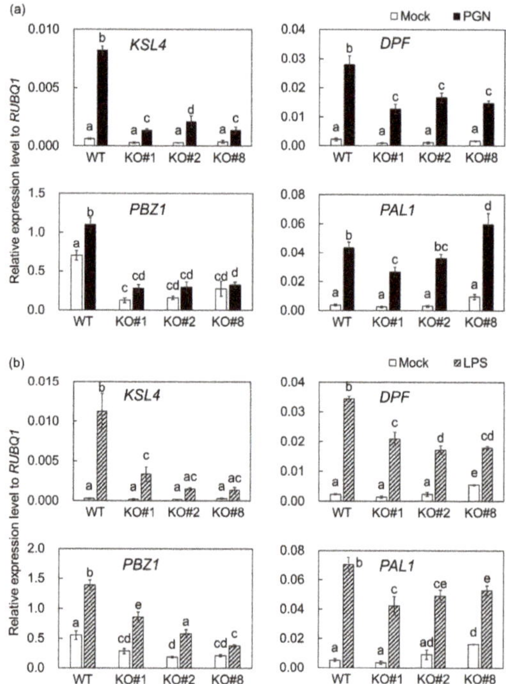

Figure 2. MAMP-induced transcript levels of defense-related genes were suppressed in *BSR1*-knockout suspension-cultured rice cells. The *PBZ1*, *PAL1*, *KSL4*, and *DPF* transcript levels at 3-h post treatment with peptidoglycan (**a**) and LPS (**b**) were normalized against the *RUBQ1* internal control levels. Values are presented as the means ± standard deviations of three biological replicates. Experiments were conducted two times with similar results. Different letters indicate significant differences (Tukey's test; $p < 0.05$). PGN, peptidoglycan; KO, knockout line; KO#1, *bsr1-1#13-1*; KO#2, *bsr1-2#16-2*; KO#8, *bsr1-8#5-1*; WT, wild-type; LPS, lipopolysaccharide.

Knocking out *BSR1* suppressed the elicitation of *KSL4* and *DPF* by LPS (Figure 2b). The significant suppression of *PBZ1* and *PAL1* were not reproducibly detected. As shown in Figures 1 and 2, BSR1 appears to function in defense responses after the plant perceives peptidoglycan and LPS.

2.3. BSR1 Overexpression Enhances Oxidative Bursts in Suspension-Cultured Cells

Contrary to the *BSR1* disruption phenotype, the overexpression of BSR1 is assumed to enhance defense responses. Whether the overexpression affects the robustness of the oxidative bursts and transcriptional activation was investigated. Rice plants overexpressing HA–PreScission–Biotin (HPB)-tagged BSR1 and GUS (BSR1-HPB:OX and GUS-HPB:OX, respectively) were generated. The GUS-HPB:OX line was used as a control. The integrities of the inserted constructs were confirmed by western analysis with an anti-HA antibody (Supplementary Materials Figure S1a). The overexpression of BSR1-HPB conferred resistance to rice blast, indicating that BSR1-HPB is functional (Supplementary Materials Figure S1b). Suspension-cultured cells were prepared from wild-type, GUS-HPB:OX, and two independent BSR1-HPB:OX lines. The transcript levels of *BSR1* and HPB-tagged transgenes in suspension-cultured cells were ascertained using qRT-PCR (Supplementary Materials Figure S1c).

In response to peptidoglycan treatments, suspension-cultured cells derived from two BSR1-HPB:OX lines produced H_2O_2 more rapidly than GUS-HPB:OX (Figure 3a; Supplementary Materials Figure S2a). At 60 min after treatment, the overexpression of BSR1 resulted in increased H_2O_2 concentrations to 1.6–2.0 times that of the control (Figure 3a). Transcript level of a defense-related gene *PAL1* was increased in BSR1-HPB:OX cells compared with GUS-HPB:OX control, while no significant changes in transcript levels of *PBZ1* and *KSL4* were detected (Figure 3b). Transcript levels in GUS-HPB:OX did not necessarily agree with those in WT, indicating that the responses would be slightly altered by overexpression of transgenes. BSR1-HPB:OX cells produced enhanced H_2O_2 bursts in response to LPS as well as peptidoglycan (Supplementary Materials Figure S3).

Interestingly, before the MAMP treatment, the overexpression of BSR1-HPB resulted in a slight but statistically significant increase in H_2O_2 concentrations compared with GUS-HPB in cell cultures. Comparisons between the untreated conditions (Figure 3a, 0 min) showed that there were significant differences between BSR1-HPB:OX lines and the GUS-HPB:OX line ($p < 0.001$ for BSR1-HPB:OX17 and BSR1-HPB:OX39, Student's *t*-test). These phenotypes were common to all the replicated experiments (Figure 4; Supplementary Materials Figure S3). BSR1 overexpression did not increase transcript levels of *RbohB*, encoding a NADPH oxidase related to ROS burst (Supplementary Materials Figure S4). These results suggest that an excess of BSR1 protein could constitutively promote NADPH oxidase activity of RBOH proteins but not their transcription.

Because of requirement of BSR1 in chitin oligomer-induced defense responses [39], we assessed the oxidative bursts after a chitin hexamer treatment. The amount of H_2O_2 produced by BSR1-HPB:OX cells significantly exceeded that of the control at each measured time point (Figure 4a; Supplementary Materials Figure S2b). At 60 min after treatment, BSR1-HPB:OX cells produced a 1.8–1.9-fold greater H_2O_2 concentration than GUS-HPB:OX cells (Figure 4a). The chitin-induced transcriptional activation of *PAL1*, but not *KSL4* and *PBZ1*, were enhanced by the overexpression of BSR1-HPB (Figure 4b). These comparisons of BSR1-HPBs with GUS-HPB clearly showed that BSR1 overexpression enhanced oxidative bursts and transcriptional activation of, at least, *PAL1* in response to multiple MAMPs.

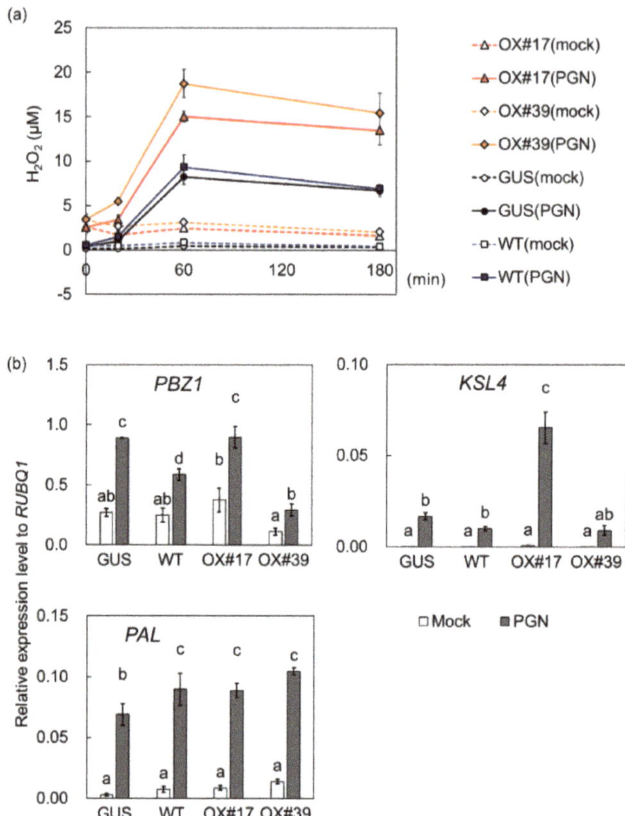

Figure 3. The overexpression of BSR1-HPB enhanced peptidoglycan-induced oxidative bursts in suspension-cultured rice cells. Cells treated with peptidoglycan were analyzed for H_2O_2 production accompanying oxidative bursts (**a**) and the transcript levels of defense-related genes (**b**). Values are presented as the means ± standard deviations of three biological replicates. In (**a**), H_2O_2 concentrations were measured before treatment and at 20, 60, and 180 min after treatment. The statistical analysis was performed as shown in Figure S2a. Experiments were conducted three times with similar results. In (**b**), the *PBZ1*, *PAL1*, and *KSL4* transcript levels were normalized against the *RUBQ1* internal control levels. Experiments were conducted two times with similar results. Different letters indicate significant differences (Tukey's test; $p < 0.05$). PGN, peptidoglycan; OX, overexpressing line; HPB, HA–PreScission–Biotin; OX#17, BSR1-HPB:OX#17; OX#39, BSR1-HPB:OX#39; GUS, GUS-HPB:OX; WT, wild-type.

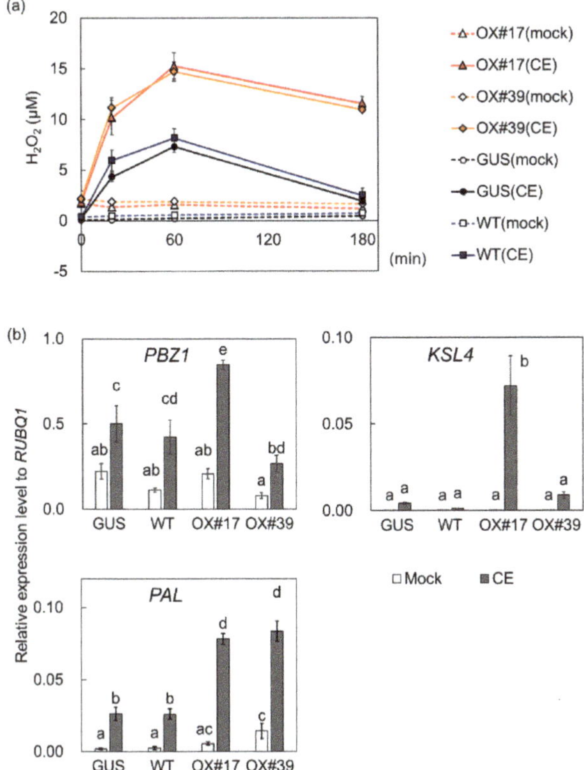

Figure 4. The overexpression of BSR1-HPB enhanced chitin-induced defense responses in suspension-cultured rice cells. Values are presented as the means ± standard deviations of three biological replicates. (**a**) H_2O_2 concentrations were measured before treatment and at 20, 60, and 180 min after treatment. The statistical analysis was performed as shown in Figure S2b. Experiments were conducted three times with similar results. (**b**) The *PBZ1*, *PAL1*, and *KSL4* transcript levels were normalized against the *RUBQ1* internal control levels. Experiments were conducted twice with similar results. Different letters indicate significant differences (Tukey's test; $p < 0.05$). CE, chitin elicitor; OX, overexpressing line; HPB, HA–PreScission–Biotin; OX#17, BSR1-HPB:OX#17; OX#39, BSR1-HPB:OX#39; GUS, GUS-HPB:OX; WT, wild-type.

2.4. Oxidative Bursts against Blast Fungus Are Enhanced in Plants Overexpressing BSR1

We speculated that H_2O_2 production in plant leaves as a response to pathogen challenges is increased by the overexpression of BSR1, as observed in suspension-cultured cells. To test the hypothesis, strips from leaf blades were quantitatively analyzed for H_2O_2 production after being treated with conidia of the blast fungus, which had been autoclaved to eliminate any biological activity. Before the treatment, the H_2O_2 concentration in water containing leaf strips of BSR1-HPB:OX#17 was slightly greater than that of GUS-HPB:OX (Figure 5). After exposure to autoclaved conidia, leaf strips of BSR1-HPB:OX plants produced far greater H_2O_2 concentrations than those of GUS-HPB:OX plants (Figure 5a). Taking into consideration the difference between untreated conditions, we calculated changes in H_2O_2 concentrations during the experiment. By 180 min after treatment, the overexpression of BSR1-HPB resulted in a ~4.2-fold increase in changes in H_2O_2 concentration, compared with GUS-HPB (Supplementary Materials Figure S5a). Autoclaved conidia-induced H_2O_2 hyperproduction was also detected in leaf strips of BSR1-HPB:OX#39, an another BSR1-overexpressing line (Supplementary Materials Figure S6). Thus, *BSR1* overexpression enhanced oxidative bursts in leaf blades.

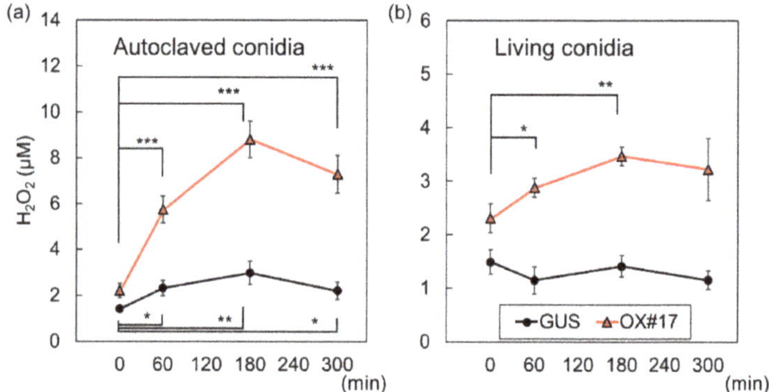

Figure 5. Rice leaf strips derived from BSR1-HPB:OX plants caused an enhanced burst of H_2O_2 when exposed to conidia of the compatible blast fungus. Leaf strips were cultivated with 8×10^4 mL^{-1} autoclaved conidia (**a**) or 8×10^3 mL^{-1} living conidia (**b**) in wells of a 12-well plate. H_2O_2 concentrations in wells were measured before treatment and at 60, 180, and 300 min after treatment. Values are presented as the means ± standard deviations of three biological replicates. Asterisks indicate significant differences between the untreated condition (0 min) values and the values at the indicated times in the same line (Student's *t*-test; * $p < 0.05$, ** $p < 0.01$, and *** $p < 0.001$). Experiments were conducted twice with similar results. OX, overexpressing line; HPB, HA–PreScission–Biotin; OX#17, BSR1-HPB:OX#17; GUS, GUS-HPB:OX.

To assess the importance of the enhanced H_2O_2 bursts in host–microbe interactions, we also examined the oxidative bursts after a living conidia treatment. A treatment with 8×10^4 mL^{-1} conidia depressed H_2O_2 levels in BSR1-HPB:OX leaves and GUS-HPB:OX leaves (Supplementary Materials Figure S7). This result corroborated a previous report that suspensions of conidia contain H_2O_2-degrading enzymes [19]. In order to avoid that abnormally strong ROS-degrading activity obscures the difference, we used lower concentration (8×10^3 mL^{-1}) of conidia. Considering that the H_2O_2-degrading activity increased with the conidial concentration, comparisons between H_2O_2 levels were performed only under the same co-cultivation conditions. When co-cultivated with 8×10^3 mL^{-1} conidia, no elevation in the H_2O_2 level was detected in GUS-HPB:OX leaves (Figure 5b; Supplementary Materials Figure S5b). Thus, the addition of this concentration of conidia completely suppressed MAMP-induced oxidative bursts in the control line. In contrast, when co-cultivated with BSR1-HPB:OX leaves, the H_2O_2 level significantly increased, compared with before the conidial inoculation (Figure 5b). These co-cultivation experiments revealed that rice plants overexpressing BSR1 produced large amounts of H_2O_2 that overwhelmed the ROS degradation caused by pathogens.

3. Discussion

BSR1, a RLCK-VII member, has a protein kinase activity and is important for the initiation of defense responses against chitin oligomers, known as a fungus-derived MAMP [39,40]. BSR1 is implicated in resistance to bacteria, as well as fungi, in wild-type and overexpressing rice lines [36,38,40], suggesting that BSR1 is also involved in responses triggered by bacterium-derived MAMPs. In this study, our experiments on suspension-cultured rice cells showed a correlation between BSR1 and the response to two bacterial elicitors, peptidoglycan and LPS. Knocking out *BSR1* significantly suppressed the elicitation of oxidative bursts and the transcript levels of defense-related genes caused by these bacterial MAMPs (Figures 1 and 2). In some experiments, variations of the H_2O_2 concentration and the transcript levels were observed among knockout lines (Figures 1 and 2). Since the variations among knockout lines were not reproducible, they were considered as the influence of the experimental manipulation. The suppression of immune responses by *BSR1* knockout is in accordance with the significant contribution of BSR1 to chitin-induced responses [39]. Rice recognizes chitin through receptor complexes containing OsCERK1 and CEBiP [6]. OsCERK1, but not CEBiP, possesses a protein kinase activity to phosphorylate cytoplasmic signaling factors [42]. The perception of peptidoglycan and LPS was mostly mediated by OsCERK1 complex [4,7,8]. Thus, in response to peptidoglycan and LPS exposure, as well as chitin, OsCERK1 would transmit a signal directly or indirectly to BSR1 to regulate its protein kinase activity (Figure 6).

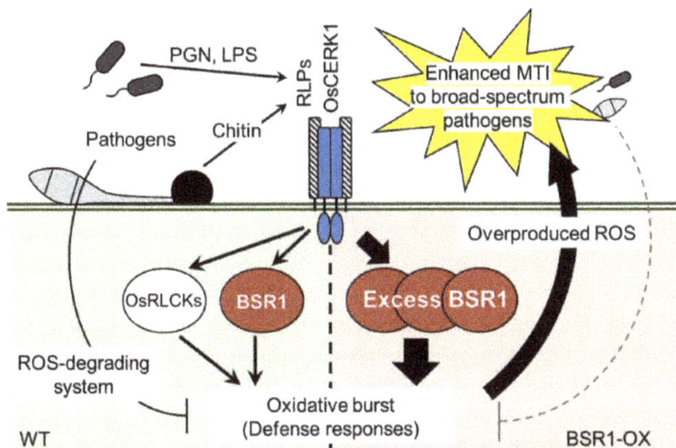

Figure 6. Proposed model in which BSR1 regulates defense responses, such as oxidative bursts, after the perception of MAMPs in wild-type (WT; **left**) and BSR1-overexpressing rice lines (BSR1-OX; **right**). PGN, peptidoglycan; LPS, lipopolysaccharide; RLPs, receptor-like proteins; ROS, reactive oxygen species; MTI, MAMP-triggered immunity.

Knocking out *BSR1* did not make rice cells nonresponsive to MAMPs (Figures 1 and 2), indicating the existence of functionally redundant factor(s) for BSR1. In *A. thaliana*, RLCK-VII members function in MAMP-induced defense responses with a robust functional redundancy [43]. The participation of other rice RLCK-VII members in PTI have been studied [27]. OsRLCK57, OsRLCK107, OsRLCK118, OsRLCK176, and OsRLCK185 positively regulate chitin- and peptidoglycan-induced responses [7,32,42]. No rice RLCKs, except for BSR1, have been reported to mediate LPS-induced oxidative bursts. However, known interactors of the LPS-(co)receptor OsCERK1, such as OsRLCK176 and OsRLCK185, may mediate LPS-signaling. To take into consideration of functional redundancy for BSR1, these RLCK-VII members could act downstream of LPS as well as peptidoglycan.

We compared BSR1-HPB:OX lines with GUS-HPB:OX control line to assess the effects of BSR1 overexpression on MAMP-triggered responses. In the absence of MAMPs, H_2O_2 levels in cell cultures and leaf strips derived from BSR1-HPB:OX lines were slightly greater than those of the control (Figures 3–5). Where this H_2O_2 originates from is unknown. Under peptidoglycan-, LPS-, and chitin-treated conditions, BSR1-HPB:OX suspension-cultured cells produced a greater amount of H_2O_2 than control cells (Figures 3 and 4; Supplementary Materials Figures S2 and S3), while the overexpression of BSR1 facilitated the transcriptional activation of *PAL1* but not *PBZ1* and *KSL4* (Figures 3b and 4b). There were variations in transcript levels of *PBZ1* and *KSL4* between two BSR1 overexpression lines, BSR1-HPB:OX#17 and BSR1-HPB:OX#39. Since transcript levels of defense-marker gene *PBZ1* in MAMP-treated and untreated BSR1-HPB:OX#39 were even lower than those in the control line, the line may have contained mutations which decrease the transcript levels of these defense-related genes. Alternatively, the condition to culture BSR1-HPB:OX#39 line may have given stress to the cells, resulting in slight increase in the expression of internal control *RUBQ1*, which encodes protein turnover factor. That is because the environment surrounding the cells, such as cell concentration, cannot be completely uniformized. In accordance with results using cell culture, BSR1-HPB:OX leaf blade tissues displayed remarkably greater oxidative bursts against MAMPs extracted from autoclaved conidia (Figure 5a). In rice and *Arabidopsis*, OsRLCK118 and *A. thaliana* BIK1, two RLCK-VII members, directly and positively regulate RBOH proteins whose NADPH oxidase activities generate ROS and cause oxidative bursts [30,31,33]. Highly expressed BIK1 leads to enhanced ROS production in response to MAMP [44]. Excess BSR1 protein also could hyperactivate RBOHs, directly or indirectly, resulting in the enhancement of oxidative bursts. Recent work showed that *BIK1* overexpression does not enhance fungal disease resistance, although deletions of negative regulators for PTI signaling result in the accumulation of BIK1 and the strong disease resistance in *A. thaliana* [45]. Unlike BIK1, BSR1 overexpression confers the robust disease resistance in rice and *A. thaliana* [36], indicating that functions of BSR1 would be quite different from those of BIK1.

A time course of the H_2O_2 levels under co-cultivation conditions revealed how the overexpression of BSR1 acts during the early phase of host–microbe interactions. Under our co-cultivation conditions, suspensions of living conidia of the blast fungus did not elicit host-derived H_2O_2 production in control leaf strips (Figure 5b; Supplementary Materials Figure S5b). These results reconfirmed previous reports that the supernatants of conidial suspensions contain ROS-degrading activities that mostly depend on CPXB, a catalase-peroxidase secreted by the blast fungus [19]. Under the same co-cultivation conditions, the overexpression of BSR1-HPB resulted in leaf blade tissues producing detectable amounts of H_2O_2 (Figure 5), indicating that oxidative bursts in BSR1-overexpressing plants are intense enough to overcome the inhibition caused by the infecting blast fungus. The enhanced responses against peptidoglycan and LPS, as well as chitin resulting from the overexpression (Figures 3 and 4; Supplementary Materials Figure S3) strongly suggested that pathogenic bacterial challenges would elicit the same responses.

In host–microbe interactions, host-derived ROS is regarded as an antimicrobial substance and a diffusible second messenger that contributes to immunity [11,16]. Indeed, host-derived ROS detoxification should be required for pathogenicity. For example, ROS-degrading activities are present in the supernatants of *P. oryzae* conidial suspensions and contribute to lesion formation when exogenously added [20]. The deletion of *P. oryzae DES1*, which is required for extracellular peroxidase activity, causes the accumulation of host-derived ROS and the induction of defense-related genes, resulting in non-pathogenicity [23]. Pathogens could not completely abolish overproduced ROS (Figure 5b), and therefore do not show full virulence in plants overexpressing BSR1 (Figure 6). Our data support the model which host-derived ROS is critical for plant interactions with pathogens.

In conclusion, we propose that the broad-spectrum disease resistance could be achieved by the enhancement of MAMP-triggered oxidative bursts and following transcriptional activation. To date, many RLCKs have been characterized as signaling factors in PTI [27]. However, no RLCK, other than *BSR1*, can enhance oxidative bursts and disease resistance when overexpressed in rice. It is unclear what allows the functional enhancement. This report clearly showed that hyperactivated MAMP-triggered immune responses could be used for broad-spectrum disease control.

4. Materials and Methods

4.1. Plant and Microbial Materials and Inoculation

Rice (*Oryza sativa* L. cv. Nipponbare) was used as the wild-type (WT) plant material. The *BSR1*-knockout lines *bsr1-1*#13-1 (KO#1), *bsr1-2*#16-2 (KO#2), and *bsr1-8*#5-1 (KO#8) generated with the clustered regularly interspaced short palindromic repeats (CRISPR)/CRISPR-associated 9 (Cas9) system in our previous study [39,46] were used. Rice calli were prepared from dehusked seeds and cultivated on N6D medium containing 0.4% gellan gum [47]. The *Pyricularia oryzae* isolate Kyu89-246 (MAFF101506, race 003.0), which is compatible with Nipponbare rice plants, was used to prepare the elicitor and for inoculations. *Xanthomonas oryzae* pv. *oryzae* (isolate T7174) was used to prepare the elicitor fraction. The culturing methods and *P. oryzae* and *X. oryzae* pv. *oryzae* inoculation techniques were as previously described [38].

4.2. Plasmid Construction and Transformation

A DNA fragment containing the maize *Ubiquitin-1* promoter was excised from pRiceFOX-GateA-*SG1* [48] with HindIII and KpnI. pMDC32-HPB [49] was digested with HindIII and KpnI, and the fragment containing the 2× 35S promoter was replaced with the HindIII–KpnI fragment containing the maize *Ubiquitin-1* promoter. The resulting plasmid was named pMDC32-Mubi-HPB. The open reading frame sequences of *BSR1* and *GUS* were amplified from the full-length cDNA (AK070024) and pBI221 (AF502128), respectively. These DNA fragments were ligated into pENTR/D-TOPO (Invitrogen, Carlsbad, CA, USA) to construct HPB-tagged *BSR1* and *GUS* (*BSR1-HPB* and *GUS-HPB*, respectively) in pMDC32-Mubi-HPB using the Gateway LR Clonase II Plus Enzyme (Invitrogen). pMDC32-Mubi-HPB containing *BSR1-HPB* or *GUS-HPB* was introduced into Nipponbare using the *Rhizobium radiobacter*-mediated transformation method [47]. BSR1-HPB:OX17 (OX#17), BSR1-HPB:OX39 (OX#39), and GUS-HPB:OX6 (GUS) transgenic lines were analyzed as two BSR1-overexpressing lines and a control line, respectively.

4.3. Measurement of H_2O_2

Rice suspension-cultured cells were prepared using a previously published method [5,39]. Modified liquid N6 medium (30 g L^{-1} sucrose; 4.1 mg L^{-1} N6 salt (Wako, Osaka, Japan); 2 mg L^{-1} glycine; 0.5 mg L^{-1} nicotinic acid; 0.5 mg L^{-1} pyridoxine HCl; 1 mg L^{-1} thiamine HCl; 100 mg L^{-1} myo-inositol; 1 mg L^{-1} 2,4-dichlorophenoxyacetic acid; 23.4 mg L^{-1} $MnSO_4 \cdot 4H_2O$; pH 5.8) was used for liquid cultivation. We treated 1 mL media containing 100 mg suspension-cultured cells with 10 μg mL^{-1} peptidoglycan from *Bacillus subtilis* (Sigma-Aldrich, St. Louis, MO, USA), 50 μg mL^{-1} LPS from *Pseudomonas aeruginosa* 10 purified by phenol extraction (Sigma-Aldrich), 10 nM *N*-acetylchitohexaose (chitin elicitor (CE)), or an autoclaved suspension of *X. oryzae* pv. *oryzae* (OD_{600} = 0.3).

To prepare leaf strips, the sixth leaves of GUS-HPB:OX and BSR1-HPB:OX17 plants at the 6–6.5-leaf stage were used. Two fragments of the leaf blades (8-mm length and 6-mm width) that were slit using bundled razor blades at approximately 0.5-mm intervals were placed in a well of a 12-well plate. Leaf strips were floated on sterile water and incubated at 28 °C for 14–15 h with shaking at 90 rpm, followed by a 1-h incubation in new water. Conidia of rice blast fungus were scraped from the gel surface with sterile water and filtered through a Kimwipe. The conidial concentration in the filtrate was calculated using a hemocytometer. A suspension of autoclaved or living conidia was poured into the wells to the indicated final concentration and incubated at 28 °C. The H_2O_2 concentration was determined at the indicated time using a previously described luminol-dependent chemiluminescence assay [5]. The statistical analyses were carried out using Dunnett's test for the experiments with cultured cells and Student's *t*-test for the experiments with leaf strips.

4.4. Quantitative Reverse Transcription (qRT)-PCR

Cultured rice cells were frozen in liquid nitrogen after a 3-h treatment with 10 μg mL^{-1} peptidoglycan, 50 μg mL^{-1} LPS, 10 nM *N*-acetylchitohexaose, an autoclaved suspension of *X. oryzae* pv. *oryzae* (OD_{600} = 0.3), or sterile water. Total RNA extraction and the qRT-PCR were performed as previously described [39]. Transcript levels were analyzed using the comparative C_T ($2^{-\Delta\Delta Ct}$) method with rice *Ubiquitin1* (*RUBQ1*; Os06g0681400) as an internal control [50,51]. The statistical analysis was carried out with Tukey's test. The primers presented in Supplementary Materials Table S1 were used for the qRT-PCR analyses.

4.5. Western Blot Analysis

Protein was extracted from 50 mg leaf blades with 300 μL SDS–urea buffer (8 M urea, 5% SDS, 0.1 mM EDTA, 2% 2-mercaptoethanol, 1 mM phenylmethanesulfonylfluoride (PMSF), 2 × complete inhibitor mix, EDTA-free (Roche, Basel, Switzerland), 40 mM Tris-HCl: pH 6.8) according to a previously described method [52]. An equal volume of each SDS–urea sample was used for the western analysis. HPB-tagged protein was detected using anti-HA antibody (Anti-HA.11, Mouse-Mono 16B12; BAB).

Supplementary Materials: The following are available online at http://www.mdpi.com/1422-0067/20/22/5523/s1.

Author Contributions: Conceptualization, Y.K., Y.N., T.K., and M.M.; resources H.N.; formal analysis, Y.K.; writing—original draft preparation, Y.K.; writing—review and editing, H.N., Y.N., and M.M.; supervision, Y.N., T.K., and M.M.; project administration, M.M.

Funding: This research received no external funding.

Acknowledgments: We thank Fumiaki Katagiri (University of Minnesota) for providing pMDC32-HPB. We thank Shoji Sugano and Satoru Maeda (NIAS, Japan) for assistance in western analyses and fungal blast infection, respectively. We thank Eiichi Minami (NIAS, Japan) for providing a luminometer, and Naoto Shibuya and Yoshitake Desaki (Meiji University) for providing the bundled razor blades used in the H$_2$O$_2$ concentration experiments. We also thank Lois Ishizaki and Yuka Yamazaki (NIAS, Japan) for their help during the rice transformation experiments and for their overall technical assistance. We thank Lesley Benyon, from Edanz Group (www.edanzediting.com/ac) for editing a draft of this manuscript.

Conflicts of Interest: The authors declare no conflict of interest.

References

1. Boller, T.; Felix, G. A renaissance of elicitors: Perception of microbe-associated molecular patterns and danger signals by pattern-recognition receptors. *Annu. Rev. Plant Biol.* **2009**, *60*, 379–406. [CrossRef] [PubMed]

2. Monaghan, J.; Zipfel, C. Plant pattern recognition receptor complexes at the plasma membrane. *Curr. Opin. Plant Biol.* **2012**, *15*, 349–357. [CrossRef] [PubMed]

3. Kaku, H.; Nishizawa, Y.; Ishii-Minami, N.; Akimoto-Tomiyama, C.; Dohmae, N.; Takio, K.; Minami, E.; Shibuya, N. Plant cells recognize chitin fragments for defense signaling through a plasma membrane receptor. *Proc. Natl. Acad. Sci. USA* **2006**, *103*, 11086–11091. [CrossRef]

4. Kouzai, Y.; Mochizuki, S.; Nakajima, K.; Desaki, Y.; Hayafune, M.; Miyazaki, H.; Yokotani, N.; Ozawa, K.; Minami, E.; Kaku, H.; et al. Targeted gene disruption of OsCERK1 reveals its indispensable role in chitin perception and involvement in the peptidoglycan response and immunity in rice. *Mol. Plant Microbe Interact.* **2014**, *27*, 975–982. [CrossRef]

5. Kouzai, Y.; Nakajima, K.; Hayafune, M.; Ozawa, K.; Kaku, H.; Shibuya, N.; Minami, E.; Nishizawa, Y. CEBiP is the major chitin oligomer-binding protein in rice and plays a main role in the perception of chitin oligomers. *Plant Mol. Biol.* **2014**, *84*, 519–528. [CrossRef] [PubMed]

6. Shimizu, T.; Nakano, T.; Takamizawa, D.; Desaki, Y.; Ishii-Minami, N.; Nishizawa, Y.; Minami, E.; Okada, K.; Yamane, H.; Kaku, H.; et al. Two LysM receptor molecules, CEBiP and OsCERK1, cooperatively regulate chitin elicitor signaling in rice. *Plant J.* **2010**, *64*, 204–214. [CrossRef] [PubMed]

7. Ao, Y.; Li, Z.; Feng, D.; Xiong, F.; Liu, J.; Li, J.F.; Wang, M.; Wang, J.; Liu, B.; Wang, H.B. OsCERK1 and OsRLCK176 play important roles in peptidoglycan and chitin signaling in rice innate immunity. *Plant J.* **2014**, *80*, 1072–1084. [CrossRef]

8. Desaki, Y.; Kouzai, Y.; Ninomiya, Y.; Iwase, R.; Shimizu, Y.; Seko, K.; Molinaro, A.; Minami, E.; Shibuya, N.; Kaku, H.; et al. OsCERK1 plays a crucial role in the lipopolysaccharide-induced immune response of rice. *New Phytol.* **2018**, *217*, 1042–1049. [CrossRef]

9. Kawasaki, T.; Yamada, K.; Yoshimura, S.; Yamaguchi, K. Chitin receptor-mediated activation of MAP kinases and ROS production in rice and Arabidopsis. *Plant Signal. Behav.* **2017**, *12*, e1361076. [CrossRef]

10. Macho, A.P.; Zipfel, C. Plant PRRs and the activation of innate immune signaling. *Mol. Cell* **2014**, *54*, 263–272. [CrossRef]

11. Waszczak, C.; Carmody, M.; Kangasjärvi, J. Reactive Oxygen Species in Plant Signaling. *Annu. Rev. Plant Biol.* **2018**, *69*, 209–236. [CrossRef] [PubMed]

12. Bradley, D.J.; Kjellbom, P.; Lamb, C.J. Elicitor-and wound-induced oxidative cross-linking of a proline-rich plant cell wall protein: A novel, rapid defense response. *Cell* **1992**, *70*, 21–30. [CrossRef]

13. Chen, S.X.; Schopfer, P. Hydroxyl-radical production in physiological reactions. A novel function of peroxidase. *Eur. J. Biochem.* **1999**, *260*, 726–735. [CrossRef] [PubMed]

14. Lu, H.; Higgins, V. The effect of hydrogen peroxide on the viability of tomato cells and of the fungal pathogen Cladosporium fulvum. *Physiol. Mol. Plant Pathol.* **1999**, *54*, 131–143. [CrossRef]

15. Peng, M.; Kuc, J. Peroxidase-Generated Hydrogen-Peroxide as a Source of Antifungal Activity in Vitro and on Tobacco Leaf-Disks. *Phytopathology* **1992**, *82*, 696–699. [CrossRef]

16. Wrzaczek, M.; Brosché, M.; Kangasjärvi, J. ROS signaling loops—Production, perception, regulation. *Curr. Opin. Plant Biol.* **2013**, *16*, 575–582. [CrossRef] [PubMed]

17. Jones, J.D.; Dangl, J.L. The plant immune system. *Nature* **2006**, *444*, 323–329. [CrossRef]

18. Molina, L.; Kahmann, R. An Ustilago maydis gene involved in H2O2 detoxification is required for virulence. *Plant Cell* **2007**, *19*, 2293–2309. [CrossRef]

19. Tanabe, S.; Ishii-Minami, N.; Saitoh, K.; Otake, Y.; Kaku, H.; Shibuya, N.; Nishizawa, Y.; Minami, E. The role of catalase-peroxidase secreted by *Magnaporthe oryzae* during early infection of rice cells. *Mol. Plant Microbe Interact.* **2011**, *24*, 163–171. [CrossRef]

20. Tanabe, S.; Nishizawa, Y.; Minami, E. Effects of catalase on the accumulation of H_2O_2 in rice cells inoculated with rice blast fungus, *Magnaporthe oryzae*. *Physiol. Plant.* **2009**, *137*, 148–154. [CrossRef]

21. Fernandez, J.; Wilson, R.A. Characterizing roles for the glutathione reductase, thioredoxin reductase and thioredoxin peroxidase-encoding genes of *Magnaporthe oryzae* during rice blast disease. *PLoS ONE* **2014**, *9*, e87300. [CrossRef] [PubMed]

22. Huang, K.; Czymmek, K.J.; Caplan, J.L.; Sweigard, J.A.; Donofrio, N.M. HYR1-mediated detoxification of reactive oxygen species is required for full virulence in the rice blast fungus. *PLoS Pathog.* **2011**, *7*, e1001335. [CrossRef] [PubMed]

23. Chi, M.H.; Park, S.Y.; Kim, S.; Lee, Y.H. A novel pathogenicity gene is required in the rice blast fungus to suppress the basal defenses of the host. *PLoS Pathog.* **2009**, *5*, e1000401. [CrossRef]

24. Fernandez, J.; Marroquin-Guzman, M.; Nandakumar, R.; Shijo, S.; Cornwell, K.M.; Li, G.; Wilson, R.A. Plant defence suppression is mediated by a fungal sirtuin during rice infection by *Magnaporthe oryzae*. *Mol. Microbiol.* **2014**, *94*, 70–88. [CrossRef] [PubMed]

25. Shiu, S.H.; Karlowski, W.M.; Pan, R.; Tzeng, Y.H.; Mayer, K.F.; Li, W.H. Comparative analysis of the receptor-like kinase family in Arabidopsis and rice. *Plant Cell* **2004**, *16*, 1220–1234. [CrossRef] [PubMed]

26. Vij, S.; Giri, J.; Dansana, P.K.; Kapoor, S.; Tyagi, A.K. The receptor-like cytoplasmic kinase (OsRLCK) gene family in rice: Organization, phylogenetic relationship, and expression during development and stress. *Mol. Plant* **2008**, *1*, 732–750. [CrossRef] [PubMed]

27. Liang, X.; Zhou, J.M. Receptor-Like Cytoplasmic Kinases: Central Players in Plant Receptor Kinase-Mediated Signaling. *Annu. Rev. Plant Biol.* **2018**, *69*, 267–299. [CrossRef]

28. Lu, D.; Wu, S.; Gao, X.; Zhang, Y.; Shan, L.; He, P. A receptor-like cytoplasmic kinase, BIK1, associates with a flagellin receptor complex to initiate plant innate immunity. *Proc. Natl. Acad. Sci. USA* **2010**, *107*, 496–501. [CrossRef]

29. Zhang, J.; Li, W.; Xiang, T.; Liu, Z.; Laluk, K.; Ding, X.; Zou, Y.; Gao, M.; Zhang, X.; Chen, S.; et al. Receptor-like cytoplasmic kinases integrate signaling from multiple plant immune receptors and are targeted by a Pseudomonas syringae effector. *Cell Host Microbe* **2010**, *7*, 290–301. [CrossRef]

30. Kadota, Y.; Sklenar, J.; Derbyshire, P.; Stransfeld, L.; Asai, S.; Ntoukakis, V.; Jones, J.D.; Shirasu, K.; Menke, F.; Jones, A.; et al. Direct regulation of the NADPH oxidase RBOHD by the PRR-associated kinase BIK1 during plant immunity. *Mol. Cell* **2014**, *54*, 43–55. [CrossRef]

31. Li, L.; Li, M.; Yu, L.; Zhou, Z.; Liang, X.; Liu, Z.; Cai, G.; Gao, L.; Zhang, X.; Wang, Y.; et al. The FLS2-associated kinase BIK1 directly phosphorylates the NADPH oxidase RbohD to control plant immunity. *Cell Host Microbe* **2014**, *15*, 329–338. [CrossRef] [PubMed]

32. Li, Z.; Ao, Y.; Feng, D.; Liu, J.; Wang, J.; Wang, H.B.; Liu, B. OsRLCK 57, OsRLCK107 and OsRLCK118 Positively Regulate Chitin- and PGN-Induced Immunity in Rice. *Rice* **2017**, *10*, 6. [CrossRef] [PubMed]

33. Fan, J.; Bai, P.; Ning, Y.; Wang, J.; Shi, X.; Xiong, Y.; Zhang, K.; He, F.; Zhang, C.; Wang, R.; et al. The Monocot-Specific Receptor-like Kinase SDS2 Controls Cell Death and Immunity in Rice. *Cell Host Microbe* **2018**, *23*, 498–510. [CrossRef] [PubMed]

34. Wang, C.; Wang, G.; Zhang, C.; Zhu, P.; Dai, H.; Yu, N.; He, Z.; Xu, L.; Wang, E. OsCERK1-Mediated Chitin Perception and Immune Signaling Requires Receptor-like Cytoplasmic Kinase 185 to Activate an MAPK Cascade in Rice. *Mol. Plant* **2017**, *10*, 619–633. [CrossRef]

35. Yamada, K.; Yamaguchi, K.; Yoshimura, S.; Terauchi, A.; Kawasaki, T. Conservation of Chitin-Induced MAPK Signaling Pathways in Rice and Arabidopsis. *Plant Cell Physiol.* **2017**, *58*, 993–1002. [CrossRef]

36. Dubouzet, J.G.; Maeda, S.; Sugano, S.; Ohtake, M.; Hayashi, N.; Ichikawa, T.; Kondou, Y.; Kuroda, H.; Horii, Y.; Matsui, M.; et al. Screening for resistance against Pseudomonas syringae in rice-FOX Arabidopsis lines identified a putative receptor-like cytoplasmic kinase gene that confers resistance to major bacterial and fungal pathogens in Arabidopsis and rice. *Plant Biotechnol. J.* **2011**, *9*, 466–485. [CrossRef]

37. Kondou, Y.; Higuchi, M.; Takahashi, S.; Sakurai, T.; Ichikawa, T.; Kuroda, H.; Yoshizumi, T.; Tsumoto, Y.; Horii, Y.; Kawashima, M.; et al. Systematic approaches to using the FOX hunting system to identify useful rice genes. *Plant J.* **2009**, *57*, 883–894. [CrossRef]

38. Maeda, S.; Hayashi, N.; Sasaya, T.; Mori, M. Overexpression of BSR1 confers broad-spectrum resistance against two bacterial diseases and two major fungal diseases in rice. *Breed. Sci.* **2016**, *66*, 396–406. [CrossRef]

39. Kanda, Y.; Yokotani, N.; Maeda, S.; Nishizawa, Y.; Kamakura, T.; Mori, M. The receptor-like cytoplasmic kinase BSR1 mediates chitin-induced defense signaling in rice cells. *Biosci. Biotechnol. Biochem.* **2017**, *81*, 1497–1502. [CrossRef]

40. Sugano, S.; Maeda, S.; Hayashi, N.; Kajiwara, H.; Inoue, H.; Jiang, C.J.; Takatsuji, H.; Mori, M. Tyrosine phosphorylation of a receptor-like cytoplasmic kinase, BSR1, plays a crucial role in resistance to multiple pathogens in rice. *Plant J.* **2018**, *96*, 1137–1147. [CrossRef]

41. Yamamura, C.; Mizutani, E.; Okada, K.; Nakagawa, H.; Fukushima, S.; Tanaka, A.; Maeda, S.; Kamakura, T.; Yamane, H.; Takatsuji, H.; et al. Diterpenoid phytoalexin factor, a bHLH transcription factor, plays a central role in the biosynthesis of diterpenoid phytoalexins in rice. *Plant J.* **2015**, *84*, 1100–1113. [CrossRef] [PubMed]

42. Yamaguchi, K.; Yamada, K.; Ishikawa, K.; Yoshimura, S.; Hayashi, N.; Uchihashi, K.; Ishihama, N.; Kishi-Kaboshi, M.; Takahashi, A.; Tsuge, S.; et al. A receptor-like cytoplasmic kinase targeted by a plant pathogen effector is directly phosphorylated by the chitin receptor and mediates rice immunity. *Cell Host Microbe* **2013**, *13*, 347–357. [CrossRef] [PubMed]

43. Rao, S.; Zhou, Z.; Miao, P.; Bi, G.; Hu, M.; Wu, Y.; Feng, F.; Zhang, X.; Zhou, J.M. Roles of Receptor-Like Cytoplasmic Kinase VII Members in Pattern-Triggered Immune Signaling. *Plant Physiol.* **2018**, *177*, 1679–1690. [CrossRef] [PubMed]

44. Monaghan, J.; Matschi, S.; Shorinola, O.; Rovenich, H.; Matei, A.; Segonzac, C.; Malinovsky, F.; Rathjen, J.; MacLean, D.; Romeis, T.; et al. The Calcium-Dependent Protein Kinase CPK28 Buffers Plant Immunity and Regulates BIK1 Turnover. *Cell Host Microbe* **2014**, *16*, 605–615. [CrossRef]

45. Wang, J.; Grubb, L.E.; Wang, J.; Liang, X.; Li, L.; Gao, C.; Ma, M.; Feng, F.; Li, M.; Li, L.; et al. A Regulatory Module Controlling Homeostasis of a Plant Immune Kinase. *Mol. Cell* **2018**, *69*, 493–504. [CrossRef]

46. Mikami, M.; Toki, S.; Endo, M. Comparison of CRISPR/Cas9 expression constructs for efficient targeted mutagenesis in rice. *Plant Mol. Biol.* **2015**, *88*, 561–572. [CrossRef]

47. Toki, S.; Hara, N.; Ono, K.; Onodera, H.; Tagiri, A.; Oka, S.; Tanaka, H. Early infection of scutellum tissue with Agrobacterium allows high-speed transformation of rice. *Plant J.* **2006**, *47*, 969–976. [CrossRef]

48. Nakagawa, H.; Tanaka, A.; Tanabata, T.; Ohtake, M.; Fujioka, S.; Nakamura, H.; Ichikawa, H.; Mori, M. Short grain1 decreases organ elongation and brassinosteroid response in rice. *Plant Physiol.* **2012**, *158*, 1208–1219. [CrossRef]

49. Qi, Y.; Katagiri, F. Purification of low-abundance Arabidopsis plasma-membrane protein complexes and identification of candidate components. *Plant J.* **2009**, *57*, 932–944. [CrossRef]

50. Jiang, C.J.; Shimono, M.; Sugano, S.; Kojima, M.; Yazawa, K.; Yoshida, R.; Inoue, H.; Hayashi, N.; Sakakibara, H.; Takatsuji, H. Abscisic acid interacts antagonistically with salicylic acid signaling pathway in rice-Magnaporthe grisea interaction. *Mol. Plant Microbe Interact.* **2010**, *23*, 791–798. [CrossRef]

51. Livak, K.J.; Schmittgen, T.D. Analysis of relative gene expression data using real-time quantitative PCR and the $2^{-\Delta\Delta CT}$ Method. *Methods* **2001**, *25*, 402–408. [CrossRef] [PubMed]

52. Matsushita, A.; Inoue, H.; Goto, S.; Nakayama, A.; Sugano, S.; Hayashi, N.; Takatsuji, H. Nuclear ubiquitin proteasome degradation affects WRKY45 function in the rice defense program. *Plant J.* **2013**, *73*, 302–313. [CrossRef] [PubMed]

International Journal of
Molecular Sciences

Article

Rice *OsAAA-ATPase1* is Induced during Blast Infection in a Salicylic Acid-Dependent Manner, and Promotes Blast Fungus Resistance

Xinqiong Liu [1,*,†], Haruhiko Inoue [2,†], Xianying Tang [1], Yanping Tan [1], Xin Xu [1], Chuntai Wang [1] and Chang-Jie Jiang [2,*]

[1] College of Life Science, South-Central University for Nationalities, Wuhan 430074, China; xytang@mail.scuec.edu.cn (X.T.); yanptan@mail.scuec.edu.cn (Y.T.); xinxu@mail.scuec.edu.cn (X.X.); wangchuntai@mail.scuec.edu.cn (C.W.)

[2] Institute of Agrobiological Sciences (NIAS), National Agriculture and Food Research Organization (NARO), Tsukuba 305-8602, Japan; haruhiko@affrc.go.jp

* Correspondence: liuxinqiong@mail.scuec.edu.cn (X.L.); cjjiang@affrc.go.jp(C.-J.J.);
  Tel.: +86-189-7122-9082 (X.L.); +81-298-838-8385(C.-J.J.)

† These authors contributed equally to this work.

Received: 29 January 2020; Accepted: 18 February 2020; Published: 20 February 2020

Abstract: Fatty acids (FAs) have been implicated in signaling roles in plant defense responses. We previously reported that mutation or RNAi-knockdown (*OsSSI2*-kd) of the rice *OsSSI2* gene, encoding ta stearoyl acyl carrier protein FA desaturase (SACPD), remarkably enhanced resistance to blast fungus *Magnaporthe oryzae* and the leaf-blight bacterium *Xanthomonas oryzae* pv. *oryzae* (*Xoo*). Transcriptomic analysis identified six AAA-ATPase family genes (hereafter *OsAAA-ATPase1–6*) upregulated in the *OsSSI2*-kd plants, in addition to other well-known defense-related genes. Here, we report the functional analysis of *OsAAA-ATPase1* in rice's defense response to *M. oryzae*. Recombinant OsAAA-ATPase1 synthesized in *Escherichia coli* showed ATPase activity. *OsAAA-ATPase1* transcription was induced by exogenous treatment with a functional analogue of salicylic acid (SA), benzothiadiazole (BTH), but not by other plant hormones tested. The transcription of *OsAAA-ATPase1* was also highly induced in response to *M. oryzae* infection in an SA-dependent manner, as gene induction was significantly attenuated in a transgenic rice line expressing a bacterial gene (*nahG*) encoding salicylate hydroxylase. Overexpression of *OsAAA-ATPase1* significantly enhanced pathogenesis-related gene expression and the resistance to *M. oryzae*; conversely, RNAi-mediated suppression of this gene compromised this resistance. These results suggest that *OsAAA-APTase1* plays an important role in SA-mediated defense responses against blast fungus *M. oryzae*.

Keywords: AAA-ATPase; salicylic acid; fatty acid; rice; *Magnaporthe oryzae*; disease resistance

---

1. Introduction

Salicylic acid (SA) plays an important signaling role in plant defense activation against pathogens. In response to pathogen attack, SA activates a battery of defense-related genes, including pathogenesis-related (PR) genes, throughout the plant, resulting in both local and systemic resistance to the pathogen [1]. In *Arabidopsis*, NPR1 (non-pathogenesis related 1) has been demonstrated to play a master role in SA-mediated defense activation [2,3]. A loss of NPR1 function (npr1) results in loss of PR gene induction, and hypersensitivity to diseases [4]. In rice, meanwhile, it has been shown that SA signaling is mediated by two downstream factors, OsNPR1 and WRKY45, acting in parallel [5,6].

In *Arabidopsis*, genetic screening for mutations that can suppress *npr1* phenotypes (based on their ability to restore SA-induced PR expression to npr1-5 plants) resulted in isolation of several *npr1* suppressor mutants (*ssi*: suppressor of SA insensitivity), which exhibit constitutive defense activation [7]. Map-based cloning of one of the *ssi* mutants (*ssi2*) revealed that the corresponding gene (*SSI2*) encodes a stearoyl-ACP desaturase, which desaturates stearoyl (18:0)-ACP into oleoyl-ACP, and finally, into oleic acid (18:1) [7]. Disruption of this gene in *ssi2* results in a ten-fold increase in 18:0 fatty acid (FA) content, indicating involvement of FAs in plant defense reactions [7]. The *ssi2* mutant plants accumulate high levels of SA and display constitutive PR gene expression and enhanced resistance to *Peronospora parasitica*, *Pseudomonas syringae* [8], and *Cucumber mosaic virus* [7,9,10].

The orthologs of *SSI2* have also been identified in soybean (*GmSACPD-A/-B*) [11], rice (*OsSSI2*) [12], and wheat (*TaSSI2*) [13,14]. Similar to *Arabidopsis ssi2*, suppression of these ortholog genes enhanced resistance to multiple pathogens: *Pseudomonas syringae* pv. *glycinea* and *Phytophthora sojae* in soybean [11]; blast fungus *Magnaporthe oryzae* and leaf-blight bacteria *Xanthomonas oryzae* pv. *oryzae* (*Xoo*) in rice [12]; and powdery mildew bacteria *Blumeria graminis* f. sp. *tritici* and Fusarium head blight fungus *Fusarium graminearum* in wheat [13,14]. These results demonstrate a common function of *SSI2* and its orthologs in defense activation in diverse plant species.

The molecular mechanisms whereby the *SSI2* family genes participate in defense reactions in plants remain to be fully elucidated. In rice, a DNA microarray analysis revealed several hundred genes differentially expressed between the wild-type and *OsSSI2*-suppressed transgenic (*OsSSI2*-kd) plants [12]. Among them was a group of six genes for AAA-ATPase (AAA: ATPases associated with diverse cellular activities) highly upregulated in *OsSSI2*-kd plants, in addition to the well-known defense-related genes, such as *WRKY45*, *PR1b*, and *PBZ1*, and a *thaumatin-like* gene [12]. These results suggest that the AAA-ATPase family genes may play important roles in defense activation in rice plants.

The AAA-ATPase family occurs in all life forms, including eukaryotes, prokaryotes, and archaebacteria, and is implicated in a variety of cellular activities, including proteolysis, protein folding, membrane trafficking, cytoskeletal regulation, organelle biogenesis, DNA replication, and immune responses [15–17]. Structurally, these proteins contain one or several conserved motifs, including the Walker A and Walker B motifs, which are, respectively, required for ATP binding and hydrolysis; they also contain a highly conserved amino acid sequence, referred to as the second region of homology (SRH) [16]. In plants, it has been reported that AAA-ATPase genes from *Nicotiana tabacum* (*NtAAA1*) [18,19] and *Arabidopsis* (*AtOM66*) [20] are, respectively, negatively or positively involved in the SA-signaling pathway and in the hypersensitive response (HR) upon pathogen infections. Moreover, in rice, map-based cloning of the *lesion mimic resembling* (*lmr*) mutant/*lesion resembling disease* (*lrd6-6*) mutant revealed that the corresponding gene (*LMR/LRD6-6, Os06g0130000*) encodes an AAA-ATPase, and is negatively involved in HR and disease resistance [21,22].

In this study, we conducted a functional analysis of *OsAAA-ATPase1*, one of the six AAA-ATPase genes upregulated in *OsSSI2*-kd rice plants [12]. We show that *OsAAA-ATPase1* is transcriptionally regulated by SA, and positively involved in resistance to blast fugus *M. oryzae*.

2. Results

In our previous study, a group of six AAA-ATPase family genes (hereafter *OsAAA-ATPase1–6*; (Table 1; Figure 1) was found to be significantly upregulated in *OsSSI2*-kd rice plants [12], implicating these genes in rice defense activation. From among them, we chose *OsAAA-ATPase1* for more detailed functional characterization in this study, because it showed SA-induced (Figure 2) and SA-dependent blast-induced (Figure 3) transcription responses.

Table 1. Genes and primer sequences used for qRT-PCR analysis.

Gene Name (RAP+DB ID)	Primer Sequences (5′→3′)	References
OsAAA-ATPase1 Os03g0802500	AGTGGTTGCTAGCTTCTCGT ACAACATGTGGTCAAATTATTCCA	[12]
OsAAA-ATPase2 Os01g0297200	CTAGGTTCTGCGATGGACAC CTCCTTTGCAATTGTTCCAC	[12]
OsAAA-ATPase3 Os06g0697600	GTTGTGATCGTGTCATGGTTGCG CAGAAAGCCACACACCATTGC	[12]
OsAAA-ATPase4 Os02g0697600	TTGCCTGAACGGCCAGGTGAT CCCATGTAAGGGTAAGGATTGC	[12]
OsAAA-ATPase5 Os02g0706500	GTTCCATCTCTTTGCCTGTAGC CATGCGCATCTCAGTCTTACC	[12]
OsAAA-ATPase6 Os07g0517600	TCAGTGGCCTCGTCGAGTTC CTACTTGCCTGCTTCACACAT	[12]
OsPR1b Os01g0382000	ACGGGCGTACGTACTGGCTA CTCGGTATGGACCGTGAAG	[23]
PBZ1 Os12g0555000	GCGTTTGAGTCCGTGAGAGT TCACCCATTGATGAAGCAAA	[24]
Rubq1 Os06g0681400	GGAGCTGCTGCTGTTCTAGG TTCAGACACCATCAAACCAGA	[25]
M. oryzae 28S rDNA	ACGAGAGGAACCGCTCATTCAGATAATT TCAGCAGATCGTAACGATAAAGCTACTC	[26]

2.1. *OsAAA-ATPase1 Encodes an AAA-ATPase Family Protein*

OsAAA-ATPase1 was predicted to encode a protein of 520 amino acids with a predicted mass of 58.1 kDa. OsAAA-ATPase1 shared 50%, 37.8%, and 22.3% amino acid sequence identity with previously reported defense-related AAA-ATPase proteins, namely, tobacco NtAAA1 [18], *Arabidopsis* AtOM66 [20], and rice LMR/LRD6-6 [21,22], respectively. Structural analysis revealed that OsAAA-ATPase1 contains consensus motifs that are typical of the AAA-ATPase family; these include the Walker A, Walker B, and SRH motifs (pfam00004; E-value = 4.95×10^{-17}) (Figure 1a).

To assess how *OsAAA-ATPase1–6* are related within the AAA-ATPase gene family, we performed a phylogenic comparison of the proteins predicted to be encoded by *OsAAA-ATPase1–6* and several known rice AAA-ATPase proteins, including LMR/LRD6-6 [21,22], OsCDC48 [27], RuvBL1a [28], RLS3 [29], OsSKD1 [30], OsFtsH5 [31], and RFC5 [32], together with NtAAA1 [18,19] and AtOM66 [20]. OsAAA-ATPase1 was grouped within a subclade of proteins related to plant defense, including OsAAA-ATPase2–6, NtAAA1, and AtOM66, but, unexpectedly, distally with LMR/LRD6-6 (Figure 1b).

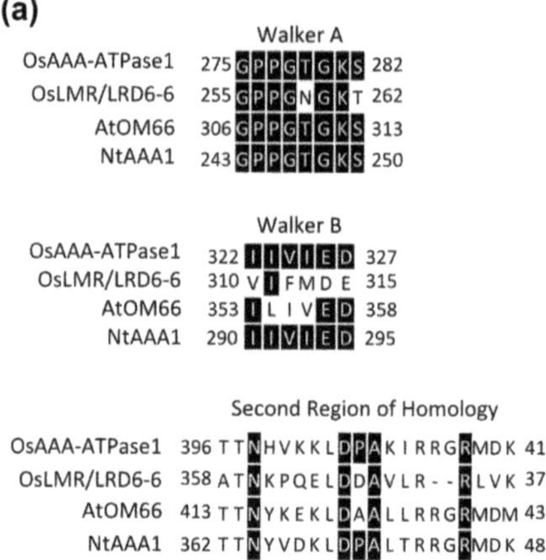
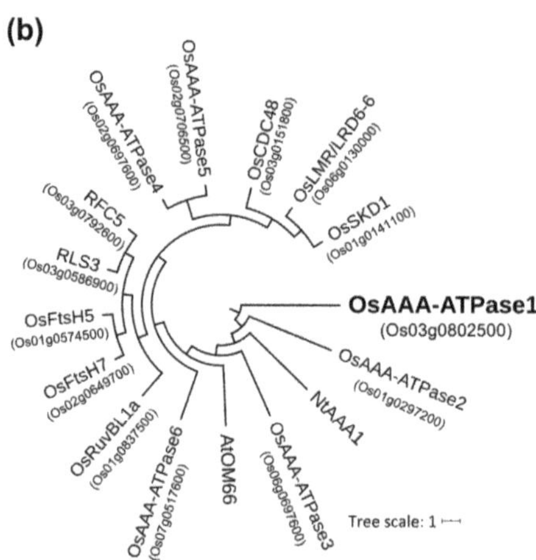

Figure 1. Amino acid sequence alignment of AAA-type ATPase proteins, and phylogenetic analysis. (**a**) Alignment of typical consensus motifs of the AAA-ATPase protein family, including the Walker A, Walker B, and SRH motifs. (**b**) Phylogeny of the AAA-ATPase proteins from rice (LMR/LRD6-6, OsCDC48, RuvBL1a, RLS3, OsSKD1, OsFtsH5, and RFC5), tobacco (NtAAA1), and *Arabidopsis* (AtOM66).

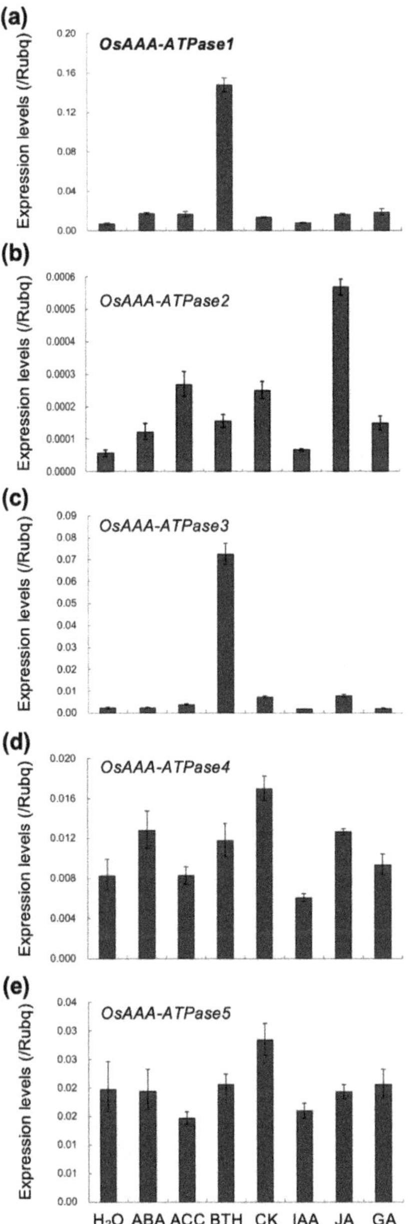

Figure 2. Expression analysis of *OsAAA-ATPase1–5* (**a–e**) in response to the plant hormones abscisic acid (ABA), ethylene (ACC, an ethylene precursor), benzothiadiazole (BTH, a functional analogue of SA), kinetin (CK, a synthetic cytokinin), auxin (IAA), jasmonic acid (JA), and gibberellic acid (GA), in Nipponbare rice seedlings. Data are represented as means ± SDs.

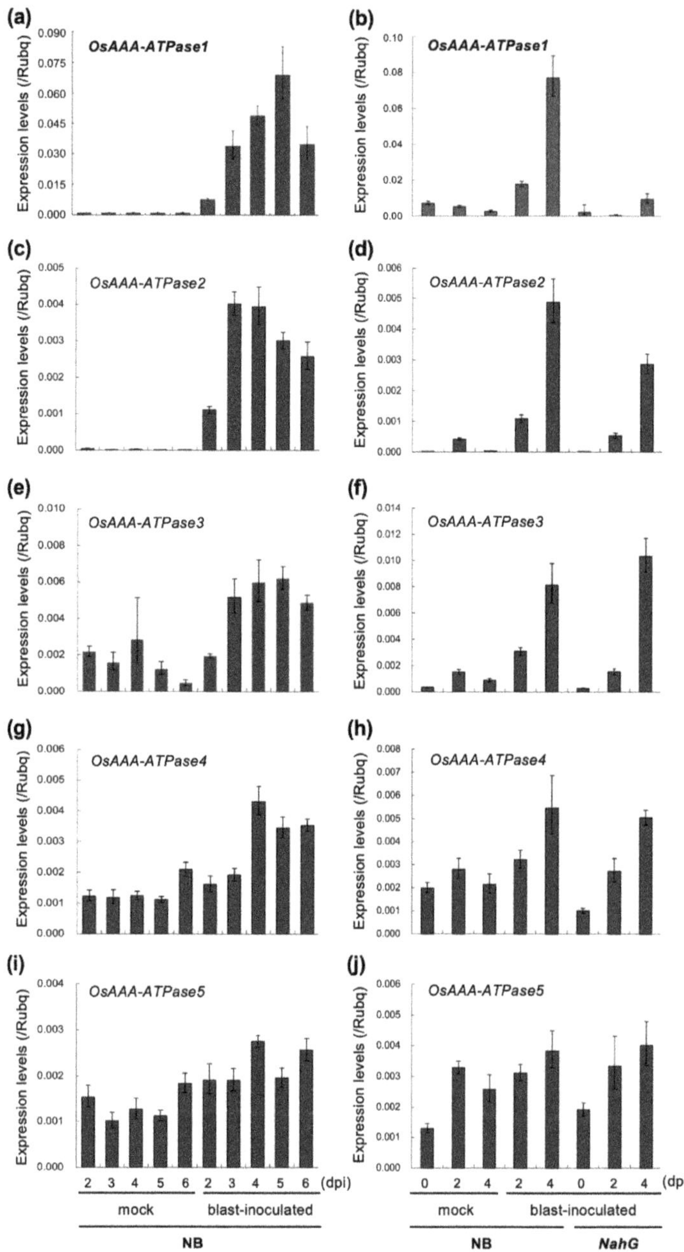

Figure 3. Expression analysis of *OsAAA-ATPase1–5* (**a–j**) in response to blast inoculation in Nipponbare (NB) and *narG* rice seedlings. Rice seedlings at the four-leaf stage (three true leaves) were subjected to mock treatment (mock) or blast inoculation (blast-inoculated), and the fourth leaf blades were sampled at indicated days post inoculation (dpi). Data are represented as means ± SDs.

2.2. *OsAAA-ATPase1 is Induced by SA Treatment*

Plant hormones have been demonstrated to play important roles in interactions between plants and pathogens. Hence, we examined the transcriptional responses of *OsAAA-ATPase1–6* to the plant hormones abscisic acid (ABA), ACC (an ethylene precursor), BTH (a functional analogue of SA), kinetin (CK, a synthetic cytokinin), auxin (IAA), jasmonic acid (JA), and gibberellic acid (GA).

OsAAA-ATPase1 (Figure 2a) and *OsAAA-ATPase3* (Figure 2c) were induced specifically by BTH treatment, and *OsAAA-ATPase2* (Figure 2b) was induced by JA treatment. Meanwhile, *OsAAA-ATPase4* (Figure 2d) and *OsAAA-ATPase5* (Figure 2e) were not specifically induced by any of the hormones, and *OsAAA-ATPase6* (*Os07g0517600*) had no detectable transcription.

2.3. *OsAAA-ATPase1 is Induced in Response to Blast Infection in An SA-Dependent Manner*

Rice seedlings of non-transformant Nipponbare rice (NB) and of NB expressing the *nahG* gene (*nahG*-rice), at the four-leaf stage, were subjected to blast inoculation. At 2–6 days post inoculation (dpi) of the blast, the fourth leaves were sampled to examine the expression of *OsAAA-ATPase1–5*.

All of the tested *OsAAA-ATPase* genes clearly showed transcriptional induction in response to blast inoculation (Figure 3c,e,g,i); in particular, *OsAAA-ATPase1* (Figure 3a) and *OsAAA-ATPase2* (Figure 3c) showed a high-fold transcriptional increase from the very low basal levels in the mock treatment. The induction of the genes became evident from 2 dpi and peaked at ca. 3–5 dpi.

In *nahG*-rice plants, in contrast, the induction of *OsAAA-ATPase1* was mostly attenuated relative to its induction in NB plants in response to blast inoculation (Figure 3b), demonstrating that the induction of this gene depends on the SA-signaling pathway. No appreciable attenuation of gene induction was observed for the other genes (Figure 3d,f,h,j).

2.4. *OsAAA-ATPase1 is Positively Involved in Blast Resistance*

To gain some insight into the role of *OsAAA-ATPase1* in disease resistance, we generated transgenic rice lines that either overexpressed the gene under maize ubiquitin promoter (*OsAAA-ATPase1*-ox; Figure 4a) or RNAi-suppressed *OsAAA-ATPase1* (*OsAAA-ATPase1*-kd; Figure 5a), and subjected these lines to blast inoculation. In order to reveal the potentially compromised resistance in *OsAAA-ATPase1*-kd plants, a half density of conidia (5×10^4/mL) was used, so as to cause blast disease moderately in NB, but more severely in OsAAA-ATPase1 plants.

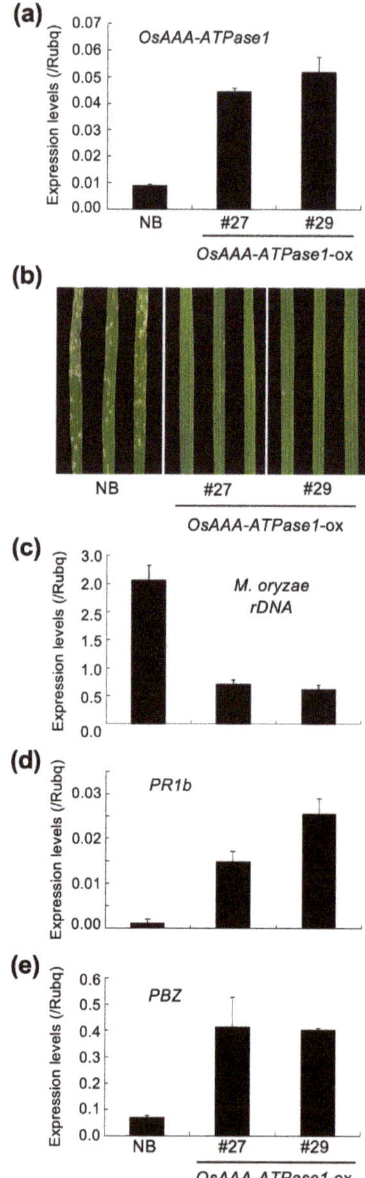

Figure 4. Blast resistance of *OsAAA-ATPase1*-ox plants. (**a**) Expression of *OsAAA-ATPase1*, (**b**) blast lesions on leaf blades, and (**c**) relative fungal growth (*Magnaporthe oryzae rDNA*). (**d**,**e**) Expression of *PR1b* (**d**) and *PBZ1* (**e**), in Nipponbare (NB) and *OsAAA-ATPase1*-ox lines (#27 and #29), respectively, at 7 days post inoculation (dpi). Data are represented as means ± SDs in (**a**,**c**–**e**).

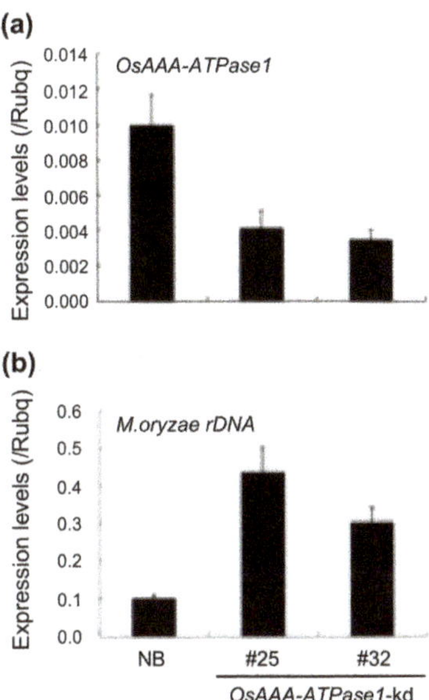

Figure 5. Compromised blast resistance in *OsAAA-ATPase1*-kd plants (#25 and #32). (**a**) Expression of *OsAAA-ATPase1*. (**b**) Relative fungal growth (*M. oryzae rDNA*) in Nipponbare (NB) and *OsAAA-ATPase1*-kd lines (#25 and #32) respectively. Data are represented as means ± SDs.

Compared with the non-transgenic control plants (NB), *OsAAA-ATPase1*-ox plants (lines #27 and #29) exhibited significantly higher resistance to blast disease, as evidenced by the fact that few susceptible blast lesions appeared on their leaf blades (Figure 4b), and that they had ca. 4-fold less fungal growth (Figure 4c). The enhanced resistance of the *OsAAA-ATPase1*-ox plants is consistent with the large increases in the expression levels of the PR genes, *OsPR1* and *PBZ1* (Figure 4d,e).

Conversely, blast resistance was significantly compromised in *OsAAA-ATPase1*-kd plants (lines #25 and #32): they had ca. 2-fold more fungal growth than the NB control plants (Figure 5b).

2.5. *OsAAA-ATPase1 Has ATPase Activity and Is Localized in the Cytosol*

To assess whether OsAAA-ATPase1 protein has ATPase activity, *OsAAA-ATPase1* N-terminal was fused to a His-tag and expressed in *Escherichia coli*, and purified using a high affinity Ni-resin. OsAAA-ATPase1 protein showed an ATPase activity level that was comparable to that of the positive control (potato ATPase) (Figure 6).

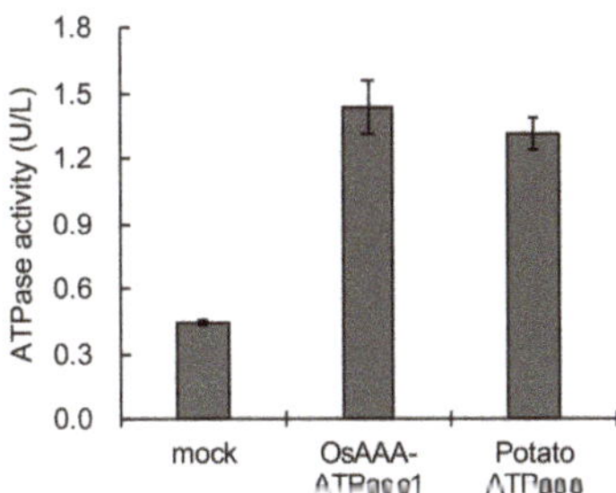

Figure 6. ATPase activity of recombinant OsAAA-ATPase1 protein. Elution buffer of Ni-resin (mock treatment), and an ATPase protein from potatoes, were used negative and positive controls, respectively.

To determine the subcellular localization of OsAAA-ATPase1 in rice cells, the EGFP-OsAAA-ATPase1 fusion protein in the rice protoplast was examined under a confocal microscope. As shown in Figure 7, EGFP-OsAAA-ATPase1 protein was co-localized with a cytosol marker, mCherry signals, indicating that OsAAA-ATPase is predominantly distributed in the cytosol.

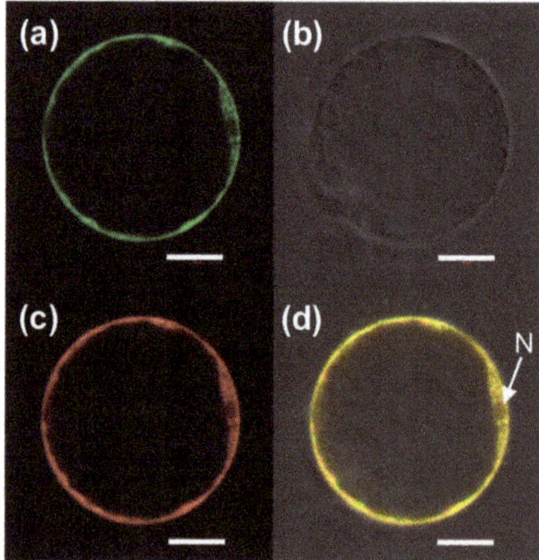

Figure 7. Subcellular localization of OsAAA-ATPase1 in rice protoplasts. (**a**) EGFP-OsAAA-ATPase1, (**b**) blight field image, (**c**) mCherry (cytoplasmic localization), and (**d**) combined image of (**a**–**c**). N, nucleus; bar, 20 μm.

3. Discussion

AAA-type ATPases constitute a large protein family in a diverse range of organisms, and thus exhibit multiple and diverse cellular functions [15,33]. In plants, AAA-ATPase genes have been implicated in proteolysis [33], male meiosis [34], vacuolar maintenance [35], peroxisome biogenesis [36], morphogenesis [37], leaf senescence [29,38], and stress [28,39] and immune responses [18–22]. In this study, we present a novel rice AAA-ATPase gene member, *OsAAA-ATPase1*. The deduced amino acid sequence of OsAAA-ATPase1 contains consensus motifs that are typical of the AAA-ATPase family; these include the Walker A, Walker B, and SRH motifs (Figure 1a) [15,17,33]. Consistent with this, biochemical analysis confirmed that there was ATPase activity in the recombinant protein of OsAAA-ATPase1 (Figure 6). Phylogenetically, OsAAA-ATPase1 was grouped within a subclade of proteins related to plant defense activation (Figure 1b), which included OsAAA-ATPase2–6 [12], tobacco NtAAA1 [18,19], and *Arabidopsis* AtOM66 [20]. These results suggest that OsAAA-ATPase1 belongs to the AAA-ATPase family. Functional analysis revealed that *OsAAA-ATPase1* is transcriptionally regulated by SA in response to blast infection (Figures 2 and 3). Overexpression or RNAi-mediated suppression of *OsAAA-ATPase1* resulted, respectively, in an increase (Figure 4) or decrease (Figure 5) in blast resistance. Taken together, our results suggest that *OsAAA-ATPase1* plays a positive role in the SA-mediated disease resistance in rice plants.

In relation to plant immune responses, several studies have shown important roles for AAA-ATPase genes. *NtAAA1* was isolated as an HR-induced gene in *Nicotiana tabacum* [18]; was found to be under the control of *N*-gene, ethylene, and jasmonate; and was localized in the cytoplasm. It was also negatively involved in the SA-signaling pathway and pathogen resistance [18,19]. In contrast, *AtOM66* (outer mitochondrial membrane protein of 66 kDa) is a stress-induced gene; overexpression of this gene increased SA content, accelerated cell death rates, and enhanced resistance to the biotrophic pathogen *Pseudomonas syringae* [20]. Recently, rice *LMR* and *LRD6-6* were map-based cloned from lesion mimic mutants *lmr* and *lrd6-6*, respectively, and were found to be the same gene (*Os06g0130000*).

LMR/LRD6-6 was shown to be localized in the multivesicular bodies (MVBs) and was negatively involved in rice immunity and cell death [21,22]. Mutation in this gene (*lmr* and *lrd6-6*) resulted in constitutive expression of *PR1* and *PBZ1*, and enhanced resistance to rice blast and bacterial blight diseases; however, no difference in SA content was determined [21,22]. By comparison, it seems that OsAAA-ATPase1 plays a role distinct from those previously reported, with respect to its association with SA-regulation and HR, its subcellular localization, and its promotion of disease resistance. Thus, our findings provide novel insights into SA-regulated defense activation in rice. Meanwhile, OsAAA-ATPase1 showed a close phylogenetic association with AtOM66 (Figure 1b); both proteins play a positive role in the SA-signaling pathway, suggesting that they may share a common cellular function.

Plants produce a variety of FAs and their derivatives, some of which have been shown to play important roles in defense activation [40,41]. In the *Arabidopsis ssi2* mutant, disruption of *SSI2*, which encodes an FA desaturase, results in an increase in the 18:0 FA content, which in turn remarkably increases SA content, PR gene expression, and resistance against multiple pathogens [42]. Similar defense-related phenotypes were observed following suppression of *SSI2*-orthologs in soybean (*GmSACPD-A/-B*) [11], rice (*OsSSI2*) [12], and wheat (*TaSSI2*) [13,14]. These results strongly suggest that *SSI2* and its orthologs serve as valuable susceptibility gene (*S* gene) resources for the development of crop cultivars with resistance to multiple pathogens, by employing targeted mutation and genome editing technologies [43–45]. In order to make such successful use of these genes in resistance breeding, it is important to understand the molecular mechanisms underlying the defense activation. In *Arabidopsis*, a mutation in the GTPase nitric oxide associated 1 (*NOA1*) gene partially restored the *ssi2* phenotype, whereas double mutations in NOA1 and either one of the two nitrate reductase isoforms (NIA1 and NIA2) completely restored the *ssi2* phenotypes; this indicates that nitric oxide (NO) is required for constitutive defense in the *ssi2* mutant [46,47]. Nevertheless, little has been reported regarding the molecular basis of defense activation in *OsSSI2*-kd rice plants. We previously identified a group of six AAA-ATPase genes (*OsAAA-ATPase1–6*) that were upregulated in *OsSSI2*-kd rice plants [12]. In this study, all of these genes tested were induced in response to blast inoculation (Figure 3), suggesting that they each play a role in resistance to blast fungus. In contrast, *OsAAA-ATPase1–5* each exhibited a distinct induction pattern in response to different plant hormone treatments (Figure 2); *OsAAA-ATPase1* and *OsAAA-ATPase3* were induced by SA, *OsAAA-ATPase2* mainly by JA, and *OsAAA-ATPase4* and *OsAAA-ATPase5* slightly by the CK treatment. These results suggest that there is functional differentiation among the *OsAAA-ATPase1–6* genes downstream of *OsSSI2* in disease resistance. Moreover, although both *OsAAA-ATPase1* and *OsAAA-ATPase3* were induced by SA treatment, only the induction of *OsAAA-ATPase1* was attenuated following blast infection in *nahG*-rice plants (Figure 3). One possible explanation for this is that *OsAAA-ATPase3* may be more sensitive to SA, allowing it to be induced even by a residual increase in the SA-signaling level in *nahG*-rice plants.

4. Materials and Methods

4.1. Plant Materials and Growth Conditions

Japonica type rice cultivar Nipponbare (*Oryza sativa* L.) was grown in commercial nursery soil (Bonsol Number 2; Sumitomo Chemical Corp., Tokyo, Japan) in a greenhouse at 28 °C (day)/23 ° C (night) with ca. 50% relative humidity.

4.2. Plasmid DNA Construction and Rice Transformation

The cDNA clone for *OsAAA-ATPase1* was provided by the Rice Genome Resource Center, Japan (accession number: AK070731). To construct a plasmid for constitutive expression of *OsAAA-ATPase1* under the maize ubiquitin promoter, a DNA fragment containing a 91 bp upstream sequence followed by the full coding sequence of *OsAAA-ATPase1* (nucleotides 2–1655) was amplified by PCR and cloned into the pUCAP/Ubi-NT vector, as previously described [5]. To construct a plasmid for *OsAAA-ATPase1*

RNAi (*OsAAA-ATPase1*-kd), part of the 3′-UTR (nucleotides 1543–1845) of *OsAAA-ATPase1* cDNA was amplified by PCR and cloned into the pANDA vector, as previously described [48,49].

Nipponbare rice plants were transformed by an *Agrobacterium tumefaciens* (strain EHA105) mediated technique, as described earlier [50]. Transgenic rice lines expressing *nahG* from *Pseudomonas putida* under the control of a double 35S promoter (*nahG*-rice) were generated using the plant expression construct previously described in Yang et al. [51].

4.3. Chemical Treatments

All stock solutions were prepared at a concentration of 100 mmol/L. Indole-3-acetic acid (IAA; Sigma, St. Louis, MO, USA), gibberellin A3 (GA$_3$; Wako, Osaka, Japan), abscisic acid ((±)-*cis-trans*, ABA; Sigma), and methyl jasmonate (ME-JA; Wako, Saitama, Japan) were dissolved in ethanol. Kinetin (Sigma) and benzothiadiazole *S*-methyl ester (BTH; Wako) were dissolved in dimethyl sulfoxide (DMSO); and 1-aminocyclopropane-1-carboxylic acid (ACC; Sigma) and sodium salicylate (SA; Nacalai Tesque, Tokyo, Japan) were dissolved in H$_2$O. The solvents did not exceed a final concentration of 0.1% in the solutions used for plant treatments, and had no effect on the expression of the rice genes examined in this study.

For plant treatments, rice seedlings at the four-leaf stage (three true leaves) were transferred to a container containing each of the plant hormone solutions at 50 μM. The rice seedlings were further grown for 1 day, and fourth leaf blades were stored in liquid nitrogen for RNA preparation.

4.4. Protein Expression, Purification, and ATPase Assay

The *OsAAA-ATPase1* sequence was amplified by PCR and cloned into the sites between BglI and HidIIIp of pET32a (Novagen) as a His-tag fusion protein; then, that was transfected into the *Escherichia coli* Origami strain BL21(Lys). The set of primers used was as follows: OsAAA1BglII 5′-GTAGATCTCTTGAGACAAATGGAGGCGACG-3′; OsAAA1HindIII 5′-GCTAAGCTTCTACTTATCCTTCCCGACCAC-3′. Expression of the protein was induced for 4 h at 25 °C with 0.5 mmol/L isopropyl β-D-1-thiogalactopyranoside. *Escherichia coli* cells were pelleted by centrifugation, resuspended in lysis buffer (20 mmol/L Tris-HCl pH 7.4, 0.1 M NaCl, 10 mmol/L imidazole), and sonicated. After the cell debris was removed by centrifugation (12,000 × g, 10 min, 4 °C), the supernatant was loaded onto a High Affinity Ni-Charged Resin (GE Healthcare, Buckinghamshire, UK), washed with washing buffer (20 mmol/L Tris-HCl pH 7.4, 0.1 M NaCl, 10 mmol/L imidazole), and eluted with elution buffer (20 mmol/L Tris-HCl pH 7.4, 0.1 M NaCl, 180 mmol/L imidazole). ATPase activity was measured by the malachite green-based colorimetric method using the QuantiChrom^{TM} ATPase/GTPase activity assay kit (Sigma-Aldrich, St. Louis, MO, USA). The elution buffer was used as the negative control, and an ATPase from potatoes (Sigma-Aldrich, St. Louis, MO, USA) was used as the positive control. One unit is defined as the amount of enzyme that catalyzes the production of 1 μM of free phosphate per minute under the assay conditions.

4.5. Subcellular Localization

For subcellular localization of OsAAA-ATPase1, the plasmid pSAT6-AFP-C1-OsAAA1 was transformed into protoplasts prepared from etiolated seedlings as previously described [52]. As a control for cytoplasmic localization, the pSAT-mCherry construct was co-transformed. Fluorescence was examined under a confocal microscope (Leica Microsystems, Wetzlar, Germany) 16 h after transformation.

4.6. Pathogen Culture and Inoculations

Culture and inoculation of the blast fungus *M. oryzae* (compatible race 007.0) was conducted essentially as previously described [12], with slight modifications. Briefly, the fungus was grown on an oatmeal agar medium (30 g/L oatmeal, 5 g/L sucrose, and 16 g/L agar) at 26 °C for 10–12 day. After removing the aerial hyphae by washing with distilled water and a brush, conidia formation was induced by irradiation under continuous black blue light (FL15BLB; Toshiba, Osaka, Japan) at 24 °C for 3 day. The conidia were suspended in 0.02% Silwet L-77 (a non-ionic surfactant; Nihon Unica, Tokyo, Japan) at a density of 10^5/mL, and were sprayed onto rice plants at the four-leaf stage. After incubation in a dew chamber at 24 °C for 24 h, the rice plants were moved back to the greenhouse.

Disease development was evaluated by determining the *M. oryzae* genomic 28S *rDNA* [26] by qRT-PCR [5,6], 6–7 dpi. At least 20 plants were used for each disease assay.

4.7. RNA Analyses

Total RNA was isolated from leaf blades of the 4th leaves of rice seedlings using the Trizol reagent (Invitrogen, Carlsbad, CA, USA). Quantitative RT-PCR (qRT-PCR) was performed on a Thermal Cycler Dice TP800 system (Takara Bio, Tokyo, Japan) using SYBR premix Ex Taq mixture (Takara Bio) as previously described [5]. The primer sequences used for qRT-PCR are listed in Table 1.

4.8. Amino Acid Sequence Alignment and Phylogenetic Analysis

The protein sequences were retrieved from the rice annotation project database (rap-db) and aligned using Clustal-X software, and the tree was constructed using iTOL software [53].

5. Conclusions

In this study, we present a novel AAA-ATPase member gene, *OsAAA-ATPase1*, one of the six AAA-ATPase genes upregulated in *OsSSI2*-kd rice plants [12]. Functional analysis revealed that *OsAAA-ATPase1* is transcriptionally regulated by SA, and plays a positive role in the SA-mediated disease resistance in rice plants. Our findings provide novel insights into SA-regulated defense activation in rice, and the molecular basis of defense activation in *OsSSI2*-kd rice plants.

Author Contributions: Conceived and designed the experiments: X.L. and C.-J.J. Performed the experiments and analyzed the data: X.L., H.I., X.T., Y.T., X.X., and C.W. Interpreted and wrote the manuscript, H.I. and C.-J.J. All authors have read and agreed to the manuscript as written.

Funding: This work was supported by the National Natural Science Foundation of China (31370306), and grants from Japan's Society for the Promotion of Science (JSPS) KAKENHI to C.J. (19K06065) and to H.I. (17K07678).

Acknowledgments: We thank the NARO Genebank for providing the cDNA clone of *OsAAA-ATPase1* (*Os03g0802500*).

Conflicts of Interest: The authors declare no conflict of interest.

References

1. Klessig, D.F.; Choi, H.W.; Dempsey, D.M.A. Systemic acquired resistance and salicylic acid: Past, present, and future. *Mol. Plant Microbe Interact.* **2018**, *31*, 871–888. [CrossRef] [PubMed]
2. Dong, X. NPR1, all things considered. *Curr. Opin. Plant Biol.* **2004**, *7*, 547–552. [CrossRef]
3. Wang, D.; Amornsiripanitch, N.; Dong, X. A genomic approach to identify regulatory nodes in the transcriptional network of systemic acquired resistance in plants. *Plos Pathog.* **2006**, *2*, e123. [CrossRef]
4. Cao, H.; Glazebrook, J.; Clarke, J.D.; Volko, S.; Dong, X. The Arabidopsis NPR1 gene that controls systemic acquired resistance encodes a novel protein containing ankyrin repeats. *Cell* **1997**, *88*, 57–63. [CrossRef]
5. Shimono, M.; Sugano, S.; Nakayama, A.; Jiang, C.J.; Ono, K.; Toki, S.; Takatsuji, H. Rice *WRKY45* plays a crucial role in benzothiadiazole-inducible blast resistance. *Plant Cell* **2007**, *19*, 2064–2076. [CrossRef]
6. Sugano, S.; Jiang, C.J.; Miyazawa, S.; Masumoto, C.; Yazawa, K.; Hayashi, N.; Shimono, M.; Nakayama, A.; Miyao, M.; Takatsuji, H. Role of OsNPR1 in rice defense program as revealed by genome-wide expression analysis. *Plant Mol. Biol.* **2010**, *74*, 549–562. [CrossRef]
7. Kachroo, P.; Shanklin, J.; Shah, J.; Whittle, E.J.; Klessig, D.F. A fatty acid desaturase modulates the activation of defense signaling pathways in plants. *Proc. Natl. Acad. Sci. USA* **2001**, *98*, 9448–9453. [CrossRef]
8. Shah, J.; Kachroo, P.; Nandi, A.; Klessig, D.F. A recessive mutation in the *Arabidopsis SSI2* gene confers SA- and *NPR1*-independent expression of *PR* genes and resistance against bacterial and oomycete pathogens. *Plant J.* **2001**, *25*, 563–574. [CrossRef] [PubMed]

9. Kachroo, P.; Kachroo, A.; Lapchyk, L.; Hildebrand, D.; Klessig, D.F. Restoration of defective cross talk in *ssi2* mutants: Role of salicylic acid, jasmonic acid, and fatty acids in *SSI2*-mediated signaling. *Mol. Plant Microbe Interact.* **2003**, *16*, 1022–1029. [CrossRef] [PubMed]

10. Sekine, K.T.; Nandi, A.; Ishihara, T.; Hase, S.; Ikegami, M.; Shah, J.; Takahashi, H. Enhanced resistance to *Cucumber mosaic virus* in the *Arabidopsis thaliana ssi2* mutant is mediated via an SA-independent mechanism. *Mol. Plant Microbe Interact.* **2004**, *17*, 623–632. [CrossRef]

11. Kachroo, A.; Fu, D.Q.; Havens, W.; Navarre, D.; Kachroo, P.; Ghabrial, S.A. An oleic acid-mediated pathway induces constitutive defense signaling and enhanced resistance to multiple pathogens in soybean. *Mol. Plant Microbe Interact.* **2008**, *21*, 564–575. [CrossRef] [PubMed]

12. Jiang, C.J.; Shimono, M.; Maeda, S.; Inoue, H.; Mori, M.; Hasegawa, M.; Sugano, S.; Takatsuji, H. Suppression of the rice fatty-acid desaturase gene *OsSSI2* enhances resistance to blast and leaf blight diseases in rice. *Mol. Plant Microbe Interact.* **2009**, *22*, 820–829. [CrossRef] [PubMed]

13. HU, L.-q.; MU, J.-j.; SU, P.-s.; WU, H.-y.; YU, G.-h.; WANG, G.-p.; Liang, W.; Xin, M.; LI, A.-f.; WANG, H.-w. Multi-functional roles of *TaSSI2* involved in Fusarium head blight and powdery mildew resistance and drought tolerance. *J. Integr. Agr.* **2018**, *17*, 368–380. [CrossRef]

14. Song, N.; Hu, Z.; Li, Y.; Li, C.; Peng, F.; Yao, Y.; Peng, H.; Ni, Z.; Xie, C.; Sun, Q. Overexpression of a wheat stearoyl-ACP desaturase (SACPD) gene *TaSSI2* in *Arabidopsis ssi2* mutant compromise its resistance to powdery mildew. *Gene* **2013**, *524*, 220–227. [CrossRef] [PubMed]

15. Vale, R.D. AAA proteins. Lords of the ring. *J. Cell Biol.* **2000**, *150*, F13–F19. [CrossRef] [PubMed]

16. Yedidi, R.S.; Wendler, P.; Enenkel, C. AAA-ATPases in protein degradation. *Front. Mol. Biosci.* **2017**, *4*, 42. [CrossRef]

17. Snider, J.; Houry, W.A. AAA+ proteins: Diversity in function, similarity in structure. *Biochem. Soc. Trans.* **2008**, *36*, 72–77. [CrossRef]

18. Sugimoto, M.; Yamaguchi, Y.; Nakamura, K.; Tatsumi, Y.; Sano, H. A hypersensitive response-induced ATPase associated with various cellular activities (AAA) protein from tobacco plants. *Plant Mol. Biol.* **2004**, *56*, 973–985. [CrossRef]

19. Lee, M.H.; Sano, H. Attenuation of the hypersensitive response by an ATPase associated with various cellular activities (AAA) protein through suppression of a small GTPase, ADP ribosylation factor, in tobacco plants. *Plant J.* **2007**, *51*, 127–139. [CrossRef]

20. Zhang, B.; Van Aken, O.; Thatcher, L.; De Clercq, I.; Duncan, O.; Law, S.R.; Murcha, M.W.; Van der Merwe, M.; Seifi, H.S.; Carrie, C. The mitochondrial outer membrane AAA ATPase AtOM66 affects cell death and pathogen resistance in *Arabidopsis thaliana*. *Plant J.* **2014**, *80*, 709–727. [CrossRef]

21. Zhu, X.; Yin, J.; Liang, S.; Liang, R.; Zhou, X.; Chen, Z.; Zhao, W.; Wang, J.; Li, W.; He, M. The multivesicular bodies (MVBs)-localized AAA ATPase LRD6-6 inhibits immunity and cell death likely through regulating MVBs-mediated vesicular trafficking in rice. *Plos Genet.* **2016**, *12*. [CrossRef] [PubMed]

22. Fekih, R.; Tamiru, M.; Kanzaki, H.; Abe, A.; Yoshida, K.; Kanzaki, E.; Saitoh, H.; Takagi, H.; Natsume, S.; Undan, J.R. The rice (*Oryza sativa* L.) *LESION MIMIC RESEMBLING*, which encodes an AAA-type ATPase, is implicated in defense response. *Mol. Genet. Genom.* **2015**, *290*, 611–622. [CrossRef] [PubMed]

23. Agrawal, G.K.; Rakwal, R.; Jwa, N.S. Rice (*Oryza sativa* L.) *OsPR1b* gene is phytohormonally regulated in close interaction with light signals. *Biochem. Biophys. Res. Commun.* **2000**, *278*, 290–298. [CrossRef] [PubMed]

24. Nakashita, H.; Yoshioka, K.; Takayama, M.; Kuga, R.; Midoh, N.; Usami, R.; Horikoshi, K.; Yoneyama, K.; Yamaguchi, I. Characterization of PBZ1, a probenazole-inducible gene, in suspension-cultured rice cells. *Biosci. Biotechnol. Biochem.* **2001**, *65*, 205–208. [CrossRef] [PubMed]

25. Wang, J.; Jiang, J.; Oard, J.H. Structure, expression and promoter activity of two polyubiquitin genes from rice (*Oryza sativa* L.). *Plant Sci.* **2000**, *156*, 201–211. [CrossRef]

26. Qi, M.; Yang, Y. Quantification of *Magnaporthe grisea* during infection of rice plants using real-time polymerase chain reaction and northern blot/phosphoimaging analyses. *Phytopathology* **2002**, *92*, 870–876. [CrossRef]

27. Huang, Q.N.; Shi, Y.F.; Zhang, X.B.; Song, L.X.; Feng, B.H.; Wang, H.M.; Xu, X.; Li, X.H.; Guo, D.; Wu, J.L. Single base substitution in OsCDC48 is responsible for premature senescence and death phenotype in rice. *J. Integ. Plant Biol.* **2016**, *58*, 12–28. [CrossRef]

28. Saifi, S.K.; Passricha, N.; Tuteja, R.; Tuteja, N. Stress-induced Oryza sativa RuvBL1a is DNA-independent ATPase and unwinds DNA duplex in 3′ to 5′ direction. *Protoplasma* **2018**, *255*, 669–684. [CrossRef]

29. Lin, Y.; Tan, L.; Zhao, L.; Sun, X.; Sun, C. RLS3, a protein with AAA+ domain localized in chloroplast, sustains leaf longevity in rice. *J. Integr. Plant Biol.* **2016**, *58*, 971–982. [CrossRef]

30. Xia, Z.; Wei, Y.; Sun, K.; Wu, J.; Wang, Y.; Wu, K. The maize AAA-type protein SKD1 confers enhanced salt and drought stress tolerance in transgenic tobacco by interacting with Lyst-interacting protein 5. *PLoS ONE* **2013**, *8*. [CrossRef]

31. Wang, F.; Liu, J.; Chen, M.; Zhou, L.; Li, Z.; Zhao, Q.; Pan, G.; Zaidi, S.-H.-R.; Cheng, F. Involvement of abscisic acid in PSII photodamage and D1 protein turnover for light-induced premature senescence of rice flag leaves. *PLoS ONE* **2016**, *11*. [CrossRef] [PubMed]

32. Furukawa, T.; Ishibashi, T.; Kimura, S.; Tanaka, H.; Hashimoto, J.; Sakaguchi, K. Characterization of all the subunits of replication factor C from a higher plant, rice (*Oryza sativa* L.), and their relation to development. *Plant Mol. Biol.* **2003**, *53*, 15–25. [CrossRef] [PubMed]

33. Santos, L. Molecular mechanisms of the AAA proteins in plants. *Adv. Agril. Food Biotechnol.* **2006**, *37*, 1–15.

34. Zhang, P.; Zhang, Y.; Sun, L.; Sinumporn, S.; Yang, Z.; Sun, B.; Xuan, D.; Li, Z.; Yu, P.; Wu, W. The rice AAA-ATPase OsFIGNL1 is essential for male meiosis. *Front. Plant Sci.* **2017**, *8*, 1639. [CrossRef]

35. Shahriari, M.; Keshavaiah, C.; Scheuring, D.; Sabovljevic, A.; Pimpl, P.; Häusler, R.E.; Hülskamp, M.; Schellmann, S. The AAA-type ATPase AtSKD1 contributes to vacuolar maintenance of *Arabidopsis thaliana*. *Plant J.* **2010**, *64*, 71–85. [CrossRef]

36. Kaplan, C.P.; Thomas, J.E.; Charlton, W.L.; Baker, A. Identification and characterisation of PEX6 orthologues from plants. *Bba-Mol. Cell Res.* **2001**, *1539*, 173–180. [CrossRef]

37. Chung, K.; Tasaka, M. RPT2a, a 26S proteasome AAA-ATPase, is directly involved in Arabidopsis CC-NBS-LRR protein uni-1D-induced signaling pathways. *Plant Cell Physiol.* **2011**, *52*, 1657–1664. [CrossRef]

38. Su, Y.; Hu, S.; Zhang, B.; Ye, W.; Niu, Y.; Guo, L.; Qian, Q. Characterization and fine mapping of a new early leaf senescence mutant es3 (t) in rice. *Plant Growth Regul.* **2017**, *81*, 419–431. [CrossRef]

39. Xu, X.; Ji, J.; Xu, Q.; Qi, X.; Weng, Y.; Chen, X. The major-effect quantitative trait locus Cs ARN 6.1 encodes an AAA ATP ase domain-containing protein that is associated with waterlogging stress tolerance by promoting adventitious root formation. *Plant J.* **2018**, *93*, 917–930. [CrossRef]

40. Siebers, M.; Brands, M.; Wewer, V.; Duan, Y.; Holzl, G.; Dormann, P. Lipids in plant-microbe interactions. *Bba-Mol. Cell Biol. L.* **2016**, *1861*, 1379–1395. [CrossRef]

41. Lim, G.H.; Singhal, R.; Kachroo, A.; Kachroo, P. Fatty Acid- and Lipid-Mediated Signaling in Plant Defense. *Annu. Rev. Phytopathol.* **2017**, *55*, 505–536. [CrossRef] [PubMed]

42. Kachroo, A.; Venugopal, S.C.; Lapchyk, L.; Falcone, D.; Hildebrand, D.; Kachroo, P. Oleic acid levels regulated by glycerolipid metabolism modulate defense gene expression in *Arabidopsis*. *Proc. Natl. Acad. Sci. USA* **2004**, *101*, 5152–5157. [CrossRef] [PubMed]

43. Delteil, A.; Zhang, J.; Lessard, P.; Morel, J.-B. Potential candidate genes for improving rice disease resistance. *Rice* **2010**, *3*, 56. [CrossRef]

44. Haque, E.; Taniguchi, H.; Hassan, M.; Bhowmik, P.; Karim, M.R.; Śmiech, M.; Zhao, K.; Rahman, M.; Islam, T. Application of CRISPR/Cas9 genome editing technology for the improvement of crops cultivated in tropical climates: Recent progress, prospects, and challenges. *Front. Plant Sci.* **2018**, *9*, 617. [CrossRef] [PubMed]

45. Yang, W.; Dong, R.; Liu, L.; Hu, Z.; Li, J.; Wang, Y.; Ding, X.; Chu, Z. A novel mutant allele of SSI2 confers a better balance between disease resistance and plant growth inhibition on Arabidopsis thaliana. *Bmc Plant Biol.* **2016**, *16*, 208. [CrossRef] [PubMed]

46. Mandal, M.K.; Chandra-Shekara, A.; Jeong, R.-D.; Yu, K.; Zhu, S.; Chanda, B.; Navarre, D.; Kachroo, A.; Kachroo, P. Oleic acid–dependent modulation of NITRIC OXIDE ASSOCIATED1 protein levels regulates nitric oxide–mediated defense signaling in *Arabidopsis*. *Plant Cell* **2012**, *24*, 1654–1674. [CrossRef] [PubMed]

47. Moreau, M.; Lee, G.I.; Wang, Y.; Crane, B.R.; Klessig, D.F. AtNOS/AtNOA1 is a functional *Arabidopsis thaliana* cGTPase and not a nitric-oxide synthase. *J. Biol. Chem.* **2008**, *283*, 32957–32967. [CrossRef]

48. Miki, D.; Itoh, R.; Shimamoto, K. RNA silencing of single and multiple members in a gene family of rice. *Plant Physiol.* **2005**, *138*, 1903–1913. [CrossRef]

49. Miki, D.; Shimamoto, K. Simple RNAi vectors for stable and transient suppression of gene function in rice. *Plant Cell Physiol.* **2004**, *45*, 490–495. [CrossRef]

50. Toki, S.; Hara, N.; Ono, K.; Onodera, H.; Tagiri, A.; Oka, S.; Tanaka, H. Early infection of scutellum tissue with Agrobacterium allows high-speed transformation of rice. *Plant J.* **2006**, *47*, 969–976. [CrossRef]

51. Yang, Y.; Qi, M.; Mei, C. Endogenous salicylic acid protects rice plants from oxidative damage caused by aging as well as biotic and abiotic stress. *Plant J.* **2004**, *40*, 909–919. [CrossRef] [PubMed]

52. Inoue, H.; Hayashi, N.; Matsushita, A.; Liu, X.; Nakayama, A.; Sugano, S.; Jiang, C.-J.; Takatsuji, H. Blast resistance of CC-NB-LRR protein Pb1 is mediated by WRKY45 through protein–protein interaction. *Proc. Natl. Acad. Sci. USA* **2013**, *110*, 9577–9582. [CrossRef] [PubMed]

53. Letunic, I.; Bork, P. Interactive Tree Of Life (iTOL) v4: Recent updates and new developments. *Nucleic Acids Res.* **2019**, *47*, W256–W259. [CrossRef] [PubMed]

International Journal of
Molecular Sciences

Article

Bulked Segregant Analysis Coupled with Whole-Genome Sequencing (BSA-Seq) Mapping Identifies a Novel *pi21* Haplotype Conferring Basal Resistance to Rice Blast Disease

Tingmin Liang [1,2,†], Wenchao Chi [1,†], Likun Huang [3], Mengyu Qu [1], Shubiao Zhang [3], Zi-Qiang Chen [2], Zai-Jie Chen [2], Dagang Tian [2], Yijie Gui [2], Xiaofeng Chen [1], Zonghua Wang [1,4], Weiqi Tang [1,*] and Songbiao Chen [1,*]

[1] Marine and Agricultural Biotechnology Laboratory, Institute of Oceanography, Minjiang University, Fuzhou 350108, China; LTM0521@163.com (T.L.); chiwenchao@mju.edu.cn (W.C.); mengyu_qu@163.com (M.Q.); chenxf@mju.edu.cn (X.C.); wangzh@fafu.edu.cn (Z.W.)

[2] Biotechnology Research Institute, Fujian Academy of Agricultural Sciences, Fuzhou 350003, China; czq@fjage.org (Z.-Q.C.); czj@fjage.org (Z.-J.C.); tdg@fjage.org (D.T.); gyj@fjage.org (Y.G.)

[3] College of Agriculture, Fujian Agriculture and Forestry University, Fuzhou 350002, China; likun.huang@foxmail.com (L.H.); zhangsbiao@aliyun.com (S.Z.)

[4] State Key Laboratory of Ecological Pest Control for Fujian and Taiwan Crops, Fujian Agriculture and Forestry University, Fuzhou 350002, China

[*] Correspondence: tangweiqi@jointgene.com (W.T.); sbchen@fjage.org (S.C.)

[†] These authors contributed equally to this article.

Received: 6 December 2019; Accepted: 19 March 2020; Published: 21 March 2020

Abstract: Basal or partial resistance has been considered race-non-specific and broad-spectrum. Therefore, the identification of genes or quantitative trait loci (QTLs) conferring basal resistance and germplasm containing them is of significance in breeding crops with durable resistance. In this study, we performed a bulked segregant analysis coupled with whole-genome sequencing (BSA-seq) to identify QTLs controlling basal resistance to blast disease in an F_2 population derived from two rice varieties, 02428 and LiXinGeng (LXG), which differ significantly in basal resistance to rice blast. Four candidate QTLs, *qBBR-4*, *qBBR-7*, *qBBR-8*, and *qBBR-11*, were mapped on chromosomes 4, 7, 8, and 11, respectively. Allelic and genotypic association analyses identified a novel haplotype of the durable blast resistance gene *pi21* carrying double deletions of 30 bp and 33 bp in 02428 (*pi21-2428*) as a candidate gene of *qBBR-4*. We further assessed haplotypes of *Pi21* in 325 rice accessions, and identified 11 haplotypes among the accessions, of which eight were novel types. While the resistant *pi21* gene was found only in *japonica* before, three Chinese *indica* varieties, ShuHui881, Yong4, and ZhengDa4Hao, were detected carrying the resistant *pi21-2428* allele. The *pi21-2428* allele and *pi21-2428*-containing rice germplasm, thus, provide valuable resources for breeding rice varieties, especially *indica* rice varieties, with durable resistance to blast disease. Our results also lay the foundation for further identification and functional characterization of the other three QTLs to better understand the molecular mechanisms underlying rice basal resistance to blast disease.

Keywords: rice; blast disease; partial resistance; *pi21*; haplotype

1. Introduction

Rice is one of the most important staple crops for more than half of the population in the world [1]. Rice blast, caused by the fungus *Magnaporthe oryzae*, is one of the most devastating diseases of rice, causing yield losses of 10%–30% annually [2]. The development and use of resistant varieties appears to be the most economical and environmentally sustainable way to control rice blast [3]. Identification of genes or genetic loci conferring resistance to blast disease could accelerate breeding programs for resistant rice varieties.

Genetically, disease resistance in plants can be categorized into two types, qualitative and quantitative [4,5]. Qualitative resistance is mainly mediated by a single resistance gene (*R* gene), which confers complete but race-specific resistance through the recognition of pathogen effectors [6,7]. *R* gene-mediated resistance has been widely deployed in crop breeding programs. However, *R* gene-mediated resistance is often not durable, as most pathogens are able to rapidly evolve new virulent races lacking the corresponding avirulence effectors to evade recognition by the cognate R protein. Quantitative resistance is mediated by multiple genes or quantitative trait loci (QTLs), providing partial or basal resistance associated with delayed and reduced development of disease lesions [8,9]. Although quantitative resistance has only partial effects, it has been considered race-non-specific and broad-spectrum, and is therefore of particular interest for breeding crops with durable resistance [4,10].

To date, more than 100 rice blast *R* genes have been identified and at least 28 *R* genes have been cloned [11,12]. All the cloned major *R* genes encode nucleotide-binding site leucine-rich repeat (NBS-LRR) proteins, with the exception of *Pid2*, encoding a B-lectin kinase [13], and *Ptr*, encoding an Armadillo repeat protein [14]. Several hundred QTLs associated with blast resistance have been identified [15]. However, only a limited number of QTLs for blast resistance have been cloned [16–20]. The cloned QTLs encode proteins that are diverse in their structure and function: *pi21* encodes a proline-rich protein with loss-of-function deletions [16]; *Pb1* encodes an atypical coiled-coil (CC)-NBS-LRR protein [17]; *Pi35* and *Pi63* encode NBS-LRR proteins [18,19]; whereas *bsr1-d1* encodes a C2H2-type transcription factor with a single nucleotide change in the promoter [20]. These findings indicate that quantitative resistances are controlled by diverse molecular mechanisms.

With the rapid development of next-generation sequencing, approaches based on bulked segregant analysis coupled with whole-genome sequencing (BSA-Seq) have been developed for the mapping of agronomically important loci in rice [21–23], including major genes or QTLs responsible for blast resistance [22,24,25]. In the present study, we apply BSA-seq to rapidly map four QTLs, *qBBR-4*, *qBBR-7*, *qBBR-8*, and *qBBR-11*, responsible for basal resistance to blast disease, and identify a novel haplotype of the durable blast resistance gene *pi21* as a candidate gene of *qBBR-4* on chromosome 4 in a *japonica* variety 02428 (*pi21-2428*). While the resistant *pi21* gene was found only in *japonica* before [16], we identify three Chinese *indica* varieties carrying the resistant *pi21-2428* allele in 325 accessions. Therefore, the novel *pi21-2428* allele and the *pi21-2428*-containing rice varieties identified in the present study provide valuable resources for breeding rice varieties, especially *indica* rice, which are durably resistant to blast disease. Our results also lay the foundation for further identification and functional characterization of the other three QTLs for a better understanding of rice basal resistance to blast disease.

2. Results

2.1. Evaluation of 02428 and LXG in Basal Resistance to Rice Blast Disease

In our earlier evaluations, the rice variety 02428 was observed to possess high basal resistance to the rice blast fungus *M. oryzae* under natural nursery conditions (data not shown). We further performed artificial inoculations on 02428 seedlings using three virulent isolates of *M. oryzae* in this study. The results showed that 02428 was moderately susceptible to isolates 501-3 and Guy11, and was moderately resistant to isolate RB22 (Figure 1A). Most lesions on leaves of 02428 were limited in size. In contrast, LiXinGeng (LXG) was highly susceptible to all three isolates. These results suggest that 02428 possesses high basal resistance, preventing blast disease development.

An F_2 population of 02428 × LXG with 626 individuals was inoculated with RB22. The results show that the frequency distribution of disease severity in the F_2 population of 02428 × LXG exhibited continuous variation (Figure 1B), indicating that the resistance to RB22 is likely controlled by multiple genes.

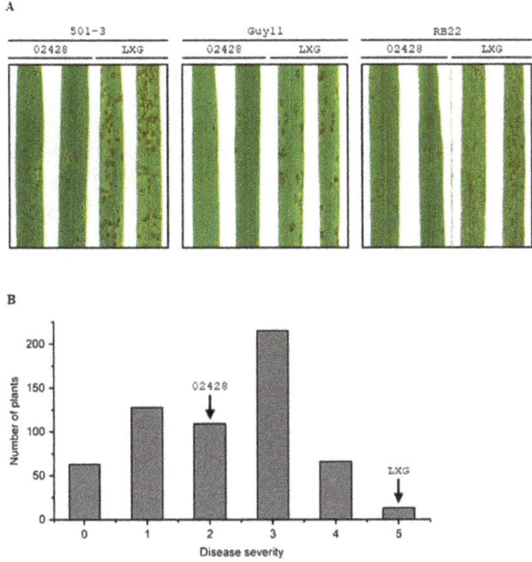

Figure 1. Resistance reaction of rice varieties 02428, LiXinGeng (LXG) and their F_2 population to rice blast disease. (**A**) Phenotypes of 02428 and LXG inoculated with *M. oryzae* isolates 501-3, Guy11, and RB22. (**B**) The frequency distribution of disease severity in the F_2 population of 02428 × LXG inoculated with *M. oryzae* isolate RB22. Disease severity was assessed following a 0–5 scale (0–1: resistant, 2: moderately resistant, 3: moderately susceptible, 4–5: severely susceptible).

2.2. SNP and Short InDel Polymorphism Profiling

Whole-genome sequencing of extremely resistant (ER) and extremely susceptible (ES) pools derived from the F_2 population of 02428 × LXG and the two parental lines 02428 and LXG generated about 90.7 to 158.9 million reads for each sub-pool or parental line (Supplementary Table S1). After filtering, a total of 469,512 bi-allelic single-nucleotide polymorphisms (SNPs), and a total of 65,766 bi-allelic short insertions and deletions (InDels) were identified (Table 1). The average densities of SNP and short InDel markers were about 1.26 SNP/kb (average every 795 bp exists a SNP) and 0.18 InDel/kb (average every 5,675 kb exists a short InDel) (Table 1), respectively. The polymorphic markers were sufficiently distributed across the whole genome, except for one region of about 6.5 Mb on chromosome 3 containing relatively fewer markers (Supplementary Figure S1).

Table 1. Chromosome-wise distribution of the identified single-nucleotide polymorphisms (SNPs) and short InDels.

Chr.	Length	SNP		Short InDel	
		Number	Density (per kb)	Number	Density (per kb)
Chr1	43,270,923	48,669	1.12	7735	0.18
Chr2	35,937,250	67,302	1.87	9069	0.25
Chr3	36,413,819	20,399	0.56	3336	0.09
Chr4	35,502,694	42,156	1.19	5558	0.16
Chr5	29,958,434	67,386	2.25	9189	0.31
Chr6	31,248,787	34,848	1.12	5309	0.17
Chr7	29,697,621	25,681	0.86	4040	0.14
Chr8	28,443,022	56,474	1.99	7152	0.25
Chr9	23,012,720	23,331	1.01	2965	0.13
Chr10	23,207,287	36,435	1.57	4062	0.18
Chr11	29,021,106	31,309	1.08	4000	0.18
Chr12	27,531,856	15,522	0.56	2671	0.10
Total	373,245,519	469,512	1.26	65,766	0.18

2.3. QTL Mapping and Heritability Estimation

Calculation results of the average allele frequency (AF) value for each marker showed that most of the identified SNP and short InDel markers had an expected AF value of around 0.5 in the four sub-pools ER-1, ER-2, ES-1, and ES-2 (Supplementary Figure S2), indicating no severe segregation distortion of the markers as a whole. Allele frequency difference (AFD) value between ER (ER-1 + ER-2) and ES (ES-1 + ES-2) pools was calculated, and four positive AFD peaks were detected in the fitted curve and exceeded the threshold (0.165 at the overall significance level of $p < 0.05$) (Figure 2A). Subsequently, unpaired t-tests were performed for the two replicated sub-pools for ER and ES. The p-values of each marker were estimated, and the peaks of negative logarithmic p value (NLP) (Figure 2B) were consistent with the peaks in AFD curve. These results suggest that there were four candidate QTLs located in these regions. The confidence intervals of the four QTLs located on chromosome 4, 7, 8, and 11 were estimated (Figures 2A and 3, Table 2), and the four QTLs were named *qBBR-4*, *qBBR-7*, *qBBR-8*, and *qBBR-11*, respectively.

Heritability estimation showed that the additive heritability and dominance heritability of each QTL varied by 0.89%–2.09% and 0.01%–7.82%, respectively (Table 2). Among the four QTLs, *qBBR-4* had the largest effect, with the biggest additive heritability at about 2.09% (Table 2), suggesting that *qBBR-4* is a major QTL involved in basal resistance to blast disease.

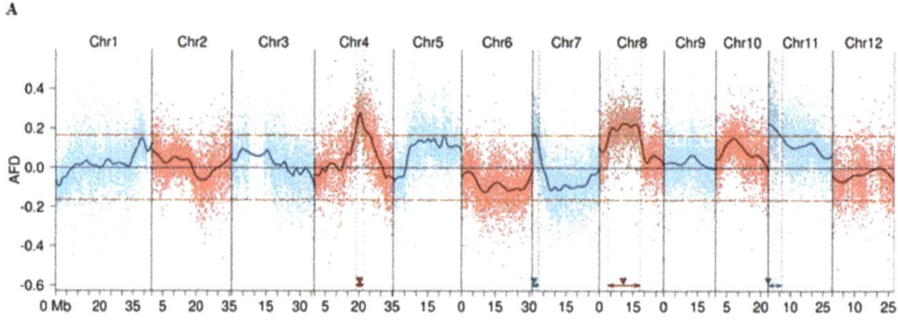

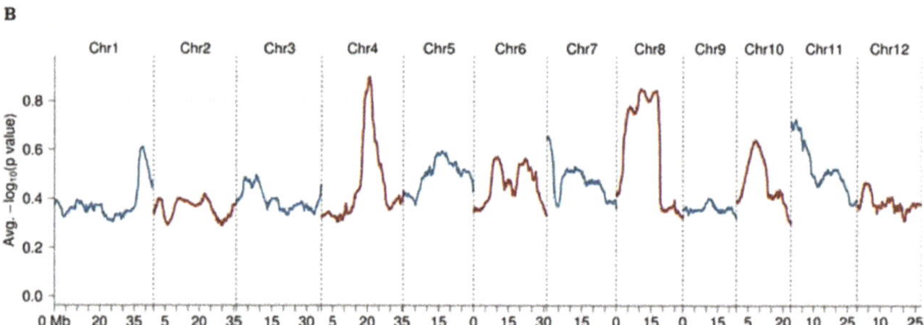

Figure 2. BSA-seq-based identification of four candidate quantitative trait loci (QTLs) conferring basal resistance to rice blast disease. (**A**) Allele frequency difference (AFD) graph from BSA-seq analysis. The horizontal orange dashed lines indicate the threshold (±0.165) at the overall significance level of $p < 0.05$. QTL positions estimated are indicated by filled triangles. AFD was obtained by subtraction of allele frequency (AF) of the extremely susceptible (ES) pool from that of the extremely resistant (ER) pool. (**B**) *T*-test verification for the two replicated sub-pools for the ER pool and the ES pool. The average negative logarithmic p-values of the markers were smoothed by sliding window (size = 3000 kb and step = 10 kb) across each chromosome. The peaks of the p-value graph were consistent with those in the AFD graph.

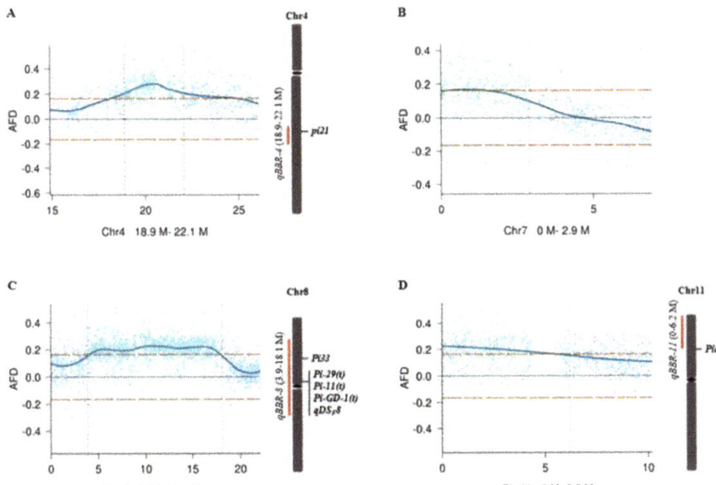

Figure 3. Estimates of confidence intervals of the four QTLs, *qBBR-4* (**A**), *qBBR-7* (**B**), *qBBR-8* (**C**), and *qBBR-11* (**D**). The horizontal orange dashed lines indicate the threshold (0.165) at the overall significance level of *p* < 0.05. The light green areas indicate 95% confidence intervals of the QTLs. Previously reported *R* genes or QTLs within or closely near the confidence intervals of *qBBR-4*, *qBBR-8*, and *qBBR-11* are indicated on the right. AFD: allele frequency difference.

Table 2. Estimates of position and heritability of the four identified QTLs.

QTL	Chr.	AFD Value [a]	Pos. (Mb) [b]	Interval (Mp) [c]	Max. NLP [d]	h_A^2 (%) [e]	h_D^2 (%) [f]
qBBR-4	4	0.278	20.46	18.90–22.10	0.90	2.09	7.82
qBBR-7	7	0.172	0.86	0–2.91	0.65	0.89	0.26
qBBR-8	8	0.227	10.63	3.89–18.09	0.85	1.56	0.01
qBBR-11	11	0.226	0.02	0–6.21	0.73	1.54	3.50

[a] Maximum value of the peak of the AFD curve; [b] Chromosome position of the peak of the AFD curve; [c] Estimated based on the 95% confidence; [d] The most significant *p*-value of the peak, which was converted to negative logarithmic *p* (NLP) value; [e] Heritability attributed to additive effect of the QTL; [f] Heritability attributed to dominance effect of the QTL.

2.4. Identification of a New Haplotype of pi21 as a Candidate Gene of qBBR-4

To further refine candidate genes involved in basal resistance to blast disease, we searched previously reported *R* genes or QTLs within the confidence intervals of the four QTLs (Figure 3). While no previously reported *R* genes or QTLs were identified within or near the confidence interval of *qBBR-7* (Figure 3B), a previously identified *Pia* [26] gene was located near the confidence interval of *qBBR-11* (Figure 3D), and there were several *R* genes or QTL, including *Pi-11(t)* [27], *Pi-29(t)* [28], *Pi33* [29], *Pi-GD-1(t)* [30], and *qDS_F8* [3], in the *qBBR-8* region (Figure 3C). Interestingly, *qBBR-4* was observed to be co-localized with a cloned recessive durable blast disease resistance QTL *pi21* (Figure 3A). *Pi21* has been found to have at least 12 variants (haplotypes A to L) based on InDel polymorphisms in the proline-rich region. While haplotype L containing double deletions of 21 bp and 48 bp (*pi21(-21/-48)*) resulting in deletions of the core motif "PxxPxxP" in the proline-rich region (Figure 4A) was identified, conferring durable blast disease resistance due to loss of function of *Pi21*, the other haplotypes carrying one of the two deletions or two smaller deletions did not confer high basal resistance to blast disease [16].

Sequence analysis of the *Pi21* alleles in LXG and 02428 showed that while the allele in LXG (*Pi21-LXG*) had a 27 bp deletion, the *pi21-2428* allele had double deletions of 30 bp and 33 bp, resulting in deletions of 10 aa and 11 aa of the core "PxxPxxP" motif, the same as in the *pi21(-21/-48)* allele (Figure 4A,B). Both *Pi21-LXG* and *pi21-2428* were different from the 12 identified haplotypes. Genotypic analysis of 61 individual ER and 59 individual ES plants showed that all *pi21-2428* homozygous seedlings showed resistance to blast disease. In contrast, about 79% (41 out of 52) of the *Pi21-LXG* homozygous, and 39% (18 out of 46) of *Pi21-LXG/pi21-2428* heterozygous seedlings showed susceptible to blast disease (Figure 4C). These results support *pi21-2428* as a candidate gene of *qBBR-4*, and suggest that while the 27 bp deletion did not affect *Pi21-LXG* function, double deletions of the 30 bp and 33 bp sequences could cause a defect in *pi21-2428* function, leading to high basal resistance to blast disease.

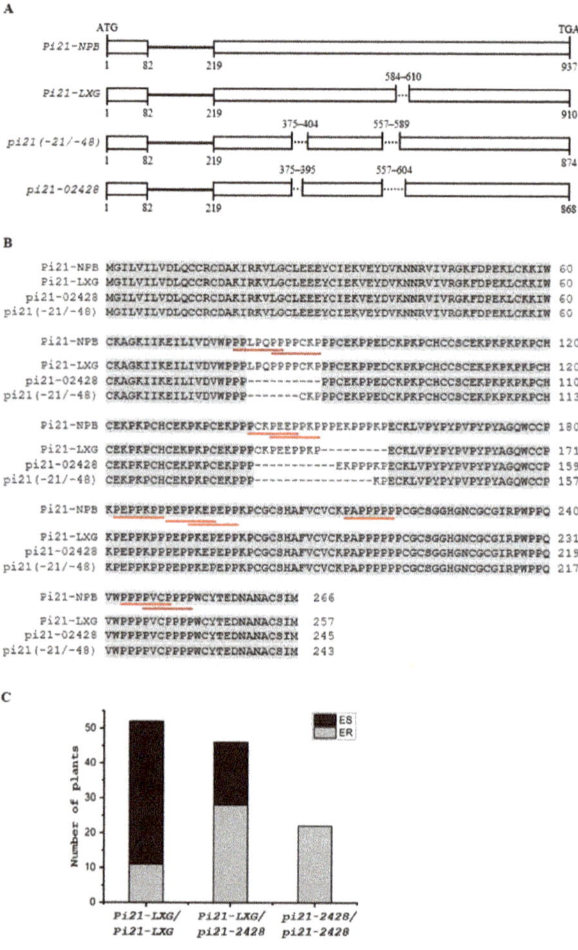

Figure 4. Identification of a novel haplotype of *pi21* as a candidate gene of *qBBR-4*. (**A**) Schematic diagram of the genomic coding region of the *Pi21* alleles in Nipponbare (*Pi21-NPB*), LXG (*Pi21-LXG*), and 02428 (*pi21-2428*), and the resistant allele of haplotype L containing double deletions of 21 bp and 48 bp (*pi21(-21/-48)*) [16]. Open boxes represent exons, lines represent introns, and dotted lines represent deletions. Numbers above the diagrams indicate positions of deletions corresponding to the *Pi21-NPB* allele, and numbers below the diagrams represent the start and end nucleotide positions of exons of each allele. (**B**) Alignment of the deduced amino acid sequences of Pi21-NPB, Pi21-LXG, pi21-2428, and pi21(-21/-48). Putative proline-rich motifs (PxxPxxP) for protein–protein interaction are underlined in red. (**C**) Genotype-phenotype correlation analysis of 61 ER individuals and 59 ES individuals. All the homozygous *pi21-2428* individuals belonged to the ER bulk.

2.5. Assessment of the pi21-2428 Haplotype in 325 Rice Accessions

Multiple sequence alignment of the *Pi21* alleles in 325 rice accessions revealed a total of 11 haplotypes among the tested accessions (Figure 5A, Supplementary Table S2). However, none of the accessions carried the resistant *pi21(-21/-48)* allele. The haplotypes were named based on insertion/deletion patterns. Among the 11 haplotypes, *Pi21-NPB* (the *Pi21* allele in Nipponbare), *Pi21(-9)*, and *Pi21(-24/-15)* were identical to the previously identified haplotypes B, C, and H, respectively (Figure 5A, Supplementary Table S2) [16], and the rest eight were novel types.

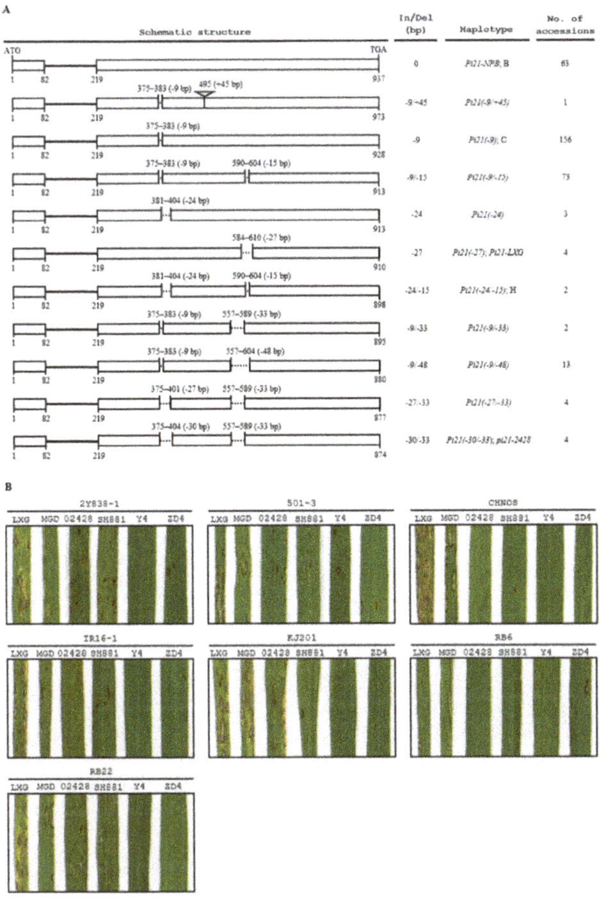

Figure 5. Assessment of haplotypes of *Pi21* in 325 rice accessions and blast inoculation of the *pi21-2428*-containing varieties. (**A**) A total of 11 haplotypes were identified among 325 rice accessions. The haplotypes were named based on insertion/deletion patterns. *Pi21-NPB*, *Pi21(-9)*, and *Pi21(-24/-15)* were identical to the previously identified haplotypes B, C, and H, respectively [16]. Open boxes represent exons, lines represent introns, and dotted lines represent deletions. Numbers above the diagrams indicate positions of deletions corresponding to the *Pi21-NPB* allele, and numbers below the diagrams represent the start and end nucleotide positions of exons of each allele. (**B**) Phenotypes of the *pi21-2428*-containing varieties 02428, ShuHui881 (SH881), Yong4 (Y4), and ZhengDa4Hao (ZD4) inoculated with *M. oryzae* isolates 2Y838-1, 501-3, CHNOS, IR16-1, KJ201, RB6, and RB22, respectively. Rice varieties LXG and MengGuDao (MGD) were used as highly susceptible controls.

Among the 325 rice accessions, in addition to 02428, three Chinese *indica* varieties ShuHui881, Yong4, and ZhengDa4Hao were detected carrying the resistant *pi21-2428* allele. The *pi21-2428*-containing varieties were inoculated with seven *M. oryzae* isolates. As shown in Figure 5B, the two susceptible varieties LXG and MengGuDao (MGD) were highly susceptible to most of the *M. oryzae* isolates. The four *pi21-2428*-containing varieties showed complete resistance, moderate resistance, or moderate susceptibility to the isolates. For example, the *indica* variety ZhengDa4Hao was highly resistant to isolates KJ201 and RB22, and moderately resistant to isolates CHNOS and IR16-1, suggesting that there might be some *R* genes conferring the high resistance in ZhengDa4Hao against the four isolates. ZhengDa4Hao also showed susceptibility to isolates 2Y838-1, 501-3, and RB6, but the lesions on leaves were limited in size and much less than those on leaves of LXG or MGD. Overall, none of the *pi21-2428*-containing varieties showed highly susceptible to the seven *M. oryzae* isolates, suggesting that the four varieties possessed high basal resistance with significantly delayed and reduced development of disease lesions.

3. Discussion

When compared with *R* gene-mediated, race-specific resistance, basal resistance has been presumed to be more durable [4,10]. Therefore, the identification of genes or QTLs conferring basal resistance is of significance in breeding crops with long-lasting resistance to plant diseases. In the present study, we employed BSA-Seq to identify four candidate QTLs *qBBR-4*, *qBBR-7*, *qBBR-8*, and *qBBR-11*, conferring basal resistance to rice blast disease on rice chromosomes 4, 7, 8, and 11, respectively. Building on advances in next-generation sequencing, the BSA-Seq method took advantage of pooled sequencing, which does not require the laborious process of genotyping of each individual from a large mapping population, allowing for rapid identification of candidate genes or QTLs controlling important agronomic traits [24,31].

Among the four QTLs detected in this study, *qBBR-4* had the largest additive effect (Table 2). The *qBBR-4* locus is located in a region of chromosome 4 with a known recessive blast disease resistant QTL *pi21*, identified in rice variety Owarihatamochi [16]. *Pi21* encodes a proline-rich protein, consisting of a putative heavy metal-binding domain and protein-protein interaction motifs. While the dominant *Pi21* appears to slow the plant's defense responses, loss-of-function of *Pi21* (double deletions of 21 bp and 48 bp in the proline-rich region; haplotype L) confers durable resistance to blast disease in rice [16]. Sequence analysis revealed that the *pi21* allele in 02428, the parental line with high basal resistance, had double deletions of 30 bp and 33 bp, resulting in deletions of 10 aa and 11 aa in the proline-rich region, which house the core motif "PxxPxxP" (Figure 4A,B) for protein–protein interaction in multicellular organisms [16,32,33]. We further performed genotype-phenotype correlation analysis of 61 ER individuals and 59 ES individuals, and the results showed that all the homozygous *pi21-2428* individuals belonged to the ER bulk (Figure 4C). Taken together, these results suggest that double deletions of the 30 bp and 33 bp sequences of *Pi21* led to high basal resistance to blast disease, and that *pi21-2428* was the candidate gene of *qBBR-4*.

Previously, sequence analysis of the *Pi21* locus revealed 12 haplotypes (A to L) among 80 Asian cultivated rice varieties. Eleven of the haplotypes (A to K) carried an insertion or smaller deletions compared with the resistant *pi21(-21/-48)* allele, and did not confer resistance to blast disease [16]. In the present study, we detected a total of 11 haplotypes of *Pi21* among 325 rice accessions (Figure 5A, Supplementary Table S2), of which three were identical to the previously identified haplotypes B, C, and H [16], and eight were novel. Interestingly, the DNA variations of all the 20 detected haplotypes identified by Fukuoka et al. [16] and in this study were found to result in amino acid insertion/deletions, but not to cause premature termination of the predicted *Pi21* product, implying that the *Pi21* or *pi21* alleles maintain certain functions important for rice [34]. Besides 02428, we identified three more varieties, ShuHui881, Yong4, and ZhengDa4Hao, possessing the resistant *pi21-2428* allele. Inoculation testing with seven *M. oryzae* isolates indicated that, while the susceptible varieties LXG and MGD were highly susceptible to most of the *M. oryzae* isolates, the four *pi21-2428*-containing varieties showed complete resistance moderate resistance or moderate susceptibility to the *M. oryzae* isolates (Figure 5B). The results suggest that there should be some *R* genes other than *pi21-2428* conferring the complete or high resistance in the four *pi21-2428*-containing varieties. On the other hand, the moderately susceptible reactions with limited lesion size and number, to virulent *M. oryzae* isolates indicated that ShuHui881, Yong4, ZhengDa4Hao, as well as 02428, possessed high basal resistance to blast disease (Figure 5B). It is worth noting that while the resistant *pi21(-21/-48)* allele was found only in *japonica* rice [16], the three *pi21-2428*-containing varieties ShuHui881, Yong4, and ZhengDa4Hao were *indica* rice (Supplementary Table S2). Therefore, the varieties identified in the present study provide valuable resources for breeding rice varieties, especially *indica* varieties, with durable resistance to blast disease.

Transgressive segregation was observed in the F_2 population of 02428 × LXG, where some F_2 segregants showed more resistance than both parents (Figure 1B). This phenomenon implies that favorable alleles from both resistant and susceptible parents could be combined in the progeny, leading to a higher resistance than in the parents. In the present study, in addition to *pi21-2428*, we detected three other QTLs, *qBBR-7*, *qBBR-8*, and *qBBR-11*, on chromosomes 7, 8, and 11, respectively. Further identification and functional characterization of the three QTLs should be helpful to better understand the mechanisms underlying rice basal resistance to blast disease. Furthermore, the finding of transgressive segregation for blast resistance in the F_2 population of 02428 × LXG indicates that pyramiding more basal resistance genes or QTL alleles with *pi21-2428* would be an effective approach to enhance durable resistance to rice blast disease [10,35].

4. Materials and Methods

4.1. Plant Materials and Blast Isolates

Rice cvs. 02428 and LXG were used to construct an F_2 population to evaluate the segregation of blast disease reaction and for QTL mapping. Besides 02428 and LXG, a collection of 323 rice accessions consisting mainly of Chinese rice varieties or breeding materials were used for haplotype assessment in this study (Supplementary Table S2). The *M. oryzae* isolates CHNOS, Guy11, and KJ201 were kindly provided by Dr. Guo-Liang Wang (Department of Plant Pathology, Ohio State University, Ohio, USA), and isolates 2Y838-1, 501-3, IR16-1, RB6, and RB22 were collected from Fujian province, China.

4.2. Rice Blast Inoculations

Rice blast inoculations were carried out following a previously described spraying method [36]. Rice seedlings were grown in a greenhouse for about 12–14 days and were inoculated with *M. oryzae* spores at a concentration of 5×10^5 spores mL^{-1}. After inoculation, the seedlings were grown at 25 °C under high humidity for 4–5 days. Blast disease reactions were scored following a 0–5 scale (0–1: resistant, 2: moderately resistant, 3: moderately susceptible, 4–5: severely susceptible) [37].

4.3. Bulking, DNA Extraction, and Whole-Genome Resequencing

To understand the genetic basis of basal resistance to rice blast disease in 02428, a total of 626 F_2 plants derived from a cross between 02428 and LXG were inoculated by RB22, and were investigated for the segregation of disease reaction. For bulked segregant analysis, about 10,000 individuals of the F_2 population of 02428 × LXG were inoculated with the *M. oryzae* isolate RB22. About 126 highly resistant and 120 highly susceptible F_2 individuals were screened to generate the ER and ES bulks, respectively. Both the ER and ES bulks were divided into two replicates, ER-1 (56 individuals) and ER-2 (70 individuals) for the ER bulk, and ES-1 (60 individuals) and ES-2 (60 individuals) for the ES bulk. The genomic DNAs of the four bulked samples were extracted and were mixed with equal amounts. DNA samples of the two parents, 02428 and LXG, and the four pools were subjected to whole-genome resequencing using the Illumina HiSeq X Ten platform, followed by standard paired-end 150 bp sequencing library construction protocols.

4.4. Analysis of Reads and Variants

The raw reads were cleaned and trimmed using BBDuk program of BBTools (http://jgi.doe.gov/data-and-tools/bbtools/). The paired reads were mapped to the IRGSP-1.0 reference rice genome (http://rapdb.dna.affrc.go.jp) by using Burrows-Wheeler Aligner based on the Maximal Exact Matches algorithm (BWA MEM) and the alignments were processed by SAMTools [38–40]. Freebayes was used to call SNPs and InDels, with default parameters [41]. To obtain reliable polymorphic markers, variant filtering was performed by custom perl scripts: firstly, SNPs or short InDels exhibiting polymorphism between the two parents were screened; secondly, to further avoid severe segregation distortion, SNPs or short InDels with AF values from 0.3 to 0.7 were retained. These markers were annotated by snpEff [42].

4.5. QTL Analysis

The marker set was employed to map QTLs. The replicated sub-pools ER-1, ER-2, and ES-1, ES-2 were firstly incorporated into one ER pool and one ES pool, respectively. AFD value between the ER and ES pools was calculated and then smoothed by block regression, following the Block Regression Mapping methodology [43]. The block size used for the regression was set to be 20 kb. The AFD curve threshold at the overall significance level of 0.05 was estimated under the assumption of theoretical allele frequency (= 0.5) in the F_2 population. For each significant AFD peak (candidate QTL), the 95% confidence interval was estimated. In addition, unpaired t-tests based on the two biological replicates of the ER and ES pools were performed to validate the candidate QTLs following the X-QTL-seq method [44]. According to the peak AFD value, the heritability of each QTL was estimated using the method of Pooled QTL Heritability Estimator (PQHE) [45].

4.6. Detection of Haplotypes of Pi21

Genomic DNAs of the rice accessions (Supplementary Table S2) were subjected to genotyping for haplotypes of *Pi21*. PCR products were amplified with primers P21-F (5′-CAAGGCTAATCAG CAGTGT-3′) and P21-R (5′-TTGGCGTTGTCCTCGGTGT-3′). DNA sequences of PCR products were aligned using Clustal W [46].

Supplementary Materials: Supplementary materials can be found at http://www.mdpi.com/1422-0067/21/6/2162/s1.

Author Contributions: W.T., Z.W. and S.C. conceived and designed the experiments. T.L., S.Z. and S.C. constructed the F_2 population. T.L., Z.-Q.C., Z.-J.C., D.T., X.C. and S.C. conducted rice blast inoculation. W.C., L.H. and W.T. analyzed the sequencing data and performed QTL mapping. T.L., M.Q. and Y.G. conducted haplotype analysis. W.T. and S.C. wrote and revised the manuscript. All authors read and approved the final manuscript.

Acknowledgments: This work was supported by grant from National Natural Science Foundation of China (U1405212), grants from Fujian Provincial Science and Technology Program (2017R1019-1, 2018R1019-9), and open project of State Key Laboratory of Ecological Pest Control for Fujian and Taiwan Crops (SKL018002).

Conflicts of Interest: The authors declare no conflicts of interest.

References

1. Khush, G.S. What it will take to feed 5.0 billion rice consumers in 2030. *Plant Mol. Biol.* **2005**, *59*, 1–6. [CrossRef] [PubMed]
2. Talbot, N.J. On the trail of a cereal killer: Exploring the biology of *Magnaporthe grisea*. *Annu. Rev. Microbiol.* **2003**, *57*, 177–202. [CrossRef] [PubMed]
3. Sun, P.; Liu, J.; Wang, Y.; Jiang, N.; Wang, S.; Dai, Y.; Gao, J.; Li, Z.; Pan, S.; Wang, D.; et al. Molecular mapping of the blast resistance gene *Pi49* in the durably resistant rice cultivar Mowanggu. *Euphytica* **2013**, *192*, 45–54. [CrossRef]
4. Poland, J.A.; Balint-Kurti, P.J.; Wisser, R.J.; Pratt, R.C.; Nelson, R.J. Shades of gray: The world of quantitative disease resistance. *Trends Plant Sci.* **2009**, *14*, 21–29. [CrossRef]
5. Kou, Y.; Wang, S. Broad-spectrum and durability: Understanding of quantitative disease resistance. *Curr. Opin. Plant Biol.* **2010**, *13*, 181–185. [CrossRef]

6. Chisholm, S.T.; Coaker, G.; Day, B.; Staskawicz, B.J. Host-microbe interactions: Shaping the evolution of the plant immune response. *Cell* **2006**, *124*, 803–814. [CrossRef]

7. Jones, J.D.; Dangl, J.L. The plant immune system. *Nature* **2006**, *444*, 323–329. [CrossRef]

8. Collins, N.C.; Thordal-Christensen, H.; Lipka, V.; Bau, S.; Kombrink, E.; Qiu, J.L.; Hückelhoven, R.; Stein, M.; Freialdenhoven, A.; Somerville, S.C.; et al. SNARE-protein-mediated disease resistance at the plant cell wall. *Nature* **2003**, *425*, 973–977. [CrossRef]

9. Aghnoum, R.; Marcel, T.C.; Johrde, A.; Pecchioni, N.; Schweizer, P.; Niks, R.E. Basal host resistance of barley to powdery mildew: Connecting quantitative trait Loci and candidate genes. *Mol. Plant-Microbe Interact* **2010**, *23*, 91–102. [CrossRef]

10. Yasuda, N.; Mitsunaga, T.; Hayashi, K.; Koizumi, S.; Fujita, Y. Effects of pyramiding quantitative resistance genes *Pi21*, *Pi34*, and *Pi35* on rice leaf blast disease. *Plant Dis.* **2015**, *99*, 904–909. [CrossRef]

11. Tian, D.; Yang, L.; Chen, Z.; Chen, Z.; Wang, F.; Zhou, Y.; Lou, Y.; Yang, L.; Chen, S. Proteomic analysis of the defense response to *Magnaporthe oryzae* in rice harboring the blast resistance gene *Piz-t*. *Rice* **2018**, *11*, 47. [CrossRef] [PubMed]

12. Li, C.; Wang, D.; Peng, S.; Chen, Y.; Su, P.; Chen, J.; Zheng, L.; Tan, X.; Liu, J.; Xiao, Y.; et al. Genome-wide association mapping of resistance against rice blast strains in South China and identification of a new *Pik* allele. *Rice* **2019**, *12*, 47. [CrossRef] [PubMed]

13. Chen, X.; Shang, J.; Chen, D.; Lei, C.; Zou, Y.; Zhai, W.; Liu, G.; Xu, J.; Ling, Z.; Cao, G.; et al. A B-lectin receptor kinase gene conferring rice blast resistance. *Plant J.* **2006**, *46*, 794–804. [CrossRef] [PubMed]

14. Zhao, H.; Wang, X.; Jia, Y.; Minkenberg, B.; Wheatley, M.; Fan, J.; Jia, M.H.; Famoso, A.; Edwards, J.D.; Wamishe, Y.; et al. The rice blast resistance gene *Ptr* encodes an atypical protein required for broad-spectrum disease resistance. *Nat. Commun.* **2018**, *9*, 2039. [CrossRef] [PubMed]

15. Zhu, D.; Kang, H.; Li, Z.; Liu, M.; Zhu, X.; Wang, Y.; Wang, D.; Wang, Z.; Liu, W.; Wang, G.L. A genome-wide association study of field resistance to *Magnaporthe oryzae* in rice. *Rice* **2016**, *9*, 44. [CrossRef]

16. Fukuoka, S.; Saka, N.; Koga, H.; Ono, K.; Shimizu, T.; Ebana, K.; Hayashi, N.; Takahashi, A.; Hirochika, H.; Okuno, K.; et al. Loss of function of a proline-containing protein confers durable disease resistance in rice. *Science* **2009**, *325*, 998–1001. [CrossRef]

17. Hayashi, N.; Inoue, H.; Kato, T.; Funao, T.; Shirota, M.; Shimizu, T.; Kanamori, H.; Yamane, H.; Hayano-Saito, Y.; Matsumoto, T.; et al. Durable panicle blast-resistance gene *Pb1* encodes an atypical CC-NBS-LRR protein and was generated by acquiring a promoter through local genome duplication. *Plant J.* **2010**, *64*, 498–510. [CrossRef]

18. Fukuoka, S.; Yamamoto, S.I.; Mizobuchi, R.; Yamanouchi, U.; Ono, K.; Kitazawa, N.; Yasuda, N.; Fujita, Y.; Nguyen, T.T.T.; Koizumi, S.; et al. Multiple functional polymorphisms in a single disease resistance gene in rice enhance durable resistance to blast. *Sci. Rep.* **2014**, *4*, 4550. [CrossRef]

19. Xu, X.; Hayashi, N.; Wang, C.T.; Fukuoka, S.; Kawasaki, S.; Takatsuji, H.; Jiang, C.J. Rice blast resistance gene *Pikahei-1(t)*, a member of a resistance gene cluster on chromosome 4, encodes a nucleotide-binding site and leucine-rich repeat protein. *Mol. Breed.* **2014**, *34*, 691–700. [CrossRef]

20. Li, W.; Zhu, Z.; Chern, M.; Yin, J.; Yang, C.; Ran, L.; Cheng, M.; He, M.; Wang, K.; Wang, J.; et al. A natural allele of a transcription factor in rice confers broad-spectrum blast resistance. *Cell* **2017**, *170*, 114–126. [CrossRef]

21. Abe, A.; Kosugi, S.; Yoshida, K.; Natsume, S.; Takagi, H.; Kanzaki, H.; Matsumura, H.; Yoshida, K.; Mitsuoka, C.; Tamiru, M.; et al. Genome sequencing reveals agronomically important loci in rice using MutMap. *Nat. Biotechnol.* **2012**, *30*, 174–178. [CrossRef]

22. Takagi, H.; Abe, A.; Yoshida, K.; Kosugi, S.; Natsume, S.; Mitsuoka, C.; Uemura, A.; Utsushi, H.; Tamiru, M.; Takuno, S.; et al. QTL-seq: Rapid mapping of quantitative trait loci in rice by whole genome resequencing of DNA from two bulked populations. *Plant J.* **2013**, *74*, 174–183. [CrossRef]

23. Zhang, H.; Wang, X.; Pan, Q.; Li, P.; Liu, Y.; Lu, X.; Zhong, W.; Li, M.; Han, L.; Li, J.; et al. QTG-Seq accelerates QTL fine mapping through QTL partitioning and whole-genome sequencing of bulked segregant samples. *Mol. Plant* **2019**, *12*, 426–437. [CrossRef]

24. Zheng, W.; Wang, Y.; Wang, L.; Ma, Z.; Zhao, J.; Wang, P.; Zhang, L.; Liu, Z.; Lu, X. Genetic mapping and molecular marker development for *Pi65 (t)*, a novel broad-spectrum resistance gene to rice blast using next-generation sequencing. *Theor. Appl. Genet.* **2016**, *129*, 1035–1044. [CrossRef]

25. Wu, S.; Qiu, J.; Gao, Q. QTL-BSA: A bulked segregant analysis and visualization pipeline for QTL-seq. *Interdiscip. Sci. Comput. Life Sci.* **2019**. [CrossRef]

26. Okuyama, Y.; Kanzaki, H.; Abe, A.; Yoshida, K.; Tamiru, M.; Saitoh, H.; Fujibe, T.; Matsumura, H.; Shenton, M.; Galam, D.C.; et al. A multifaceted genomics approach allows the isolation of the rice *Pia*-blast resistance gene consisting of two adjacent NBS-LRR protein genes. *Plant J.* **2011**, *66*, 467–479. [CrossRef]

27. Zhu, L.; Chen, Y.; Xu, Y.; Xu, J.; Cai, H.; Ling, Z. Construction of a molecular map of rice and gene mapping using a double-haploid population of a cross between *indica* and *japonica* varieties. *Rice Genet. Newsl.* **1993**, *10*, 132–134.

28. Sallaud, C.; Lorieux, M.; Roumen, E.; Tharreau, D.; Berruyer, R.; Svestasrani, P.; Garsmeur, O.; Ghesquiere, A.; Notteghem, J.L. Identification of five new blast resistance genes in the highly blast-resistant rice variety IR64 using a QTL mapping strategy. *Appl. Genet.* **2003**, *106*, 794–803. [CrossRef]

29. Berruyer, R.; Adreit, H.; Milazzo, J.; Gaillard, S.; Berger, A.; Dioh, W.; Lebrun, M.H.; Tharreau, D. Identification and fine mapping of *Pi33*, the rice resistance gene corresponding to the *Magnaporthe grisea* avirulence gene *ACE1*. *Appl. Genet.* **2003**, *107*, 1139–1147. [CrossRef]

30. Liu, B.; Zhang, S.; Zhu, X.; Yang, Q.; Wu, S.; Mei, M.; Mauleon, R.; Leach, J.; Mew, T.; Leung, H. Candidate defense genes as predictors of quantitative blast resistance in rice. *Mol. Plant-Microbe Interact* **2004**, *17*, 1146–1152. [CrossRef]

31. Gao, J.; Dai, G.; Zhou, W.; Liang, H.; Huang, J.; Qing, D.; Chen, W.; Wu, H.; Yang, X.; Li, D.; et al. Mapping and identifying a candidate gene *Plr4*, a recessive gene regulating purple leaf in rice, by using bulked segregant and transcriptome analysis with next-generation sequencing. *Int. J. Mol. Sci.* **2019**, *20*, 4335. [CrossRef]

32. Pawson, T. Protein modules and signalling networks. *Nature* **1995**, *373*, 573–580. [CrossRef]

33. Ball, L.J.; Kuhne, R.; Schneider-Mergener, J.; Oschkinat, H. Recognition of proline-rich motifs by protein-protein-interaction domains. *Angew. Chem. Int. Ed. Engl.* **2005**, *44*, 2852–2869. [CrossRef]

34. Fukuoka, S.; Okuno, K. Strategies for breeding durable resistance to rice blast using *pi21*. *Crop Breed. Genet. Genom.* **2019**, *1*, e190013.

35. Fukuoka, S.; Saka, N.; Mizukami, Y.; Koga, H.; Yamanouchi, U.; Yoshioka, Y.; Hayashi, N.; Ebana, K.; Mizobuchi, R.; Yano, M. Gene pyramiding enhances durable blast disease resistance in rice. *Sci. Rep.* **2015**, *5*, 7773. [CrossRef]

36. Tian, D.; Chen, Z.; Chen, Z.; Zhou, Y.; Wang, Z.; Wang, F.; Chen, S. Allele-specific marker-based assessment revealed that the rice blast resistance genes *Pi2* and *Pi9* have not been widely deployed in Chinese *indica* rice cultivars. *Rice* **2016**, *9*, 19. [CrossRef]

37. Mackill, D.J.; Bonman, J.M. Inheritance of blast resistance in near-isogenic lines of rice. *Phytopathology* **1992**, *82*, 746–749. [CrossRef]

38. Kawahara, Y.; de la Bastide, M.; Hamilton, J.P.; Kanamori, H.; McCombie, W.R.; Ouyang, S.; Schwartz, D.C.; Tanaka, T.; Wu, J.; Zhou, S.; et al. Improvement of the *Oryza sativa* Nipponbare reference genome using next generation sequence and optical map data. *Rice* **2013**, *6*, 4. [CrossRef]

39. Li, H.; Durbin, R. Fast and accurate short read alignment with Burrows—Wheeler transform. *Bioinformatics* **2009**, *25*, 1754–1760. [CrossRef]

40. Li, H.; Handsaker, B.; Wysoker, A.; Fennell, T.; Ruan, J.; Homer, N.; Marth, G.; Abecasis, G.; Durbin, R. The sequence alignment/map format and SAMtools. *Bioinformatics* **2009**, *25*, 2078–2079. [CrossRef]

41. Garrison, E.; Marth, G. Haplotype-based variant detection from short-read sequencing. *arXiv* **2012**, arXiv:1207.3907.

42. Cingolani, P.; Platts, A.; Wang, L.; Coon, M.; Nguyen, T.; Wang, L.; Land, S.J.; Lu, X.; Ruden, D.M. A program for annotating and predicting the effects of single nucleotide polymorphisms, SnpEff: SNPs in the genome of *Drosophila melanogaster* strain w^{1118}; iso-2; iso-3. *Fly* **2012**, *6*, 80–92. [CrossRef] [PubMed]

43. Huang, L.; Tang, W.; Bu, S.; Wu, W. BRM: A statistical method for QTL mapping based on bulked segregant analysis by deep sequencing. *Bioinformatics* **2019**. [CrossRef]

44. Ehrenreich, I.M.; Torabi, N.; Jia, Y.; Kent, J.; Martis, S.; Shapiro, J.A.; Gresham, D.; Caudy, A.A.; Kruglyak, L. Dissection of genetically complex traits with extremely large pools of yeast segregants. *Nature* **2010**, *464*, 1039–1042. [CrossRef]

45. Tang, W.; Huang, L.; Bu, S.; Zhang, X.; Wu, W. Estimation of QTL heritability based on pooled sequencing data. *Bioinformatics* **2018**, *34*, 978–984. [CrossRef]

46. Thompson, J.D.; Higgins, D.G.; Gibson, T.J. CLUSTAL W: Improving the sensitivity of progressive multiple sequence alignment through sequence weighting, position-specific gap penalties and weight matrix choice. *Nucleic Acids Res.* **1994**, *22*, 4673–4680. [CrossRef]

Int. J. Mol. Sci. **2020**, *21*, 2162

International Journal of
Molecular Sciences

Article

Pyramiding Bacterial Blight Resistance Genes in Tainung82 for Broad-Spectrum Resistance Using Marker-Assisted Selection

Yu-Chia Hsu [1], Chih-Hao Chiu [2], Ruishen Yap [1], Yu-Chien Tseng [1] and Yong-Pei Wu [2,*]

[1] Department of Agronomy, National Chiayi University, Chiayi 60004, Taiwan; hsuychia@mail.ncyu.edu.tw (Y.-C.H.); ruishen0506@gmail.com (R.Y.); yct@mail.ncyu.edu.tw (Y.-C.T.)

[2] Department of Agronomy, Chiayi Agricultural Experiment Station, Taiwan Agricultural Research Institute, Chiayi 60014, Taiwan; robinchiu3310@gmail.com

* Correspondence: wuypei@dns.caes.gov.tw; Tel.: +886-5-2753156

Received: 3 December 2019; Accepted: 12 February 2020; Published: 14 February 2020

Abstract: Tainung82 (TNG82) is one of the most popular *japonica* varieties in Taiwan due to its relatively high yield and grain quality, however, TNG82 is susceptible to bacterial blight (BB) disease. The most economical and eco-friendly way to control BB disease in *japonica* is through the utilization of varieties that are resistant to the disease. In order to improve TNG82's resistance to BB disease, five bacterial blight resistance genes (*Xa4*, *xa5*, *Xa7*, *xa13* and *Xa21*) were derived from a donor parent, IRBB66 and transferred into TNG82 via marker-assisted backcrossing breeding. Five BB-resistant gene-linked markers were integrated into the backcross breeding program in order to identify individuals possessing the five identified BB-resistant genes (*Xa4*, *xa5*, *Xa7*, *xa13* and *Xa21*). The polymorphic markers between the donor and recurrent parent were used for background selection. Plants having maximum contribution from the recurrent parent genome were selected in each generation and crossed with the recipient parent. Selected BC_3F_1 plants were selfed in order to generate homozygous BC_3F_2 plants. Nine pyramided plants, possessing all five BB-resistant genes, were obtained. These individuals displayed a high level of resistance against the BB strain, XF89-b. Different BB gene pyramiding lines were also inoculated against the BB pathogen, resulting in more than three gene pyramided lines that exhibited high levels of resistance. The five identified BB gene pyramided lines exhibited yield levels and other desirable agronomic traits, including grain quality and palatability, consistent with TNG82. Bacterial blight-resistant lines possessing the five identified BB genes exhibited not only higher levels of resistance to the disease, but also greater yield levels and grain quality. Pyramiding multiple genes with potential characteristics into a single genotype through marker-assisted selection can improve the efficiency of generating new crop varieties exhibiting disease resistance, as well as other desirable traits.

Keywords: rice; pyramiding; bacterial blight; marker-assisted selection; foreground selection; background selection

---

1. Introduction

As a carbohydrate-rich staple of more than half the world's diet, rice (*Oryza sativa* L.) is one of the most important food crops on the planet. The Food and Agriculture Organization of the United Nations (FAO) estimates that by 2050, overall global agricultural production may need to be increased by up to 70% to meet the dietary requirements of the world's projected population of nine billion [1]. In order to satisfy the demand corresponding to the FAO's projected population in 2050, global rice production would have to increase by nearly 42% over present-day levels [2]. Bacterial blight (BB) caused by *Xanthomonas oryzae* pv. *oryzae* (*Xoo*) is a disease that poses one of the greatest threats to rice production worldwide. In Asia, BB has proven to be capable of reducing crop yields by as much as 50% [3] to 80% [4]. The disease, being systemic, affects the photosynthetic areas of plants, which results in a drastically lower yield. Although BB can be managed through the use of fungicides, enhancing the genetic resistance in rice is the most effective and ecological method of overcoming the threat posed by the disease.

To date, 42 BB resistance genes have been identified from diverse sources, of which *Xa4*, *xa5*, *Xa7*, *xa13* and *Xa21* are most frequently utilized in BB resistance breeding programs [5–7]. Although the BB resistance genes *xa5* and *xa13* are recessive in nature, abundant molecular marker resources allow for molecular marker-assisted breeding [8–11]. The *xa5* gene encodes a mutated gamma subunit of basal transcription factor IIA 5 (TFIIAγ5), and along with the dominant resistance gene *Xa7*, has shown strong resistance to a virulent BB strain, Z-173, in China [11,12]. Another broad-spectrum recessive gene, *xa13*, was correlated with a plasma membrane protein conferring recessive resistance to PXO99 [13]. *Xa21*, which encodes a leucine-rich repeat (LRR) receptor kinase-type gene, was identified from *O. longistaminata*; it is one of the most effective genes utilized in breeding programs designed to enhance the BB resistance of rice cultivars [8,14].

Conventional backcross breeding embedded with marker-assisted selection (MAS) has been successfully employed in developing crop varieties exhibiting agronomically important traits. The utility of MAS in pyramiding several resistance genes to develop a variety possessing broad-spectrum durable resistance has been successfully demonstrated against numerous pathotypes [6,15,16]; *Jalmagna*, a high-yield, deep-water rice variety, was improved for BB resistance by pyramiding three resistance genes, *xa5* + *xa13* + *Xa21* [6]; a Korean elite *japonica* variety, *Mangeumbyeo*, improved with the introgression of the *Xa4* + *xa5* + *Xa21* genes, which were shown to possess a wide range of resistance to BB [16]. Recently, *xa5*, *xa13* and *Xa21* genes were introgressed into the hybrid rice maintainer lines CO2B, BO23B and CO24B through MAS, which can form the basis to develop new, widely adaptable heterotic hybrids possessing resistance against the destructive diseases to which rice is vulnerable [17]. In addition, there have been several examples of MAS being utilized to successfully incorporate different genes which provide higher resistance to various biotic and abiotic stresses (for example, the pyramiding of QTLs of submergence tolerance (*Sub1A*), leaf/neck blast (*qBL1* and *qBL11*), brown planthopper (*Bph3*) and BB (*xa5* and *Xa21*) in high-yielding and aromatic rice variety 'Pink3' [18]).

According to the annual report of the Council of Agriculture, of the 271,000 hectares of rice paddy fields in Taiwan, approximately 7% are affected by BB per year. Most Taiwanese *japonica* rice cultivars lack BB resistance genes [10], resulting in significant yield loss in fields severely infected by the disease. Pyramiding multiple *R* genes by MAS provides a rapid and precise way to develop a variety with wide-spectrum and durable resistance [19]. A set of 17 near-isogenic lines (NILs) in IR24 background, having single or two to four pyramided *Xa* genes, were included in the panel to serve as controls of known disease reactions [20]. IRBB66, carrying *Xa4*, *xa5*, *Xa7*, *xa13* and *Xa21*, in an *indica* rice IR24 genetic background, conferred strong resistance to races of BB. In the present study, five BB resistance genes were introgressed from IRBB66 into an elite *japonica* variety, 'Tainung82' (TNG82), using marker-assisted backcrossing (MAB) and marker-assisted background analysis of selected backcross progenies using SSR markers. The aims of this study were to (i) develop five gene pyramiding lines using MAB, (ii) evaluate the effects of BB-resistant lines carrying different R genes after inoculation with BB strain, (iii) select individuals possessing agronomic traits and grain quality performance from the resulting BB-resistant lines. The development of BB-resistant lines with more than three genes pyramided has a promising future in molecular breeding of durable BB-resistant rice cultivars.

2. Results

2.1. Development of BC₃F₄ Pyramided Lines Using Marker-Assisted Breeding

Tainung82 is one of the most widely cultured elite *japonica* varieties in Taiwan, but it exhibits a high susceptibility to bacterial blight disease. In order to develop a BB-resistant *japonica* cultivar, TNG82 was used as the recurrent parent to backcross with IRBB66 for three generations, and then self-crossed to produce a BC_3F_4 population. The polymorphism was detected between donor parent IRBB66 and recurrent parent TNG82 with the markers Xa4F/4R, RM604F/604R, Xa7F/7-1R/7-2R, Xa13F/13R and Xa21F/21R for *Xa4*, *xa5*, *Xa7*, *xa13* and *Xa21*, respectively. In addition, the parents were screened with 216 rice microsatellite markers, of which 143 were polymorphic and 117 were used for background selection (Figure S1). The breeding scheme using molecular markers for the selection of the five BB-resistant genes is shown in Figure 1. During the breeding procedure, functional marker selection was practiced from the F_1 generation until the BC_3F_2 generation. The plants possessing all five resistance genes were selected in each stage, of which only two progenies were advanced to the next generation. A total of two plants having all five BB resistance genes (*Xa4*, *xa5*, *Xa7*, *xa13* and *Xa21*) were screened from 960 F_2 plants and confirmed by lined molecular markers [10]. The two F_2 plants were backcrossed to TNG82. A total of 53 of 147 BC_1F_1 plants containing different BB resistance genes were selected by MAS. The percentages of recurrent parent genome (%RPG) of BC_1F_1 ranged from 60% to 85%, with an average of 73.8% (Figure 2). Ten BC_1F_1 plants containing both the five BB resistance genes, as well as a high %RPG (average of 81.7%), were used for further backcrossing with TNG82.

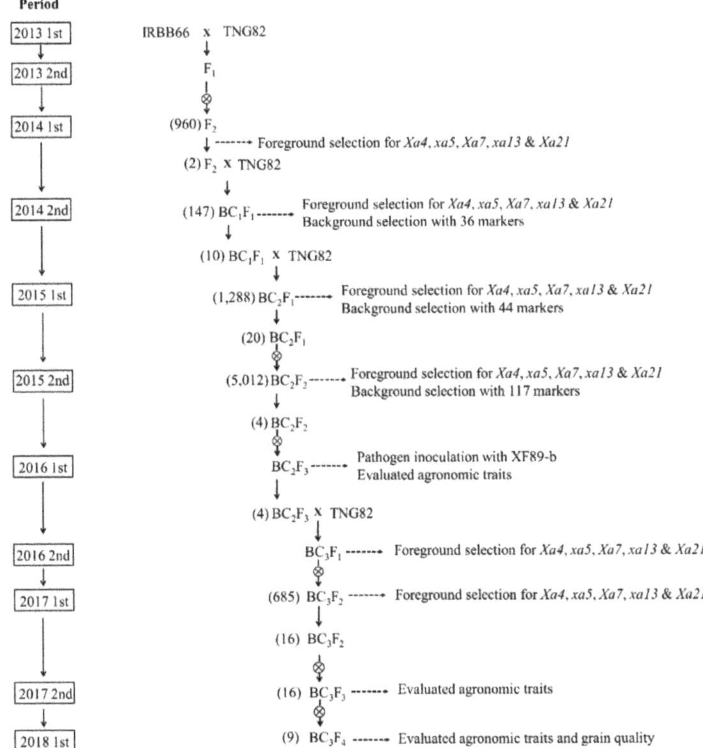

Figure 1. Schematic diagram for pyramiding bacterial blight resistance genes into Taiwanese *japonica* rice cultivar, TNG82, using marker-assisted selection and number of plants selected in every generation.

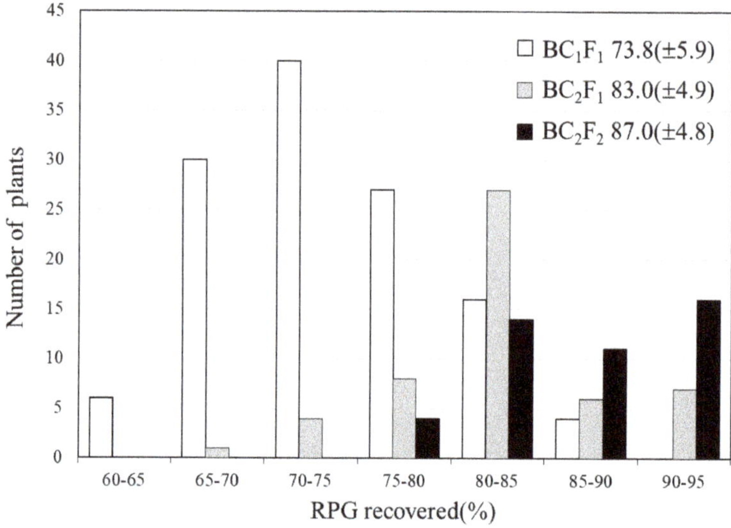

Figure 2. The frequency distribution of recurrent parent genome (RPG) recovered rate using marker-assisted backcrossing in BC$_1$F$_1$, BC$_2$F$_1$ and BC$_2$F$_2$ populations derived from the backcross of IRBB66/TNG82. The numbers inside the right side of frame indicate the mean values (SD) of RPG recovered.

A total of 50 of 1228 BC$_2$F$_1$ plants containing different BB resistance genes possessed the recurrent genome content of TNG82, ranging from 72% to 94%, with an average of 83% (Figure 2). The 20 selected BC$_2$F$_1$ plants, heterozygous for all five BB resistance genes and possessing a high %RPG (average of 87.3%), were selfed to obtain the BC$_2$F$_2$ population. The plants homologous for all five target genes were segregated with a Mendelian pattern (homozygous preference genotype = 1/4^n). The four BC$_2$F$_2$ plants carrying five positive homozygous alleles of the donor genes, including *Xa4*, *xa5*, *Xa7*, *xa13* and *Xa21*, were screened from 5012 BC$_2$F$_2$ plants. Four BC$_2$F$_2$ plants showed recurrent genome content of TNG82 with %RPG of 92.05% (29), 84.3% (18), 83.1% (5) and 79.4% (43), with an average of 84.71% (Table S1). In the BC$_2$F$_3$ generation, 17 plants containing different BB resistance genes were used to confirm resistance reaction by inoculation with *Xoo* isolate XF89-b and evaluated for agronomic performance. Four BC$_2$F$_3$ plants with the five BB resistance genes were backcrossed to TNG82. In the BC$_3$F$_2$ generation, 16 of 685 plants containing the five BB resistance genes were identified and grown as BC$_3$F$_3$. These 16 five-gene-pyramided genotypes were selfed and evaluated for agronomic performance. The nine BC$_3$F$_4$ lines containing five BB resistance genes, *Xa4*, *xa5*, *Xa7*, *xa13* and *Xa21* (Figure 3) were selected and evaluated for agronomic performance in the field, as well as analyzed for grain quality.

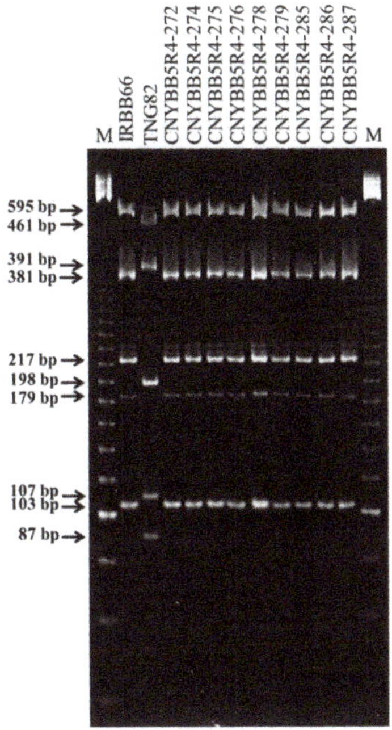

Figure 3. Multiplex PCR amplification of five bacterial blight resistance genes, *Xa4*, *xa5*, *Xa7*, *xa13* and *Xa21*. The five expected band sizes of 217, 106, 179, 381 and 595 bp, correlated with *Xa4*, *xa5*, *Xa7*, *xa13* and *Xa21*, respectively, were amplified in IRBB66 and nine five-gene-pyramided lines using multiplex PCR. P1:IRBB66, P2:TNG82. DNA products were separated by 6% polyacrylamide gel in 0.5 × TBE at 100 v for 60 min. M: DNA ladder marker.

2.2. Development of BC₃F₄ Pyramided Lines Using Marker-Assisted Breeding

The BC_2F_3 pyramided rice genotypes were evaluated for their resistance to BB in the field conditions using the Taiwanese *Xanthomonase oryzae* strain isolate, XF89-b. The resistance donor IRBB66, containing five BB resistance genes, showed shorter lesion lengths (mean lesion length of 0.43 cm), while the susceptible checks, TN1, TCS10, IR24 and TNG82, exhibited a range of longer lesion lengths, between 6.75 and 12.56 cm (Table 1, Figure 4). The genotypes having either BB resistance genes alone or more than two genes pyramided were shown to be moderately resistant, resistant, and highly resistant to the BB disease (Figure 5). In addition, the five-gene-pyramided BC_2F_3 genotypes exhibited a range of shorter lesion lengths, between 0.37 and 0.46 cm (Table 1). The five-gene-pyramided lines displayed higher levels of disease resistance and a broader resistance spectrum compared to both the parental rice variety, TNG82 and the genotypes possessing a single gene.

Table 1. The results of TNG82/IRBB66 BC_2F_3 lines after incubating the *Xoo* strains XF89-b in the field at the first crop season in 2016.

No.	Lines	Genotypes	Lesion Length [†] (cm)	Resistance Scale [¥]
1	TN1		12.56 ± 2.98 [a]	S
2	TCS10		9.85 ± 2.07 [b]	MS
3	IR24		11.75 ± 1.80 [a]	S
4	TNG82		6.75 ± 2.54 [c,d]	MS
5	IRBB66	*Xa4 + xa5 + Xa7 + xa13 + Xa21*	0.43 ± 0.70 [i]	HR
6	CNYBB0R01	Without resistance gene	7.30 ± 1.50 [d]	MS
7	CNYBB0R02	Without resistance gene	5.91 ± 2.04 [c]	MR
8	CNYBB1R01	*Xa4*	2.67 ± 1.00 [e,f]	R

Table 1. *Cont.*

No.	Lines	Genotypes	Lesion Length [†] (cm)	Resistance Scale [¥]
9	CNYBB1R02	*xa5*	2.27 ± 0.73 [e,f,g]	R
10	CNYBB1R03	*Xa7*	2.77 ± 1.04 [e,f]	R
11	CNYBB1R04	*xa13*	5.55 ± 2.28 [d]	MR
12	CNYBB1R05	*Xa21*	1.48 ± 1.48 [f,g,h,i]	R
13	CNYBB2R03	*xa13 + Xa21*	0.77 ± 0.45 [h,i]	HR
14	CNYBB2R04	*Xa4 + Xa21*	0.68 ± 0.24 [h,i]	HR
15	CNYBB2R05	*Xa4 + xa5*	1.24 ± 0.98 [g,h,i]	R
16	CNYBB2R06	*xa5 + xa13*	1.82 ± 0.54 [f,g,h]	R
17	CNYBB2R01	*xa5 + Xa7*	0.75 ± 0.33 [h,i]	HR
18	CNYBB2R07	*Xa7 + xa13*	3.25 ± 0.73 [e]	MR
19	CNYBB2R02	*Xa7 + Xa21*	0.76 ± 0.19 [h,i]	HR
20	CNYBB3R03	*Xa4 + xa13 + Xa21*	0.45 ± 0.17 [h,i]	HR
21	CNYBB3R04	*Xa4 + xa5 + xa13*	1.25 ± 0.74 [g,h,i]	R
22	CNYBB3R05	*Xa4 + xa5 + Xa21*	0.46 ± 0.17 [h,i]	HR
23	CNYBB3R01	*Xa4 + Xa7 + Xa21*	0.56 ± 0.16 [h,i]	HR
24	CNYBB3R06	*xa5 + xa13 + Xa21*	0.44 ± 0.08 [h,i]	HR
25	CNYBB3R07	*xa5 + Xa7 + xa13*	0.71 ± 0.45 [h,i]	HR
26	CNYBB3R02	*xa5 + Xa7 + Xa21*	0.44 ± 0.12 [h,i]	HR
27	CNYBB4R03	*Xa4 + xa5 + xa13 + Xa21*	0.43 ± 0.12 [i]	HR
28	CNYBB4R01	*Xa4 + xa5 + Xa7 + Xa21*	0.48 ± 0.17 [h,i]	HR
29	CNYBB4R02	*Xa4 + Xa7 + xa13 + Xa21*	0.68 ± 0.50 [h,i]	HR
30	CNYBB5R01	*Xa4 + xa5 + Xa7 + xa13 + Xa21*	0.37 ± 0.13 [i]	HR
31	CNYBB5R02	*Xa4 + xa5 + Xa7 + xa13 + Xa21*	0.38 ± 0.12 [i]	HR
32	CNYBB5R03	*Xa4 + xa5 + Xa7 + xa13 + Xa21*	0.46 ± 0.14 [h,i]	HR
33	CNYBB5R04	*Xa4 + xa5 + Xa7 + xa13 + Xa21*	0.38 ± 0.12 [i]	HR

[†] Mean ± standard error. [¥] HR = highly resistant (lesion length < 1 cm); R = resistant (1 cm < lesion length < 3 cm); MR = moderately resistant (3 cm < lesion length < 6 cm); MS = moderately susceptible (6 cm < lesion length < 10 cm); S = susceptible (10 cm < lesion length). Means with none or the same letter of a row are not significantly different at 5% level by least significant difference (LSD) test.

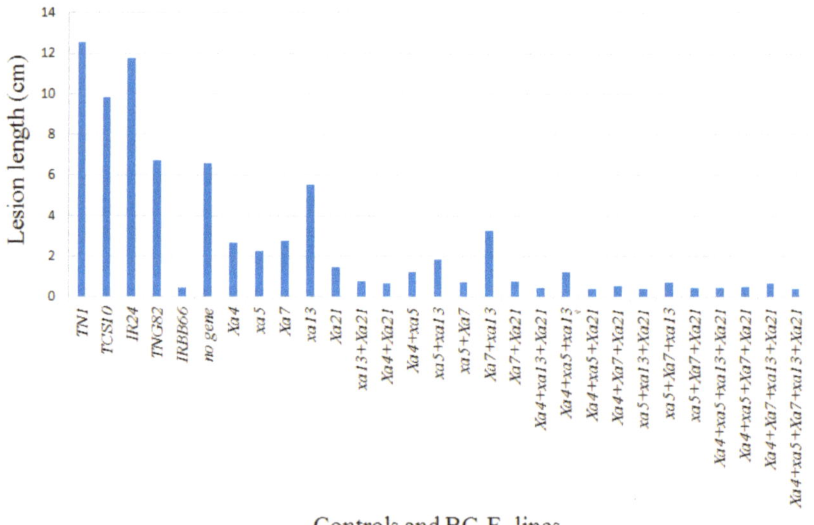

Figure 4. The leaf lesion length of TNG82/IRBB66 BC_2F_3 genotypes after 21 days inoculum of bacterial blight pathogen XF89-b in the field at the first crop season in 2016. Susceptible cultivar: TCS10, IR24, and TN1; Parental: TNG82 and IRBB66.

Figure 5. The leaf lesion photo of TNG82/IRBB66 BC_2F_3 genotypes after 21 days inoculum of bacterial blight pathogen XF89-b in the field at the first crop season in 2016. Susceptible cultivar check: TCS10, IR24, and TN1; Parents: TNG82 and IRBB66.

2.3. Development of BC_3F_4 Pyramided Lines Using Marker-Assisted Breeding

Nine five-gene-pyramided lines at the BC_3F_4 generation, along with the recurrent and donor parents, were evaluated in the first crop season of 2018 at Taiwan Agricultural Research Institute (TARI), Taiwan. Significant variances were observed between the pyramided lines and parental rice genotypes for plant height, days to 50% flowering, panicle length, panicles/plant, panicle weight, number of grains/panicle, 1000-seed weight, and single plant yield (Table 2). The recurrent parent, TNG82, recorded mean grain yield of 36.8 g/plant, while the donor parent, IRBB66, was 30.1 g/plant. Six of the nine five-gene-pyramided lines, CNYBB5R4-275, -276, -278, -279, -285 and -287, produced significantly higher grain yields per plant than the recurrent parent, which ranged from 37.1 to 44.5 g/plant, and displayed a similar phenotype to the donor parent TNG82 (Figure 6).

Table 2. Agro-morphologic traits of parental and five-gene-pyramided BC_3F_4 genotypes.

Pyramided Lines	Plant Height (cm) ($n = 20$)	Days to 50% Flowering ($n = 20$)	Panicle Length (cm) ($n = 20$)	Panicles/Plant ($n = 20$)	Panicle Weight (g) ($n = 20$)	No. of Grains/Panicle ($n = 20$)	1000-Seed Weight (g) ($n = 20$)	Single Plant Yield (g) ($n = 20$)
TNG82	105.4 ± 1.1	90	20.3 ± 0.2	14 ± 0.7	3.3 ± 0.1	102 ± 2.5	30.8 ± 0.1	36.8 ± 0.1
IRBB66	90.2 ± 1.4	95	22.6 ± 0.1	17 ± 0.5	3.7 ± 0.1	108 ± 1.0	24.2 ± 0.1	30.1 ± 0.7
CNYBB5R4-272	120.4 ± 1.3	92	22.8 ± 1.8	16 ± 0.8	3.6 ± 0.4	122 ± 16.7	26.0 ± 0.1	36.3 ± 7.3
CNYBB5R4-274	115.9 ± 2.7	91	21.4 ± 1.6	15 ± 0.5	2.8 ± 0.1	95 ± 8.2	28.4 ± 0.4	32.6 ± 4.4
CNYBB5R4-275	116.3 ± 3.7	91	20.1 ± 0.6	16 ± 2.1	3.2 ± 0.3	113 ± 17.8	25.2 ± 0.4	41.2 ± 3.3
CNYBB5R4-276	116.3 ± 3.2	91	20.4 ± 0.7	16 ± 0.5	2.8 ± 0.2	100 ± 9.9	26.4 ± 0.4	43.7 ± 0.9
CNYBB5R4-278	115.8 ± 2.4	91	21.7 ± 0.2	12 ± 0.8	3.9 ± 0.3	135 ± 10.3	26.2 ± 0.1	44.5 ± 1.3
CNYBB5R4-279	112.7 ± 4.2	91	18.3 ± 0.1	15 ± 1.7	3.1 ± 0.1	120 ± 4.7	25.2 ± 0.2	39.2 ± 3.0
CNYBB5R4-285	121.8 ± 3.0	92	21.6 ± 0.1	14 ± 0.8	3.1 ± 0.1	106 ± 0.8	28.2 ± 0.2	38.0 ± 0.3
CNYBB5R4-286	113.7 ± 6.8	91	20.2 ± 1.0	15 ± 1.0	2.9 ± 0.2	100 ± 12.1	27.6 ± 0.1	35.9 ± 2.6
CNYBB5R4-287	115.3 ± 6.6	91	21.7 ± 0.6	14 ± 0.2	2.8 ± 0.4	103 ± 16.5	26.6 ± 1.6	37.1 ± 8.1
LSD ($p = 0.05$)	9.2	0.6	2.6	3.1	0.7	24.0	1.7	12.1

LSD, least significant difference at 5% probability level.

Figure 6. Phenotype of the five-gene-pyramided BC_3F_4 genotypes compared with recurrent parental variety TNG82.

A significant difference was noted between the parental rice varieties and pyramided genotypes in grain quality traits (Table 3). The palatability among pyramided BC_3F_4 genotypes varied between 69.8 (CNYBB5R4-275) and 74.5 (CNYBB5R4-276). The protein content among pyramided BC_3F_4 genotypes varied between 6.4 (CNYBB5R4-276 and CNYBB5R4-286) and 7.4 (CNYBB5R4-272). The brown rice ratio for the five-gene-pyramided genotypes varied from 72.8% to 79.3%. The four genotypes, CNYBB5R4-272, -275, -276 and -278, were found to have higher head rice ratios, however, the amount of total milled rice was not significantly different from the recurrent parent, TNG82. The evaluation of agronomic traits in BC_3F_3 and BC_3F_4 provided us with an important selection criteria, which can select candidate lines with stable agronomic performances and resistance to disease.

Table 3. Grain quality of parental and five-gene-pyramided BC_3F_4 genotypes.

Pyramided Lines	Palatability ($n = 20$)	Protein ($n = 20$)	Amylose ($n = 20$)	Brown Rice (%) ($n = 20$)	Head Rice (%) ($n = 20$)	Total Milled Rice (%) ($n = 20$)
TNG82	76.3 ± 3.2	6.0 ± 0.5	16.0 ± 0.1	77.6 ± 0.2	55.0 ± 0.1	61.5 ± 0.7
IRBB66	60.3 ± 1.1	8.7 ± 0.3	18.1 ± 0.1	72.3 ± 0.1	42.3 ± 0.1	45.7 ± 0.4
CNYBB5R4-272	69.0 ± 2.1	7.4 ± 0.4	14.5 ± 0.9	74.1 ± 0.5	50.5 ± 0.7	57.8 ± 0.8
CNYBB5R4-274	70.0 ± 3.5	7.3 ± 0.7	15.3 ± 0.2	72.8 ± 0.6	31.2 ± 1.8	48.6 ± 0.1
CNYBB5R4-275	69.8 ± 1.1	7.2 ± 0.2	16.0 ± 0.4	75.1 ± 2.3	53.8 ± 1.5	58.7 ± 1.3
CNYBB5R4-276	74.5 ± 2.8	6.4 ± 0.5	15.4 ± 0.1	77.2 ± 0.9	50.1 ± 2.4	59.0 ± 1.0
CNYBB5R4-278	72.3 ± 1.8	6.7 ± 0.3	15.6 ± 0.2	77.5 ± 0.7	55.8 ± 0.3	62.1 ± 0.1
CNYBB5R4-279	72.3 ± 0.4	6.8 ± 0.1	15.3 ± 0.2	77.2 ± 1.8	49.3 ± 2.6	58.1 ± 3.6
CNYBB5R4-285	72.0 ± 1.4	7.0 ± 0.7	15.2 ± 0.1	77.0 ± 1.4	37.0 ± 1.9	54.4 ± 2.2
CNYBB5R4-286	74.3 ± 1.4	6.4 ± 0.3	15.3 ± 0.2	79.3 ± 0.5	48.4 ± 2.4	61.2 ± 0.1
CNYBB5R4-287	74.3 ± 0.4	6.8 ± 0.1	15.4 ± 0.1	75.5 ± 0.5	42.0 ± 9.0	54.7 ± 0.6
LSD ($p = 0.05$)	4.6	0.9	1.0	3.5	9.8	4.5

LSD, least significant difference at 5% probability level.

3. Discussion

Conventional backcross breeding is the primary method used to develop highly BB-resistant rice cultivars, but it cannot accurately transfer multiple genes into the cultivar by phenotypic screening and the process requires a significant amount of time [21,22]. Modified backcross pyramid breeding, combined with molecular marker-assisted selection, has already been demonstrated to increase the precision and efficiency of breeding [23–25]. Due to the relatively large amount of work involved with the MAS process, the conventional backcross breeding approach has been widely adopted in breeding programs designed to breed for BB resistance [10,26–28].

To date, many rice cultivars with broad-spectrum resistance against *Xoo* isolates have been developed; Singh et al. (2001) pyramided three *R* genes, *xa5*, *Xa13* and *Xa21*, in the *indica* rice cultivars PR106 and Jalmagna using MAS to enhance the bacterial blight resistance [6,15]; the four genes *Xa4*, *xa5*, *Xa13* and *Xa21* were introgressed into the recurrent parent lines Jyothi, IR50, Mahsuri, PRR78, KMR3 and Pusa 6B [26,29,30]; different BB-resistant genes, *Xa7*, *Xa21*, *Xa22* and *Xa23*, were also transferred to an elite hybrid rice restorer line, Huahui 1035, in order to improve BB resistance and enhance rice yield [31].

In Taiwan, many *japonica* rice cultivars lack BB resistance, resulting in significant yield loss in severely infected fields. One such variety is Tainung82, which was released in Taiwan for commercial cultivation in 2006. TNG82 is described as a popular *japonica* rice variety, with high-yield potential (6–7 t/ha), excellent grain quality, various culinary applications, and relatively low grain protein content (4.5%–5.5%). As an extremely valuable yet BB-susceptible variety, TNG82 was selected as the focus of this study to increase BB resistance through the introgression of five BB-resistant genes, *Xa4*, *xa5*, *Xa7*, *xa13* and *Xa21*.

The primary purpose of backcross breeding is to transfer one or multiple genes of interest, linked to desirable traits, from donor parents into a base variety for improvement, a process which typically requires six to eight backcrosses to recover the recurrent parent's phenotype [32]. However, in the MAS scheme, three to four generations of backcrossing is generally enough to achieve more than 99% of the recurrent parent genome [33]. The theoretical %RPG of each generation, BC_1, BC_2, BC_3 and BC_4, were 75%, 87.5%, 93.8% and 96.9%, respectively. Furthermore, the %RPG can be improved by using MAS for background selection [16,34]. The 80% and 89% recovery rates following two and three backcrosses were obtained from three-BB-gene-pyramided BC_2 and BC_3 genotypes, via MAS [35]. Balachiranjeevi et al. (2015) transferred the BB gene, *Xa21* and rice blast-resistant gene, *Pi54*, to DRR17A and were able to recover 73.4%, 84.8% and 93.4% RPG in the BC_1, BC_2 and BC_3 generations, respectively.

In this study, the recurrent parent genome recovery rates in BC_1F_1, BC_2F_1 and BC_2F_2 were 73.8%, 83% and 84.7% (Figure 2), respectively. Compared with the theoretical %RPG, a relatively low background recovery rate was obtained, however, the results were consistent with those found in previous studies [36,37]. Marker-assisted backcrossing can accelerate the breeding process and facilitate a speedy recovery for most of the recurrent genome within a few generations [38], however, the population size of each backcross generation, linkage drag, number of background markers used and genetic background between two parents are considered to be factors that reduce the efficiency of MAB and %RPG [32].

Bacterial blight is one of the most destructive diseases affecting rice productivity in Asia. In Taiwan, rice production is frequently affected by BB in the second crop season, resulting in substantial yield loss. In recent years, BB has become a more prevalent threat, due to climate change [39]. XF89-b, a strong and stable Taiwanese epidemic pathogen, has also been used for genetic analysis and the mapping of BB-related resistance genes [40]. In our bioassays, artificial screening of BC_2F_3 progenies revealed that all genotypes containing at least one BB-resistant gene displayed a degree of increased resistance (Table 1, Figure 4). The BC_2F_3 progenies that pyramided more than three BB-resistant genes exhibited a very high level of BB resistance against the XF89-b strain, compared to parental lines (Figure 5). The lesion lengths were measured between 0.37 and 1.25 cm (Table 1). The data indicated that multiple BB-resistant genes pyramided in rice can improve resistance to *Xoo*. The BB pyramiding lines are expected to enhance the adaptability and durability necessary to provide resistance against the dynamic nature of the pathogen. In addition, the results suggest that the gene combinations containing the *Xa21* gene were most resistant, as evidenced by shorter lesions lengths, followed by *Xa4* + *Xa21*, *Xa7* + *Xa21* and *xa13* + *Xa21*, while lines with *Xa4* + *xa5*, *xa5* + *xa13* and *Xa7* + *xa13* were less effective. These results are consistent with previous studies, which have shown the presence of *Xa21* to be correlated with high levels of persistent resistance against BB disease in rice [6,14,15,17,25]. *Xa21* is the cell surface receptor, kinase, which is able to provide resistance to *Xoo* infections; *Xa21* not only suppresses *Xoo* growth, but also triggers broad perturbation in rice transcriptomes and mediated signaling pathways, preventing *Xoo* infections [14].

The agronomic performance evaluation of BC_3F_4 derived in the genetic background of TNG82 revealed that all pyramided lines for most of the agro-morphological traits were, in general, similar to the recipient parent, TNG82. However, six candidate lines, CNYBB5R4-275, -276, -278, -279, -285 and -287, produced significantly higher grain yields per plant than the recurrent parent, which was further confirmed by the multilocation evaluation. In addition, three candidate lines, CNYBB5R4-276, -278 and -286, were not significantly different in palatability, protein, amylose, brown rice ratio, head rice ratio or total milled rice ratio, indicating that the BC_3F_4 pyramiding lines had grain quality consistent with TNG82. The data showed that there were no yield or grain quality reductions, but rather improvements, due to the pyramiding of the five BB-resistant genes.

4. Materials and Methods

4.1. Plant Materials

The donor parent, IRBB66, contained five resistance genes, *Xa4*, *xa5*, *Xa7*, *xa13* and *Xa21*, which were introgressed from wild species in the background of IR24. IRBB66 was provided as a courtesy by the Genetic Resources Center (GRC) of the International Rice Research Institute (IRRI). The recurrent parent was TNG82, an elite japonica cultivar with low protein content and good grain quality, but susceptible to bacterial blight disease. A cross was made between TNG82 and IRBB66, with F_1 plants backcrossed thrice with TNG82 to obtain BC_3F_1 plants, which were selfed to obtain the BC_3F_4 progeny. Selection based on foreground, background and agronomic traits were practiced from BC_1F_1 to BC_2F_2 as a means of identifying lines similar to the recurrent parent.

4.2. Evaluation of Bacterial Blight Resistance

The parental varieties (IRBB66 and TNG82), susceptible varieties (Taichung Native 1 (TN1), Taichung sen 10 (TCS10)), BC_2F_2 and BC_2F_3 generation genotypes were pyramided with the five BB-resistant genes, with IR24 as control. Different combinations were evaluated for BB resistance under greenhouse and field conditions with the pathogen, *X. oryzae* pv. *oryzae*. Pathogen inoculation was performed at the maximum tillering stage in the field through the modified leaf clipping method, as previously described [41]. A strong Taiwanese epidemic pathogen isolate, XF89-b, was used in this study. The isolate was grown in 523 medium [42] with agitation at room temperature for two days. After adjusting the optical bacterial density to 10^9 CFU/mL with distilled water, the cultures were used to screen the rice plants for BB resistance. Approximately six leaves from one plant were clipped from the top 2–3 cm simultaneously. All inoculation was completed within 1 h following the preparation of bacterial suspensions. Lesion length for BB was scored after inoculation when the lesion of the susceptible variety, TN1, reached approximately 3/4 of overall leaf length (approximately 21–28 days). The resistance reaction was classified as highly resistant (HR), resistant (R), moderately resistant (MR), moderately susceptible (MS), and susceptible (S) when the values of lesion length were recorded as 0–1 cm, 1.1–3 cm, 3.1–6 cm, 6.1–10 cm, and more than 10 cm, respectively [43,44].

4.3. Evaluation of Agronomic Traits

During the second and first crop season of 2017 and 2018, respectively, the 30-day-old seedlings of the BC_3F_3 and BC_3F_4 pyramided lines and both the parents were transplanted into three rows, with 27 plants per row, per entry, at 15 × 25 cm spacing, under a randomized complete-block design, with two replications at the Taiwan Agricultural Research Institute's Chiayi Agricultural Experiment Station Farm. Ten plants from each entry were recorded as one data replication. Single plant yield was recorded for the 16 BC_3F_3 genotypes as a basis for selection. In BC_3F_4, variables for agronomic traits were recorded for nine pyramided lines, including: plant height (cm), days to 50% flowering, panicle length (cm), panicles/plant, panicle weight (g), number of grains/panicle and 1000-seed weight (g), while single plant yield (g) was recorded on a whole-plot basis. In addition, the grain quality, including palatability, protein, amylose, brown rice ratio, head rice ratio and total milled rice ratio, was investigated and analyzed. For palatability analyses, the rice grains were hulled and ground into a fine flour. Approximately 33 g of rice flour was used for the palatability evaluation, which was performed by using a palatability analyzer system (Toyo Taste Meter, Model MA-30), in accordance with the manufacturer's operation manual (TRCM Co., Toyo Rice Polishing Machine Factory, Japan), as previously described [45]. Protein and amylose contents were measured with a near-infrared spectrometer (AN820, Kett Electric Laboratory Co. Ltd., Tokyo, Japan) (Near Infrared Spectrometer, Foss Japan Co. Ltd., Tokyo, Japan). Statistical analysis was performed with independent samples using least significant difference (LSD).

4.4. DNA Isolation and PCR Amplification

A rice genomic DNA extraction, with modification, was adopted for minipreparation [45]. Approximately 0.05 g of fresh leaf tissue from 6- to 8-week-old seedlings was homogenized with 300 μL extraction buffer (100 mM Tis-HCl, pH 9.0; 40 mM EDTA-2Na, pH 8.0; 1.67% SDS) at 30 1/s for 2 min by use of TissueLyser (Qiagen Retsch GmbH, Haan, Germany). A total of 150 μL benzyl chloride was added to the homogenized tissue and vortexed. After incubation in a 50 °C water bath for 15 min, 150 μL of 3 M sodium acetate (pH 5.2) was added. Supernatants were saved after centrifugation at 15,000 rpm for 15 min at 4 °C, and 300 μL of ice-cold isopropanol was added to precipitate DNA. After centrifugation at 15,000 rpm for 10 min, DNA pellets were saved and washed with 70% ethanol, air-dried and dissolved in 50 μL TE buffer.

A 10 μL PCR reaction containing 20 ng genomic DNA, 0.2 μM forward and reverse primers, 5 μL Multiplex PCR Master Mix (QIAGEN, Inc., Valencia, CA), and 1 μL Q-Solution (QIAGEN, Inc., Valencia, CA) was performed by use of a thermocycler (GeneAmp PCR System 9700, PerkinElmer Corp., Norwalk, CT, USA) at 95 °C for 15 min for 1 cycle; 94 °C for 30 s, 57 °C for 2 min, 72 °C for 2 min for 30 cycles; and 60 °C for 30 min for 1 cycle. Following PCR, 2 μL of amplified DNA products was separated by 6% polyacrylamide gel (PAGE) in 0.5 × TBE at 100 v (Dual Triple-Wide Mini-Vertical System, C. B. S. Scientific, CA, USA) for 60 min.

4.5. Marker Analysis

Five gene-specific primers, Xa4F/4R, RM604F/604R, Xa7F/7-1R/7-2R, Xa13F/13R, and Xa21F/21R, tightly linked to the resistance genes *Xa4*, *xa5*, *Xa7*, *xa13* and *Xa21*, respectively, were used to confirm the presence of the R genes in each generation. All markers in this study were published in the previous report [10]. In addition, a total of 36 and 44 markers of known chromosomal positions were used for genotyping in BC_1F_1 and BC_2F_1, respectively. In BC_2F_2, 117 markers, including 57 SSRs, 9 STS, and 51 InDel, distributed evenly on the 12 chromosomes with an average marker interval of 12.76 cM, were used in a genome-wide survey to identify the chromosome segment substitution locations. These polymorphic markers were used for background selection in order to select plants having maximum recovery of the recurrent parent genome. The genotypes from polymorphic bands are recorded as A (IRBB66), B (TNG82) and H (IRBB66/TNG82). The Graphical Geno Types (GGT) Version 2.0 [46] software program was used for the assessment of the recurrent parent genome (%RPG) in the selected recombinants, based on marker data.

5. Conclusions

The use of marker assisted selection in backcross breeding is an effective and reliable approach for pyramiding BB-resistant genes in rice. In this study, the pyramiding lines that possess resistance against BB strains, high potential yields, and high grain quality were both developed and improved.

The BB-pyramided breeding lines containing all five genes, *Xa4*, *xa5*, *Xa7*, *xa13* and *Xa21*, can serve as donors to introgress the resistance genes into other elite rice cultivars in order to accelerate the improvement of rice for disease resistance in Taiwan. These BB-pyramided lines are expected to have a high impact on domestic rice production stability, and also reduce the need for pesticides.

Supplementary Materials: The following are available online at http://www.mdpi.com/1422-0067/21/4/1281/s1. Figure S1: The polymorphic markers used for background selection of TNG82/IRBB66 backcross population. Table S1: The genome composition of BC_2F_2 derived from TNG82/IRBB66 by 117 polymorphic markers used for MAS.

Author Contributions: Conceptualization, Y.-C.H. and Y.-P.W.; methodology, Y.-P.W.; validation, Y.-P.W.; formal analysis, C.-H.C.; investigation, Y.-C.H., C.-H.C., R.Y., and Y.-P.W.; data curation, Y.-C.H. and R.Y.; writing—original draft preparation, Y.-C.H.; writing—review and editing, Y.-C.T. and Y.-P.W.; supervision, Y.-C.H. and Y.-P.W. All authors have read and agreed to the published version of the manuscript.

Funding: This work was supported by the Ministry of Science and Technology (MOST 108-2313-B-415-007) of Taiwan. We are grateful to National Plant Resources Center for providing the rice materials.

Acknowledgments: The authors are highly grateful to the technical support from Plant Protection Division, Chiayi Agricultural Experiment Station, Taiwan Agricultural Research Institute.

Conflicts of Interest: The authors declare that they have no conflict of interests.

References

1. FAO. *Global Agriculture Towards 2050*; FAO: Rome, Italy, 2009.
2. Ray, D.K.; Mueller, N.D.; West, P.C.; Foley, J.A. Yield trends are insufficient to double global crop production by 2050. *PLoS ONE* **2013**, *8*, e66428. [CrossRef] [PubMed]
3. Khush, G.S.; Mackill, D.J.; Sidhu, G.S. Breeding rice for resistance to bacterial leaf blight. In *IRRI Bacterial Blight of Rice*; IRRI: Manila, Philippines, 1989; pp. 207–217.
4. Singh, G.P.; Singh, S.R.; Singh, R.V.; Singh, R.M. Variation and qualitative losses caused by bacterial blight in different rice varieties. *Indian Phytopathol.* **1997**, *30*, 180–185.
5. Perez, L.M.; Redoña, E.D.; Mendioro, M.S.; Vera Cruz, C.M.; Leung, H. Introgression of *Xa4*, *Xa7* and *Xa21* for resistance to bacterial blight in thermosensitive genetic male sterile rice (*Oryza sativa* L.) for the development of two-line hybrids. *Euphytica* **2008**, *164*, 627–636. [CrossRef]
6. Pradhan, S.K.; Nayak, D.K.; Mohanty, S.; Behera, L.; Barik, S.R.; Pandit, E.; Lenka, S.; Anandan, A. Pyramiding of three bacterial blight resistance genes for broad-spectrum resistance in deepwater rice variety, Jalmagna. *Rice* **2015**, *8*, 51. [CrossRef]
7. Prahalada, G.D.; Ramkumar, G.; Hechanova, S.L.; Vinarao, R.; Jena, K.K. Exploring key blast and bacterial blight resistance genes in genetically diverse rice accessions through molecular and phenotypic evaluation. *Crop Sci.* **2017**, *57*, 1881–1892. [CrossRef]
8. Song, W.Y.; Wang, G.L.; Chen, L.L.; Kim, H.S.; Pi, L.Y.; Holsten, T.; Gardner, J.; Wang, B.; Zhai, W.X.; Zhu, L.H.; et al. A receptor kinase-like protein encoded by the rice disease resistance gene, Xa21. *Science* **1995**, *270*, 1804–1806. [CrossRef]
9. Sun, X.; Yang, Z.; Wang, S.; Zhang, Q. Identification of a 47-kb DNA fragment containingXa4, a locus for bacterial blight resistance in rice. *Theor. Appl. Genet.* **2003**, *106*, 683–687. [CrossRef]
10. Yap, R.; Hsu, Y.-C.; Wu, Y.P.; Lin, Y.R.; Kuo, C.W. Multiplex PCR genotyping for five bacterial blight resistance genes applied to marker-assisted selection in rice (*Oryza sativa*). *Plant Breed.* **2016**, *135*, 309–317. [CrossRef]
11. Iyer, A.S.; McCouch, S.R. The rice bacterial blight resistance gene *xa5* encodes a novel form of disease resistance. *Mol. Plant Microbe Interact.* **2004**, *17*, 1348–1354. [CrossRef]
12. Wang, Y.H.; Liu, S.J.; Ji, S.L.; Zhang, W.W.; Wang, C.M.; Jiang, L.; Wan, J.M. Fine mapping and marker-assisted selection (MAS) of a low glutelin content gene in rice. *Cell Res.* **2005**, *15*, 622–630. [CrossRef]
13. Chu, Z.; Fu, B.; Yang, H.; Xu, C.; Li, Z.; Sanchez, A.; Park, Y.J.; Bennetzen, J.L.; Zhang, Q.; Wang, S. Targetingxa13, a recessive gene for bacterial blight resistance in rice. *Theor. Appl. Genet.* **2006**, *112*, 455–461. [CrossRef] [PubMed]
14. Peng, H.; Chen, Z.; Fang, Z.; Zhou, J.; Xia, Z.; Gao, L.; Chen, L.; Li, L.; Li, T.; Zhai, W.; et al. Rice Xa21 primed genes and pathways that are critical for combating bacterial blight infection. *Sci. Rep.* **2015**, *5*, 12165. [CrossRef] [PubMed]

15. Singh, S.; Sidhu, J.S.; Huang, N.; Vikal, Y.; Li, Z.; Brar, D.S.; Dhaliwal, H.S.; Khush, G.S. Pyramiding three bacterial blight resistance genes (Xa5, Xa13 and Xa21) using marker-assisted selection into indica rice cultivar PR106. *Theor. Appl. Genet.* **2001**, *102*, 1011–1015. [CrossRef]

16. Suh, J.P.; Jeung, J.U.; Noh, T.H.; Cho, Y.C.; Park, S.H.; Park, H.S.; Shin, M.S.; Kim, C.K.; Jena, K.K. Development of breeding lines with three pyramided resistance genes that confer broad-spectrum bacterial blight resistance and their molecular analysis in rice. *Rice* **2013**, *6*, 5. [CrossRef] [PubMed]

17. Ramalingam, J.; Savitha, P.; Alagarasan, G.; Saraswathi, R.; Chandrababu, R. Functional marker assisted improvement of stable cytoplasmic male sterile lines of rice for bacterial blight resistance. *Front. Plant Sci.* **2017**, *8*, 1131. [CrossRef]

18. Ruengphayak, S.; Chaichumpoo, E.; Phromphan, S.; Kamolsukyunyong, W.; Sukhaket, W.; Phuvanartnarubal, E.; Korinsak, S.; Korinsak, S.; Vanavichit, A. Pseudo-backcrossing design for rapidly pyramiding multiple traits into a preferential rice variety. *Rice* **2015**, *8*, 7. [CrossRef]

19. Guvvala, L.; Koradi, P.; Shenoy, V.; Marella, L. Making an Indian traditional rice variety Mahsuri, bacterial blight resistant using marker-assisted selection. *J. Crop Sci. Biotech.* **2013**, *16*, 111–121. [CrossRef]

20. Huang, N.; Angeles, E.R.; Domingo, J.; Magpantay, G.; Singh, S.; Zhang, G.; Kumaravadivel, N.; Bennett, J.; Khush, G.S. Pyramiding of bacterial blight resistance genes in rice: Marker-assisted selection using RFLP and PCR. *Theor. Appl. Genet.* **1997**, *95*, 313–320. [CrossRef]

21. Crossa, J.; Perez-Rodriguez, P.; Cuevas, J.; Montesinos-Lopez, O.; Jarquin, D.; de Los Campos, G.; Burgueno, J.; Gonzalez-Camacho, J.M.; Perez-Elizalde, S.; Beyene, Y.; et al. Genomic selection in plant breeding: Methods, models, and perspectives. *Trends Plant Sci.* **2017**, *22*, 961–975. [CrossRef]

22. Mehta, S.; Singh, B.; Dhakate, P.; Rahman, M.; Islam, M.A. Rice, Marker-Assisted Breeding, and Disease Resistance. In *Disease Resistance in Crop Plants: Molecular, Genetic and Genomic Perspectives*; Wani, S.H., Ed.; Springer International Publishing: Cham, Switzerland, 2019; pp. 83–111.

23. Abhilash Kumar, V.; Balachiranjeevi, C.H.; Bhaskar Naik, S.; Rambabu, R.; Rekha, G.; Harika, G.; Hajira, S.K.; Pranathi, K.; Anila, M.; Kousik, M.; et al. Development of gene-pyramid lines of the elite restorer line, RPHR-1005 possessing durable bacterial blight and blast resistance. *Front Plant Sci.* **2016**, *7*, 1195. [CrossRef]

24. Abhilash Kumar, V.; Balachiranjeevi, C.H.; Bhaskar Naik, S.; Rekha, G.; Rambabu, R.; Harika, G.; Pranathi, K.; Hajira, S.K.; Anila, M.; Kousik, M.; et al. Marker-assisted pyramiding of bacterial blight and gall midge resistance genes into RPHR-1005, the restorer line of the popular rice hybrid DRRH-3. *Mol. Breed.* **2017**, *37*, 86. [CrossRef]

25. Yugander, A.; Sundaram, R.M.; Singh, K.; Senguttuvel, P.; Ladhalakshmi, D.; Kemparaju, K.B.; Madhav, M.S.; Prasad, M.S.; Hariprasad, A.S.; Laha, G.S. Improved versions of rice maintainer line, APMS 6B, possessing two resistance genes, Xa21 and Xa38, exhibit high level of resistance to bacterial blight disease. *Mol. Breed.* **2018**, *38*, 100. [CrossRef]

26. Bharathkumar, S.; Paulraj, R.S.D.; Brindha, P.V.; Kavitha, S.; Gnanamanickam, S.S. Improvement of bacterial blight resistance in rice cultivars Jyothi and IR50 via marker-assisted backcross breeding. *J. Crop Improv.* **2008**, *21*, 101–116. [CrossRef]

27. Joseph, M.; Krishnan, S.; Sharma, R.; Singh, V.P.; Singh, A.; Singh, N.; Mohapatra, T. Combining bacterial blight resistance and Basmati quality characteristics by phenotypic and molecular marker-assisted selection in rice. *Molecular Breed.* **2004**, *13*, 377–387. [CrossRef]

28. Luo, Y.; Sangha, J.S.; Wang, S.; Li, Z.; Yang, J.; Yin, Z. Marker-assisted breeding of *Xa4*, *Xa21* and *Xa27* in the restorer lines of hybrid rice for broad-spectrum and enhanced disease resistance to bacterial blight. *Molecular Breed.* **2012**, *30*, 1601–1610. [CrossRef]

29. Lalitha Devi, G.; Koradi, P.; Shenoy, V.; Shanti, L. Improvement of resistance to bacterial blight through marker assisted backcross breeding and field validation in rice (*Oryza sativa*). *Res. J. Biol.* **2013**, *1*, 52–66.

30. Shanti, M.L.; Shenoy, V.V.; Lalitha Devi, G.; Mohan Kumar, V.; Premalatha, P.; Naveen Kumar, G.; Shashidhar, H.E.; Zehr, U.B.; Freeman, W.H. Marker-assisted breeding for resistance to bacterial leaf blight in popular cultivar and parental lines of hybrid rice. *J. Plant Pathol.* **2010**, *92*, 495–501.

31. Huang, B.; Xu, J.Y.; Hou, M.S.; Ali, J.; Mou, T.M. Introgression of bacterial blight resistance genes *Xa7*, *Xa21*, *Xa22* and *Xa23* into hybrid rice restorer lines by molecular marker-assisted selection. *Euphytica* **2012**, *187*, 449–459. [CrossRef]

32. Hasan, M.M.; Rafii, M.Y.; Ismail, M.R.; Mahmood, M.; Rahim, H.A.; Alam, M.A.; Ashkani, S.; Malek, M.A.; Latif, M.A. Marker-assisted backcrossing: A useful method for rice improvement. *Biotechnol. Biotechnol. Equip.* **2015**, *29*, 237–254. [CrossRef]

33. Jiang, G.L. Molecular Markers and Marker–Assisted Breeding in Plants. In *Plant Breeding from Laboratories to Fields*; Andersen, S.B., Ed.; IntechOpen: Croatia, 2013; pp. 45–83.

34. Rajpurohit, D.; Kumar, R.; Kumar, M.; Paul, P.; Awasthi, A.; Osman Basha, P.; Puri, A.; Jhang, T.; Singh, K.; Dhaliwal, H.S. Pyramiding of two bacterial blight resistance and a semidwarfing gene in Type 3 Basmati using marker-assisted selection. *Euphytica* **2011**, *178*, 111–126. [CrossRef]

35. Sundaram, R.M.; Vishnupriya, M.R.; Biradar, S.K.; Laha, G.S.; Reddy, G.A.; Rani, N.S.; Sarma, N.P.; Sonti, R.V. Marker assisted introgression of bacterial blight resistance in Samba Mahsuri, an elite indica rice variety. *Euphytica* **2008**, *160*, 411–422. [CrossRef]

36. Balachiranjeevi, C.; Bhaskar, N.S.; Abhilash, V.; Akanksha, S.; Viraktamath, B.C.; Madhav, M.S.; Hariprasad, A.S.; Laha, G.S.; Prasad, M.S.; Balachandran, S.; et al. Marker-assisted introgression of bacterial blight and blast resistance into DRR17B, an elite, fine-grain type maintainer line of rice. *Mol. Breed.* **2015**, *35*, 151. [CrossRef]

37. Singh, A.; Singh, V.K.; Singh, S.P.; Pandian, R.T.P.; Ellur, R.K.; Singh, D.; Bhowmick, P.K.; Gopala Krishnan, S.; Nagarajan, M.; Vinod, K.K.; et al. Molecular breeding for the development of multiple disease resistance in Basmati rice. *AoB Plants* **2012**, *2012*. [CrossRef] [PubMed]

38. Cobb, J.N.; Biswas, P.S.; Platten, J.D. Back to the future: Revisiting MAS as a tool for modern plant breeding. *Theor. Appl. Genet.* **2019**, *132*, 647–667. [CrossRef]

39. Tseng, H.Y.; Lin, D.G.; Hsieh, H.Y.; Tseng, Y.J.; Tseng, W.B.; Chen, C.W.; Wang, C.S. Genetic analysis and molecular mapping of QTLs associated with resistance to bacterial blight in a rice mutant, SA0423. *Euphytica* **2015**, *205*, 231–241. [CrossRef]

40. Wang, C.S.; Lin, D.G. The Application of Genomic Approaches in Studying a Bacterial Blight–Resistant Mutant in Rice. In *Advances in International Rice Research*; Li, J.Q., Li, J., Eds.; IntechOpen: London, UK, 2017.

41. Kauffman, H.E.; Reddy, A.P.K.; Hsieh, S.P.Y.; Merca, S.D. An improved technique for evaluating resistance of rice varieties to *Xanthomonas oryzae*. *Plant Dis. Rep.* **1973**, *57*, 537–541.

42. Kado, C.I.; Heskett, M.G. Selective media for isolation of Agrobacterium, Corynebacterium, Erwinia, Psedumonoas and Xanthomonas. *Phytopathology* **1970**, *60*, 969–976. [CrossRef]

43. IRRI. *Standard Evaluation System*; In International Rice Research Institute: Manila, Philippines, 1996; p. 52.

44. Latif, M.A.; Badsha, M.A.; Tajul, M.I.; Kabir, M.S.; Rafii, M.Y.; Mia, M.A.T. Dentification of genotypes resistant to blast, bacterial leaf blight, sheath blight and tungro and efficacy of seed treating fungicides against blast disease of rice. *Sci. Res. Essays* **2011**, *6*, 2804–2811.

45. Hsu, Y.C.; Tseng, M.C.; Wu, Y.P.; Lin, M.Y.; Wei, F.J.; Hwu, K.K.; Hsing, Y.I.; Lin, Y.R. Genetic factors responsible for eating and cooking qualities of rice grains in a recombinant inbred population of an inter-subspecific cross. *Mol. Breed.* **2014**, *34*, 655–673. [CrossRef]

46. Van Berloo, R. GGT: Software for the display of graphical genotypes. *J. Hered.* **1999**, *90*, 328–330. [CrossRef]

International Journal of
Molecular Sciences

Article

Genetic Dissection of Germinability under Low Temperature by Building a Resequencing Linkage Map in *japonica* Rice

Shukun Jiang [1,*,†], Chao Yang [2,†], Quan Xu [3], Lizhi Wang [1], Xianli Yang [1], Xianwei Song [2], Jiayu Wang [3], Xijuan Zhang [1], Bo Li [1], Hongyu Li [4], Zhugang Li [1,*] and Wenhua Li [1,*]

[1] Crop Cultivation and Tillage Institute of Heilongjiang Academy of Agricultural Sciences, Heilongjiang Provincial Key Laboratory of Crop Physiology and Ecology in Cold Region, Heilongjiang Provincial Engineering Technology Research Center of Crop Cold Damage, Harbin 150086, China; wanglizhi0451@163.com (L.W.); aiwei.ni@163.com (X.Y.); xijuanzhang@163.com (X.Z.); blnky@163.com (B.L.)

[2] Institute of Genetics and Developmental Biology, Chinese Academy of Sciences, Beijing 100101, China; chaoyang@genetics.ac.cn (C.Y.); xwsong@genetics.ac.cn (X.S.)

[3] Rice Research Institute of Shenyang Agricultural University, Shenyang 110866, China; kobexu34@syau.edu.cn (Q.X.); ricewjy@126.com (J.W.)

[4] College of Agronomy, Heilongjiang Bayi Agricultural University, Daqing 163000, China; ndrice@163.com

* Correspondence: sk.jiang@hotmail.com (S.J.); lizhugang@163.com (Z.L.); nkylwh@163.com (W.L.)

† These authors have contributed equally to this work.

Received: 25 January 2020; Accepted: 13 February 2020; Published: 14 February 2020

Abstract: Among all cereals, rice is highly sensitive to cold stress, especially at the germination stage, which adversely impacts its germination ability, seed vigor, crop stand establishment, and, ultimately, grain yield. The dissection of novel quantitative trait loci (QTLs) or genes conferring a low-temperature germination (LTG) ability can significantly accelerate cold-tolerant rice breeding to ensure the wide application of rice cultivation through the direct seeding method. In this study, we identified 11 QTLs for LTG using 144 recombinant inbred lines (RILs) derived from a cross between a cold-tolerant variety, Lijiangxintuanheigu (LTH), and a cold-sensitive variety, Shennong265 (SN265). By resequencing two parents and RIL lines, a high-density bin map, including 2,828 bin markers, was constructed using 123,859 single-nucleotide polymorphisms (SNPs) between two parents. The total genetic distance corresponding to all 12 chromosome linkage maps was 2,840.12 cm. Adjacent markers were marked by an average genetic distance of 1.01 cm, corresponding to a 128.80 kb physical distance. Eight and three QTL alleles had positive effects inherited from LTH and SN265, respectively. Moreover, a pleiotropic QTL was identified for a higher number of erected panicles and a higher grain number on Chr-9 near the previously cloned *DEP1* gene. Among the LTG QTLs, *qLTG3* and *qLTG7b* were also located at relatively small genetic intervals that define two known LTG genes, *qLTG3-1* and *OsSAP16*. Sequencing comparisons between the two parents demonstrated that LTH possesses *qLTG3-1* and *OsSAP16* genes, and SN-265 owns the *DEP1* gene. These comparison results strengthen the accuracy and mapping resolution power of the bin map and population. Later, fine mapping was done for *qLTG6* at 45.80 kb through four key homozygous recombinant lines derived from a population with 1569 segregating plants. Finally, *LOC_Os06g01320* was identified as the most possible candidate gene for *qLTG6*, which contains a missense mutation and a 32-bp deletion/insertion at the promoter between the two parents. LTH was observed to have lower expression levels in comparison with SN265 and was commonly detected at low temperatures. In conclusion, these results strengthen our understanding of the impacts of cold temperature stress on seed vigor and germination abilities and help improve the mechanisms of rice breeding programs to breed cold-tolerant varieties.

Keywords: *japonica* rice; cold stress; germinability; high-density linkage map; QTLs

---

1. Introduction

Rice (*Oryza sativa* L.), which is a staple food and nutritional source for many countries, fulfills the nutritional requirements for over half of the world's population and is cultivated across the globe, except in a few areas where icy conditions prevail during most of the year [1]. Since rice originated in tropical and sub-tropical climates, it is one of the most sensitive cereals to cold stress [2], which limits its growth, development, and yield formation, especially when cold stress prevails at the germination stage [3]. Cold stress impacts all growth stages of rice, including tillering, booting, flowering, and grain-filling, but if stress dominates at the germination stage, it proves adverse for rice development at later growth stages. Low-temperature stress during the germination stages of rice affects seedling vigor and produces poor seedling emergence and an uneven stand establishment with a lower growth rate, which delays panicle development and enhances spikelet sterility [4]. Across China's mainland, most of the rice cultivation areas are affected by frequent cold stresses. The Chinese agricultural sector suffers from an average loss of rice of about 3–5 million tons of rice every year due to these frequent cold stresses [5]. Two kinds of cold stresses occur in Chinese rice-growing regions. The (1) "cold spring" and (2) "cold autumn wind" often cause severe yield losses in double-cropping rice regions across the Yangtze River in China. In Northeast China (NEC), commonly considered a rice region at a high latitude, and the Yunnan-Guizhou Plateau, considered a rice cultivation region at a low latitude, severe cold summer damage was observed, with an average of three to four years of cold stress. These areas are expected to encounter more severe damage in the near future due to low temperature stress [6].

Traditional genetic and molecular analyses on Arabidopsis, rice, and other model plants have revealed that C-repeat binding factors (CBFs) are mainly involved in the cold signaling pathways. Recent studies have further revealed that the protein kinases and transcription factors are also involved in cold signaling in plants [7]. Additionally, genetic research on rice has detected numerous quantitative trait loci that control cold tolerance on nearly all 12 chromosomes [8]. Among these loci, only a few quantitative trait loci (QTLs) have been thoroughly researched and cloned, while the functional mechanisms of most are still largely unknown. Among all the QTLs for low-temperature germination in rice, only *qLTG3-1* and *OsSAP16* were cloned. *qLTG3-1* encodes a protein with glycine-rich and lipid trans-protein domain structures [9], and *OsSAP16* encodes a zinc-finger protein that positively regulates germination under low temperatures [10]. The QTLs *qCTS7*, *LTG1*, *COLD1*, *qCTS9*, *bZIP73*, *qPSR10*, and *HAN1* control the pathways for cold tolerance in rice. *qCTS7* increases cold tolerance at the seedling stage due to its overexpression [11]. *LTG1* encodes a casein kinase that plays a role in regulating the rice growth rate under cold stress [4]. A regulator of Ca^{2+} signaling in the plasma membrane and endoplasmic reticulum is encoded by *COLD1* [12], whereas a novel protein that interacts with Brassinosteroid Insensitive-1 is encoded by *qCTS9* [13]. There is a functional interaction between *bZIP73* and *bZIP71* that makes rice seedlings tolerant to greater cold [14]. A single-nucleotide polymorphism (SNP), SNP^{2G}, at position 343 in *qPSR10*, is responsible for conferring cold tolerance during the seedling stage [15]. *HAN1* encodes an oxidase that provides functional contributions to the Jasmonic acid mediated cold response in temperate *japonica* rice [16]. The other three QTLs, *Ctb1*, *CTB4a*, and *bZIP73*, control cold tolerance at the booting stage. The first encodes an F-box protein [17], the second encodes a leucine rich repeat kinase that enhances seed setting through increased ATP-synthase activity under low temperature stress [3], and the third increases the cold tolerance rate by enhancing the soluble sugar transport from anthers to pollens at the booting stage [18].

Advances in genome-wide sequencing technology have provided an effective method to detect DNA sequencing differences among closely related rice materials and to ensure the presence of sufficient markers for a genetic mapping analysis. A genotype calling method for RILs that utilizes resequencing data was developed [19], which determined the construction of resequencing bin maps and accelerated genetics-based studies for many crops, including cereals [19–27]. Based on the above discussions, many important advances have been achieved in the study of rice cold stress, but we still need to use high-throughput sequencing technology to mine further cold-tolerance genes from *japonica* rice, especially cold tolerance genes at the germination stage for breeding practice.

The current study was arranged with the following objectives: (1) constructing a high density bin map by re-sequencing a set of 144 RILs with large differences in germination abilities under cold stress; (2) identifying QTLs for LTG in RIL populations by using the built linkage map; and (3) creating an accurate map of *qLTG6*, with a high low-temperature germinability (LOD) score and relatively small genetic intervals.

2. Results

2.1. Phenotypic Variation among the Parent and RIL Populations

After incubating at 28 °C for three days, the germination rates of LTH and SN265 were 100%, indicating that both parents share a similar germination rate at a normal temperature, as presented in Figure 1A. However, when they were incubated at 15 °C for six days to determine their low temperature germinability, LTH and SN265 showed broad differences in their germination rates, as shown in Figure 1A,B. There was a delay of about two days in germination after incubating LTH under low temperature conditions, and, thereafter, the germination rate reached up to 90% after five days of incubation. Comparatively, for SN265, germination started after five days of incubation and, thereafter, took five more days to reach 90% of the germination of LTH, which shows a clear difference in the time taken for germination by LTH and SN265 (Figure 1B). A range of five to eight days was chosen for a proper and dynamic comparison of the low-temperature germinability among the parents and RILs (Figure 1C–F). At all four time points, the distributions of the germination percentages of the RIL populations were continuous, indicating that low-temperature germinability is controlled by QTLs. Generally, it is already known that the longer the germination time, the higher the germination percentage. Therefore, on the basis of the larger differences in germination between the parents after six days (LTH and SN265 showed 86.5% and 15.4%, respectively), those data were used for the subsequent QTL mapping.

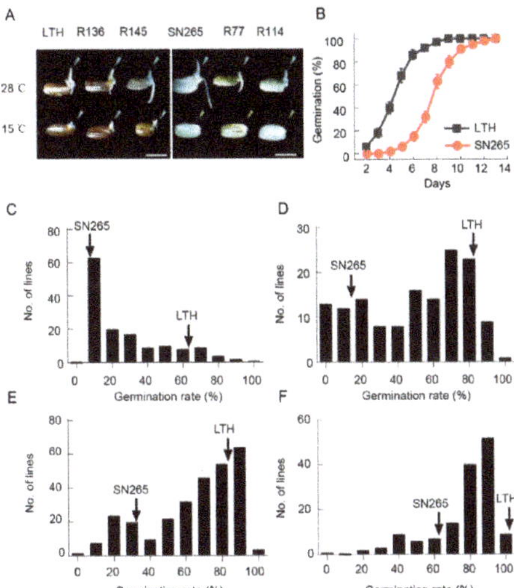

Figure 1. The seed germination of two parents and their recombinant inbred lines (RILs) at normal (28 °C) and low (15 °C) temperatures. (**A**). The germination phenotype of the two parents, Lijiangxintuanheigu (LTH) and Shennong265 (SN265), as well as the four RIL lines incubated for 3 days at 28 °C and for 6 days at 15 °C. R136 and R145 have 11 positive QTLs, whereas R77 and R144 have no positive QTLs. The bars = 5 mm; (**B**). The germination behavior of LTH (black) and SN265 (red) under low temperatures at 15 °C. The means are shown in triplicate; (**C–F**). The frequency distribution of low-temperature germinability in the RILs incubated for 5 days (**C**), 6 days (**D**), 7 days (**E**), and 8 days (**F**). Arrowheads indicate LTH or SN265.

2.2. Bin Map Construction and Comparison of the Physical Map to the Genetic Map

For proper identification of the SNP between the two parents as molecular markers, deep resequencing was done for LTH and SN265. The effective sequencing depths of LTH and SN265 were about 19-fold and 17-fold, respectively. After resequencing, a total of 123,859 SNPs were produced between the two parents. Construction of the genetic linkage map for the RILs was carried out by re-sequencing the 144 RIL lines, which were already derived from LTH and SN265. In this way, these 144 RIL lines produced a range of reads between about 7,448,879 and 12,118,209, with a mean value of 9,660,250. The overall effective depth coverage of these RIL lines ranged from 5.98-fold to 9.73-fold, with an average depth of 7.75-fold. About 58,738 recombination breakpoints were used to construct the fine bin map. Each RIL line was comprised of breakpoints ranging from 236 to 1007 breakpoints, with an average value of 405. The 144 RIL lines were merged into a high-density bin map comprising 2,818 recombination bins of the 12 rice chromosomes, including most recombination events (Figure 2A). The average of the physical intervals for the adjacent bins was observed between 15.00 kb and 3.60 Mb, with a mean value of 128.80 kb, where most of the bins with physical intervals less than 100.00 kb were found to be around 68.60%. Overall, only 32 bins exceeded a physical interval of 1.00 Mb, most of which were in centromeric regions. The average physical distance between the bin markers was 98.23 kb and 169.03 kb on chromosome 10 and chromosome 1, respectively. Then, we constructed 12 chromosomes with a total genetic distance of 2,840.12 cm (Figure 2B). The average genetic distance observed between the two markers was nearly 1.01 cm. Among all chromosomes under consideration, chromosome 1 was the longest, with 254 bin markers, and its genetic distance was 338.65 cm. In contrast, chromosome 9 was seen to be the shortest one, with 188 bin markers encompassing a genetic distance of 111.32 cm, as given in Table 1. Comparing the genetic distance between chromosome 9 and 11, the average values were 0.59 cm and 1.74 cm, respectively.

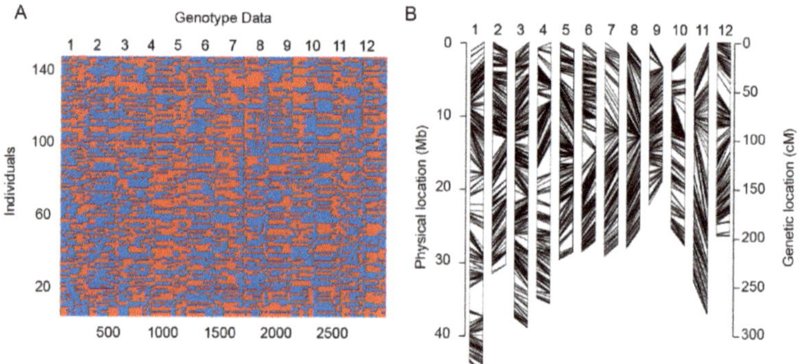

Figure 2. High-resolution genotyping and comparison of the physical maps and genetic maps of RILs. (**A**). Aligned recombination maps of 144 RILs from a cross between LTH and SN265. The two genotypes are indicated by blue (SN265) and red (LTH); (**B**). Comparison of the physical maps and genetic maps. Left: physical map. Right: genetic map.

2.3. The Quality and Accuracy of the Bin Map

QTL mapping for the typical panicle trait named *Dense and Erect Panicle1* (*DEP1*) was performed to estimate the accuracy and mapping resolution ability of the respective bin maps and the RILs. Careful observations were taken for the panicle curvature of the 144 RILs lines 25 days after the start of heading, whereas the erect-type panicle lines remained erect. One QTL was identified on chromosome 9, with a peak interval of 16.4 Mb, as shown in Figure 3C. The QTL peak was located on the previously characterized and cloned *DEP1/EP1* gene. Sequence comparisons of the *DEP1* region between LTH and SN265 illustrated the replacement of a 637-bp in the middle of exon-5 by a 12-bp sequence (Figure 3F), which caused a frame shift mutation, as described previously [28].

Table 1. Characteristics of the high-density genetic map derived from a cross between LTH and SN265.

Chr. [a]	No. Markers [b]	Genetic Distance (cm)	Physical Distance (Mb)	Avg Distance between Markers (cm/kb)	<1 Mb Gap	Min. Gap (kb)	Max. Gap (Mb)
1	254	338.65	42.93	1.33/169.03	246	15.27	3.16
2	254	227.11	31.61	0.89/124.46	252	15.47	3.45
3	313	284.80	36.14	0.91/115.45	310	15.06	1.57
4	281	300.18	34.29	1.07/122.03	275	15.63	1.91
5	239	269.15	29.68	1.13/124.19	235	15.32	1.08
6	247	181.10	28.63	0.73/115.92	244	15.53	3.63
7	187	217.94	29.23	1.17/156.34	184	17.18	1.46
8	238	193.86	28.12	0.82/118.15	237	15.02	0.99
9	188	111.32	22.30	0.59/118.61	185	15.89	1.92
10	234	199.84	22.99	0.85/98.23	230	15.77	1.27
11	186	323.32	28.49	1.74/153.18	183	15.01	3.48
12	207	192.85	26.70	0.93/130.43	202	15.63	2.33

[a] Chr., indicates chromosome; [b] No. markers, the number of markers on the chromosome.

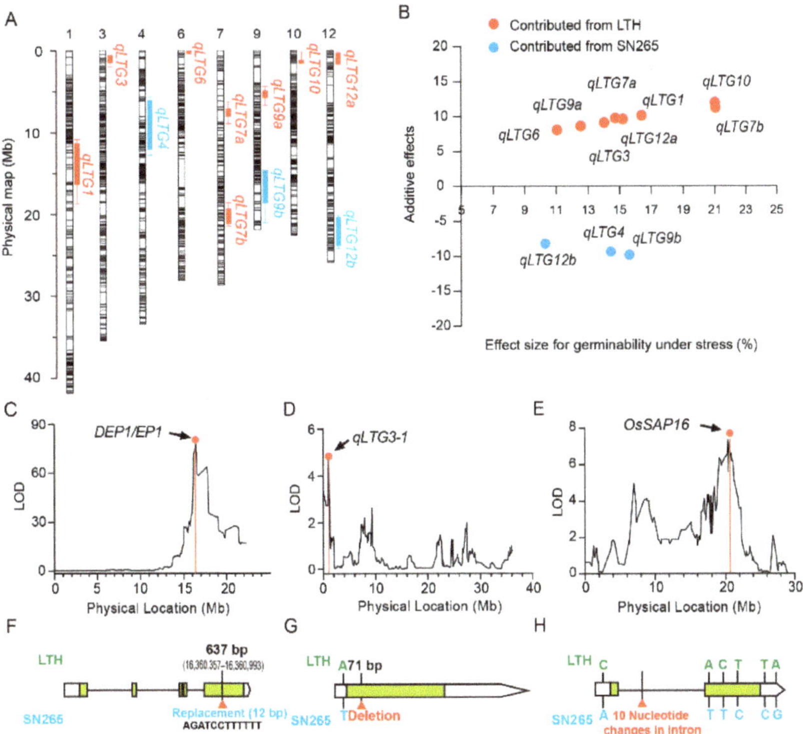

Figure 3. (**A**). QTL scan using whole-genome sequencing and a sequence comparing the panicle curvature and low-temperature germinability QTLs near the previously identified genes. Curves indicate the chromosome locations (Mb) and the low-temperature germinability (LOD) values of the detected QTLs. Arrowheads represent the relative genetic positions of the candidate genes. (**A**). Genomic locations of the 11 QTLs with strong effects for low-temperature germinability (LOD > 3) identified in the RIL population; (**B**). Plots of the additive effect and the allele effect of the 11 QTLs. (**C**). A plot of the LOD values of the panicle curvature. (**D**). A plot of the LOD values of the low-temperature germinability on chromosome 3. (**E**). A plot of the LOD values of low-temperature germinability on chromosome 7. (**F**). Sequence comparisons of the *DEP1* region between LTH and SN265. (**G**). Sequence comparisons of the *qLTG3-1* region between LTH and SN265. (**H**). Sequence comparisons of the *OsSAP16* region between LTH and SN265.

2.4. QTL Analysis of Low-Temperature Germinability

After combining the bin-map and phenotyping data of each RIL line, scanning of the LTG QTL was carried out. A total of 11 QTLs were identified on chromosome 1 (11.48–19.34 Mb), chromosome 3 (0.00–1.31 Mb), chromosome 4 (6.10–11.41 Mb), chromosome 6 (0.34–0.74 Mb), chromosome 7 (6.98–9.21 Mb; 18.93–21.76 Mb), chromosome 9 (5.91–6.83 Mb; 14.91–21.38 Mb), chromosome 10 (1.20–1.60 Mb), and chromosome 12 (0.86–2.35 Mb; 21.05–25.16 Mb). The confidence intervals associated with these QTLs spanned the genomic physical position from 0.07 to 7.86 Mb compared with the reference genome of rice (Table 2, Figure 3A). Five QTLs showed relatively small confidence intervals of less than 1.5 Mb, including *qLTG-3* (1.31 Mb), *qLTG6* (0.40 Mb), *qLTG-9a* (0.92 Mb), *qLTG-10* (0.40 Mb), and *qLTG-12a* (1.49 Mb). Among these QTLs, phenotypic variation ranged from 10.30% to 21.04% (Table 2, Figure 3B). Overall, five QTLs depicted the explained phenotypic variation exceeding 15%, and two major QTLs were detected: *qLTG7b*, found on chromosome 7, with 21.04% phenotypic variation and 7.39 of the LOD value; and *qLTG10*, located on chromosome 10 and comprised of 21.00% phenotypic variation with a 7.38 LOD value. Furthermore, eight QTL alleles, *qLTG1*, *qLTG3*, *qLTG6*, *qLTG7a*, *qLTG7b*, *qLTG9a*, *qLTG10*, and *qLTG12a*, had a positive effect inherited from LTH. The other three QTL alleles, *qLTG4*, *qLTG9b*, and *qLTG12b*, had a positive effect inherited from SN265 (Figure 3B). Based on the high-resolution bin map, two QTLs, *qLTG3* and *qLTG7b*, were localized to the chromosome intervals that subsumed the cloned LTG genes. The QTL named *qLTG3* was located in the interval of the LTG gene described as *qLTG3-1*, with a 4.71 LOD value and 14.02% phenotype variance [9]. The QTL named *qLTG7b* had a 7.39 LOD value and 21.04% phenotypic variation and was located on *OsSAP16*, a rice LTG gene that has already been identified by the association study method [10]. PCR amplification and sequencing were conducted to screen out the causal polymorphisms of *qLTG3-1* and *OsSAP-16*, which determined that LTH has two cold-tolerant genes (Figure 3E,F). Additionally, these results also indicate that the data used for scanning QTLs are entirely effective (Figure 3D,E,G,H).

Table 2. Characteristics of the high-density genetic map derived from a cross between LTH and SN265.

QTL	Chr. [a]	Peak		QTL Interval			LOD [c]	Var (%) [d]	Add. [e]	Positive Allele
		Pos. (cm)	Pos. (Mb) [b]	Linkage (cm)	Physical (Mb)	Location Interval (cm/Mb)				
qLTG1	1	116.34	16.90	103.19–128.81	11.48–19.34	25.62/7.86	5.59	16.40	10.16	LTH
qLTG3	3	3.95	1.10	0.00–10.78	0.00–1.31	10.78/1.31	4.71	14.02	9.18	LTH
qLTG4	4	54.58	6.27	36.64–77.67	6.10–11.41	41.03/5.31	4.85	14.43	−9.30	SN265
qLTG6	6	0.18	1.34	0.18–1.07	0.34–0.74	0.89/0.40	3.64	11.05	8.13	LTH
qLTG7a	7	104.99	8.88	87.28–107.84	6.98–9.21	20.56/2.23	4.98	14.72	9.80	LTH
qLTG7b	7	152.12	20.32	148.13–162.80	18.93–21.76	14.67/2.84	7.39	21.04	11.26	LTH
qLTG9a	9	4.31	6.07	4.35–8.38	5.91–6.83	4.03/0.92	4.17	12.53	8.68	LTH
qLTG9b	9	79.24	15.27	71.69–102.85	14.91–21.38	31.16/6.47	5.29	15.60	−9.76	SN265
qLTG10	10	13.93	1.60	8.55–14.06	1.20–1.60	5.51/0.40	7.38	21.00	12.00	LTH
qLTG12a	12	6.17	0.87	4.72–16.25	0.86–2.35	11.53/1.49	5.14	15.20	9.65	LTH
qLTG12b	12	131.25	22.79	111.25–145.46	21.05–25.16	34.21/4.11	3.41	10.30	−8.12	SN265

[a] Chr., chromosome; [b] Positions in the linkage map (unit: Mb); [c] Logarithm (base 10) of the odds for the corresponding QTL peak; [d] Percentage of the phenotypic variation explained by the corresponding QTL; [e] Additive effect of the corresponding QTL.

2.5. Fine Mapping and Candidate Gene Prediction for qLTG6

To identify the possible novel genes for low germination within the QTLs, *qLTG6* was thoroughly analyzed because it has a high LOD value and relatively small genetic intervals (Figure 4A, Table 2). *qLTG6* was first identified within a 400 kb interval. Then, further high-resolution mapping of *qLTG6* was carried out using four key homozygous recombination lines from 1,569 segregating population plants and newly developed markers, as shown in Table 3. Finally, *qLTG6* was localized to a region of a 45.8 kb physical interval between markers M002 and M008 via the progeny testing of key homozygous recombinant lines (Figure 4B). Overall, there was the prediction of about seven genes in the target region (Table 4). A re-sequencing data analysis of the delimited region only detected differences in the gene *LOC_Os06g01320* (Figure 4C). To confirm and estimate these differences, PCR-based sequencing was conducted to analyze the gene body and the 3 kb promoter of *LOC_Os06g01320*, which revealed a 32 bp deletion and a T–C transition in the promoter. In the gene body, a C–A transition was also detected in the 18th exon, causing a substitution of Thr to Asn in LTH (Figure 4D).

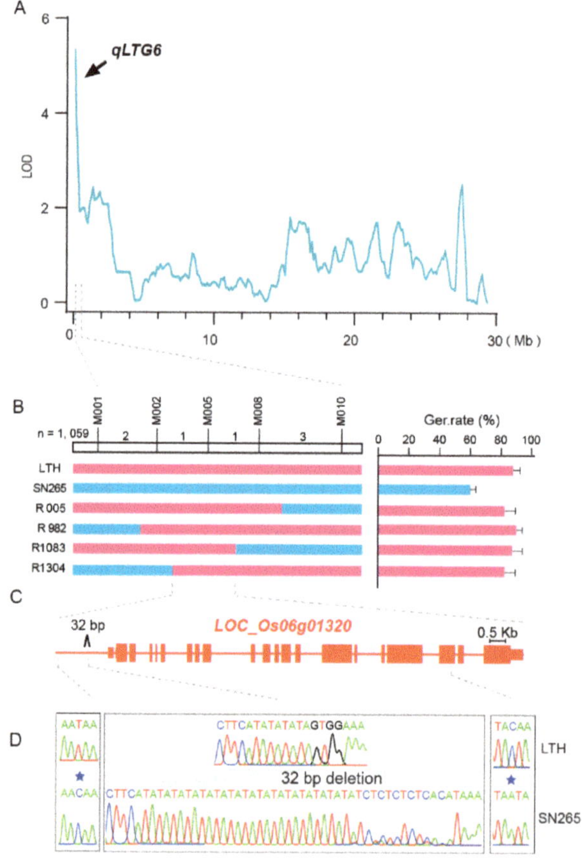

Figure 4. Identification of the candidate gene of the QTL, *qLTG6*. (**A**). Linkage analyses of the QTL, *qLTG6*; (**B**). Graphical representation of recombinants in the RILs of germinability under low temperatures, refining the location of *qLTG6* in an interval defined by bin markers; (**C**). The identification of the candidate gene, *LOC_Os06g01320*, which encodes a chromodomain, helicase/ATPase, and DNA-binding domain (CHD)-related (CHR) proteins, CHR723; (**D**). Validation of the mutation of *LOC_Os06g01320* by PCR sequencing.

Table 3. The markers developed for the fine mapping of *qLTG6* on chromosome 6.

Molecular Marker	Primer Sequence (5′→′)
M001	CTTCGCACTCCAGTCGCTCTCC GTTGAGGAGGTGTATGGGCTTGG
M002	AGCTCACCAGGGACAACATCAAGG TTAACCAGCTCCGCCAGCATCC
M005	CGCCACTGATCGATCTCCTCTCC CGAGCTGGCCTTCTTCCTTGG
M008	AATTGATGCAGGTTCAGCAAGC GGAAATGTGGTTGAGAGTTGAGAGC
M010	TGTTGGATTGGAATCGGAAAGC CTCTGCTGTGCTGTGCTGCTAGG

Table 4. Candidate genes in the 45.8 kb target region corresponding to *qLTG6*.

Name	Location	Protein
LOC_Os06g01250	163205–165539	Cytochrome P450
LOC_Os06g01260	167364–174331	Glutathione gamma-glutamylcysteinyltransferase 1
LOC_Os06g01270	178580–178343	Expressed protein
LOC_Os06g01280	180215–181423	Retrotransposon protein
LOC_Os06g01290	182104–184623	Expressed protein
LOC_Os06g01304	185692–191452	Spotted leaf 11
LOC_Os06g01320	195018–208583	Chromodomain, helicase/ATPase, and DNA-binding domain (CHD) proteins

2.6. Expression Analysis of LOC_Os06g01320

Both an SNP and deletion were detected at the promoter of *LOC_Os06g01320* in LTH, which suggests that expression might vary between parents. For further investigations on these variants, qPCR analyses were executed to evaluate the expression patterns of *LOC_Os06g01320* in two parents. During seed germination under optimal conditions at 28 °C, the expression levels of *LOC_Os06g01320* decreased in the LTH and SN265 from 0 h to 1 d (Figure 5), which indicates that *LOC_Os06g01320* is necessarily required to be repressed during seed germination. Under the conditions of low-temperature stress with a temperature range around 15 °C, the expression levels of *LOC_Os06g01320* in LTH were significantly lower than SN265, both at 12 h and at 1 d. After conducting a further refinement, these results showed that LTH has lower expression levels of *LOC_Os06g01320* than SN265, which could reduce the negative regulation of gene expression and promote seed germination under low-temperature stress.

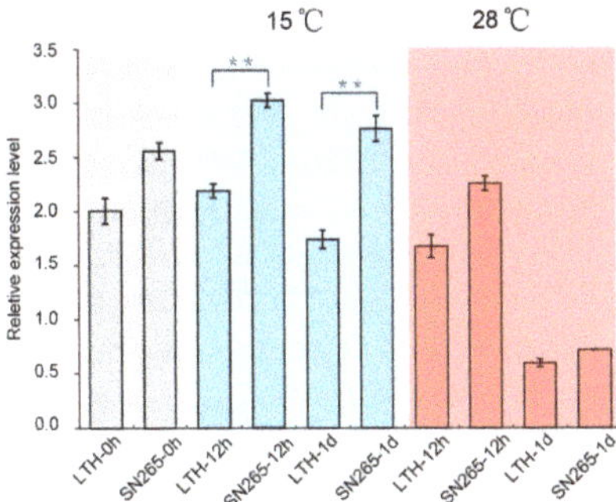

Figure 5. The expression analysis of the candidate gene of *qLTG6*. Relative expression of LOC_Os06g01320 at the germination stage under 15 °C and 28 °C in the LTH and SN265, respectively. ** $p < 0.01$, using Student's *t*-test. Bar SEM, $n = 3$.

3. Discussion

In recent years, direct-seeded rice has received much attention in Asia because of its time and labor savings and low input demand as an alternative to conventional rice systems. However, the long-term cultivation method of seedling-transplantation has led to a loss of expressions of some low-temperature-tolerant genes, which are usually expressed at the germination stage. Poor germinability remains one of the major problems [5,21,29,30]. Therefore, screening cultivars with high germination abilities under low-temperature stress is necessary to sustain rice yield and ensure the application of direct seeding cultivation in the NEC region. In the current study, the *japonica* landrace variety, LTH, a locally well-adopted cultivar originating from the far Southwest province of China, was selected since it shows a high germination rate under low-temperature conditions. The germination of LTH began two days after the start of incubation at a temperature of 15 °C and took five more days to reach a germination rate of almost 90%. Screening of the low-temperature germination ability of 135 cultivars from the NEC was then done, which revealed that no cultivar was more tolerant to low-temperature stress than LTH. Therefore, the identification of tolerant genes during germination under low-temperature stress in LTH has important scientific value for improving the low-temperature germination ability of rice in the NEC.

In this study, high-throughput genotyping was employed through whole-genome resequencing, and bin map construction was carried out for QTL mapping. A total of 11 QTLs were identified on Chr.1, Chr.3, Chr.4, Chr.6, Chr.7, Chr.9, Chr.10, and Chr.12. By comparing the positions of the QTLs, 19 previously identified LTG QTLs were found to be near 10 QTLs, except for *qLTG9a*, as presented in Table 5. The *qLTG-1* was mapped near the *qCTGERM1-5* region for LTG [31]. The QTL location of *qLTG-3* was very close to that of the cloned QTL *qLTG3-1* [9]. The QTLs of *qLTG4*, *qLTG6*, and *qLTG10* were as also found by Teng et al. [32], Ji et al. [33] and Shakiba et al. [31] to enhance low-temperature germinability. *qLTG7a* was mapped near *qCTGERM7-1* [31], *qLTG-7* [34], *qGR-7*, and *qGI-7* [35]. Furthermore, *qLTG7b* was found to be mapped near *qCTGERM7-4* [31], *qLTG-7* [36], and *OsSAP16* [10]. At the germination stage, there was a correlation between the cold tolerance of *qLTG9b* and the *qLTG-9* region [32,33]. *qLTG12a* was mapped near *qLTG-12a* [37] and *qLTG-12* [38]. *qLTG12b* was identified near the region of *qCTGERM12-1* [31], *qGR-12* [10], *qLTG-12* [36], and *qLTG-12* [39]. The above comparison results reflect the accuracy of this study, as well as the complexity of cold tolerance during germination. Moreover, this study also suggests that LTH's strong low-temperature germination ability was acquired by accumulating more cold-tolerant genes. These results not only strengthen the findings of previous studies but also reflect the complexity of low-temperature germination in accelerating the breeding programs for enhanced cold-tolerance among rice cultivars.

Among the 11 QTLs identified in this study, *qLTG-6* was ultimately narrowed down to the 45.80 kb region (Figure 4B). Currently, seven genes have been observed in the target region (Table 5) where *LOC_Os06g01250* is found to encode the protein named cytochrome P450. *LOC_Os06g01260* has functional activities in encoding Glutathione gamma-glutamyl cysteinyl transferase-1. Moreover, *LOC_Os06g01280* encodes a retrotransposon protein. *LOC_Os06g01270* and *LOC_Os06g01290* encode a protein that has yet to be discovered. *LOC_Os06g01304* encodes Spotted Leaf 11, whereas *LOC_Os06g01320* encodes chromodomain, helicase/ATPase, and DNA-binding domain proteins. Considering the organ specificity in gene expression and molecular function information, it is difficult to ensure that a gene is a target gene. According to the sequencing data, this study identified only the C–A transition in the 18th exon of *LOC_Os06g01320*, which is predicted to result in the substitution of Thr to Asn in LTH. We also found a 32 bp deletion and a T–C transition in the promoter region. In addition, the expression level of *LOC_Os06g01320* in LTH was found to be lower than that of SN265 under lower temperatures. This lower expression level could be associated with seed germination at low-temperature stress. The sequence and gene expression data suggest that *LOC_Os06g01320* might be the most plausible prospect for *qLTG-6*, but the current evidence remains insufficient and will require us to carry out genetic modification complementation or gene editing verification.

Table 5. Comparison of QTL positions on the rice genome.

QTL	Chr. [a]	QTL interval	Prior near QTLs Location	Reference
		Physical (Mb)	Physical (Mb)	
qLTG1	1	11.48–19.34	*qCTGERM1-5* (12.71)	[31]
qLTG3	3	0.00–1.31	*qLTG3-1* (0.22)	[9]
qLTG4	4	6.10–11.41	*qLTG-4* (6.58–13.64)	[32]
qLTG6	6	0.34–0.74	*qLTG-6* (0.65–2.69)	[33]
qLTG7a	7	6.98–9.21	*qCTGERM7-1* (10.46–10.65)	[31]
qLTG7b	7	18.93–21.76	*qLTG7* (20.16–22.55) *qGR-7* and *qGI-7*(20.35–21.59) *qCTGERM7-4* (19.59–20.26) *qLTG-7* (16.88–22.52) *OsSAP16* (22.93)	[34] [35] [31] [36] [10]
qLTG9a	9	5.91–6.83		
qLTG9b	9	14.91–21.38	*qLTG-9* (12.29–18.90) *qLTG-9* (11.81–15.32)	[32] [33]
qLTG10	10	1.20–1.60	*qCTGERM10-1* (1.40–1.53)	[31]
qLTG12a	12	0.86–2.35	*qLTG12a* (0.75) *qLTG-12* (2.43–3.19)	[37] [38]
qLTG12b	12	21.05–25.16	*qCTGERM12-1* (24.89–24.90) *qGR-12*(22.78–25.15) *qLTG12* (24.52–25.08) *qLTG-12* (24.52–25.08)	[31] [10] [36] [39]

[a] Chr., chromosome.

4. Materials and Methods

4.1. Plant Materials

The 144 *japonica* rice RIL population was built via the single-seed descent method from a cross between Shennong265 (SN265) and Lijiangxintuanheigu (LTH). The SN265 is a locally well-adopted and super high yielding cultivar in Liaoning province, NEC, whereas LTH is a landrace from the Yunnan Province, in the far Southwest of China. These RILs were planted at the agricultural farm of institute of Crop Cultivation and Tillage, Heilongjiang Academy of Agricultural Sciences. The DNA of F11 RIL generations was isolated for genotyping with the specified protocols. One residual heterozygous plant was selected in the F6 population for the fine mapping of *qLTG6*. It had already been observed through the genotyping of 114 DNA markers that the *qLTG6* region of this selected plant is heterozygous, and other chromosome regions are homozygous [40]. A total of 1,569 segregating individuals were developed from the selected heterozygous plant.

4.2. Preparation of Seeds for the Germination Test

The RILs and their parents were grown in the experimental fields in 2017, where the nursery sowing date was 15th April, and transplanting was done on 16th May, with one seedling per hill. Sixty plants of each line were planted in 4 rows, with a plant to plant distance of 13.3 × 30 cm. The planting of the segregated *qLTG6* population was performed during the rice-growing seasons of 2018. The sowing and transplanting dates were 18th April and 21st May, respectively. The transplanting standards were kept consistent with those of the RILs. Field management practices were done according to the most followed agricultural practices of local farmers. The nitrogen (N), phosphorus (P), and potassium (K) fertilizers, in the form of Urea, single superphosphate, and Murate of Potash, were applied at rates of 120, 60, and 60 kg/ha, respectively. Rice harvesting was done on 30 September, and the plants were retained to dry them for three months. Immature and unfilled grains were removed to obtain high quality grain, which was stored at 5 °C to maintain the relative humidity around 10%.

4.3. Evaluation of Germinability under Cold Stress

The evaluation of the germination ratio under cold stress was performed following standard protocol. The seeds broke out of dormancy at 50 °C after 48 h. Further matured and fully-filled grains were sterilized with 0.1% mercury chloride solution for about 10 min, rinsed with tap water 3 to 4 times, and deionization was performed with deionized water 3 to 4 times. One hundred grains per line were placed on a filter paper in a petri dish, and 10 mL of distilled water was added. The petri dishes were then put in an incubator at the recommended temperature of 15 °C. To avoid any kind of reactions and contaminations, the tap water was changed every two days. The germination of the seeds was noted carefully every day. The germination rate on the 6th day was used for QTL mapping because of the large differences observed between the two parents. The germination was noted three times for each line.

4.4. DNA Extraction, Re-Sequencing, and SNP Calling

The extraction of the genomic DNA of the two parents and each RIL was performed using a modified CTAB method [41]. Biomarker Technologies were used for the re-sequencing of the parents and RILs. The procedure was performed according to Jiang et al. [41]. The short read alignment was done as described by Li and Durbin [42]. Straining of the low-quality data was performed to produce better-quality mapping. The clean data were then aligned to the Nipponbare reference genome (Os-Nipponbare-Reference-IRGSP-1.0) [43] using the BWA software [42]. The calculations of the sequencing coverage and depth were performed through Samtools [44]. Then, the Genome Analysis Toolkit (GATK) was used to detect the SNPs with default parameters [45]. The accession number raw sequence data obtained in our study have been deposited in the NCBI Short Read Archive, with accession number PRJNA587802.

4.5. Genotyping and Construction Bin Map

The estimation of genotype calling parameters, determination of the recombination breakpoint, and construction of the bin map were carried out with minor modifications via the method of Sliding Window, as described by Huang et al. [19]. The genotype of each window was determined by the SNP ratio between the two parents; if the SNP ratio of SN265 to LTH was 15:5 or higher, the genotype was considered to be a homozygous SN265 genotype, if the ratio was lower, then it was designated as a homozygous LTH genotype. Moreover, if the SNP ratio between the two parents was between 5:15 and 15:5, the genotype was recognized as heterozygous. The genotype calling parameters were performed according to the methods described by Song et al. [22]. Adjacent 15 kb segments with the same genotypes were merged as one bin marker [46]. The linkage map was built using the *est.map* function with the R/qtl software [47].

4.6. QTL Mapping for Low-Temperature Germinability

In this study, we used the composite interval method (CIM) in the R/qtl package [47] to detect the QTLs for LTG. The threshold level of CIM was ensured by 1000 permutations, and the QTL confidence interval was estimated by using a 1.5 LOD-drop from the peak LOD. The germination rate (%) after 6 days under controlled temperature conditions of 15 °C was used for the QTL analysis.

4.7. qRT-PCR and Expression Analysis

The extraction of the total RNA was carried out with a TRIzol reagent. The first-strand of cDNA was reverse transcribed by using a TransScript II First-Strand cDNA Synthesis SuperMix kit (Transgen). The qRT–PCR analysis was executed by using the kit named SYBR FAST qPCR (KAPA). The qPCR primer pairs sequences for *LOC_Os06g01320* were 5′-CAAAAAAAAAGACAATAAGGTGGA-3′ (forward) and 5′-CAGACATTGCTTACCCTTATTTATTTT-3′ (reverse). EF-1 alpha was used as an internal control, and the sequencing was 5′-GCACGCTCTTCTTGCTTTCACTCT-3′ (forward) and 5′-AAAGGTCACCACCATACCAGGCTT-3′ (reverse).

Author Contributions: S.J., L.W., X.Y., J.W. conducted the phenotyping and genotyping of parental lines and individuals of the RIL population. C.Y. and S.J. performed the SNP calling, data analyses, and bin-map construction. X.Z., X.S., Q.X., B.L., H.L., and W.L. participated in the phenotyping and data analysis. S.J., Z.L., and W.L. designed the experiment. S.J., C.Y., Q.X. and X.S. drafted the manuscript. All authors have read and agreed to the published version of the manuscript.

Funding: This work was supported by a grant from the National Key Research and Development Program of China (2016YFD0300104), the National Natural Science Foundation of China (31661143012), the Natural Science Foundation of Heilongjiang Province of China (C2016050, YQ2019C020), Heilongjiang Postdoctoral Financial Assistance (LBH-Q15133), Provincial funding for the National Key Research and Development Program in Heilongjiang Province (768001), and Heilongjiang Province Agricultural Science and Technology Innovation Project (2018CQJC002, 2019CQJC002).

Conflicts of Interest: The authors declare no conflict of interest. The funders had no role in the design of the study; in the collection, analyses, or interpretation of data; in the writing of the manuscript, or in the decision to publish the results.

References

1. Qian, Q. Genomics assisted germplasm improvement. *J. Integr. Plant Biol.* **2018**, *60*, 82–84. [CrossRef]

2. Zhang, Q.; Chen, Q.; Wang, S.; Hong, Y.; Wang, Z. Rice and cold stress methods for its evaluation and summary of cold tolerance-related quantitative trait loci. *Rice* **2014**, *7*, 24. [CrossRef] [PubMed]

3. Zhang, Z.; Li, J.; Pan, Y.; Li, J.; Zhou, L.; Shi, H.; Zeng, Y.; Guo, H.; Yang, S.; Zheng, W.; et al. Natural variation in *CTB4a* enhances rice adaptation to cold habitats. *Nat. Commun.* **2017**, *8*, 14788. [CrossRef] [PubMed]

4. Lu, G.; Wu, F.; Wu, W.; Wang, H.; Zheng, X.; Zhang, Y.; Chen, X.; Zhou, K.; Jin, M.; Cheng, Z.; et al. Rice *LTG1* is involved in adaptive growth and fitness under low ambient temperature. *Plant J.* **2014**, *78*, 468–480. [CrossRef] [PubMed]

5. Liu, C.; Wang, W.; Mao, B.; Chu, C. Cold stress tolerance in rice: Physiological changes, molecular mechanism, and future prospects. *Hereditas (Beijing)* **2018**, *40*, 171–185.

6. Wang, Z.; Wang, J.; Wang, F.; Bao, Y.; Wu, Y.; Zhang, H. Genetic control of germination ability under cold stress in rice. *Rice Sci.* **2009**, *16*, 173–180. [CrossRef]

7. Shi, Y.; Ding, Y.; Yang, S. Molecular regulation of CBF signaling in cold acclimation. *Trends Plant Sci.* **2018**, *23*, 623–637. [CrossRef]

8. Yonemaru, J.; Yamamoto, T.; Fukuoka, S.; Uga, Y.; Hori, K.; Yano, M. Q-TARO: QTL annotation rice online database. *Rice* **2010**, *3*, 194. [CrossRef]

9. Fujino, K.; Sekiguchi, H.; Matsuda, Y.; Sugimoto, K.; Ono, K.; Yano, M. Molecular identification of a major quantitative trait locus, *qLTG3-1*, controlling low-temperature germinability in rice. *Proc. Natl. Acad. Sci. USA* **2008**, *105*, 12623–12628. [CrossRef]

10. Wang, X.; Zou, B.; Shao, Q.; Cui, Y.; Lu, S.; Zhang, Y.; Huang, Q.; Huang, J.; Hua, J. Natural variation reveals that *OsSAP16* controls low-temperature germination in rice. *J. Exp. Bot.* **2018**, *69*, 413–421. [CrossRef]

11. Liu, F.; Xu, W.; Song, Q.; Tan, L.; Liu, J.; Zhu, Z.; Fu, Y.; Su, Z.; Sun, C. Microarray-assisted fine-mapping of quantitative trait loci for cold tolerance in rice. *Mol. Plant* **2013**, *6*, 757–767. [CrossRef] [PubMed]

12. Ma, Y.; Dai, X.; Xu, Y.; Luo, W.; Zheng, X.; Zeng, D.; Pan, Y.; Lin, X.; Liu, H.; Zhang, D.; et al. *COLD1* confers chilling tolerance in rice. *Cell* **2015**, *160*, 1209–1221. [CrossRef] [PubMed]

13. Zhao, J.; Zhang, S.; Dong, J.; Yang, T.; Mao, X.; Liu, Q.; Wang, X.; Liu, B. A novel functional gene associated with cold tolerance at the seedling stage in rice. *Plant Biotechnol. J.* **2017**, *15*, 1141–1148. [CrossRef] [PubMed]

14. Liu, C.; Ou, S.; Mao, B.; Tang, J.; Wang, W.; Wang, H.; Cao, S.; Schläppi, M.; Zhao, B.; Xiao, G.; et al. Early selection of *bZIP73* facilitated adaptation of *japonica* rice to cold climates. *Nat. Commun.* **2018**, *9*, 3302. [CrossRef] [PubMed]

15. Xiao, N.; Gao, Y.; Qian, H.; Gao, Q.; Wu, Y.; Zhang, D.; Zhang, X.; Yu, L.; Li, Y.; Pan, C.; et al. Identification of genes related to cold tolerance and a functional allele that confers cold tolerance. *Plant Physiol.* **2018**, *177*, 1108–1123. [CrossRef]

16. Mao, D.; Xin, Y.; Tan, Y.; Hu, X.; Bai, J.; Liu, Z.; Yu, Y.; Li, L.; Peng, C.; Fan, T.; et al. Natural variation in the *HAN1* gene confers chilling tolerance in rice and allowed adaptation to a temperate climate. *Proc. Natl. Acad. Sci. USA* **2019**, *116*, 3494–3501. [CrossRef]

17. Saito, K.; Hayanosaito, Y.; Kuroki, M.; Sato, Y. Map-based cloning of the rice cold tolerance gene *ctb1*. *Plant Sci.* **2010**, *179*, 97–102. [CrossRef]

18. Liu, C.; Schläppi, M.; Mao, B.; Wang, W.; Wang, A.; Chu, C. The bZIP73 transcription factor controls rice cold tolerance at the reproductive stage. *Plant Biotechnol. J.* **2019**, *17*, 1834–1849.

19. Huang, X.; Feng, Q.; Qian, Q.; Zhao, Q.; Wang, L.; Wang, A.; Guan, J.; Fan, D.; Weng, Q.; Huang, T.; et al. High-throughput genotyping by whole-genome resequencing. *Genome Res.* **2009**, *19*, 1068–1076. [CrossRef]

20. Chen, Z.; Wang, B.; Dong, X.; Liu, H.; Ren, L.; Chen, J.; Hauck, A.; Song, W.; Lai, J. An ultra-high density bin-map for rapid QTL mapping for tassel and ear architecture in a large F2 maize population. *BMC Genom.* **2014**, *15*, 433. [CrossRef]

21. Jiang, N.; Shi, S.; Shi, H.; Khanzada, H.; Wassan, G.; Zhu, C.; Peng, X.; Yu, Q.; Chen, X.; He, X.; et al. Mapping QTL for seed germinability under low temperature using a new high-density genetic map of rice. *Front. Plant Sci.* **2007**, *8*, 1223. [CrossRef] [PubMed]

22. Song, W.; Wang, B.; Hauck, A.; Dong, X.; Li, J.; Lai, J. Genetic dissection of maize seedling root system architecture traits using an ultra-high density bin-map and a recombinant inbred line population. *J. Integr. Plant Biol.* **2016**, *58*, 266–279. [CrossRef] [PubMed]

23. Wang, B.; Liu, H.; Liu, Z.; Dong, X.; Guo, J.; Li, W.; Chen, J.; Gao, C.; Zhu, Y.; Zheng, X.; et al. Identification of minor effect QTLs for plant architecture related traits using super high density genotyping and large recombinant inbred population in maize (*Zea mays*). *BMC Plant Biol.* **2018**, *18*, 17. [CrossRef] [PubMed]

24. Wang, B.; Zhu, Y.; Zhu, J.; Liu, Z.; Liu, H.; Dong, X.; Guo, J.; Li, W.; Chen, J.; Gao, C.; et al. Identification and Fine-Mapping of a Major Maize Leaf Width QTL in a Re-sequenced Large Recombinant Inbred Lines Population. *Front. Plant Sci.* **2018**, *9*, 101. [CrossRef] [PubMed]

25. Wang, L.; Wang, A.; Huang, X.; Zhao, Q.; Dong, G.; Qian, Q.; Sang, T.; Han, B. Mapping 49 quantitative trait loci at high resolution through sequencing-based genotyping of rice recombinant inbred lines. *Theor. Appl. Genet.* **2010**, *122*, 327–340. [CrossRef]

26. Zhou, Z.; Zhang, C.; Zhou, Y.; Hao, Z.; Wang, Z.; Zeng, X.; Di, H.; Li, M.; Zhang, D.; Yong, H.; et al. Genetic dissection of maize plant architecture with an ultra-high density bin map based on recombinant inbred lines. *BMC Genom.* **2016**, *17*, 178. [CrossRef]

27. Zou, G.; Zhai, G.; Feng, Q.; Yan, S.; Wang, A.; Zhao, Q.; Shao, J.; Zhang, Z.; Zou, J.; Han, B.; et al. Identification of QTLs for eight agronomically important traits using an ultra-high-density map based on SNPs generated from high-throughput sequencing in sorghum under contrasting photoperiods. *J. Exp. Bot.* **2012**, *63*, 5451–5462. [CrossRef]

28. Zhao, M.; Sun, J.; Xiao, Z.; Cheng, F.; Xu, H.; Tang, L.; Chen, W.; Xu, Z.; Xu, Q. Variations in *DENSE AND ERECT PANICLE 1* (*DEP1*) contribute to the diversity of the panicle trait in high-yielding *japonica* rice varieties in northern China. *Breed. Sci.* **2016**, *66*, 599–605. [CrossRef]

29. Lee, J.; Lee, W.; Kwon, S. A quantitative shotgun proteomics analysis of germinated rice embryos and coleoptiles under low-temperature conditions. *Proteome Sci.* **2015**, *13*, 27. [CrossRef]

30. Sasaki, T.; Kinoshita, T.; Takahashi, M. Estimation of the number of genes in the germination ability at low temperature in rice genetical. *J. Fac. Agric. Hokkaido Univ.* **1974**, *57*, 301–312.

31. Shakiba, E.; Edwards, J.; Jodari, F.; Duke, S.; Baldo, A.; Korniliev, P.; McCouch, S.; Eizenga, G. Genetic architecture of cold tolerance in rice (*Oryza sativa*) determined through high resolution genome-wide analysis. *PLoS ONE* **2017**, *12*, e0172133. [CrossRef] [PubMed]

32. Teng, S.; Zeng, D.; Qian, Q.; Yasufumi, K.; Huang, D.; Zhu, L. QTL analysis of rice low temperature germinability. *Chin. Sci. Bull.* **2001**, *46*, 1800–1803. [CrossRef]

33. Ji, S.; Jiang, L.; Wang, Y.; Liu, S.; Liu, X.; Zhai, H.; Wan, J. Detection and analysis of QTL for germination rate at low temperature in rice (*Oryza sativa* L.). *J. Nanjing Agric. Uni.* **2007**, *30*, 1–6.

34. Iwata, N.; Fujino, K. Genetic effects of major QTLs controlling low-temperature germinability in different genetic backgrounds in rice (*Oryza sativa* L.). *Genome* **2010**, *53*, 763–768. [CrossRef] [PubMed]

35. Ji, S.; Jiang, L.; Wang, Y.; Zhang, W.; Liu, X.; Liu, S.; Chen, L.; Zhai, H.; Wan, J. Quantitative trait loci mapping and stability for low temperature germination ability of rice. *Plant Breed.* **2009**, *128*, 387–392. [CrossRef]

36. Hou, M.; Wang, C.; Jiang, L.; Wan, J.; Hideshi, Y.; Atsushi, Y. Inheritance and QTL mapping of low temperature germinability in rice (*Oryza sativa* L.). *Acta. Genet. Sin.* **2004**, *31*, 701–706.

37. Fujino, K.; Obara, M.; Shimizu, T.; Koyanagi, K.-O.; Ikegaya, T. Genome-wide association mapping focusing on a rice population derived from rice breeding programs in a region. *Breed. Sci.* **2015**, *65*, 403–410. [CrossRef]

38. Li, L.; Liu, X.; Xie, K.; Wang, Y.; Liu, F.; Lin, Q.; Wang, W.; Yang, C.; Lu, B.; Liu, S.; et al. *qLTG-9*, a stable quantitative trait locus for low-temperature germination in rice (*Oryza sativa* L.). *Theor. Appl. Genet.* **2013**, *126*, 2313–2322. [CrossRef]

39. Ji, S.; Jiang, L.; Wang, Y.; Liu, S.; Liu, X.; Zhai, H.; Yoshimura, A.; Wan, J. QTL and epistasis for low temperature germinability in rice. *Acta. Agron. Sin.* **2008**, *34*, 551–556.

40. Jiang, S.; Zhang, X.; Zhang, F.; Xu, Z.; Chen, W.; Li, Y. Identification and Fine Mapping of *qCTH4*, a quantitative trait loci controlling the chlorophyll content from tillering to heading in rice (*Oryza sativa* L.). *J. Hered.* **2012**, *103*, 720–726. [CrossRef]

41. Jiang, S.; Sun, S.; Bai, L.; Ding, G.; Wang, T.; Xia, T.; Jiang, H.; Zhang, X.; Zhang, F. Resequencing and variation identification of whole genome of the japonica rice variety "Longdao24" with high yield. *PLoS ONE* **2017**, *12*, e0181037. [CrossRef] [PubMed]

42. Li, H.; Durbin, R. Fast and accurate long-read alignment with Burrows-Wheeler transform. *Bioinformatics* **2010**, *26*, 589–595. [CrossRef] [PubMed]

43. Kawahara, Y.; De la Bastide, M.; Hamilton, J.; Kanamori, H.; McCombie, W.; Ouyang, S.; Schwartz, D.; Tanaka, T.; Wu, J.; Zhou, S.; et al. Improvement of the *Oryza sativa* Nipponbare reference genome using next generation sequence and optical map data. *Rice* **2013**, *6*, 4. [CrossRef] [PubMed]

44. Li, H.; Handsaker, B.; Wysoker, A.; Fennell, T.; Ruan, J.; Homer, N.; Marth, G.; Abecasis, G.; Durbin, R.; Proc, G. The Sequence Alignment/Map format and SAMtools. *Bioinformatics* **2009**, *25*, 2078–2079. [CrossRef] [PubMed]

45. McKenna, A.; Hanna, M.; Banks, E.; Sivachenko, A.; Cibulskis, K.; Kernytsky, A.; Garimella, K.; Altshuler, D.; Gabriel, S.; Daly, M.; et al. The Genome Analysis Toolkit: A MapReduce framework for analyzing next-generation DNA sequencing data. *Genome Res.* **2010**, *20*, 1297–1303. [CrossRef]

46. Han, K.; Jeong, H.; Yang, H.; Kang, S.; Kwon, J.; Kim, S.; Choi, D.; Kang, B. An ultra-high-density bin map facilitates high-throughput QTL mapping of horticultural traits in pepper (*Capsicum annuum*). *DNA Res.* **2016**, *23*, 81–91. [CrossRef]

47. Arends, D.; Prins, P.; Jansen, R.; Broman, K. R/qtl: High-throughput multiple QTL mapping. *Bioinformatics* **2010**, *26*, 2990–2992. [CrossRef]

International Journal of
Molecular Sciences

Article

Genetic Dissection of Seed Dormancy using Chromosome Segment Substitution Lines in Rice (*Oryza sativa* L.)

Shaowen Yuan [1], Yuntong Wang [2,3], Chaopu Zhang [1,2], Hanzi He [1,*] and Sibin Yu [1,2,*]

[1] College of Plant Science and Technology, Huazhong Agricultural University, Wuhan 430070, Hubei, China; wenshaoY@webmail.hzau.edu.cn (S.Y.); zchaopu@163.com (C.Z.)
[2] National Key Laboratory of Crop Genetic Improvement, Huazhong Agricultural University, Wuhan 430070, Hubei, China; wangyt@biomarker.com.cn
[3] Biomarker Technologies Corporation, Beijing 101300, China
[*] Correspondence: hzhe@mail.hzau.edu.cn (H.H.); ysb@mail.hzau.edu.cn (S.Y.)

Received: 14 January 2020; Accepted: 14 February 2020; Published: 17 February 2020

Abstract: Timing of germination determines whether a new plant life cycle can be initiated; therefore, appropriate dormancy and rapid germination under diverse environmental conditions are the most important features for a seed. However, the genetic architecture of seed dormancy and germination behavior remains largely elusive. In the present study, a linkage analysis for seed dormancy and germination behavior was conducted using a set of 146 chromosome segment substitution lines (CSSLs), of which each carries a single or a few chromosomal segments of Nipponbare (NIP) in the background of Zhenshan 97 (ZS97). A total of 36 quantitative trait loci (QTLs) for six germination parameters were identified. Among them, *qDOM3.1* was validated as a major QTL for seed dormancy in a segregation population derived from the *qDOM3.1* near-isogenic line, and further delimited into a genomic region of 90 kb on chromosome 3. Based on genetic analysis and gene expression profiles, the candidate genes were restricted to eight genes, of which four were responsive to the addition of abscisic acid (ABA). Among them, LOC_Os03g01540 was involved in the ABA signaling pathway to regulate seed dormancy. The results will facilitate cloning the major QTLs and understanding the genetic architecture for seed dormancy and germination in rice and other crops.

Keywords: seed dormancy; quantitative trait locus; ABA; seed germination; chromosome segment substitution lines; linkage mapping

---

1. Introduction

Seed dormancy is an important evolutionary trait. It can optimize the distribution and timing of germination over time in nature [1]. Seed dormancy also plays an important role in agricultural production. Extremely strong dormancy leads to a low germination rate in the field, the irregular emergence of seedlings and an impact on sowing time, and can even affect the final yield. On the contrary, too weak dormancy leads to pre-harvest sprouting (PHS), especially in the high temperature and rainy environment during seed maturity. The economic loss caused by pre-harvest sprouting has become an important factor restricting the yield of cereal crops (such as rice, wheat, maize, etc.) [2], and seriously affects the sowing quality and processing quality of crops, and even causes the change of storage quality [3]. Therefore, crop seeds require a well-balanced level of dormancy to ensure a high rate of germination and to control pre-harvest sprouting in the field.

Seed dormancy is a highly complex trait and largely influenced by genetic and environmental factors [4]. Recent progress in plant genomics and various genetic populations has facilitated the identification of quantitative trait loci (QTLs) for seed dormancy in many species, for example, in *Arabidopsis* [5], *Lepidium sativum* [6], oilseed rape [7], sorghum [8], barley [9], and wheat [10–12].

By using different genetic populations constructed from cultivated rice, wild rice, and weedy rice, more than 160 QTLs have been identified that affect the germination or dormancy of rice (available online: https://archive.gramene.org/qtl/). For example, five seed dormancy QTLs were detected by BC1 and F_2 populations constructed from rice variety N22 with strong dormancy and two weak dormant varieties [13]. Three seed dormancy QTLs were mapped using the chromosomal segment substitution line (CSSL) population and their derived F_2 populations constructed by the strong dormant variety Nona Bokra and the weak dormant variety Koshihikari [14]. By using a population of recombinant inbred lines (RILs), nine seed dormancy QTLs were identified in three developmental stages [15]. In addition, besides biparental genetic populations, genome-wide association analysis across natural accession also revealed genetic variation of seed dormancy among rice natural populations, and the analysis can be used to identify new candidate genes related to seed dormancy [16,17].

In 2010, Japanese scientists isolated the first rice seed dormancy gene *Seed dormancy 4 (Sdr4)* by map-based cloning [18]. *Sdr4* can be positively regulated by the seed maturation-related gene *OsVP1* and two *Arabidopsis* dormancy gene *Delay of Germination 1 (DOG1)* [19] homologous genes, thereby enhancing seed dormancy. After that, several rice seed dormancy genes were identified and proved to be involved in the hormonal regulation of seed dormancy. By rice mutant screening, *PHS8* was isolated to be a starch debranching enzyme named isoamylase1, and it determined seed dormancy and germination by affecting abscisic acid (ABA) signaling [20]. The rice *GERMINATION DEFECTIVE 1* regulated seed germination by integrating gibberellin acid (GA) and carbohydrate metabolism [21]. A weedy red rice dormancy QTL (*SD7-1/Rc*) was identified as a basic helix-loop-helix transcription factor that controls ABA synthesis, influencing red pericarp color and seed dormancy [22]. By map-based cloning, a gibberellin synthesis gene *OsGA20ox2* was identified within QTL *Seed Dormancy1-2 (qSD1-2)* [23]. *OsGA20ox2* (the green revolutionary gene *SD1*) is involved in the biosynthesis of GA, regulating the development of endosperm-imposed dormancy in rice.

Hormonal regulation may be a highly conserved mechanism of seed dormancy among many species. The balance of ABA and GA or other hormones plays crucial roles in the regulation of seed dormancy and germination [24,25]. ABA is an essential positive regulator of both dormancy induction during seed maturation and maintenance of the dormant state after imbibition [26–28]. In *Arabidopsis*, *DOG1* was the first cloned dormancy QTL, and encoded a protein with unknown functional domain [19] and had conserved function throughout many species. In recent years, *DOG1* was reported to play a regulatory role in ABA signaling. It encodes for a plant-specific protein that enhances ABA signaling through its binding to protein phosphatase 2C (PP2C) ABA HYPERSENSITIVE GERMINATION1 (AHG1) and AHG3 [29,30]. In addition, *DOG1* may mediate a conserved seed coat dormancy mechanism in the temperature- and GA-dependent pathways [31]. Besides *DOG1*, the previously mentioned seed dormancy genes, such as *PHS8* and *SD7-1/Rc*, were involved in either ABA metabolism or signaling pathway. Thus, it is worthwhile to investigate whether there are more genes involved in the hormonal regulation of seed dormancy.

Here, we presented the identification of QTLs for seed dormancy in a set of genome-wide single nucleotide polymorphism (SNP) genotyped chromosomal segment substitution lines (CSSLs) by backcrossing and marker-assisted selection, in which *japonica* Nipponbare (NIP) was the donor parent and the recurrent parent was *indica* Zhenshan 97 (ZS97). The CSSL population, which comprised 146 lines, was developed and genotyped in a previous study [32,33] and has not been used for a seed dormancy study.

Therefore, the objectives of the present study were to dissect the genetic base of seed dormancy and germination performances in the CSSL population, and to fine map the major QTLs using the CSSL-derived population. Moreover, we investigated how the candidate gene was involved in the ABA regulation of seed dormancy.

2. Results

2.1. Seed Dormancy Variation in Parents

The germination assay for the freshly harvested seed of the parental lines, which were *indica* variety Zhenshan97 (ZS97) and *japonica* variety Nipponbare (NIP), was performed. The number of germinated seeds was counted daily for seven consecutive days and the germination percentage was calculated each day to construct cumulative germination curves. ZS97 and NIP exhibited significant difference in seed dormancy, in which ZS97 had significantly higher germination percentage (100%) than NIP (37%) at seven days (168 h of germination) (Figure 1A). Dry seeds were treated at 43 °C for three days to break seed dormancy (which was called after-ripened seeds) and then underwent a seed germination experiment. After seed dormancy was broken, the germination percentage of seven days (168 h) of the parental lines was almost the same (around 90%) (Figure 1B). However, the germination rate of NIP was still lower as compared to ZS97. At two days of germination (48 h), the radicle protrusion of ZS97 was 86%, whereas no radicle protrusion occurred for NIP (Figure 1B).

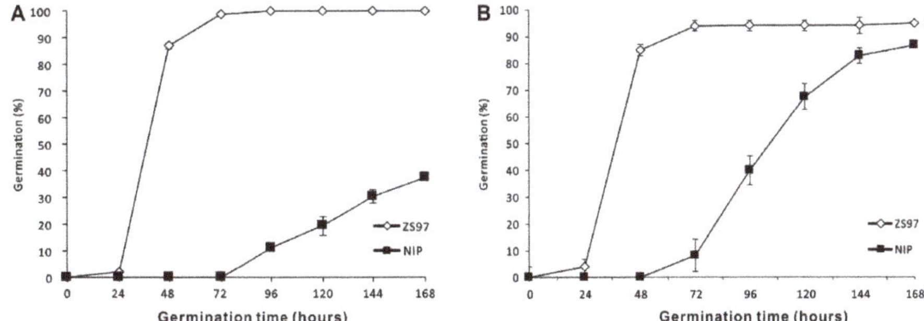

Figure 1. (**A**) Germination behavior of freshly harvested *japonica* variety Nipponbare (NIP) and *indica* variety Zhenshan97 (ZS97) seeds; (**B**) germination behavior of after-ripened NIP and ZS97 seeds.

2.2. Seed Dormancy Variation in the CSSL Population

The CSSL population, which comprised 146 lines, was developed and genotyped in a previous study [32,33] and the lines were used to dissect the genetic architecture of seed dormancy underlying this population.

Seed germination was analyzed using the six germination parameters from the Germinator package [34], which were G_{max} (maximum germination percentage at seven days); G_{3d} (germination percentage atthree3 days), T_{50} (germination speed: time to reach 50% germination of the total number of germinated seeds), U_{8416} (germination uniformity: time interval between 16% and 84% of viable seed to germinate), AUC (area under the curve), and GI (germination index). The mean performance of the CSSL population is presented in Table 1. The frequency distribution for the six germination parameters of the CSSL population showed large variation (Figure S1, Supplementary Materials). These results indicated that there might be seed dormancy QTLs in this CSSL population.

Table 1. Germination parameters of the two parents and the chromosomal segment substitution line (CSSL) population.

Index	Parents		CSSLs		
	NIP	ZS97	Mean ± SD	CV%	Min to Max
G_{max} (%)	37.7 ± 3.2**	100.0 ± 0.0	96.7 ± 7.03	7.26	36.4 to 100.0
G_{3d} (%)	0**	91.01 ± 6.7	92.3 ± 14.90	16.15	12.0 to 100.0
T_{50}	-	36.8 ± 2.5	41.6 ± 9.66	23.22	24.4 to 91.0
U_{8416}	47.9 ± 6.3**	11.2 ± 2.3	15.9 ± 11.21	70.57	4.14 to 75.4
AUC	19.2 ± 0.03**	130.3 ± 2.4	121.2 ± 15.24	12.57	45.1 to 141.9
GI	8.4 ± 0.5**	76.6 ± 2.9	23.5 ± 3.97	16.92	8.27 to 35.9

G_{max}, maximum germination percentage of seven days of germination; G_{3d}, germination percentage at three days; T_{50}, time to reach 50% germination of the total number of germinated seeds; U_{84-16}, germination uniformity, which is time interval between 84% and 16% of viable seed to germinate; AUC; area under the germination curve until 168 h; GI, germination index. SD, standard deviation; CV, coefficient variation; Min to Max, the minimum and maximum value in the CSSL population. Asterisks **, indicate significant difference between the parents at the level of 0.01.

2.3. Seed Dormancy QTL Detection in CSSL Population

To determine the genetic regions controlling seed dormancy, the CSSL population was genotyped by the RICE6K SNP array. A total of 518 bins (defined as Bin1 to Bin518) across the whole genome were obtained [33]. QTL mapping was carried out by using ridge regression analysis with the 518 bins, and it identified 9, 19, 25, 23, 17, and 21 QTLs for G_{max}, G_{3d}, T_{50}, U_{8416}, AUC, and GI, respectively (Figure 2). Detailed information about the *p*-value, phenotypic variation explained, and the effect of the identified QTLs are shown in Table 2.

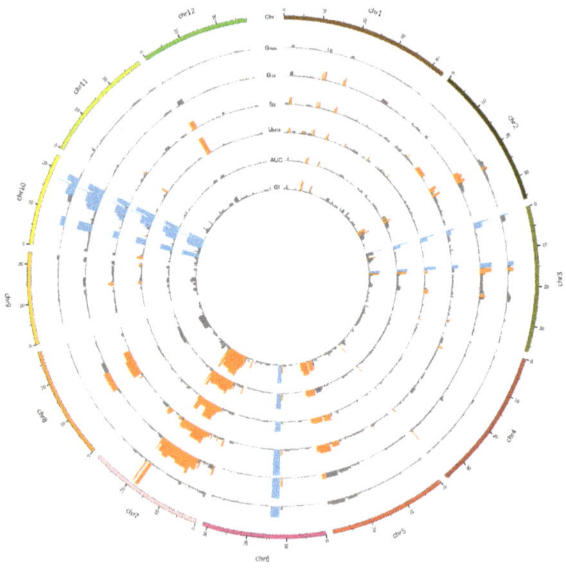

Figure 2. Circos plot illustrating the quantitative trait loci (QTLs) of six germination parameters. Chr: Size of the 12 chromosomes of *Oryza sativa*; G_{max}, maximum germination percentage of seven days germination; G_{3d}, germination percentage at three days; T_{50}, time to reach 50% germination of the total number of germinated seeds; U_{8416}, germination uniformity, which is time interval between 84% and 16% of viable seed to germinate; AUC; area under the germination curve until 168 h; GI, germination index. Blue indicates significant QTLs identified by all six germination parameters. Orange indicates significant QTLs identified in the corresponding germination parameter. Gray indicates non-significant QTLs.

Table 2. QTLs identified for six germination parameters in the NIP/ZS97 CSSL population using the single nucleotide polymorphism (SNP) bin markers.

Chr	Interval (Mb)	QTL	G_{max}[a]			G_{3d}			T_{50}			U_{8416}			AUC			GI			QTL/Gene[c]
			p[b]	PVE%	E	p	PVE%	E	p	PVE%	E	p	PVE%	E	p	PVE%	E	p	PVE%	E	
1	2.3–3.0	qDOM1.1	_[d]	-	-	-	-	-	4.4	0.9	1.59	3.0	1.7	1.46	-	-	-	-	-	-	
1	12.5–13.8	qDOM1.2	-	-	-	3.6	1.7	0.00	4.2	1.5	1.19	4.2	1.3	1.75	2.6	1.0	−0.37	3.1	1.5	−0.17	
1	19.5–20.3	qDOM1.4	-	-	-	3.3	1.8	0.00	3.4	1.7	1.49	4.0	2.1	1.85	2.7	1.2	−0.44	2.3	1.3	−0.18	
1	28.5–28.8	qDOM1.5	-	-	-	-	-	-	-	-	-	2.2	1.7	−2.03	2.5	1.2	−0.44	2.2	1.1	−0.09	
1	40.2–40.6	qDOM1.6	-	-	-	-	-	-	-	-	-	3.1	0.4	0.77	-	-	-	-	-	-	
2	6.7–6.9	qDOM2.1	-	-	-	-	-	-	3.5	3.9	0.67	3.0	2.0	0.69	-	-	-	2.2	1.6	−0.09	qDOR-2
2	12.1–23.1	qDOM2.2	-	-	-	-	-	-	2.7	1.0	−2.81	-	-	-	-	-	-	-	-	-	
2	23.9–24.9	qDOM2.3	2.2	2.1	−0.01	3.6	1.8	0.00	5.1	2.2	1.88	-	-	-	3.2	1.7	−0.53	3.0	1.8	−0.21	
3	0.2–2.1	qDOM3.1	3.8	4.9	−0.01	6.6	5.2	−0.01	4.9	5.1	2.88	9.2	8.3	4.46	6.6	6.4	−0.87	4.4	3.4	−0.30	
3	3.9–4.4	qDOM3.2	-	-	-	-	-	-	-	-	-	-	-	-	-	-	-	2.3	3.3	0.14	qSD-3
3	13.2–14.8	qDOM3.3	2.6	2.6	−0.01	4.9	3.1	0.00	4.1	3.5	1.86	7.2	5.5	2.78	4.5	3.7	−0.58	3.0	1.8	−0.19	qSD-3-2
3	14.8–16.9	qDOM3.4	2.3	2.6	−0.01	4.7	2.9	0.00	-	-	-	2.3	4.4	1.72	-	-	-	2.7	1.6	−0.18	
3	23.0–24.7	qDOM3.5	-	-	-	2.3	1.4	0.00	-	-	-	2.8	2.0	1.13	-	-	-	-	-	-	
3	36.2–36.4	qDOM3.6	-	-	-	-	-	-	-	-	-	2.3	1.1	−0.88	-	-	-	-	-	-	
4	5.5–6.0	qDOM4.1	-	-	-	-	-	-	3.1	1.4	−2.51	-	-	-	-	-	-	-	-	-	
4	34.6–4.9	qDOM4.2	-	-	-	5.4	3.0	0.00	2.5	3.6	0.90	-	-	-	3.1	1.3	−0.37	3.1	1.4	−0.15	
5	7.2–7.3	qDOM5.1	-	-	-	-	-	-	3.2	1.2	−1.13	-	-	-	-	-	-	-	-	-	
5	21.4–24.1	qDOM5.2	-	-	-	-	-	-	2.9	1.6	1.47	4.1	1.9	2.00	-	-	-	2.5	2.7	−0.19	
5	25.0–30.0	qDOM5.3	-	-	-	2.1	4.9	−0.01	4.9	3.0	3.50	3.2	2.3	3.00	2.6	5.6	−0.63	3.4	4.0	−0.33	
6	11.0–11.5	qDOM6.1	-	-	-	3.5	2.0	−0.01	-	-	-	-	-	-	2.8	1.5	−0.69	2.1	1.1	−0.26	
6	12.0–14.4	qDOM6.2	4.3	8.6	−0.02	8.2	7.4	−0.01	15.7	6.5	6.44	6.1	3.6	4.34	6.4	4.9	−1.47	4.8	3.3	−0.54	
7	0.0–0.5	qDOM7.1	-	-	-	4.0	2.4	−0.01	4.7	2.1	2.72	-	-	-	3.1	1.7	−0.72	2.3	1.2	−0.27	
7	2.4–2.8	qDOM7.2	-	-	-	-	-	-	2.7	1.1	−1.15	-	-	-	-	-	-	-	-	-	
7	4.9–5.2	qDOM7.3	-	-	-	3.9	2.2	0.00	3.2	1.7	1.10	2.4	3.1	0.96	3.0	1.5	−0.29	2.8	1.8	−0.12	
7	5.7–17.7	qDOM7.4	-	-	-	7.7	6.2	0.00	8.3	8.1	2.01	9.5	9.9	2.78	5.8	3.9	−0.55	5.6	4.7	−0.25	qSD7-1/Rc
7	17.7–18.5	qDOM7.5	10.5	9.6	−0.01	5.4	6.3	0.00	-	-	-	-	-	-	6.7	8.3	−0.54	5.9	6.7	−0.21	
7	19.2–19.5	qDOM7.6	9.3	7.2	−0.01	5.9	7.3	0.00	-	-	-	-	-	-	6.4	7.5	−0.61	4.5	3.4	−0.22	qSD.7
7	22.6–24.4	qDOM7.5	-	-	-	-	-	-	4.0	1.4	−1.36	-	-	-	-	-	-	-	-	-	Sdr4
8	8.1–19.1	qDOM8.1	-	-	-	2.4	4.6	−0.01	5.6	5.0	5.04	-	-	-	-	-	-	-	-	-	
9	13.8–15.0	qDOM9.1	-	-	-	-	-	-	3.4	3.7	1.60	2.2	1.7	1.72	-	-	-	-	-	-	qDOR-9-1
9	21.5–22.6	qDOM9.2	-	-	-	-	-	-	-	-	-	2.1	1.2	0.77	-	-	-	-	-	-	qDOR-9-2
10	0.1–1.4	qDOM10.1	-	-	-	-	-	-	-	-	-	3.8	3.6	1.38	-	-	-	-	-	-	
10	6.4–10.3	qDOM10.2	2.1	1.3	0.00	6.0	4.9	0.00	3.8	5.9	1.22	6.4	7.3	1.78	5.4	5.4	−0.43	4.7	4.7	−0.17	
10	13.2–20.3	qDOM10.3	8.2	11.6	−0.02	10.6	11.5	−0.01	15.7	10.4	5.44	15.7	12.5	6.87	10.7	12.9	−1.33	6.7	5.8	−0.44	OsFbx352
11	6.5–7.2	qDOM11.1	-	-	-	-	-	-	2.7	0.4	−1.00	4.5	0.7	1.37	-	-	-	-	-	-	
12	1.1–3.0	qDOM12.1	-	-	-	-	-	-	5.2	3.6	9.13	10.8	8.8	14.90	-	-	-	-	-	-	qSD12

a. G_{max}: maximum germination percentage of seven days of germination, G_{3d}: germination percentage at three days, T_{50}: germination speed, which is time to reach 50% germination of the total number of germinated seeds, U_{8416}: germination uniformity, which is time interval between 84% and 16% of viable seed to germinate, AUC: area under the germination curve, GI: germination index; b. When two or more consecutive bins were significant, the lowest p-value was selected and displayed as −log10(P). PVE presents the phenotypic variance explained for a given QTL. E indicates estimated additive effect by a given QTL. For G_{max}, G_{3d}, AUC, and GI, the positive E value represents the NIP alleles that increased the effect, and for T_{50} and U_{8416} the positive E value represents the ZS97 alleles that increased the effect; c. The genes near 500 kb of the most significantly associated SNP; d. - Indicates no QTLs detected.

In total, 36 QTLs were detected for seed dormancy using the six parameters in the CSSL population (Table 2). Among the six parameters, U_{8416} explained the highest phenotypic variance (77.9%) with 23 QTLs detected, while T_{50} explained 71.5% of phenotypic variance with the largest number of QTLs detected (25 QTLs). G_{3d}, AUC, and GI explained 77%, 66.3%, and 52.7% of phenotypic variance with 19, 17, and 21 QTLs detected, respectively. G_{max} only detected 9 QTLs and explained 50.5% of phenotypic variance. The 36 QTLs were distributed on each of the chromosome in which both chromosome 6 and 10 had the highest -log10(p) value (15.7) for U_{8416}, and the phenotypic variance were 6.5% and 10.4%, respectively. Among all 36 QTLs, 10 QTLs were identified or cloned previously for seed dormancy, suggesting the consistency of our QTL analysis with others. The other remaining 26 QTLs may be new ones for seed dormancy, as they do not contain any QTLs for dormancy in rice that have been described before (https://archive.gramene.org/qtl/).

Five out of thirty-six QTLs were common QTLs detected in all six parameters and distributed on chromosomes 3, 6, and 10. One common QTL (*qDOM3.3*) on chromosome 3 was identified as a seed dormancy QTL in the "Asominori×IR24" CSSL population [35], and *qDOM10.3*, which was detected in all six parameters, contained a gene, namely *OsFbx352* [36], that plays a regulatory role in the regulation of glucose-induced suppression of seed germination by targeting ABA metabolism. The other three common QTLs (*qDOM3.1*, *qDOM6.2*, and *qDOM10.2*) have not been reported before. *qDOM3.1* was detected in almost the same region on the upper end of chromosome 3 by the six germination parameters, suggesting the robustness of this QTL in the present CSSL population. U8416 of *qDOM3.1* had the highest -log10(P) value (9.2) among the six parameters and explained 8.3% of phenotypic variance.

2.4. Verification of qDOM3.1 for Seed Dormancy

To validate and fine map *qDOM3.1*, one line, namely NQ96, in the CSSL population was selected (Figure S2, Supplementary Materials). It carries a NIP substitution segment encompassing *qDOM3.1* on top of chromosome 3 in the ZS97 genetic background, with another NIP substitution segment on chromosome 9. The germination behavior of NQ96 was significantly lower and slower than ZS97 (Table 3). This indicated that the introduced NIP segment contained the QTL of seed dormancy. To confirm the genetic effect of the *qDOM3.1* on seed dormancy, we generated an F_2 segregating population comprising 338 individuals by crossing NQ96 with ZS97. The F_2 population was genotyped using ten polymorphic markers in the *qDOM3.1* region and one polymorphic marker on the other introgressed segment on chromosome 9. There was no significant difference on seed dormancy for NIP and ZS97 allele on chromosome 9 with marker RM410, denoting that the introgressed segment on chromosome 9 had no effect on seed dormancy. Thus, NQ96 only contained *qDOM3.1*, which had a genetic effect on seed dormancy.

Table 3. Confirmation of *qDOM3.1* loci by NQ96.

	G_{max}%	G_{3d}%	T_{50} (h)	U_{8416} (h)	AUC	GI
ZS97	100 ± 0	91.01 ± 6.7	36.76 ± 2.48	11.22 ± 2.26	130.29 ± 2.37	76.63 ± 2.97
NQ96	57 ± 7.07**	30 ± 8.49**	97.03 ± 9.79**	32.11 ± 11.82*	63.97 ± 10.13*	35.48 ± 6.36**
NIP	37.71 ± 3.24	0	-	47.9 ± 6.25	19.22 ± 0.03	8.42 ± 0.46

Germination behaviors that are significantly different from that of ZS97 are indicated by asterisks (* $p < 0.05$, ** $p < 0.01$). - no data available.

To determine the genetic effect of *qDOM3.1*, we performed a genetic segregation analysis of seed dormancy using the *qDOM3.1*-derived F_2 population. Frequency distribution of G_{max} in the population indicated that *qDOM3.1* from ZS97 was dominant (Figure 3).

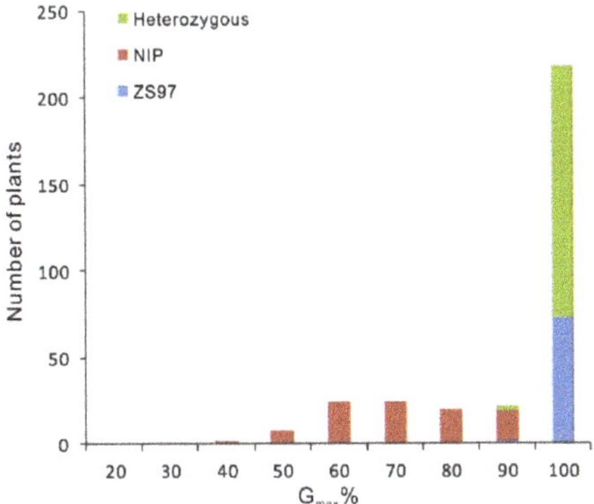

Figure 3. Frequency distribution of G_{max} in F_2 segregation population.

Afterward, based on the genotyping results of ten polymorphic markers distributed within the target region RM14238-RM14317 for the F_2 individuals, *qDOM3.1* was detected in the interval RM14238-MP030012 (approximately 252 kb) with a logarithm of the odds (LOD) score peaked around MP03008 (Figure 4), which explained 69.9%, 75.4%, 73.2%, 71.4%, 38.3%, and 46.5% of the phenotypic variance in G_{max}, G_{3d}, AUC, GI, U8416, and T50, respectively.

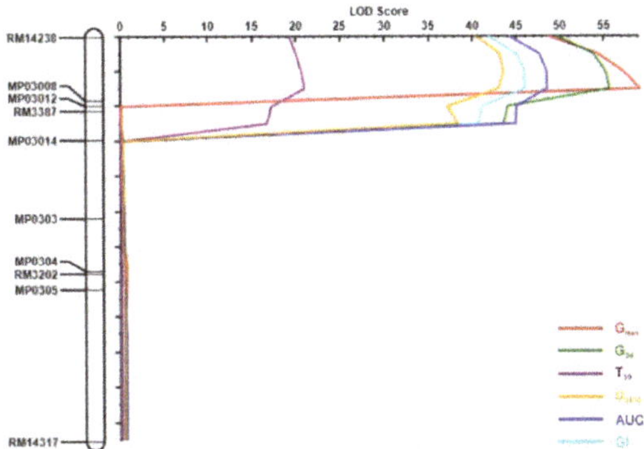

Figure 4. Verification of the QTL effect in the F_2 primary segregation population. QTL scans along chromosome 3 for the six indexes in the CSSL-derived F_2 population from the cross of NQ96 and ZS97. Logarithm of odds profile of QTL region on chromosome 3 in the F_2 population, showing a QTL (*qDOM3.1*) for G_{max}, G_{3d}, T_{50}, U_{8416}, AUC, and GI.

2.5. Fine-Mapping of qDOM3.1 for Seed Dormancy

To further fine map *qDOM3.1*, we selected the heterozygous lines in the CSSL-derived F_2 population flanked by the markers RM14238 and MP030012, and self-pollinated these heterozygous lines to generate a larger segregating population ($n = 2500$). Through genotyping with 14 additional markers, seven informative recombinants with *qDOM3.1* were identified. A progeny test of the informative recombination plants delimited *qDOM3.1* for seed dormancy to a 90 kb region (Figure 5 and Figure S3, Supplementary Materials). This region encompassed 19 open reading frames (ORFs) according to the RGAP database (available online: http://rice.plantbiology.msu.edu/, Release 7).

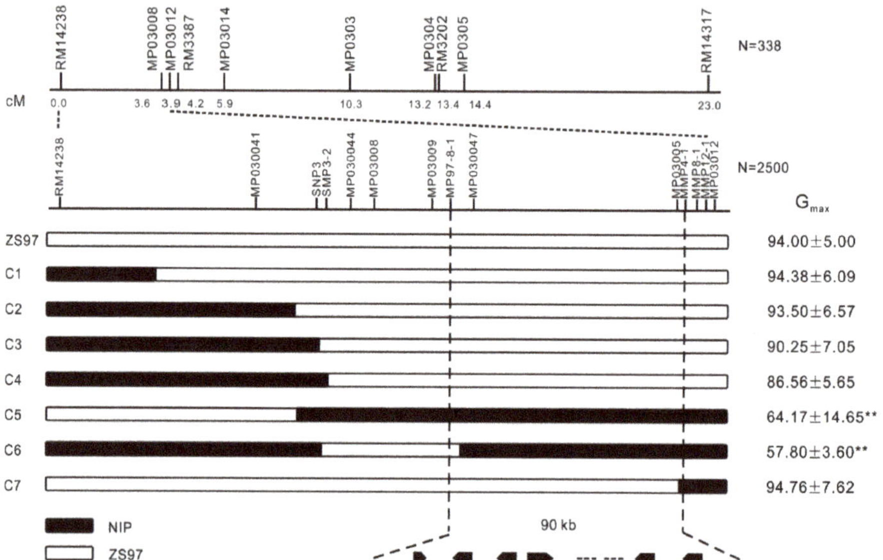

Figure 5. Fine mapping of *qDOM3.1*. The QTLs narrowed down to the region flanked by MP97-8-1 and MMP4-1 on the upper end of chromosome 3. Some important recombinant plants derived from a large F_2 group generated by selfing a single individual heterozygous at the *qDOM3.1* region and divided into 7 groups based on their genotypes. G_{max} (mean ± SD) (%) is given on the right for each genotype. The phenotypes of each recombinant individual were evaluated by germination experiments. ** Indicates significant difference at $p < 0.01$ by Dunnett's test against the control.

The 19 genes included 11 expressed proteins, 1 transposon protein, 2 retrotransposon proteins, 1 hypothetical protein, and 4 genes with functional annotation. The chromosomal synteny analysis between NIP and ZS97 showed there were 7 genes missing in the ZS97 (available online: http://rice.hzau.edu.cn/cgi-bin/gb2/gbrowse_syn/3rice_syn/) (Figure 6). As the candidate gene should be dominant in ZS97 (Figure 3), those 7 genes were ruled out from the candidate genes. Thus, only 12 genes remained, including 10 expressed proteins, 1 gene annotated as tubulin/FtsZ domain-containing protein (LOC_Os03g01530), and another annotated as DNA-binding protein (LOC_Os03g01540). According to the expression profile in the database (http://rice.plantbiology.msu.edu/expression.shtml), 4 genes (LOC_Os03g01430, LOC_Os03g01450, LOC_Os03g01460, and LOC_Os03g01520) had no expression or very low expression among all the tissues. Therefore, those 4 genes were unlikely to be our candidate genes, leaving 8 genes as candidate genes. The expression profiles were obtained from the RiceXPro website (available online: http://ricexpro.dna.affrc.go.jp/) (Figure S4, Supplementary Materials). None of the 8 genes were seed-specific expressed, except LOC_Os03g01360 that had a relatively higher expression in embryo from 7 days after flowering until 42 days after flowering. Based on the sequence comparison between NIP and ZS97, only LOC_Os03g01530 had no amino acid change in the coding region; all the other 7 genes contained at least one missense variant.

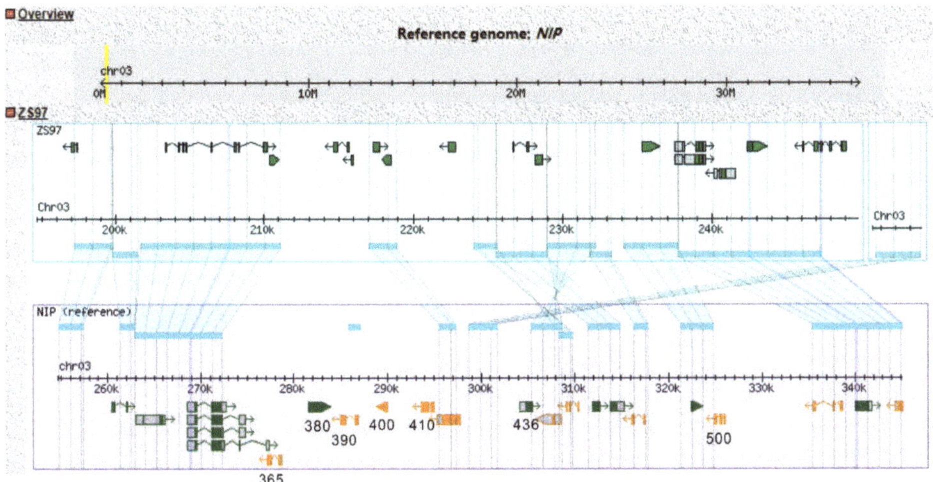

Figure 6. Chromosomal synteny analysis of ZS97 and NIP of candidate region on chromosome 3. The genes in which ZS97 was not contained are indicated with the final three numbers of the gene ID (e.g., LOC_Os03g01365 is displayed as 365). In NIP panel, green gene symbol means gene direction from left to the right and orange gene symbol means gene direction from right to the left.

2.6. qDOM3.1 Increased Seed Endogenous ABA Content and ABA Sensitivity

ABA plays an essential role in the regulation of seed dormancy [4,24,37]. The endogenous ABA level was measured in the near-isogenic line (NIL) of *qDOM3.1* (NIL-NIP) and the corresponding background line (NIL-ZS97). NIL-NIP had an ABA level almost five times higher than that of the NIL-ZS97 (Figure 7).

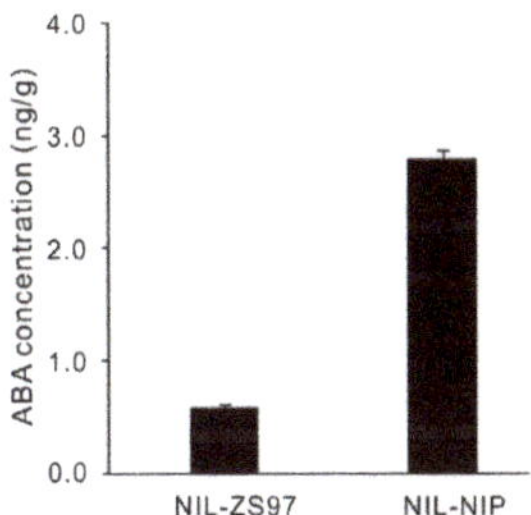

Figure 7. Endogenous ABA concentration in near-isogenic line (NIL)-ZS97 and NIL-NIP.

Subsequently, the ABA sensitivity was investigated in the near-isogenic lines (NIL-NIP and NIL-ZS97) for freshly harvested seeds (Figure 8A) and after-ripened seeds (Figure 8B). The germination percentage and germination speed were significantly lower and slower in NIL-NIP compared with NIL-ZS97 (Figure 8A). Freshly harvested seeds were treated in 43 °C for three days to break seed dormancy, and the germination behavior was almost the same for after-ripened NIL-NIP and NIL-ZS97 (Figure 8B). Then, the pair of near-isogenic lines (after-ripened) were treated in a series of ABA solution to investigate ABA sensitivity. Up to 10μM ABA had no significant effect on seed germination, whereas 20–100μM ABA significantly decreased seed germination of NIL-NIP compared with NIL-ZS97 (Figure 8C). Thus, *qDOM3.1* was sensitive to ABA treatment. Therefore, we hypothesized that the target region may contain an ABA responsive gene.

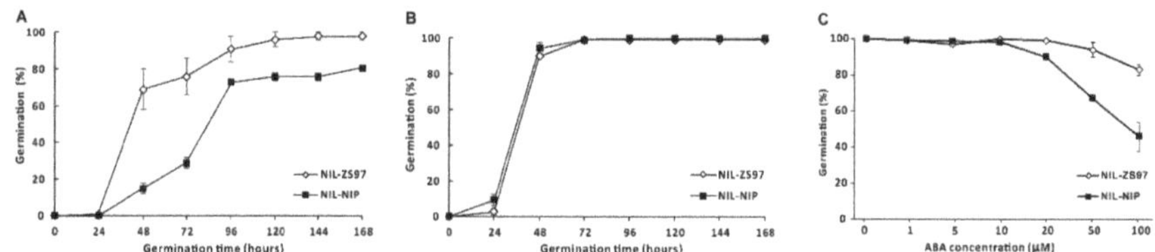

Figure 8. (**A**) Germination behavior of freshly harvested NIL-ZS97 and NIL-NIP seeds; (**B**) germination behavior of after-ripened NIL-ZS97 and NIL-NIP seeds; (**C**) ABA sensitivity of NIL-ZS97 and NIL-NIP.

2.7. Candidate Gene Expression Changes Upon ABA Treatment

For that reason, the gene expression of eight candidates was measured upon ABA treatment (20 µM), using after-ripened NIL-NIP and NIL-ZS97. In total, four out of eight candidate genes were ABA responsive genes (Figure 9). LOC_Os03g01442's expression level had no significant difference in non-treated NIL-NIP and NIL-ZS97 (CK); however, upon ABA treatment, the expression level was significantly higher in NIL-NIP than NIL-ZS97. LOC_Os03g01540 had the opposite effect, which was lower in non-treated NIL-NIP than in NIL-ZS97 and, upon ABA treatment, the expression level had no difference in NIL-NIP and NIL-ZS97. LOC_Os03g01530's expression level was significantly increased upon ABA treatment. LOC_Os03g01470's expression level was significantly higher in non-treated NIL-NIP than NIL-ZS97, and, after ABA treatment, the expression level was significantly lower in NIL-NIP compared with NIL-ZS97. The other four candidate genes had the same trend before and after the ABA treatment.

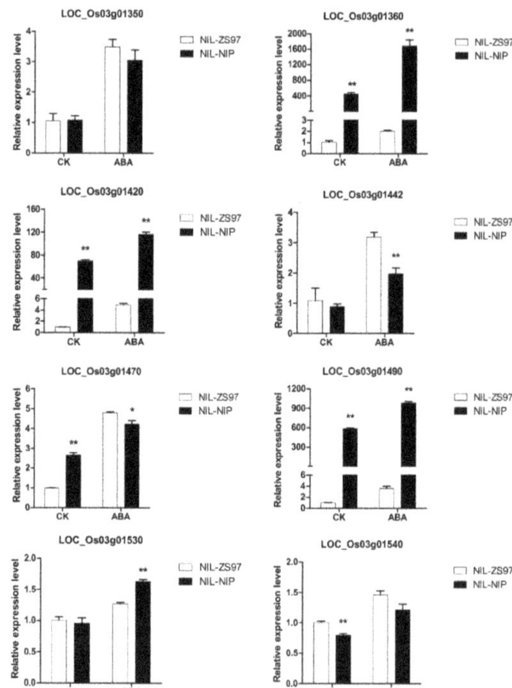

Figure 9. Relative mRNA abundance level of eight candidate genes under *qDOM3.1* using after-ripened NIL-ZS97 and NIL-NIP before (CK) and after 20 µM of abscisic acid (ABA) treatment (ABA). * and ** indicate significant differences at $p < 0.05$ and $p < 0.01$ using the Student's *t*-test of NIL-NIP against NIL-ZS97, respectively. CK of NIL-ZS97 in each figure was determined as 1.

Therefore, if the expression of the candidate gene was changed by the addition of ABA, they were unlikely to be our candidate genes. Therefore, there were four candidate genes under *qDOM3.1*, including two expressed proteins (LOC_Os03g01442 and LOC_Os03g01470), one tubulin/FtsZ domain-containing protein (LOC_Os03g01530), and one DNA-binding protein (LOC_Os03g01540). However, we cannot rule out the posttranslational modifications of the candidate gene, such as phosphorylation/dephosphorylation. More experimental evidence is needed.

3. Discussion

3.1. Seed Dormancy QTL Analysis

Seed dormancy in rice is generally a complex trait and is controlled by multiple genes. In our study, the genetic architecture of seed dormancy was examined in the NIP/ZS97 CSSL population with each line carrying one or a few different introgressed segments from NIP and otherwise sharing the uniform genetic background of ZS97. Each introgression segment was defined by high-density SNP markers. The CSSL population has several advantages over other mapping populations such as F_2, BC1, and RIL [38–40]. First, the detection power of QTLs in CSSLs was higher than that of other mapping populations reported. In total, 36 QTLs for seed dormancy were identified in this CSSL population using six germination parameters (Table 2) and the six germination parameters explained from 50.5% to 77.9% of phenotypic variance. However, only four seed dormancy QTLs were detected in a double haploid (DH) population [41]. Four [42] and nine [15] seed dormancy QTLs were identified by two different RIL populations, respectively. Second, it is easier to develop a secondary F_2 population derived from a cross between a CSSL line containing the target QTL and the recurrent parent for fine mapping [38,43]. In our study, by developing an F_2 segregation population, one of the novel seed dormancy QTLs, namely *qDOM3.1*, was delimited to 90 kb (Figure 5).

Among all 36 QTLs, 10 were detected or cloned previously for seed dormancy (Table 2). The results implied that our NIP/ZS97 population and QTL analysis method turned out to be efficient to detect seed dormancy QTLs across the whole genome. The first cloned seed dormancy gene in rice *Sdr4* was only detected in T_{50} (*qDOM7.5*), implying that our population had mild seed dormancy level. *qDOM7.4* covered the gene *SD7-1/Rc* and was detected in all six parameters except G_{max}. *SD7-1/Rc* was a pleiotropic gene that most likely controlled the dormancy and pigment traits by regulating ABA and flavonoid biosynthetic pathways, respectively [22]. *OsFbx352*, which was located under *qDOM10.3*, was detected in all six parameters. It was involved in the regulation of glucose-induced suppression of seed germination by targeting ABA metabolism [36]. Besides the QTLs co-located with the previous study, there were 26 new QTLs identified in the present study; therefore, both the plant material and abundant QTL information will facilitate the use of the dormancy alleles in other breeding programs or other research studies.

3.2. Candidate Gene Analysis

In the present study, three new QTLs were detected in all six germination parameters, namely *qDOM3.1*, *qDOM6.2*, and *qDOM10.2*. Through a CSSL-derived F_2 population, the major effect of *qDOM3.1* on seed dormancy was validated for the first time using the six seed dormancy-related parameters and delimited to a 90 kb region (Figures 4 and 5 and Figure S3), which included 19 candidate genes. As the heterozygous line had the same phenotype as ZS97 (Figure 3), the candidate gene in ZS97 should be dominant. Based on the chromosomal synteny analysis and the gene expression profile in different tissues, we deducted that eight genes left as candidate genes. For the eight candidate genes, LOC_Os03g01540 was annotated as DNA-binding protein, LOC_Os03g01530 was annotated as tubulin/FtsZ domain-containing protein, and the other six candidate genes were annotated as expressed protein (based on the RGAP database: http://rice.planthiology.mou.edu/, Release 7).

Hormonal regulation may be a highly conserved mechanism of seed dormancy among many species. ABA plays an essential role in the regulation of seed dormancy [4,24,37]. In the present study, we investigated the ABA content in freshly harvested NIL-NIP and NIL-ZS97, and found out the ABA content was almost five times higher in NIL-NIP than in NIL-ZS97 (Figure 7). Subsequently, the ABA sensitivity assay demonstrated that NIL-NIP was indeed more sensitive to high ABA solutions than NIL-ZS97 (Figure 8C). Based on those results, we assumed that our candidate gene could be responsive to ABA treatment, and the two alleles in NIP and ZS97 should respond differently. Thus, four candidate genes were selected (Figure 9), which were two expressed proteins (LOC_Os03g01442 and LOC_Os03g01470), one tubulin/FtsZ domain-containing protein (LOC_Os03g01530), and one DNA-binding protein (LOC_Os03g01540).

The DNA-binding protein (LOC_Os03g01540) had 60% of protein sequence similarity with AT-hook motif DNA-binding family protein in *Arabidopsis* and 56.35% of similarity with AT-hook protein 1 in *Oryza sativa*. A recent study showed that the DNA-binding protein AT-Hook-Like 10 (AHL10) could be dephosphorylated by a protein phosphatase Highly ABA-Induced1 (HAI1), which was involved in abiotic stress and abscisic acid signaling. AHL10 phosphorylation was crucial for hormone-related genes during drought stress [44]. Therefore, we thought our candidate gene LOC_Os03g01540 was somehow involved in the ABA signaling pathway to regulate seed dormancy. However, further research will be needed to prove our hypothesis.

LOC_Os03g01530 is one of the isotype genes controlling β-tubulin, which is a basic component of microtubules. Proteomic analysis of rice embryo showed that LOC_Os03g01530 was upregulated during seed germination, suggesting its possible role in seed germination [45]. LOC_Os03g01490 was annotated as expressed protein. A research study showed that it is a functional new chimerical gene for *Oryza sativa* ssp. *japonica* by comparing *Oryza sativa* ssp. *japonica* and its five wild progenitors [46], although its function was not revealed. Another research paper identified LOC_Os03g01490 and LOC_Os03g01470 as phosphopeptides [47]. The other four candidate genes were annotated as expressed protein, and very limited information was available based on a literature search.

4. Materials and Methods

4.1. Plant Materials

4.1.1. Experimental Design

The plant materials were planted at the experimental field of Huazhong Agricultural University at Wuhan (29.58°N, 113.41°E). The temperature during the late stage of maturity ranged from 25 °C to 30 °C, which was normal for rice growth and seed maturation. The first flowering date of each plant was recorded by the emergence of the first panicle from the leaf sheath [13].

Seeds were harvested from the individual plants at 35 days after flowering, which was defined as freshly harvested seeds and then equilibrated about 5–6 days at 15% relative humidity (called freshly harvested seeds), and then stored at −20 °C for subsequent analyses. A three-day heat treatment at 43 °C for dry seeds was used to break seed dormancy, and the seeds were called after-ripened seeds. All the analyses were performed with three biological replicates.

4.1.2. CSSL Population

The CSSL population was developed and genotyped in a previous study [32,33]. The details are as follows. A set of chromosomal segment substitution lines (CSSLs) comprising 146 lines was developed in a previous study [32], in which the donor parent was *japonica* variety Nipponbare (NIP) and the recurrent parent *indica* was variety Zhenshan97 (ZS97). The CSSL population was genotyped previously, of which 518 bins were defined with a median size of 400 kb [33].

4.1.3. CSSL Line-Derived Population

One of the CSSLs (NQ96), which contains the target QTL, was selected to backcross with ZS97 to generate an F$_2$ population (called CSSL-derived population) for QTL validation. A total of 338

F$_2$ plants and the F$_{2:3}$ families of several recombinants were screened with polymorphic markers to identify the respective genotype.

4.1.4. Plant Materials for ABA Content Measurement, ABA Sensitivity, and qRT-PCR Analysis

A pair of near-isogenic lines (NILs) containing the NIP alleles (NIL-NIP) and ZS97 alleles (NIL-ZS97), respectively, at the target QTL (*qDOM3.1*) in a common background of ZS97 were developed based on polymorphic markers. NIL-NIP and NIL-ZS97 were used for ABA content measurement, ABA sensitivity, and qRT-PCR analysis.

4.2. Seed Trait Measurement

In all, 50 seeds of each sample were spread on moistened filter paper in Petri dishes in a 25 °C growth chamber for germination experiments. The number of germinated seeds was counted daily for seven consecutive days to construct cumulative germination curves. Germination was defined as the length of the protruded radicle by 3–5 mm. Germination tests for the after-ripened seeds were also conducted as described in the above method.

Germination was scored using the Germinator package [34]. We calculated the six relevant parameters from the germination curve. The parameters included maximum germination percentage of seven days germination (G$_{max}$); germination percentage at three days (G$_{3d}$); germination speed, which is the time to reach 50% germination of the total number of germinated seeds (T$_{50}$); germination uniformity, which is the time interval between 84% and 16% of viable seed to germinate (U$_{8416}$) (16% and 84% stands for −1SD and +1SD, respectively); and area under the germination curve (AUC). Germination index (GI) was calculated by the method of Cao et al. [48]: (GI = Σ(Gt/Tt), where Gt is the number of the germinated seeds on Day t, and Tt is the time corresponding to Gt in days.

4.3. DNA Extraction and Marker Analysis

DNA was isolated from 2 cm long leaves using the cetyl trimethylammonium bromide (CTAB) method [49]. According to the sequence variation between NIP and ZS97 (available online: http: //ricevarmap.ncpgr.cn/v2/), single nucleotide polymorphism (SNP) markers and insertion/deletion (Indel) markers in the target region were developed (Table S1, Supplementary Materials). The primers used for nucleotide variation analysis were designed according to the Nipponbare reference genome by Primer 3 (available online: http://redb.ncpgr.cn/modules/redbtools/primer3.php). Polymerase chain reaction (PCR) amplification and gel electrophoresis for marker genotype analysis were conducted following the methods described previously [50]. PCR products were sequenced by Sangon Biotech (Shanghai, China) and the sequences were analyzed using Sequencer 5.0 (Gene Codes Corporation, Ann Arbor, MI, USA).

4.4. QTL Analysis and Linkage Mapping

Germination percentage (x) such as G$_{max}$ and G$_{3d}$ was transformed by arcsine(x)0.5 to make the trait mean independent from the variance.

Based on the SNP genotypes, a bin was defined by a unique overlapping substitution segment from the CSSLs and used as a marker for QTL analysis. A linear ridge regression in the R package "ridge" (available online: http://www.r-project.org/) was applied for QTL analysis in the CSSL population [33]. In addition, *p* < 0.01 was set as the significance level for the presence of a putative QTL. The most significant bin was selected if several adjacent bins showed significant *p*-values. The phenotypic variance explained by each QTL (bin) was calculated using lmg in the R package named "relaimpo".

A linkage map was constructed for the CSSL-derived F$_2$ population with an additional 10 markers on the target region. Linkage analysis and QTL validation for seed dormancy were performed using the ICIMAPPING software (version 4.1, Chinese Academy of Agriculture Sciences, Beijing, China). The presence of a QTL was declared when an LOD score was larger than 3. The additive effect and the phenotypic variation explained by each QTL were estimated by ICIMAPPING.

4.5. Quantification of Endogenous ABA

About 100 mg of embryo from fresh seeds were extracted with 750 µL of methanol/water/acetic acid (80:19:1) and 10 ng/ ml of d6-ABA as internal standard, shaking for 16 h at 300 rpm in 4 °C. After centrifuging for 10 min at 13,000 rpm, the supernatant was transported to a new tube. The precipitate was re-extracted with 450 µL methanol/water/acetic acid (80:19:1) shaking for 4 h at 300 rpm in 4 °C. After centrifuging for 10 min at 13,000 rpm, the supernatant was combined with the previous one. The extracts were filtered with 22 µm filter membrane (Bizcomr, Nylon Syringe Filter, Guangzhou, China) and dried with N_2. The residue was dissolved in 200µL of methanol and centrifuged for 15 min at 13,000 rpm under 4 °C. About 150 µL of supernatant were used for ultra-fast liquid chromatography (UFLC)/electrospray ionization/tandem mass spectrometry system (ESI/MS/MS) (Agilent 6520 QTOF, Hong Kong, China).

4.6. Exogenous ABA Treatment

A seed germination assay was performed as described above. For the ABA sensitivity assay, after-ripened seeds of NIL-NIP and NIL-ZS97 were treated with a series of ABA solutions (0, 1, 5, 10, 20, 50, and 100 µM). Dimethyl sulfoxide (DMSO) was used to dissolve ABA.

Subsequently, a 20 µM portion of ABA was chosen to treat after-ripened NIL-NIP and NIL-ZS97 to further measure the candidate gene expression. DMSO was used as control (CK). Embryos germinated for three days were used for RNA isolation.

4.7. RNA Isolation and qRT–PCR Analysis

RNA isolation: RNAPrep Pure Plant Plus Kit (TIANGEN, catalog number DP441, Beijing, China) was used to extract total RNA from embryos of germinated seeds, according to the manufacturer's instructions. IScript cDNA Synthesis Kit (Bio-Rad, Hercules, CA, USA) was used for cDNA synthesis for quantitative real-time PCR (qRT-PCR), according to the manufacturer's instructions. qRT-PCR was performed using a QuantStudio (TM) 6 Flex Real-Time PCR instrument (Applied Biosystems, Waltham, MA, USA) with iQ SYBR Green Supermix (Bio-Rad, Hercules, CA, USA). Three biological replicates were used for each sample. The data were normalized to the amplification of a rice ACTIN gene (LOC_Os03g50885). The mean value was then plotted with its standard error. Primers for real-time PCR are included in Table S2, Supplementary Materials.

5. Conclusions

Rice seed dormancy is an important agronomic trait that is crucial to the quality of rice. Currently, rice breeding programs have been focused on cultivars with a balance between pre-harvest sprouting and deep dormancy [2]. Cultivars with moderate dormancy levels have multiple advantages, including decreasing pre-harvest sprouting, improving the survival rate of direct seedling rice, increasing seedling growth uniformity, and further improving the storage quality of rice. The QTL *qDOM3.1* controlling seed dormancy identified in the present study will accelerate the breeding of new rice varieties with suitable seed dormancy. Besides *qDOM3.1*, many of the QTLs identified in the present study may also be useful in an agricultural context, providing new genes that can be used to improve crop performance under fluctuating environments.

Supplementary Materials: Supplementary materials can be found at http://www.mdpi.com/1422-0067/21/4/1344/s1.

Author Contributions: Conceptualization, H.H. and S.Y. (Sibin Yu); methodology, S.Y. (Shaowen Yuan), Y.W., and C.Z.; software, Y.W. and C.Z.; writing—original draft preparation, S.Y. (Shaowen Yuan) and Y.W.; writing—review and editing, H.H. and S.Y. (Sibin Yu) All authors have read and agreed to the published version of the manuscript.

Funding: This research was funded by the National Natural Science Foundation of China (31971864), Natural Science Foundation of Hubei Province of China, grant number 2019CFC851, and Fundamental Research Funds for the Central Universities, grant numbers 2662016QD040 and 2662018YJ025.

Conflicts of Interest: The authors declare no conflict of interest. The funders had no role in the design of the study; in the collection, analyses, or interpretation of data; in the writing of the manuscript, or in the decision to publish the results.

Abbreviations

AUC	Area Under the Germination Curve
CSSL	Chromosomal Segment Substitution Line
GI	Germination Index
G_{max}	Maximum Germination Percentage of Seven Days of Germination
G_{3d}	Germination Percentage at Three Days
NIL	Near-Isogenic Line
T_{50}	Germination Speed, which is the time to reach 50% germination of the total number of germinated seeds
U_{8416}	Germination Uniformity, which is the time interval between 84% and 16% of viable seed to germinate

References

1. Bewley, J.D. Seed germination and dormancy. *Plant Cell* **1997**, *9*, 1055–1066. [CrossRef]
2. Xie, K.; Jiang, L.; Lu, B.; Yang, C.; Li, L.; Liu, X.; Zhang, L.; Zhao, Z.; Wan, J. Identification of QTLs for seed dormancy in rice (*Oryza sativa* L.). *Plant Breeding* **2011**, *130*, 328–332. [CrossRef]
3. He, H.; de Souza Vidigal, D.; Snoek, L.B.; Schnabel, S.; Nijveen, H.; Hilhorst, H.; Bentsink, L. Interaction between parental environment and genotype affects plant and seed performance in Arabidopsis. *J. Exp. Bot.* **2014**, *65*, 6603–6615. [CrossRef]
4. Penfield, S. Seed dormancy and germination. *Curr. Biol.* **2017**, *27*, 874–878. [CrossRef]
5. Alonso-Blanco, C.; Bentsink, L.; Hanhart, C.J.; Blankestijn-de Vries, H.; Koornneef, M. Analysis of natural allelic variation at seed dormancy loci of *Arabidopsis thaliana*. *Genetics* **2003**, *164*, 711–729.
6. Graeber, K.; Linkies, A.; Müller, K.; Wunchova, A.; Rott, A.; Leubner-Metzger, G. Cross-species approaches to seed dormancy and germination: Conservation and biodiversity of ABA-regulated mechanisms and the Brassicaceae *DOG1* genes. *Plant Mol. Biol.* **2010**, *73*, 67–87. [CrossRef]
7. Schatzki, J.; Schoo, B.; Ecke, W.; Herrfurth, C.; Feussner, I.; Becker, H.C.; Mollers, C. Mapping of QTL for seed dormancy in a winter oilseed rape doubled haploid population. *Theor. Appl. Genet.* **2013**, *126*, 2405–2415. [CrossRef]
8. Benech-Arnold, R.L.; Rodriguez, M.V. Pre-harvest sprouting and grain dormancy in sorghum bicolor: What have we learned? *Front. Plant Sci.* **2018**, *9*, 811. [CrossRef]
9. Nagel, M.; Alqudah, A.M.; Bailly, M.; Rajjou, L.; Pistrick, S.; Matzig, G.; Börner, A.; Kranner, I. Novel loci and a role for nitric oxide for seed dormancy and preharvest sprouting in barley. *Plant Cell Environ.* **2019**, *42*, 1318–1327. [CrossRef]
10. Lin, M.; Zhang, D.; Liu, S.; Zhang, G.; Yu, J.; Fritz, A.K.; Bai, G. Genome-wide association analysis on pre-harvest sprouting resistance and grain color in U.S. winter wheat. *BMC genomics* **2016**, *17*, 794–810. [CrossRef]
11. Shao, M.; Bai, G.; Rife, T.W.; Poland, J.; Lin, M.; Liu, S.; Chen, H.; Kumssa, T.; Fritz, A.; Trick, H.; et al. QTL mapping of pre-harvest sprouting resistance in a white wheat cultivar Danby. *Theor. Appl. Genet.* **2018**, *131*, 1683–1697. [CrossRef]
12. Vetch, J.M.; Stougaard, R.N.; Martin, J.M.; Giroux, M.J. Review: Revealing the genetic mechanisms of pre-harvest sprouting in hexaploid wheat (*Triticum aestivum* L.). *Plant Sci.* **2019**, *281*, 180–185. [CrossRef]
13. Wan, J.; Jiang, L.; Tang, J.; Wang, C.; Hou, M.; Jing, W.; Zhang, L. Genetic dissection of the seed dormancy trait in cultivated rice (*Oryza sativa* L.). *Plant Sci.* **2006**, *170*, 786–792. [CrossRef]
14. Marzougui, S.; Sugimoto, K.; Yamanouchi, U.; Shimono, M.; Hoshino, T.; Hori, K.; Kobayashi, M.; Ishiyama, K.; Yano, M. Mapping and characterization of seed dormancy QTLs using chromosome segment substitution lines in rice. *Theor. Appl. Genet.* **2012**, *124*, 893–902. [CrossRef]
15. Cheng, J.; Wang, L.; Du, W.; Lai, Y.; Huang, X.; Wang, Z.; Zhang, H. Dynamic quantitative trait locus analysis of seed dormancy at three development stages in rice. *Mol. Breeding* **2014**, *34*, 501–510. [CrossRef]
16. Magwa, R.A.; Zhao, H.; Xing, Y. Genome-wide association mapping revealed a diverse genetic basis of seed dormancy across subpopulations in rice (*Oryza sativa* L.). *BMC Genet.* **2016**, *17*, 28–41. [CrossRef]

17. Lu, Q.; Niu, X.; Zhang, M.; Wang, C.; Xu, Q.; Feng, Y.; Yang, Y.; Wang, S.; Yuan, X.; Yu, H.; et al. Genome-wide association study of seed dormancy and the genomic consequences of improvement footprints in rice (*Oryza sativa* L.). *Front. Plant Sci.* **2017**, *8*, 2213–2226. [CrossRef]

18. Sugimoto, K.; Takeuchi, Y.; Ebana, K.; Miyao, A.; Hirochika, H.; Hara, N.; Ishiyama, K.; Kobayashi, M.; Ban, Y.; Hattori, T.; et al. Molecular cloning of *Sdr4*, a regulator involved in seed dormancy and domestication of rice. *Proc. Natl. Acad. Sci. USA* **2010**, *107*, 5792–5797. [CrossRef]

19. Bentsink, L.; Jowett, J.; Hanhart, C.J.; Koornneef, M. Cloning of *DOG1*, a quantitative trait locus controlling seed dormancy in Arabidopsis. *Proc. Natl. Acad. Sci. USA* **2006**, *103*, 17042–17047. [CrossRef]

20. Du, L.; Xu, F.; Fang, J.; Gao, S.; Tang, J.; Fang, S.; Wang, H.; Tong, H.; Zhang, F.; Chu, J.; et al. Endosperm sugar accumulation caused by mutation of *PHS8/ISA1* leads to pre-harvest sprouting in rice. *Plant J.* **2018**, *95*, 545–556. [CrossRef]

21. Guo, X.; Hou, X.; Fang, J.; Wei, P.; Xu, B.; Chen, M.; Feng, Y.; Chu, C. The rice *GERMINATION DEFECTIVE 1*, encoding a B3 domain transcriptional repressor, regulates seed germination and seedling development by integrating GA and carbohydrate metabolism. *Plant J.* **2013**, *75*, 403–416. [CrossRef]

22. Gu, X.; Foley, M.E.; Horvath, D.P.; Anderson, J.V.; Feng, J.; Zhang, L.; Mowry, C.R.; Ye, H.; Suttle, J.C.; Kadowaki, K.; et al. Association between seed dormancy and pericarp color is controlled by a pleiotropic gene that regulates abscisic acid and flavonoid synthesis in weedy red rice. *Genetics* **2011**, *189*, 1515–1524. [CrossRef]

23. Ye, H.; Feng, J.; Zhang, L.; Zhang, J.; Mispan, M.S.; Cao, Z.; Beighley, D.H.; Yang, J.; Gu, X. Map-based cloning of *Seed Dormancy1-2* identified a gibberellin synthesis gene regulating the development of endosperm-imposed dormancy in rice. *Plant Physiol.* **2015**, *169*, 2152–2165. [CrossRef]

24. Finkelstein, R.; Reeves, W.; Ariizumi, T.; Steber, C. Molecular aspects of seed dormancy. *Plant Biol.* **2008**, *59*, 387–415. [CrossRef]

25. Shu, K.; Liu, X.; Xie, Q.; He, Z. Two faces of one seed: Hormonal regulation of dormancy and germination. *Mol. Plant* **2016**, *9*, 34–45. [CrossRef]

26. Fang, J.; Chu, C. Abscisic acid and the pre-harvest sprouting in cereals. *Plant Signal. Behav.* **2008**, *3*, 1046–1048. [CrossRef]

27. Finch-Savage, W.E.; Leubner-Metzger, G. Seed dormancy and the control of germination. *New Phytol.* **2006**, *171*, 501–523. [CrossRef]

28. Rajjou, L.; Duval, M.; Gallardo, K.; Catusse, J.; Bally, J.; Job, C.; Job, D. Seed germination and vigor. *Annu. Rev. Plant Biol.* **2012**, *63*, 507–533. [CrossRef]

29. Née, G.; Xiang, Y.; Soppe, W.J.J. The release of dormancy, a wake-up call for seeds to germinate. *Curr. Opin. Plant Biol.* **2017**, *35*, 8–14. [CrossRef]

30. Nishimura, N.; Tsuchiya, W.; Moresco, J.J.; Hayashi, Y.; Satoh, K.; Kaiwa, N.; Irisa, T.; Kinoshita, T.; Schroeder, J.I.; Yates, J.R.; et al. Control of seed dormancy and germination by DOG1-AHG1 PP2C phosphatase complex via binding to heme. *Nat. Commun.* **2018**, *9*, 2132. [CrossRef]

31. Graeber, K.; Linkies, A.; Steinbrecher, T.; Mummenhoff, K.; Tarkowska, D.; Tureckova, V.; Ignatz, M.; Sperber, K.; Voegele, A.; de Jong, H.; et al. *DELAY OF GERMINATION 1* mediates a conserved coat-dormancy mechanism for the temperature- and gibberellin-dependent control of seed germination. *Proc. Natl. Acad. Sci. USA* **2014**, *111*, E3571–E3580. [CrossRef]

32. Chen, Q.; Mu, J.; Zhou, H.; Yu, S. Genetic effect of *japonica* alleles detected in *indica* candidate introgression lines. *Sci. Agri. Sin.* **2007**, *40*, 2379–2387.

33. Sun, W.; Zhou, Q.; Yao, Y.; Qiu, X.; Xie, K.; Yu, S. Identification of genomic regions and the isoamylase gene for reduced grain chalkiness in rice. *PLoS ONE* **2015**, *10*, e0122013. [CrossRef]

34. Joosen, R.V.L.; Kodde, J.; Willems, L.A.J.; Ligterink, W.; van der Plas, L.H.W.; Hilhorst, H.W.M. GERMINATOR: A software package for high-throughput scoring and curve fitting of Arabidopsis seed germination. *Plant J.* **2010**, *62*, 148–159. [CrossRef]

35. Jiang, L.; Cao, Y.J.; Wang, C.M.; Zhai, H.Q.; Wan, J.M.; Yoshimura, A. Detection and analysis of QTL for seed dormancy in rice (*Oryza sativa* L.) using RIL and CSSL population. *Yi chuan xue bao = Acta genetica Sinica* **2003**, *30*, 453–458.

36. Song, S.; Dai, X.; Zhang, W.H. A rice F-box gene, *OsFbx352*, is involved in glucose-delayed seed germination in rice. *J. Exp. Bot.* **2012**, *63*, 5559–5568. [CrossRef]

37. Liu, Y.; Fang, J.; Xu, F.; Chu, J.; Yan, C.; Schlappi, M.R.; Wang, Y.; Chu, C. Expression patterns of ABA and GA metabolism genes and hormone levels during rice seed development and imbibition: A comparison of dormant and non-dormant rice cultivars. *J. Genet. Genomics* **2014**, *41*, 327–338. [CrossRef]

38. He, Q.; Yang, H.; Xiang, S.; Tian, D.; Wang, W.; Zhao, T.; Gai, J. Fine mapping of the genetic locus L1 conferring black pods using a chromosome segment substitution line population of soybean. *Plant Breeding* **2015**, *134*, 437–445. [CrossRef]

39. Ali, M.L.; Sanchez, P.L.; Yu, S.-b.; Lorieux, M.; Eizenga, G.C. Chromosome segment substitution lines: A powerful tool for the introgression of valuable genes from Oryza wild species into cultivated rice (*O. sativa*). *Rice* **2010**, *3*, 218–234. [CrossRef]

40. Keurentjes, J.J.; Bentsink, L.; Alonso-Blanco, C.; Hanhart, C.J.; Blankestijn-De Vries, H.; Effgen, S.; Vreugdenhil, D.; Koornneef, M. Development of a near-isogenic line population of *Arabidopsis thaliana* and comparison of mapping power with a recombinant inbred line population. *Genetics* **2007**, *175*, 891–905. [CrossRef]

41. Guo, L.; Zhu, L.; Xu, Y.; Zeng, D.; Wu, P.; Qian, Q. QTL analysis of seed dormancy in rice (*Oryza sativa* L.). *Euphytica* **2004**, *140*, 155–162. [CrossRef]

42. Li, W.; Xu, L.; Bai, X.; Xing, Y. Quantitative trait loci for seed dormancy in rice. *Euphytica* **2011**, *178*, 427–435. [CrossRef]

43. Wan, X.; Wan, J.; Su, C.; Wang, C.; Shen, W.; Li, J.; Wang, H.; Jiang, L.; Liu, S.; Chen, L.; et al. QTL detection for eating quality of cooked rice in a population of chromosome segment substitution lines. *Theor. Appl. Genet.* **2004**, *110*, 71–79. [CrossRef] [PubMed]

44. Wong, M.M.; Bhaskara, G.B.; Wen, T.-N.; Lin, W.-D.; Nguyen, T.T.; Chong, G.L.; Verslues, P.E. Phosphoproteomics of *Arabidopsis* Highly ABA-Induced1 identifies AT-Hook–Like10 phosphorylation required for stress growth regulation. *Proc. Natl. Acad. Sci. USA* **2019**, *116*, 2354–2363. [CrossRef] [PubMed]

45. Han, C.; He, D.; Li, M.; Yang, P. In-depth proteomic analysis of rice embryo reveals its important roles in seed germination. *Plant & cell physiology* **2014**, *55*, 1826–1847. [CrossRef]

46. Zhang, C.; Wang, J.; Marowsky, N.C.; Long, M.; Wing, R.A.; Fan, C. High occurrence of functional new chimeric genes in survey of rice chromosome 3 short arm genome sequences. *Genome Biol Evol* **2013**, *5*, 1038–1048. [CrossRef]

47. Nakagami, H.; Sugiyama, N.; Mochida, K.; Daudi, A.; Yoshida, Y.; Toyoda, T.; Tomita, M.; Ishihama, Y.; Shirasu, K. Large-scale comparative phosphoproteomics identifies conserved phosphorylation sites in plants. *Plant physiology* **2010**, *153*, 1161–1174. [CrossRef]

48. Cao, D.; Hu, J.; Huang, X.; Wang, X.; Guan, Y.; Wang, Z. Relationships between changes of kernel nutritive components and seed vigor during development stages of F_1 seeds of sh_2 sweet corn. *J. Zhejiang Univ. Sci. B* **2008**, *9*, 964–968. [CrossRef]

49. Murray, M.; Thompson, W.F. Rapid isolation of high molecular weight plant DNA. *Nucleic Acids Res.* **1980**, *8*, 4321–4326. [CrossRef]

50. Chen, X.; Temnykh, S.; Xu, Y.; Cho, Y.; McCouch, S. Development of a microsatellite framework map providing genome-wide coverage in rice (*Oryza sativa* L.). *Theor. Appl. Genet.* **1997**, *95*, 553–567. [CrossRef]

International Journal of *Molecular Sciences*

Article

GABA-Alleviated Oxidative Injury Induced by Salinity, Osmotic Stress and their Combination by Regulating Cellular and Molecular Signals in Rice

Mohamed S. Sheteiwy [1,2], Hongbo Shao [1,3,4,*], Weicong Qi [1], Yousef Alhaj Hamoud [1], Hiba Shaghaleh [1], Nasr Ullah Khan [1], Ruiping Yang [3] and Boping Tang [3]

[1] Salt-Soil Agricultural Center, Key Laboratory of Agricultural Environment in the Lower Reaches of Yangtze River Plain, Institute of Agriculture Resources and Environment, Jiangsu Academy of Agricultural Sciences (JAAS), Nanjing 210014, China; salahco_2010@mans.edu.eg (M.S.S.); weicong_qi@126.com (W.Q.); yousef@sina.com (Y.A.H.); Hiba-shaghaleh@njfu.edu.cn (H.S.); khan@hotmail.com (N.U.K.)

[2] Department of Agronomy, Faculty of Agriculture, Mansoura University, Mansoura 35516, Egypt

[3] Jiangsu Key Laboratory for Bioresources of Saline Soils, Jiangsu Synthetic Innovation Center for Coastal Bio-agriculture, Yancheng Teachers University, Yancheng 224002, China; yangruiping@ycnu.edu.cn (R.Y.); tangbp@ycnu.edu.cn (B.T.)

[4] College of Environment and Safety Engineering, Qingdao University of Science & Technology (QUST), Qingdao 266000, China

[*] Correspondence: shaohongbochu@126.com

Received: 20 September 2019; Accepted: 11 November 2019; Published: 14 November 2019

Abstract: This study was conducted in order to determine the effect of priming with γ-aminobutyric acid (GABA) at 0.5 mM on rice (*Oryza sativa* L.) seed germination under osmotic stress (OS) induced by polyethylene glycol (30 g/L PEG 6000); and salinity stress (S, 150 mM NaCl) and their combination (OS+S). Priming with GABA significantly alleviated the detrimental effects of OS, S and OS+S on seed germination and seedling growth. The photosynthetic system and water relation parameters were improved by GABA under stress. Priming treatment significantly increased the GABA content, sugars, protein, starch and glutathione reductase. GABA priming significantly reduced Na^+ concentrations, proline, free radical and malonaldehyde and also significantly increased K^+ concentration under the stress condition. Additionally, the activities of antioxidant enzymes, phenolic metabolism-related enzymes, detoxification-related enzymes and their transcription levels were improved by GABA priming under stress. In the GABA primed-plants, salinity stress alone resulted in an obvious increase in the expression level of Calcineurin B-like Protein-interacting protein Kinases (*CIPKs*) genes such as *OsCIPK01*, *OsCIPK03*, *OsCIPK08* and *OsCIPK15*, and osmotic stress alone resulted in obvious increase in the expression of *OsCIPK02*, *OsCIPK07* and *OsCIPK09*; and OS+S resulted in a significant up-regulation of *OsCIPK12* and *OsCIPK17*. The results showed that salinity, osmotic stresses and their combination induced changes in cell ultra-morphology and cell cycle progression resulting in prolonged cell cycle development duration and inhibitory effects on rice seedlings growth. Hence, our findings suggested that the high tolerance to OS+S is closely associated with the capability of GABA priming to control the reactive oxygen species (ROS) level by inducing antioxidant enzymes, secondary metabolism and their transcription level. This knowledge provides new evidence for better understanding molecular mechanisms of GABA-regulating salinity and osmotic-combined stress tolerance during rice seed germination and development.

Keywords: rice; salinity; osmotic stress; combined stress; GABA; phenolic metabolism; *CIPKs* genes

1. Introduction

Rice is one of the most important cereal crops that serves as the staple food for almost half of the world's population. Rice is not a salt-tolerant crop, but is suited for cultivation in affected saline soils due to its highly consumption of fresh water for most of the growing season, which could dilute the salts and increase the availability of essential nutrients such as Fe, Mn, N, and P, which contribute to improving rice growth and yield. The mechanism by which rice can tolerate salinity stress is mainly related to the maintenance of ion homeostasis, predominantly low Na^+/K^+ or high K^+/Na^+ ratios, through exclusion, compartmentation, and partitioning of Na^+ [1]. In addition, rice plants can tolerate salt by ion exclusion which mainly involves Na^+ and Cl^- transport processes in roots and prevention of the excess accumulation of Na^+ and Cl^- in leaves [2], as well as osmotic stress tolerance which maintains leaf expansion and stomatal conductance [3]. However, rice productivity is affected by salinity stress, which originates from the accumulation of underground salt and is exacerbated by salt mining, deforestation and irrigation [4]. The tolerance limit of rice to saline conditions may vary among the different growth and developmental stages. In this regards, Zhu et al. [5] reported that rice is more tolerance to salinity during the germination and tillering stages, whereas it seems to be more sensitive during early vegetative and reproductive stages.

Salinity, being an important environmental factor, severely causes a significant reduction in the seed germination, seedling growth and development of rice. It has been reported that more than 800 million hectares of the global cultivated area are severely affected by salt stress [2]. A previous study has showed that rice plants experience osmotic stress in saline soil as a result of reduced osmotic potential of the soil solution, and ultimately reduced water uptake by plants [6]. Under the salinity stress condition, the photosynthetic rate decreased due to stomatal closure and resulted in limited availability of CO_2 and thus altered carbohydrate content of the leaf [7]. Plants can adapt to these conditions by accumulation of compatible solutes such as proline and starch, which function as osmoprotectants and have a vital role in plant adaptation to osmotic stress through stabilization of the tertiary structure of proteins [2]. Salinity reduces the ability of plants to take up water, which leads to increasing of osmotic substances, and causes inhibition of plant growth rate accompanied by metabolic changes similar to those induced by osmotic stress [8]. This action of salinity can induce osmotic stress, oxidative damage, stomatal closure, inhibition of photosynthesis, and damage of cellular structures, and decreased gas exchange rates [9]. Osmotic stress can cause a significant crop yield loss worldwide. It can reduce the plant productivity and seedling growth [10] by affecting the stomatal closure and photosynthesis process [9]. In the present study, PEG (6000) was used to induce osmotic stress in rice plants, being frequently used to induce osmotic stress in several plant species [11,12]. Moreover, the osmotic stress induced by PEG can reduce the photosynthetic rate and chlorophyll content by inhibiting the electron transport system.

The feedback regulation of plants to the combination of salinity and osmotic stresses is unique and cannot be directly extrapolated from the response of plants to each of the two stresses applied individually [13]. The physiological response of barley was investigated under combinations of two different abiotic stresses [8]; however, the molecular mechanism of plant adaptation to a combination of two different stresses remains a matter of debate [14]; this adaptation might require conflicting or antagonistic responses [14,15]. As such, plants can be adapted to heat stress by increasing the transpiration rate through opening stomata to recover from the high temperature of their leaves. Nevertheless, when plants are exposed to the combination of heat and osmotic stresses, plants have to close their stomata to reduce water loss under the osmotic stress condition [14].

GABA accumulates rapidly in response to biotic and abiotic stresses [16]. Despite the rapid accumulation of GABA during stresses, the specificity of the response and the specific role of GABA under these conditions are still elusive [17]. The accumulation of GABA in different plant cells under the osmotic stress condition requires a specific response which can act as a signaling molecule by modulating the activity of H^+-ATPase and regulating stomatal movement [17,18]. This action of GABA accumulation has been observed in plants under different environmental stresses such as osmotic stress, oxygen deficiency, mechanical stimulation, low temperature and pathogen attack [16,19]. As a physiological response under abiotic stress, GABA plays vital roles in plants for maintaining the C/N balance, regulating cytosolic pH and scavenging reactive oxygen species (ROS) [20]. Under salinity stress, exogenous GABA application acts as a signaling molecule and functions in phenolic compound enrichment [21], scavenging of ROS [22] and modulating antioxidant enzyme activities in nitrogen metabolic pathways [23]. Recent studies reported that GABA has been implicated in signaling processes affecting the nitrate-uptake system [24] and guidance of the pollen tube [25].

Salinity and osmotic stresses have a significant devastating effect limiting worldwide crop production. Taking into consideration the expected increasing world population and food demand, finding ways to improve crop tolerance to abiotic stress constraints is an urgent issue for further improving agricultural production and enhancing global food security. GABA metabolism could be involved in regulation of plant development under abiotic stress through regulation of C and N metabolism [26]. Therefore, the present study hypothesized that GABA could be involved in rice tolerance to salinity and osmotic stresses and their combination. Moreover, the *CIPKs* genes were frequently expressed and participated in plant growth and development during abiotic stresses such as heat, drought, salinity, and chilling stresses [27]. For this reason, we have investigated the effects of GABA on rice *CIPK* genes in responses to the combination of salinity and osmotic stresses to evaluate the potential usefulness of the stress-responsive *CIPK* genes in genetic improvement of stress tolerance. In addition, the present study hypothesized that high accumulation of secondary metabolites during salinity and osmotic stress could facilitate osmotic adjustment and ultimately increase rice tolerance to the combined stresses. This study was undertaken to elucidate the mechanism by which exogenously supplied GABA is involved in the stresses tolerance, in the context of regulation of Na^+ and K^+ balance, the photosynthetic system, antioxidant system, cell cycle development, cellular regulation, and controlling stomatal conductance which ultimately might result in improving rice growth under the stress conditions. The current study could significantly contribute to further understanding the tolerance mechanism induced by priming rice seeds with GABA under salinity and osmotic stresses.

2. Results

2.1. Effects of GABA Priming on the Morpho-Physiological Parameters under Salinity, Osmotic Stress and OS+S

The effects of the salinity, osmotic stress and their combination on the physiological parameters of rice seedlings with or without GABA priming are presented in Table 1. Salinity, osmotic stresses, and their combination caused a significant reduction in the germination percentage, germination energy, root length, shoot length, seedling fresh and dry weight, and seedling vigor index as compared with unstressed seedlings. Both the salinity and OS+S stresses induced a greater reduction than the osmotic stress alone (Table 1). However, priming with 0.5 mM GABA improved the physiological parameters under the salinity, osmotic stress and their combination as compared with unprimed plants. Regardless of the effect of salinity and osmotic stress, priming with 0.5 mM GABA improved the germination percentage, germination energy and vigor index by 1.12%, 5.47% and 8.88%, respectively as compared with the control condition (Table 1). Under the osmotic stress, salinity, and OS+S conditions, priming with 0.5 mM of GABA improved the germination percentage by 4.15%, 4.56% and 6.75%; germination energy by 2.85%, 23.79%, and 25.35%; and vigor index by 14.51%, 16.60% and 33.79%, respectively. Irrespective of the priming treatment, the effects of different stresses were different. The combination of salinity and osmotic stress was the most damaging for rice growth followed by individual salinity and osmotic stress (Table 1).

2.2. Effects of GABA Priming on the Photosynthetic and Water Relation Parameters under Salinity, Osmotic Stress and OS+S

Salinity and osmotic stresses and their combination resulted in a significant reduction in the net photosynthetic (Pn), transpiration rate (Tr), stomatal conductance (Gs), intracellular CO_2 (Ci), chlorophyll content (SPAD), water potential (Ψw), osmotic potential (Ψs) and relative water content (RWC) as compared with unstressed seedlings (Table 2). Priming with 0.5 mM GABA resulted in the highest Pn, Tr, Gs, SPAD, Ψw, Ψs, RWC and WUE as compared with unprimed seedlings; however, the unprimed seedlings resulted in highest Ci (Table 2). As compared with salinity and osmotic stress, the OS+S resulted in the lowest values of Pn, Tr, Gs, Ψw and RWC (Table 2). Interestingly, osmotic stress alone resulted in the lowest SPAD, while the salinity alone resulted in the lowest Ψs. As compared to the unprimed seeds, priming with 0.5 mM GABA improved the Pn, Tr, Gs, SPAD, Ψw, Ψs, RWC and WUE by 39.90%, 43.06%, 59.40%, 28.52%, 32.20%, 42.85%, 48.05% and 76.72%, respectively (Table 2). Under osmotic stress, salinity and OS+S conditions, priming with 0.5 mM GABA improved Pn by 38.21%, 23.86% and 32.85%; Tr by 44.62%, 54.94% and 50.56%; Gs by 16.57%, 38.96% and 49.56%; Ci by 9.39%, 34.53% and 11.44; SPAD by 27.24%, 16.87% and 16.32%; Ψw by 44.76%, 74.65% and 63.35%; Ψs by 17.64%, 36.48% and 3.84%; RWC by 55.29%, 44.90% and 57.50%; and WUE by 33.85%, 43.26% and 38.35%, respectively, as compared with unprimed seeds (Table 2). The present study suggested that application of 0.5 mM GABA to stressed plants with salinity, osmotic stress and their combination affected photosynthetic mechanisms, such as CO_2 diffusion through stomatal control. Moreover, it also affected the leaf water balance by controlling water and osmotic potential of the leaf to maintain water uptake for plant growth with relatively little water loss by the plant.

2.3. Effects of GABA Priming on the Sugar, Protein, Starch and GABA Contents under Salinity, Osmotic Stress and OS+S

The mean data concerning effects of salinity and osmotic stress and their combination on the sugars, protein, and starch contents are presented in Table 3. The results reported that osmotic stress, salinity and their combination decreased the sugar content by 49.60%, 60.32%, 66.64%; protein content by 54.85%, 57.16% and 64.51; and starch content by 49.90%, 59.12%, and 56.54%, respectively, as compared with the unstressed condition. However, priming with GABA improved the sugar content under osmotic stress, salinity and OS+S by 36.36%, 44.88% and 59.07%; protein content by 32.28%, 52.26% and 49.56% and starch content by 47.96%, 44.42% and 26.93%, respectively, as compared with unprimed seeds. The osmotic stress, salinity and OS+S resulted in a significant increase in the GABA content as compared with the unstressed condition (Table 3). Priming seeds with GABA significantly improved the GABA content under salinity and osmotic stress and their combination, and the highest GABA content was recorded under osmotic stress as compared with either salinity alone or OS+S stress. The current study emphasized that application of 0.5 mM GABA resulted in the accumulation of sugars, starch and protein (Table 3), which may serve as osmolytes to provide energy and carbon at times when photosynthesis may be inhibited under salinity and osmotic stresses.

Table 1. Effects of GABA treatment on the germination percentage (GP) germination energy (GE), root length (RL, cm), shoot length (SL, cm), seedlings fresh weight (SFW, g), seedling dry weight (SDW, g) and seedling vigor index (SVI) of rice seedlings exposed to salinity, osmotic stress and their combined stress.

Treatments	GP	GE	RL	SL	SFW	SDW	SVI
Ck	88.66 ± 0.57a	80.66 ± 3.78ab	12.40 ± 0.17b	7.63 ± 5.41a	0.160 ± 0.01b	0.062 ± 0.002b	1775 ± 459ab
Osmotic (OS)	84.66 ± 0.41b	79.33 ± 4.72bc	9.53 ± 0.25d	7.50 ± 0.30a	0.126 ± 0.07b	0.057 ± 0.002bc	1443 ± 106bc
Salinity (S)	76.66 ± 0.35de	57.66 ± 0.57e	7.60 ± 0.45f	6.56 ± 0.37a	0.096 ± 0.02b	0.049 ± 0.003de	1085 ± 53cd
OS+S	73.66 ± 0.35e	52.00 ± 2.00f	6.56 ± 0.37g	6.50 ± 0.50a	0.061 ± 0.04c	0.035 ± 0.001f	962 ± 65d
Priming (P)	89.00 ± 1.00a	85.33 ± 0.57a	14.30 ± 0.62a	7.63 ± 5.40a	0.192 ± 0.02a	0.086 ± 0.002a	1948 ± 518a
OS+P	88.33 ± 1.15a	81.66 ± 2.08ab	10.78 ± 0.36c	8.33 ± 0.11a	0.131 ± 0.03b	0.057 ± 0.006bc	1688 ± 14ab
S+P	80.33 ± 0.57c	75.66 ± 0.57c	8.50 ± 0.40e	7.70 ± 0.26a	0.114 ± 0.05b	0.053 ± 0.003cd	1301 ± 20b–d
OS+S+P	79.00 ± 0.37cd	69.66 ± 3.51d	9.40 ± 0.36d	9.00 ± 0.45a	0.092 ± 0.01b	0.044 ± 0.001e	1453 ± 36bc

Values are means ± SD ($n = 3$) and the same letters within a column indicate no significant difference at a 95% probability level ($p < 0.05$).

Table 2. Effects of GABA treatment on the net photosynthetic rate (Pn), transpiration rate (Tr), stomatal conductance (Gs), intracellular CO_2 concentration (Ci); chlorophyll content (SPAD), water potential (WP, Ψw), osmotic potential (OP, Ψs), relative water content (RWC) and water use efficiency (WUE) of rice seedlings exposed to salinity, osmotic stress and their combined stress.

Treatments	Pn	Tr	Gs	Ci	SPAD	WP (Ψw)	OP (Ψs)	RWC	WUE
Ck	6.67 ± 0.15b	3.60 ± 0.20b	0.613 ± 0.47b	417.67 ± 35.30a	11.70 ± 0.55b	−2.36 ± 0.15ab	−0.14 ± 0.01b	33.04 ± 2.15b	1.59 ± 0.19f
Osmotic (OS)	3.46 ± 0.11d	1.70 ± 0.10e	0.453 ± 0.06c	225.00 ± 5.00b	6.33 ± 0.15e	−6.68 ± 0.17e	−0.17 ± 0.01b	11.56 ± 1.48f	3.38 ± 0.23d
Salinity (S)	2.84 ± 0.09e	1.23 ± 0.05f	0.260 ± 0.03e	115.00 ± 5.00d	7.98 ± 0.45d	−9.43 ± 0.90f	−0.38 ± 0.04d	16.81 ± 1.37e	2.36 ± 0.07e
OS+S	1.86 ± 0.13f	0.88 ± 0.02g	0.173 ± 0.02f	136.67 ± 7.63cd	8.20 ± 0.36d	−13.40 ± 0.95g	−0.26 ± 0.02c	9.68 ± 0.50f	2.17 ± 0.12e
Priming (P)	11.10 ± 0.52a	5.46 ± 0.30aa	1.51 ± 0.05a	410.00 ± 65.57a	16.37 ± 0.12a	−1.60 ± 0.16a	−0.08 ± 0.01a	63.61 ± 2.60a	6.83 ± 0.10a
OS+P	5.60 ± 0.19c	3.07 ± 0.06c	0.543 ± 0.05b	248.33 ± 10.40b	8.70 ± 0.85d	−3.69 ± 0.13c	−0.14 ± 0.01b	25.86 ± 2.53c	5.11 ± 0.41b
S+P	3.73 ± 0.12d	2.73 ± 0.07d	0.426 ± 0.02c	175.67 ± 4.04c	9.60 ± 0.17c	−2.93 ± 0.06bc	−0.24 ± 0.01c	30.51 ± 1.17b	4.16 ± 0.12c
OS+S+P	2.77 ± 0.19e	1.78 ± 0.17e	0.343 ± 0.03d	154.33 ± 5.13cd	9.80 ± 0.43c	−4.91 ± 0.61d	−0.26 ± 0.01c	22.78 ± 0.67d	3.52 ± 0.07d

Values are means ± SD (n = 3) and the same letters within a column indicate no significant difference at a 95% probability level ($p < 0.05$).

Table 3. Effects of GABA treatment on the GABA content (mg/g DW), sugars (µg/g FW), protein (mg/g FW), starch (%); proline (µg/g FW); hydrogen peroxide (H_2O_2, nmol g^{-1} FW), superoxide radical (O_2^-, nmol min^{-1}g^{-1} FW), hydroxyl ion (OH$^-$, nmol g^{-1} FW); malonaldehyde (MDA, nmol mg^{-1} protein) and glutathione reductase (GR, µmol min^{-1} mg^{-1} protein) of rice seedlings exposed to salinity, osmotic stress and their combined stress.

Treatments	GABA Con.	Sugars	Protein	Starch	Proline	H_2O_2	O_2^-	OH$^-$	MDA	GR
Ck	3.53 ± 0.10g	12.50 ± 0.17a	240.0 ± 8.6b	66.37 ± 0.89b	30.35 ± 2.07c	1.64 ± 0.06de	2.59 ± 0.15c	0.103 ± 0.05d	81.75 ± 4.8e	6.72 ± 0.54f
Osmotic (OS)	5.71 ± 0.10d	6.30 ± 0.25d	108.35 ± 6.2e	33.25 ± 2.65de	48.55 ± 1.89b	3.52 ± 0.06b	2.68 ± 0.18c	0.153 ± 0.01b	105.60 ± 1.5cd	10.53 ± 0.07e
Salinity (S)	4.76 ± 0.24e	4.96 ± 0.20e	102.81 ± 6.3e	27.13 ± 1.47e	73.19 ± 2.51a	4.34 ± 0.18a	3.95 ± 0.50a	0.210 ± 0.04a	159.90 ± 3.5a	14.44 ± 0.42d
OS+S	4.52 ± 0.24e	4.17 ± 0.30f	85.17 ± 5.84f	28.84 ± 0.41e	76.44 ± 1.7a	4.15 ± 0.53a	3.49 ± 0.05b	0.156 ± 0.05b	125.46 ± 2.0b	17.31 ± 0.42c
Priming (P)	4.07 ± 0.06f	12.24 ± 0.30a	361.01 ± 12.9a	81.07 ± 7.62a	30.40 ± 1.15c	1.47 ± 0.21e	2.69 ± 0.34c	0.099 ± 0.05d	83.24 ± 2.65e	6.10 ± 0.18f
OS+P	10.45 ± 0.05a	9.90 ± 0.52b	160.0 ± 5.60d	63.90 ± 3.72b	34.56 ± 1.04c	1.98 ± 0.01d	2.67 ± 0.27c	0.126 ± 0.05b-d	83.38 ± 10.4e	18.61 ± 0.82bc
S+P	8.42 ± 0.39b	9.00 ± 0.46c	215.37 ± 5.0c	48.82 ± 3.26c	41.68 ± 1.70bc	2.69 ± 0.24c	2.73 ± 0.07c	0.136 ± 0.05bc	112.02 ± 2.7c	20.00 ± 0.21b
OS+S+P	6.60 ± 0.16c	10.19 ± 0.54b	168.86 ± 16.3d	39.47 ± 4.91d	49.18 ± 6.30b	2.95 ± 0.07c	2.55 ± 0.09c	0.116 ± 0.01cd	102.67 ± 3.8d	27.68 ± 2.00a

Values are means ± SD ($n = 3$) and the same letters within a column indicate no significant difference at a 95% probability level ($p < 0.05$).

2.4. Effects of GABA Priming on ROS, MDA, Proline and GR under Salinity, Osmotic Stress and OS+S

The results revealed that osmotic stress, salinity and OS+S improved ROS, i.e., hydrogen peroxide (H_2O_2), superoxide radical (O_2^-) and hydroxyl ion (OH^-) accumulation in the leaf tissues as compared with the control condition (Table 3). Priming treatment significantly reduced the accumulation of the H_2O_2 under osmotic stress, salinity and OS+S by 43.75%, 38.01% and 28.91%; O_2^- by 0.37%, 30.88%, and 26.93%; and OH^- by 17.64%, 22.38% and 25.64%, respectively, as compared with unprimed seeds (Table 3), and these findings were confirmed by the confocal microscopic investigation as depicted in Figure 1. Plants exposed to osmotic stress, salinity and OS+S experienced a significant increase in proline, malonaldehyde (MDA), and glutathione reductase (GR) concentrations as compared with the unstressed condition (Table 3). Proline, MDA and GR contents were not significantly affected by 0.5 mM GABA priming irrespective of the stress effects. However, the primed seedlings exposed to osmotic stress, salinity and OS+S had the lowest contents of MDA and proline, specifically under osmotic stress. Priming with 0.5 mM GABA resulted in a significant decrease of proline under osmotic stress, salinity, and OS+S, by 28.81%, 43.05% and 35.66%, respectively, relative to unprimed seeds (Table 3). By contrast, priming with 0.5 mM GABA resulted in a significant increase in the GR activity under the stress condition. It could be concluded that priming with 0.5 mM GABA improved GR activity under osmotic stress, salinity and OS+S stresses by 43.41%, 27.8% and 37.46%, respectively, relative to unprimed seeds. It could be stated that 0.5 mM GABA inhibited the release of free radicals by reducing the ROS and MDA levels and thus reducing the cell membrane damage under salinity, osmotic stress and their combination.

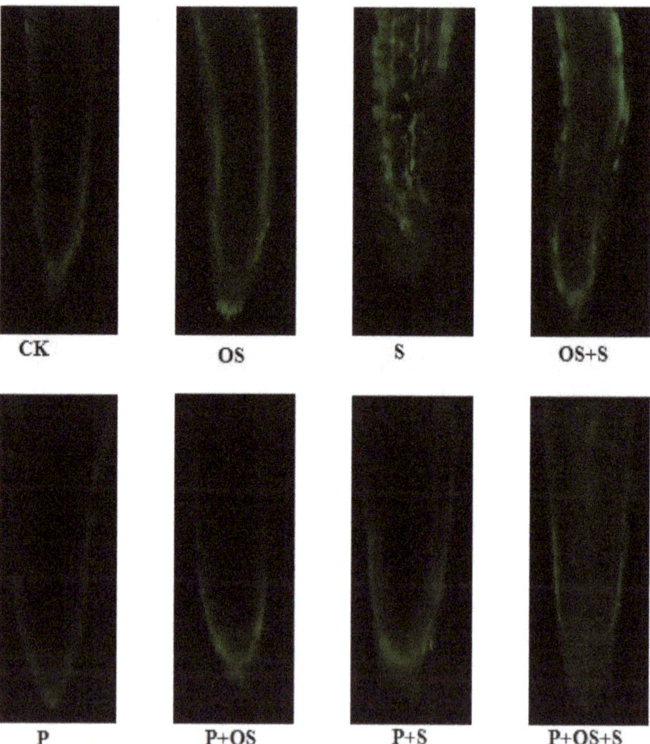

Figure 1. Effects of priming treatment on H_2O_2 accumulation in the root cells of rice seedlings exposed to salinity, osmotic stress, and their combined stress (OS+S). P (Priming) OS (Osmotic stress); S (Salinity).

2.5. Effects of GABA Priming on Ion Accumulation under Salinity, Osmotic Stress and OS+S

The obtained results showed that salinity and OS+S stresses significantly increased the Na^+ concentration in the leaf and root tissues as compared to the control condition, and a higher concentration of Na^+ was observed in the root as compared to the leaf (Figure 2). Interestingly, the Na^+ concentration was not significantly reduced by GABA priming in the leaf and root under the unstressed condition. However, under the salinity and OS+S stresses, the GABA priming resulted in a significant reduction of the Na^+ concentrations in the leaf and root as compared with unprimed seeds. Priming decreased Na^+ concentrations under osmotic stress, salinity and OS+S stresses by 2.20%, 23.72% and 48.87% in the leaf, and in the root by 0.64%, 36.39% and 31.00%, respectively. Irrespective of the salinity, osmotic stress and their combination, priming with 0.5 mM GABA significantly improved the K^+ concentration in the leaf and root tissues (Figure 2). It could be concluded that priming seeds with GABA significantly improved the K^+ concentration in the leaf by 23.84%, 43.89%, and 30.65%, and in the root by 25.96%, 31.68%, and 37.50% under the osmotic stress, salinity and OS+S stresses, respectively (Figure 2). The present study suggested that priming with 0.5 mM GABA has the potential to maintain the balance between the accumulation of Na^+ in the plant cell and the loss of K^+ under salinity, osmotic stress and their combination.

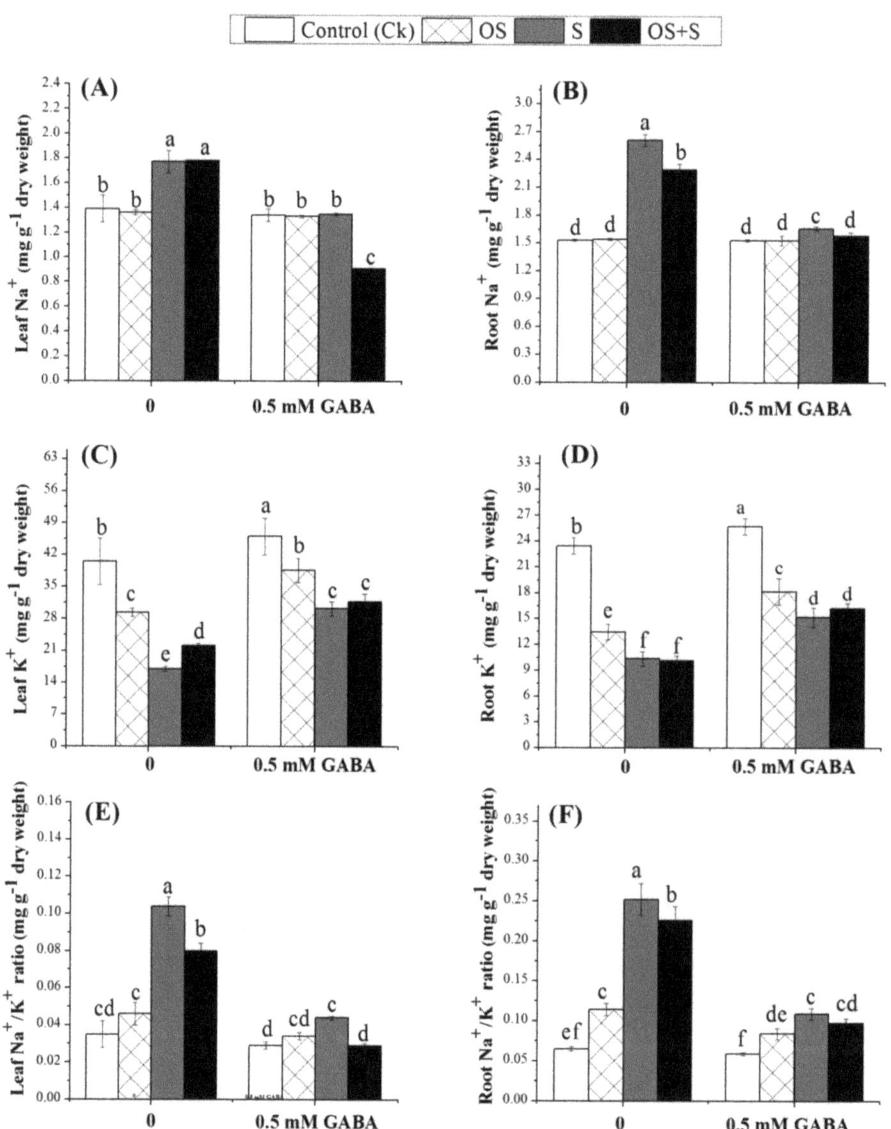

Figure 2. Effects of GABA treatment on the concentration of Na^+ in the leaf (**A**), Na^+ in the root (**B**), K^+ in the leaf (**C**), K^+ in the root (**D**), Na^+/K^+ ratio in the leaf (**E**) and Na^+/K^+ ratio in the roots (**F**) of rice seedlings exposed to salinity, osmotic stress, and their combined stress (OS+S). OS (Osmotic stress); S (Salinity).

2.6. Effects of GABA Priming on Enzyme Activities under Salinity, Osmotic Stress and OS+S

As shown in Figure 3A–C, the activities of antioxidant enzymes such as superoxide dismutase (SOD), catalase (CAT) and ascorbic peroxidase (APX) increased in the seedlings exposed to osmotic stress, salinity and OS+S stresses as compared with their respective controls. Priming seeds with 0.5 mM GABA significantly improved the activities of these enzymes under the salinity, osmotic stress and their combination, but the impact was more obvious under the salinity stress as compared with osmotic stress and the combined stress (Figure 3A–C). Compared to the unprimed seeds, SOD activity was improved in the GABA-primed seeds under osmotic stress, salinity and OS+S stresses by 36.46%, 40.64% and 28.29%; CAT activity by 36.92%, 49.06% and 30.17%, and APX activity by 39.64%, 32.89% and 27.65%, respectively (Figure 3A–C). It seems that the application of 0.5 mM GABA can regulate the antioxidant enzymes activity which played a crucial role in scavenging H_2O_2 helping to minimize excessive ROS in the stressed plants under salinity, osmotic stress and their combination.

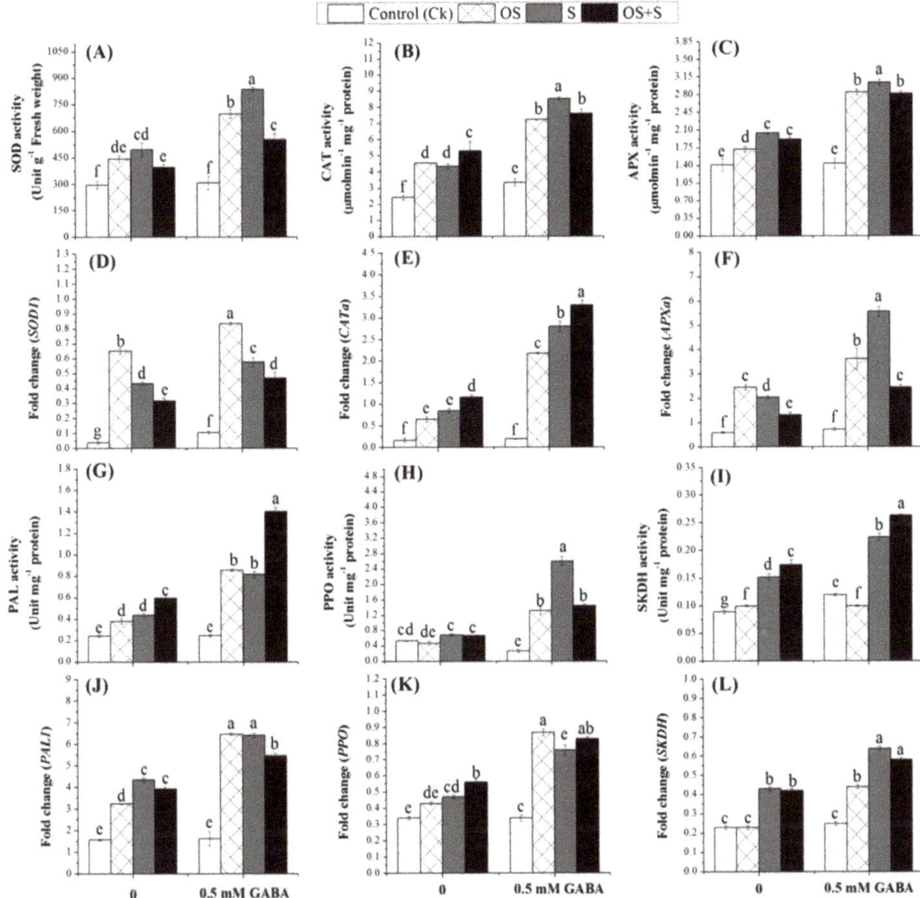

Figure 3. Effects of GABA treatment on SOD (**A**), CAT (**B**), APX (**C**) activities; *SOD1* (**D**); *CATa* (**E**); *APXa* (**F**); PAL (**G**); PPO (**H**); SKDH (**I**); *PAL1* (**J**); *PPO* (**K**) and *SKDH* (**L**) of rice seedlings exposed to salinity, osmotic stress, and their combined stress (OS+S). OS (Osmotic stress); S (Salinity).

2.7. Effects of GABA Priming on Phenolic Metabolism under Salinity, Osmotic Stress and OS+S

The results presented in Figure 3G–I showed that phenylalanine ammonia-lyase (PAL) activity significantly increased when the seedlings were exposed to the osmotic stress, salinity and OS+S relative to their controls. Priming with 0.5mM GABA resulted in a significant increase in PAL activity as compared to unprimed seeds. The priming treatment resulted in an increase of the PAL activity under osmotic stress, salinity and OS+S stress by 47.3%, 46.07% and 57.42%, respectively, relative to the unprimed seeds (Figure 3G). Similarly, polyphenol oxidase (PPO) activity was increased in the primed seedlings under osmotic stress, salinity and OS+S by 67.72%, 73.73% and 54.16%, respectively

(Figure 3H). Interestingly, the PPO activity in the unprimed seedlings under osmotic stress was lower than that in the control seedlings. The results revealed that priming with 0.5 mM GABA resulted in a significant increase in Shikimate dehydrogenase (SKDH) activity under salinity and OS+S, but there was no significant difference under the osmotic stress in primed and unprimed seedlings (Figure 3I). Cinnamyl alcohol dehydrogenase (CAD) activity significantly increased in the primed seedlings exposed to osmotic stress, salinity and OS+S, while there was no significant increase in the CAD activity in the unprimed seeds under osmotic stress (Figure 4A). GABA priming increased the CAD activity under osmotic stress, salinity and OS+S by 31.88%, 30.83% and 28.52%, respectively, relative to unprimed seeds. Hence, in the present study, priming with 0.5 mM GABA resulted in up-regulation of secondary metabolism, such as PAL, PPO and SKDH, which can generate a defense mechanism against oxidative stress induced by salinity, osmotic stress and their combination.

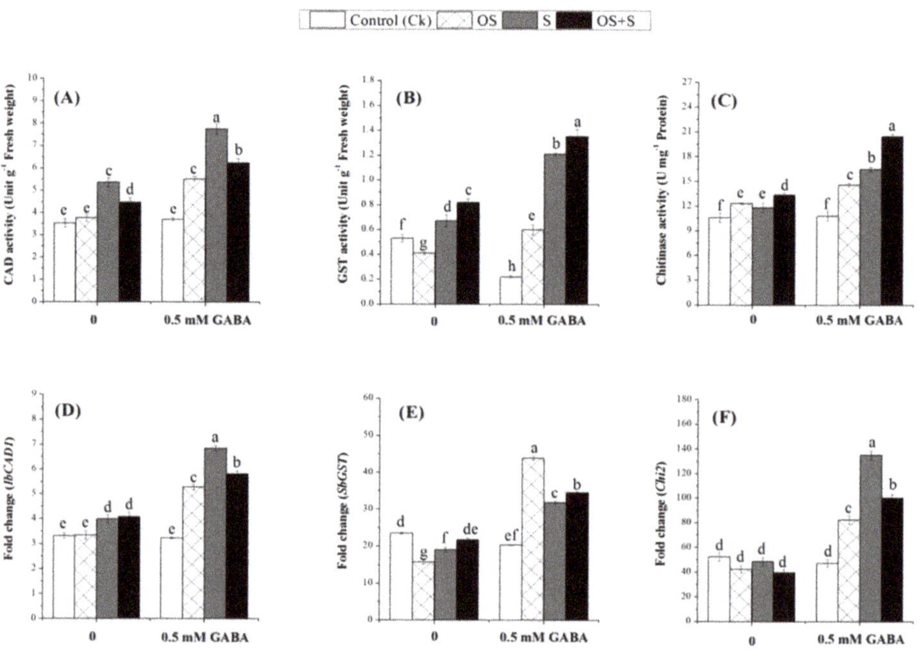

Figure 4. Effects of GABA treatment on the activities of CAD (**A**), GST (**B**) and Chitinase (**C**) and, their transcript levels (**D–F**) in rice seedlings exposed to salinity, osmotic stress, and their combined stress (OS+S). OS (Osmotic stress); S (Salinity).

2.8. Effects of GABA Priming on Detoxification-related Enzymes under Salinity, Osmotic Stress and OS+S

The activities of the detoxification-related enzyme such as Glutathione-s-transferase (GST) and chitinase are shown in Figure 4B,C. Without stress, the GST activity in the primed seedlings was significantly lower than that in the control seedlings (Figure 4B). The GST activity was significantly increased by priming with 0.5 mM GABA under osmotic stress, salinity and OS+S as compared with unprimed seeds. Priming with GABA resulted in an increase in the GST activity under the osmotic stress, salinity and OS+S by 31.66%, 44.62% and 39.25%, respectively, relative to the controls (Figure 4B). Similarly, the chitinase activity was also increased under the osmotic stress, salinity and OS+S as compared to the unstressed condition (Figure 4C). There was no significant difference in the chitinase activity in the primed and unprimed seedlings under the unstressed condition. However, priming with 0.5 mM GABA increased the chitinase activity under the osmotic stress, salinity and OS+S, by 15.54%, 28.34% and 42.09%, respectively, relative to the controls (Figure 4C).

2.9. Effects of GABA Priming on Gene Expression under Salinity, Osmotic Stress and OS+S

In order to further understand the molecular mechanism by which GABA priming could alleviate the detrimental effects of osmotic stress, salinity and their combination, we investigated the transcript levels of *APXa*, *CATa*, *SOD1* (Figure 3D–F), *PAL1*, *PPO*, *SKDH* (Figure 3J–L), *IbCAD1*, *SbGST*, *Chi2* (Figure 4D–F), *OsCIPK01*, *OsCIPK02*, *OsCIPK03*, *OsCIPK07*, *OsCIPK08*, *OsCIPK09*, *OsCIPK12*, *OsCIPK15*, *OsCIPK17* genes (Figure 5A–I). As shown in Figure 3D–F, *SOD1*, *CATa* and *APXa* were significantly up-regulated in the GABA-primed seedlings under osmotic stress, salinity and OS+S stress relative to their controls. In the primed seedlings, *SOD1* was significantly up-regulated under osmotic stress as compared with salinity and the combined stress, while *CATa* was highly significantly up-regulated under the combined stress, whereas the *APXa* was highly significantly up-regulated under the salinity stress. The transcript levels of these genes under different treatments are somewhat consistent with the activity of their corresponding enzymes activity (Figure 3A–C). Under the unstressed conditions, priming with 0.5 mM GABA up-regulated *SOD1*, *CATa* and *APXa* by 62.26, 17.00 and 19.17-fold, respectively as compared with the control condition. The transcription level of *SOD1* was up-regulated in the GABA-primed seeds under the osmotic stress, salinity and OS+S by 22.24, 24.82 and 32.33-fold, *CATa* by 70.04, 69.64, and 64.74-fold; and *APXa* by 32.13, 63.37 and 45.67-fold, respectively (Figure 3D–F). Similarly, priming seeds with 0.5 mM GABA resulted in up-regulation of the *PAL1*, *PPO* and *SKDH* genes under osmotic stress, salinity and OS+S stress relative to their controls (Figure 3J–L). The transcription level of *PAL1* and *SKDH* greatly increased under salinity and combined stress without significant differences between them. In contrast, *PPO* expression was greatly increased under the combined stress. In the absence of stress conditions, priming with 0.5 mM GABA resulted in up-regulation of the *PAL1*, *PPO* and *SKDH* genes by 3.06, 2.85 and 8.00-fold, respectively, as compared with the control condition. The relative expression level of *PAL1*, *PPO* and *SKDH* genes was up-regulated in the primed seedlings under the osmotic stress, by 50.07, 50.57 and 47.72-fold; salinity by 32.39, 38.15 and 32.81-fold, and OS+S by 28.38, 32.53 and 27.58-fold, respectively, as compared with their controls (Figure 3J–L). The transcription levels of *ibCAD1* (Figure 4D) and *Chi2* (Figure 4F) in the GABA-primed plants were highly significantly up-regulated when the seedlings were exposed to salinity stress as compared to osmotic stress and the combined stress, while the transcription level of *SbGST* was highly significantly up-regulated in the GABA-primed plants when exposed to osmotic stress as compared with salinity and combined stress (Figure 4E). Osmotic stress, salinity, and OS+S, priming with 0.5 mM GABA up-regulated *ibCAD1* by 36.55, 41.66 and 29.77-fold; *SbGST* by 64.19, 25.14, and 36.91-fold; and *Chi2* by 48.70, 64.02 and 60.46-fold, respectively, as compared with the controls (Figure 4D–F). The results reported that the *OsCIPK* genes in the GABA-primed seedlings were up-regulated under the osmotic, salinity and OS+S conditions as compared to the controls (Figure 5). Salinity stress resulted in significant increases in the expression level of the *OsCIPK01*, *OsCIPK03*, *OsCIPK08* and *OsCIPK15* genes as compared to osmotic stress and the combined stress (Figure 5A,C,E,H). Osmotic stress resulted in a greater increase in the expression levels of *OsCIPK02*, *OsCIPK07* and *OsCIPK09* genes as compared with salinity and the combined stress (Figure 5B,D,F), whereas the combined stress (OS+S) resulted in a significant up-regulation of *OsCIPK12* and *OsCIPK17* genes as compared with salinity and osmotic stress (Figure 5G,I).

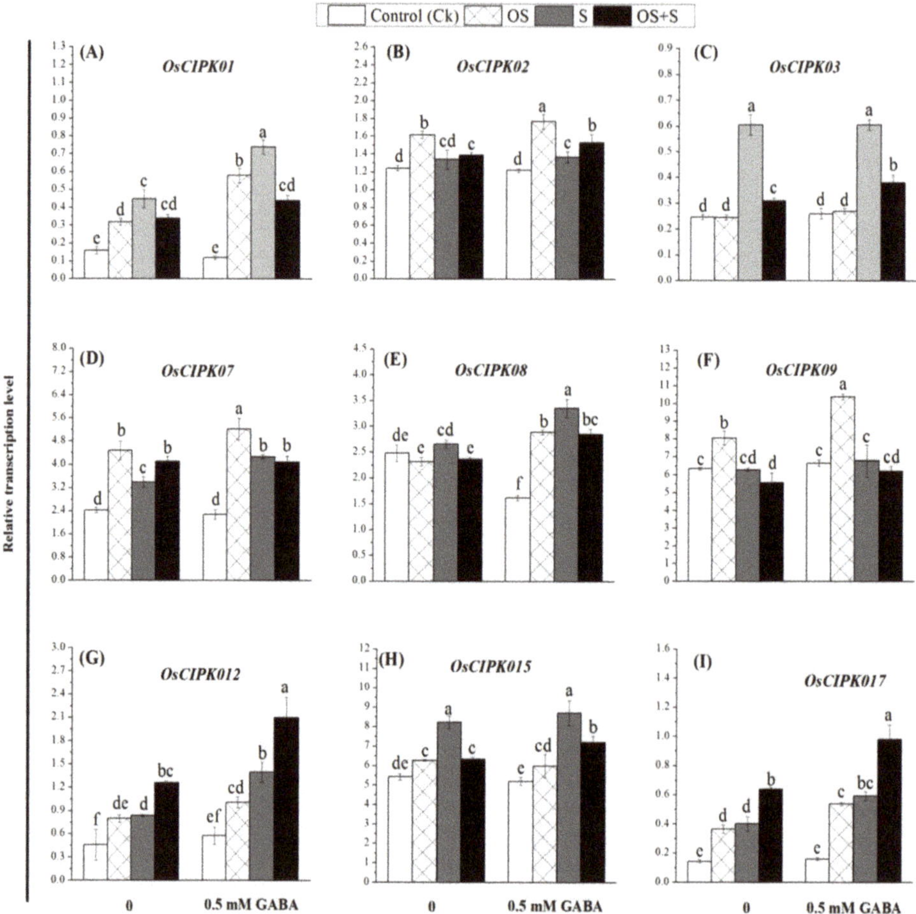

Figure 5. Effects of GABA treatment on the transcription levels of *OsCIPK01* (**A**), *OsCIPK02* (**B**), *OsCIPK03* (**C**), *OsCIPK07* (**D**), *OsCIPK08* (**E**), *OsCIPK09* (**F**), *OsCIPK012* (**G**), *OsCIPK015* (**H**) and *OsCIPK017* (**I**) genes in rice seedlings exposed to salinity, osmotic stress, and their combined stress (OS+S). OS (Osmotic stress); S (Salinity).

2.10. Effects of GABA Priming on the Nuclear DNA Content and Ultramorphology of the Cell under Salinity, Osmotic Stress and OS+S

In order to investigate whether the osmotic stress, salinity and their combination could inhibit cell cycle progression, the nuclear DNA content was analyzed using the flow cytometry technique. The results showed that seedlings without GABA priming under osmotic stress and salinity and their combination underwent changes in cell cycle progression (Figure 6A–D). Under the osmotic stress condition (Figure 6B), the cells were blocked in at the G2/M phase, but the cells showed pronounced nuclear accumulation at the G0/G1 phase. However, with GABA priming and under the osmotic stress (Figure 6F), the nuclear accumulation was more obvious in both phases of the cell cycle. The salinity and the OS+S stresses induced a pronounced cell accumulation in G2/M and a sharp inhibition in the cell progression through the G0/G1 phase irrespective of the priming treatment (Figure 6C,D,G,H). The flow cytometry analysis indicated that the G0/G1 phases were more sensitive under the salinity alone or the combined stress as compared to the osmotic stress. These findings suggested that the osmotic stress, salinity and their combination result in a longer time for cells to progress through the cell cycle. However, each treatment affects cell cycle progression in a different way. The cell ultrastructure was affected due to the priming, salinity and the osmotic stress treatments (Figure 7). Under the control condition (Figure 7A) and priming treatment (Figure 7B), the transmission electron microcopy (TEM) analysis showed clear cell walls and developed chloroplasts (Chl) with uniform thylakoids (Thy). Under the salinity and the combined stress (OS+S), in the unprimed plants (Figure 7C,E), an unclear cell wall, a lot of vacuoles and raptured chloroplasts were observed. However, under the salinity and the combined stress (Figure 7D,F), the priming treatment somewhat improved the cell structure, which was represented by a clear cell wall, developed starch grain and the absence of the vacuoles. These results suggested that the priming with 0.5 mM GABA led to changes in the cell cycle progression and cell ultramorphology as a kind of cell signaling under salinity, osmotic stress and their combination.

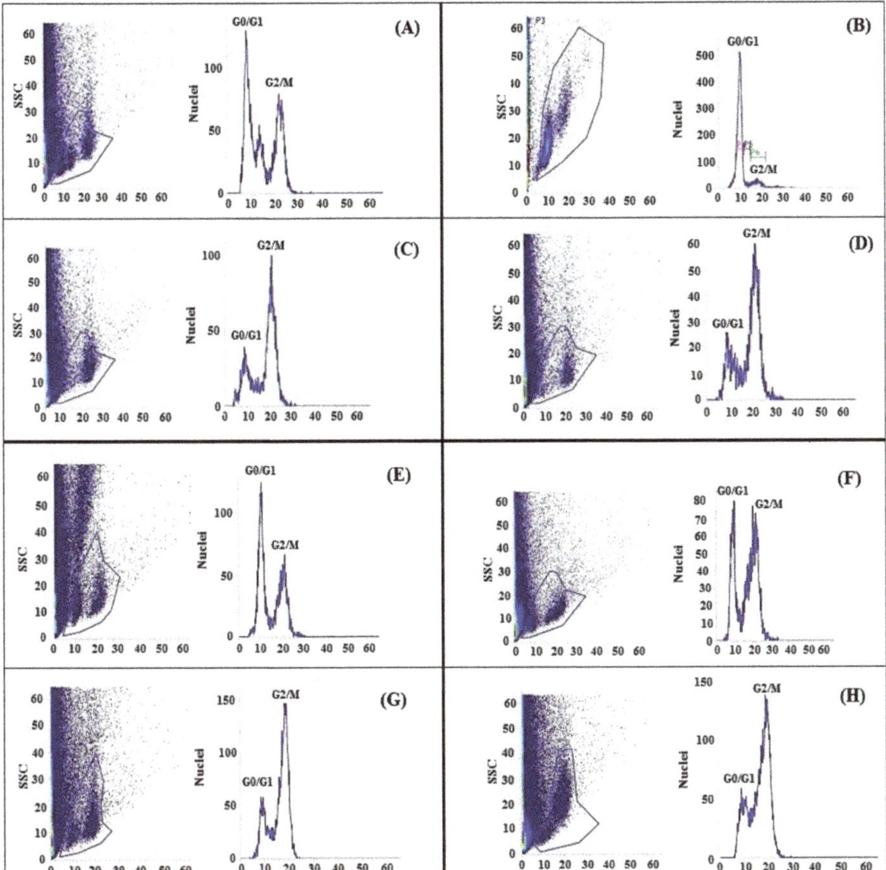

Figure 6. Flow cytometric analysis of GABA-unprimed rice (**A–D**) showing the nuclear DNA content of the root cell under control (**A**); osmotic stress (**B**); salinity (**C**) and their combined stress (**D**) and GABA-primed rice (**E–H**) under control (**E**); osmotic stress (**F**); salinity (**G**) and their combined stress (**H**).

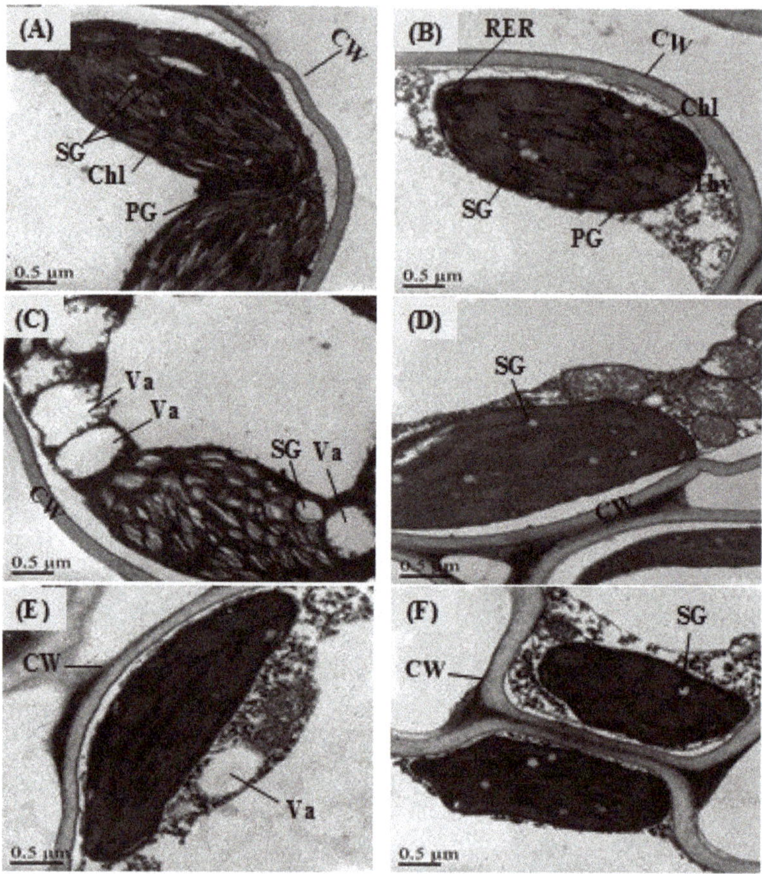

Figure 7. Transmission electron microscopic images of the leaf mesophyll of rice seedlings primed with 0.5 mM of GABA and grown under the control, salinity and the combination stress (OS+S) conditions. Control condition (**A**); primed with 0.5 mM GABA (**B**); salinity stress (**C**); salinity stress and primed with 0.5 mM GABA (**D**); combined stress (OS+S) (**E**); Combined stress (OS+S) and primed with 0.5 mM of GABA (**F**).

3. Discussion

Enhancing plant tolerance to abiotic stresses by inducing a stress-responsive gene pathway in transgenic plants is a promising approach [28]. Combined stress is a new kind of abiotic stress in plants that requires integration defense or cross-talk response, and the plants should be tested for their tolerance to a combination of different stresses prior to cultivation under field conditions. The present study reported that priming with 0.5 mM GABA improved the physiological parameters under salinity and osmotic stress and their combination as compared with unprimed plants (Table 1). During seed germination and the stress condition, GABA could improve starch catabolism and mobilization of sugar and amino acids which are necessary for seedling growth [29]. GABA can also improve the antioxidant system for mitigation of oxidative damage, and increase Na^+/K^+ transportation for osmotic regulation under salt stress [29]. Furthermore, GABA could maintain the hormones and mineral nutrients and reduce lipid peroxidation under different environmental stresses [30]. During the early stage of seed imbibition, the up-regulation of some germination-related genes contributed to the metabolic process prior to seed germination which could improve the germination and seedling growth under the stress condition [31]. The reduction in rice growth from seed germination to the maturity stage under salinity stress may be due to the increase in osmotic pressure of the root medium and ion effects [32]. Several studies have reported that osmotic stress significantly decreased fresh and dry weights of shoots [33] and roots [34]. It has been found that exogenous GABA could significantly improve the shoot length, root length, and fresh weight of maize seedlings [35] and white clover [29] under abiotic stresses. In the present study, GABA induced starch catabolism, which is of primary importance for providing available carbohydrates for seed germination and growth under osmotic stress, salinity and the combined stress. These findings are consistent with previous studies, which found that environmental stresses such as osmotic stress, salt and heat stress decreased seed germination due to the inhibition of starch catabolism under these conditions [29,36,37]. It has been found that the metabolite mobilization of starch and soluble sugars is critical for the maintenance of cell turgor and energy sources when seeds are subjected to salinity and osmotic stress [36]. Additionally, GABA could increase starch catabolism and provide available carbohydrates for seed germination and growth of white clover under salt stress [29]. Moreover, starch catabolism could be accelerated by activating α- and β-amylase activities in GABA-pretreated seeds [29]. It is well-known that both organic and inorganic osmolytes such as sugars and starch are important osmotic regulators for plant adaption to environmental stresses [38–40]. These osmolytes were decreased in the present study under salinity, osmotic stress and OS+S, and were enhanced by priming with 0.5 mM GABA (Table 3). It has been found that the accumulation of osmolytes such as free starch, soluble sugars and protein regulated the osmotic pressure of the plants under abiotic stress including salinity and osmotic stress [41]. Furthermore, plants under stress conditions may accumulate small molecular weight proteins that could be used as a source of storage nitrogen and could be rapidly mobilized when required for the alleviation of stress [42]. These proteins could also have a role in osmotic adjustment [43]. Additionally, proline can also accumulate in plants to act as a solute for adaptation of plants in response to different environmental stresses [44].

The photosynthetic response to abiotic stress, especially salinity and osmotic stress, is highly complex. It involves the interaction of stress signaling with different plant cells which can promote plant growth and development. In the present study, rice plants grown under salinity, osmotic stress and OS+S treatments showed a marked reduction in Pn, gs and Tr (Table 2). The remarkable reduction in gs may be a consequence of stomatal closure caused by the higher osmotic pressure in guard cells under OS+S [45]. Another study observed a decrease of the chlorophyll content in rice under water stress, which may be related to the inhibition of the photosynthetic system of the plant under the water stress [46], or might be due to increases in ROS and lipid peroxidation levels leading to chlorophyll damage and a change in the leaf color from green to yellow [47]. In the present study, RWC was decreased by the salinity and osmotic stress and OS+S, and enhanced by GABA priming as compared with the control condition (Table 2). A decrease in the RWC under PEG-induced osmotic stress was also reported in rice leaves [46] and in tomato [48]; this might be due to the decreased water potential under salinity and osmotic stresses [49]. Similarly, the cellular accumulation of GABA could achieve a balance in the reduction of water potential that occurs during cellular dehydration under the stress condition [50].

Rice controls the transport of salts initially by selective uptake by root cells and ions entering into the root along with water through symplastic and apoplastic routes [51]. The ratio of Na^+/K^+ can be used as a physiological index for salt response in several crop plants such as tomato (*Solanum lycopersicum*) [52], chickpea (*Cicer arietinum*) [53], barley (*Hordeum vulgare*) [54] and white clover (*Trifolium repens*) [29]. Moreover, the concentration of Na^+ is key for the salinity tolerance mechanism, interacting with K^+ homeostasis, and especially given its involvement in numerous metabolic processes, maintaining a balanced cytosolic Na^+/K^+ ratio [55]. In the present study, priming with 0.5 mM GABA under salinity, osmotic stress and OS+S caused a significant reduction in the Na^+ concentration in the leaf and root as compared with unprimed seeds (Figure 2). Under the salinity stress, plant cell loses the balance in the Na^+/K^+ ratio due to a continuously increasing Na^+ ion concentration and decrease in K^+ ion uptake [56]. In the current study, a higher concentration of Na^+ was recorded in roots as compared to the leaf under the salinity and osmotic stress and their combination (Figure 2). The high accumulation of Na^+ especially in the leaf resulted in a reduction in the photosynthesis due to stomatal and non-stomatal limitation [57]. Therefore, plants have to reduce the concentration of Na^+ in the plant leaves by either minimizing the entry from the root symplast to reduce loading, maximizing Na^+ retrieval from the xylem [58], or exporting Na^+ from the leaf to the phloem [59]. As such, the down-regulation of genes encoding Na^+ influx transporters (*OsCNGC1*) in rice root significantly contributed to the salinity tolerance, as it could avert toxic Na^+ influx [60].

In the present study, antioxidant enzymes were measured to investigate the capability of priming with 0.5 mM GABA to induce such enzymes for mitigating the oxidative stress by scavenging ROS in the cell and thus increase the rice tolerance under the salinity, osmotic stress and their combination. In the current work, salt and osmotic stress and their combination (OS+S) significantly decreased the activity of CAT, SOD and APX (Figure 3), which was consistent with that of barely exposed to salinity stress [21]. However, the decrease in these enzymes was inhibited when the seeds were primed with 0.5 mM of GABA under the stress conditions (Figure 3). Similar findings were observed in different plant species primed with GABA such as rice seedlings [61], black pepper seedlings [62], perennial ryegrass [63] and white clover [29]. The accumulation of ROS and oxidative stress signal are the main common mechanisms for the plant tolerance could be observed under the salinity and osmotic stresses. The intensive accumulation of ROS in the different plant cells resulted in significant pigment loss, a reduction in the photosynthetic system efficiency and decreased protein assimilation [64]. However, induction of the antioxidant defense system could protect plants from oxidative damage induced by ROS accumulation under stress conditions [65]. In the present study, salinity and osmotic stress and OS+S improved the ROS, i.e., H_2O_2, O_2^- and OH^- accumulation in the leaf tissue as compared with the control condition (Table 2). The present study reported that GABA-activated antioxidant enzymes could play vital roles in scavenging free radicals induced by H_2O_2 and O_2^- and reduced lipid peroxidation by up-regulating the genes (*APXa*, *CATa* and *SOD1*) involved in antioxidant enzymes during rice seed germination and seedling growth under salinity and osmotic stress and their combination (Figure 3).

The accumulation of phenolic compounds under environmental stresses can protect plants from damage caused by ROS-induced oxidative stress [66] by scavenging free radicals, breaking radical chain reactions, and decomposing peroxides [67]. In the present study, the enzymes involved in the phenolic metabolism such as PAL, PPO, SKDH and CAD were improved by priming with 0.5 mM GABA under the osmotic, salinity and OS+S stresses (Figure 3G–I and Figure 4A). This phenolic metabolism may provide an effective defense tool for plant tolerance under environmental stresses [68]. Recently, SKDH activity was increased in plants exposed to salinity stress for 3 days [69]. Moreover, the activity and the expression level of the *PAL* protein were increased in barley exposed to 200 μM of Al for 24 h [70]. The increased PAL and PPO levels might help plants to cope with oxidative stress by scavenging ROS [45,71]. In the present study, the accumulation of PPO in the leaf under salinity and osmotic stresses (Figure 3H), might be due to the induction of *PPO* genes in the leaf tissues under osmotic stress [72]. However, the PPO and CAD activities were not affected in *Matricaria chamomilla* plants exposed to salinity stress, but a higher activity of SKDH was observed in the root [68]. In the present study, the induction of the defense genes involved in secondary metabolites such as *PAL1*, *PPO*, *SKDH* and *IbCAD1* in the primed plants may increase the tolerance of plants to salinity, osmotic stress and OS+S (Figure 3J–L and Figure 4B), which is consistent with the finding of Ahammed et al. [71], who also found that plant growth regulators help in alleviating the oxidative stress in plant leaves through the induction of phenolic metabolism defense under the stress condition. The mechanism by which GABA treatment increases the tolerance to salt stress might be due to the ability of GABA to induce endogenous GABA, proline and the total phenolic content, thus enhancing the antioxidant capacity [67]. The present study revealed that the GST and chitinase activities and their transcript levels were increased by priming with 0.5 mM of GABA under salinity, osmotic stresses and their combination (Figure 4). Similarly, a previous study reported that the *Chi2* gene was up-regulated in pepper leaves under salinity and osmotic stresses which could protect plant tissues against osmotic stress via an

ABA-independent signal transduction pathway [73]. Another study reported that chitinase is involved in heavy metal stress tolerance and chilling tolerance [74]. Moreover, GST activity and expression level increased in salinity and osmotic -stressed barley plants [45].

In the present study, up-regulation of *OsCIPK* genes was observed in the GABA-primed seedlings under salinity and osmotic stress and OS+S as compared to their controls (Figure 5). The *OsCIPK* genes induced by different stresses may provide new signaling pathway to reveal the molecular mechanism of rice response to different stresses alone or in combination considering the nature of *CIPKs* as putative signaling components [75]. The signaling pathway of *OsCIPK* genes may be involved in the substantial common regulatory systems or cross talks triggered by different stresses [75]. Our findings indicated that expression patterns of *OsCIPK* genes were induced under the salinity, osmotic stress and OS+S, which is consistent with cross talk between salinity and osmotic stress as previously reported by Seki et al. [76]. The interaction between co-activated pathways is likely to be mediated at different levels under the combined stresses [13]. This pathway could include the interaction between different transcription factors and mitogen-activated protein kinase (MAPK) cascades [77], different stress hormones such as ethylene, jasmonic acid and abscisic acid [78], between calcium and/or ROS signaling [79] and between different receptors and signaling complexes [80].

The present study revealed that nuclear accumulation was inhibited under the stress condition especially under salinity and the combined stress in the G0/G1 as compared with the control or osmotic stress (Figure 6). A recent study showed that root growth was inhibited under the abiotic stress conditions, and the cell division and cell cycle regulation might be involved in this inhibition [80], or might be associated with the reduction of cell production [81]. The reduced cell production under the salinity and osmotic stress and their combination might be due to a smaller number of dividing cells such as a meristem size reduction, and the temporary inhibition of mitotic activity that allows the adaptation to the stress condition is most likely mediated by post-translational control of cyclin-dependent kinase activity (CDK) [80].

4. Materials and Methods

4.1. Plant Materials and Growth Conditions

Rice (*Oryza sativa* cv. Qian You No. 0508) seeds were purchased from the Seed Production Unit, Jiangsu Academy of Agricultural Sciences, China. Before priming, seeds were sterilized with 0.5% NaClO solution for 15 min and washed several times to remove the traces of the disinfectant. The seeds were then primed with GABA at the optimized concentration (0 and 0.5 mM) at 15 °C in darkness for 24 h. The time and concentration of the priming agent were initially selected based on a preliminary study. In the preliminary experiment, several concentrations of GABA, i.e., 0, 0.1, 0.2, 0.3, 0.4, 0.5, 0.6 and 0.7 mM were used for seed priming. GABA at 0.5 mM significantly improved rice germination and seedling growth as compared to other concentrations. The seed were primed for 24 h as this time was sufficient enough for rice seeds to trigger the activation of various metabolic processes such as the synthesis of hydrolytic enzymes which resulted in hydrolysis of reserve food into a simple available form for embryo uptake as stated recently in our previous study [82–85]. The primed seeds were dried at room temperature to maintain their original moisture content [82]. Thereafter, the primed and unprimed seeds were germinated for two weeks in a plastic germination box containing two layers of germination paper moistened with water. Fifty seeds and three replications for each treatment were used. Then, seeds were incubated in a germination chamber at 25 °C with 80 % relative humidity under alternating cycles of 16 h illumination and 8 h darkness for 14 days.

The salinity and osmotic stress and their combination were applied to 7-day-old rice seedlings, in which the salinity stress (150 mM NaCl) and the osmotic stress (30 g/L PEG, 6000) and their combination (150 mM NaCl+30 g/L PEG) were supplied to plants for one week. The seedlings without GABA priming and stress treatments were used as the control (Ck). After fourteen days, the germination percentage, germination energy, root and shoot length, seedling fresh and dry weight and seedling vigor index were measured according to the methods of Hu et al. [86].

4.2. Determination of Photosynthesis and Leaf water Relation

The physiological parameters such as Pn, Tr, Gs and Ci were measured according to our previous study [87]. The chlorophyll content was measured spectrophotometrically according the method of Sheteiwy et al. [88]. The values of Ψw, WUE and RWC were measured according to our previous study [87]. The Ψs was measured according to the method of Ahmed et al. [8].

4.3. Biochemical Analysis

For further investigation regarding the potential role of GABA priming to alleviate the oxidative stress induced by salinity and osmotic stress, the antioxidant enzymes, i.e., SOD, CAT and APX were measured according the method of Sheteiwy et al. [88]. GR and total soluble sugars were measured according the method of Sheteiwy et al. [89]. The starch content and total soluble proteins were measured according to the method of Sheteiwy et al. [90].

4.4. Analysis of MDA, Proline and ROS Contents

The MDA content of the leaf was measured according to the method of Zhou and Leul [91]. Proline concentration was determined by a spectrophotometer according to the method of Li [92]. Briefly, 100 mg of leaf was homogenized with 5 mL of 3% sulfosalicylic acid and centrifuged at $5000 \times g$ for 10 min. The supernatant was treated with acid-ninhydrin and acetic acid, after which the supernatant was boiled for 1 h at 100 °C. Absorbance was determined at 520 nm and the proline content was expressed as $\mu g\ gFW^{-1}$. For determination of the H_2O_2 content, 0.5 g of leaf was homogenized with 5.0 mL of 0.1% trichloroacetic acid (TCA) using an ice bath, and then the homogenate was centrifuged for 15 min at $12,000 \times g$ [93]. The H_2O_2 content in the supernatant was read using a spectrophotometer at 390 nm. The content of O_2^- was measured according to Jiang and Zhang's method [94]. The content of OH^- in the leaf was determined according to our previous study [89].

4.5. Determination of GABA Content

The GABA content in the leaf tissues was measured using GABAase commercial enzyme preparation (Sigma chemical Co., St. Louis, MI, USA) as previously described by Ma et al. [21].

4.6. Assay of Phenolic Metabolism-related Enzymes

The activity of PAL was determined according to the method of Zheng et al. [95] with slight modification. Fresh leaves (2g) were homogenized with 2.5 mL of solution containing 100 mM K_3PO_4 buffer, 2 mM EDTA, 1% (*m/v*) PVP, and 1 mM phenyl-methylsulfonyl fluoride (PMSF). Then, the homogenates were centrifuged at $12,000 \times g$ for 15 min at 4 °C, and the supernatant fractions were used for enzyme analysis. The absorbance was spectrophotometrically measured at 290 nm. The activities of PPO and SKDH were measured according to the method of Sheteiwy et al. [82]. The activity of CAD was measured according to the method of Wyrambik and Grisebach [96].

4.7. Measurements of Detoxification-related Enzymes

In order to determine the activity of GST, 0.3 g of leaf tissue was homogenized with 2 mL phosphate buffer solution (pH 6.5) + 1 mM EDTA. The suspension was centrifuged at $4000 \times g$ for 10 min. The GST activity was measured spectrophotometry at 412 nm following the method of Chun-hua and Ying [97]. The chitinase activity was measured according to the method of Chun-hua and Ying [97].

4.8. Measurement of Na⁺ and K⁺ Ions in the Leaf and Root Tissues

The concentrations of Na^+ and K^+ in leaf and root were measured according to the methods of Zhao et al. [98]. Briefly, the samples were washed with distilled water before the determination to remove any traces of the Na^+ from the leaf and root tissues, which were then dried at 50 °C for 4 d. Thereafter, the dried leaf and root tissues were ground into a fine powder in liquid nitrogen. The powder was then digested in 5 mL nitric acid overnight. Thereafter, the digested solution was diluted to 25 mL with double-distilled water. The concentration of Na^+ and K^+ in the acid-digested tissues was measured using a flame photometer according to the method of Zhao et al. [98].

4.9. Analysis of Gene Expression

In order to further study the mechanism by which GABA can alleviate the effects of both salinity and osmotic stresses alone or in combination on rice germination and seedling growth, the antioxidant enzymes, detoxification-related enzymes, phenolic metabolism-related enzymes and *OsCIPK* responses genes were investigated at the molecular levels. For this purpose, frozen leaf tissues (100 mg each) were ground thoroughly in liquid nitrogen using a pestle and mortar. Thereafter, the total RNA was isolated from the leaf and the concentration of the RNA was determined by a NanoDrop 2000/2000c (Thermo Scientific, Wilmington, Delaware, USA). The RNA purity was also checked by the spectrophotometer using the 260/280 nm ratio before quantitative real-time PCR. The primers of the *OsCIPK* genes presented in Supplementary Table S1 are the same as those used previously by Xiang et al. [75]. Quantitative real-time RT-PCR was performed using SYBR premix EX Taq (Takara, Japan). The PCR program used in this study is the same as that used recently by Sheteiwy et al. [88].

4.10. Ultra-structure and Flow Cytometry Analysis

The ultramorphology of the leaf was investigated according to our previous study [82]. The H_2O_2 was detected according to the method of Sheteiwy et al. [82]. Briefly, the roots were stained with 5 μM dichlorodihydrofluorescein diacetate for 15 min, and then washed with excess 20 mM sodium phosphate buffer (pH 6.1) to stop the reaction. The changes in $\Delta\Psi m$ were analyzed using a tetramethylrhodamine methyl ester assay kit (Immunochemistry Technologies, Bloomington, IN, USA) and imaged using a laser confocal scan microscope (Zeiss LSM 780, Zeiss, Germany). Then, nuclear isolation was performed according to the method of Hu et al. [99]. The root samples were cut into small pieces and then fixed with nucleus isolation buffer [10 mM $MgSO_4$, 50 mM KCl, 5 mM Hepes, 1 mg/mL dithiothreitol (Sigma, St. Louis, MI, USA) and 0.2% Triton X-100] and filtered through a 33 mm nylon mesh. The nuclei were fixed in 4% paraformaldehyde for 30 min and were then precipitated (200 g, 10 min, 4u C) and re-suspended in the isolation buffer.

4.11. Experimental Design and Statistical Analysis

The treatments were applied using a Completely Randomized Block Design (CRBD) with a factorial arrangement. All the obtained values are the means of three replicates ± standard deviation (SD). The data were analyzed using two-way analysis of variance (ANOVA) by SPSS v16.0 (SPSS, Inc., Chicago, IL, USA), and means were separated using Duncan's multiple range tests ($\alpha = 0.05$).

5. Conclusions

Priming with 0.5 mM of GABA could be an effective technique to alleviate salinity and osmotic stresses and OS+S causing inhibition of rice seed vigor. Under the stress conditions, GABA induced a balance in Na^+/K^+ accumulation and transport from the root to the leaf which could be attributed to the osmotic adjustment through the mobilization of organic osmolytes such as proline, sugars, and starch during seed germination. The fluorescence staining revealed that H_2O_2 formation was increased under the stress condition and decreased by the GABA priming treatment. These findings indicated that GABA could also act as a signal molecule under salinity, osmotic stress and their combination by increasing antioxidant enzymes, phenolic metabolism-related enzymes and detoxification-related enzyme activities and their transcript levels. The significantly improved starch and sugar contents and *CIPK* gene expression in rice seedlings by GABA treatment under the stress conditions may be the main mechanism of rice tolerance to salinity, osmotic stress and their combination. The present study elucidated the possible cross talk between the salinity and osmotic stresses when the plant are exposed to both at the same time, and thus to develop transgenic crops with enhanced tolerance to field conditions, further studies need to expand their area to include stress combinations. Current findings provide new evidence for better understanding of GABA-regulated osmotic and salinity combined stress tolerance during seed germination and development. The results showed that the different abiotic stresses induced changes in cell cycle progression resulting in inhibition in rice root cell development. Priming with 0.5 mM GABA has the potential to improve cell ultra-morphology under the stress condition.

Supplementary Materials: Supplementary materials can be found at http://www.mdpi.com/1422-0067/20/22/5709/s1.

Author Contributions: M.S.S. initiated and designed the experiment. W.Q., R.Y., B.T. and Y.A.H. helped in conducting experiments. M.S.S. performed the RNA extractions and plant hormone analysis. M.S.S., Y.A.H., W.Q., N.U.K. and H.S. (Hiba Shaghaleh) analyzed the data and results. M.S.S. monitored the experimental work and wrote the manuscript. M.S.S. and H.S. (Hongbo Shao) revised and edited the final version of the manuscript. All authors read and approved the final manuscript.

Funding: This work was supported by Jiangsu Autonomous Innovation of Agricultural Science & Technology [CX (15)1005], the Excellent Scientist Plan of JAAS, and Shuangchuang Talent Plan of Jiangsu Province.

Acknowledgments: We thank Feng Lin for technical assistance in the flow cytometry analysis of rice tissues.

Conflicts of Interest: The authors declare that they have no competing interests.

References

1. Blumwald, E. Sodium transport and salt tolerance in plants. *Curr. Opin. Cell Biol.* **2000**, *12*, 431–434. [CrossRef]

2. Munns, R.; Tester, M. Mechanisms of salinity tolerance. *Ann. Rev. Plant Biol.* **2008**, *59*, 651–681. [CrossRef] [PubMed]

3. Rajendran, K.; Tester, M.; Roy, S.J. Quantifying the three main components of salinity tolerance in cereals. *Plant Cell Environ.* **2009**, *32*, 237–249. [CrossRef] [PubMed]

4. Akbar, M. Breeding for salinity tolerance in rice. In *Salt-affected Soils of Pakistan, India and Thailand*; IRRI, Ed.; International Rice Research Institute: Manila, Philippines, 1986; pp. 39–63.

5. Zhu, G.Y.; Kinet, M.; Lutts, S. Characterization of rice (*Oryza sativa* L.) F3 populations selected for salt resistance. I. Physiological behaviour during vegetative growth. *Euphytica* **2001**, *121*, 251–263. [CrossRef]

6. Castillo, E.G.; Tuong, T.P.; Ismail, A.M.; Inubushi, K. Response to salinity in rice: Comparative effects of osmotic and ionic stresses. *Plant Prod. Sci.* **2007**, *10*, 159–170. [CrossRef]

7. Kaiser, W.M. Effects of water deficit on photosynthetic capacity. *Plant Physiol.* **2009**, *71*, 142. [CrossRef]

8. Ahmed, I.M.; Dai, H.; Zheng, W.; Cao, F.; Zhang, G.; Sun, D.; Wu, F. Genotypic differences in physiological characteristics in the tolerance to drought and salinity combined stress between Tibetan wild and cultivated barley. *Plant Physiol. Biochem.* **2013**, *63*, 49–60. [CrossRef]

9. Wang, R.; Chen, S.; Zhou, X.; Shen, X.; Deng, L.; Zhu, H.; Shao, J.; Shi, Y.; Dai, S.; Fritz, E.P.; et al. Ionic homeostasis and reactive oxygen species control in leaves and xylem sap of two poplars subjected to NaCl stress. *Tree Physiol.* **2008**, *28*, 947–957. [CrossRef]

10. Singh, N.P.; Pal, P.K.; Vaishali, S.K. Morpho-physiological characterization of Indian wheat genotypes and their evaluation under drought condition. *Afr. J. Biotechnol.* **2014**, *13*, 2022–2027.

11. Turkan, I.; Bor, M.; Ozdemir, F.; Koca, H. Differential responses of lipid peroxidation and antioxidants in the leaves of drought-tolerant P. acutifolius Gray and drought-sensitive P. vulgaris L. subjected to polyethylene glycol mediated water stress. *Plant Sci.* **2005**, *168*, 223–231. [CrossRef]

12. Wu, Y.; Yang, C. Physiological responses and expression profile of NADPH oxidase in rice (*Oryza sativa*) seedlings under different levels of submergence. *Rice* **2016**, *9*, 2–10. [CrossRef] [PubMed]

13. Mittler, R. Abiotic stress, the field environment and stress combination. *Trends Plant Sci.* **2006**, *11*, 15–19. [CrossRef] [PubMed]

14. Rizhsky, L.; Liang, H.; Shuman, J.; Shulaev, V.; Davletov, S.; Mittler, R. When defense pathways collide: The response of Arabidopsis to a combination of drought and heat stress. *Plant Physiol.* **2004**, *134*, 1683–1696. [CrossRef] [PubMed]

15. Fowler, S.; Thomashow, M.F. Arabidopsis Transcriptome profiling indicates that multiple regulatory pathways are activated during cold acclimation in addition to the CBF cold response pathway. *Plant Cell* **2002**, *14*, 1675–1690. [CrossRef]

16. Kinnersley, A.M.; Turano, F.J. Gamma aminobutyric acid (GABA) and plant responses to stress. *Crit. Rev. Plant Sci.* **2000**, *19*, 479–509. [CrossRef]

17. Mekonnen, D.W.; Flüggea, U.; Ludewig, F. Gamma-aminobutyric acid depletion affects stomata closure and drought tolerance of *Arabidopsis thaliana*. *Plant Sci.* **2016**, *245*, 25–34. [CrossRef]

18. Jordan, B.R.; Givan, C.V. Effects of light and inhibitors on glutamate metabolism in leaf discs of *Vicia faba* L. *Plant Physiol.* **1979**, *64*, 1043–1047. [CrossRef]

19. Bown, A.W.; MacGregor, K.B.; Shelp, B.J. Gamma-aminobutyrate: Defense against invertebrate pests. *Trends Plant Sci.* **2006**, *11*, 424–427. [CrossRef]

20. Bouché, N.; Fromm, H. GABA in plants: Just a metabolite. *Trends Plant Sci.* **2004**, *9*, 110–115. [CrossRef]

21. Ma, Y.; Wang, P.; Wang, M.; Sun, M.; Gu, Z.; Yang, R. GABA mediates phenolic compounds accumulation and the antioxidant system enhancement in germinated hulless barley under NaCl stress. *Food Chem.* **2019**, *270*, 593–601. [CrossRef]

22. Renault, H.; Roussel, V.; Amrani, A.E.; Arzel, M.; Renault, D.; Bouchereau, A. The Arabidopsis pop2-1 mutant reveals the involvement of GABA transaminase in salt stress tolerance. *BMC Plant Biol.* **2010**, *10*, 20. [CrossRef] [PubMed]

23. Barbosa, J.M.; Singh, N.K.; Cherry, J.H.; Locy, R.D. Nitrate uptake and utilization is modulated by exogenous gamma-aminobutyric acid in Arabidopsis thaliana seedlings. *Plant Physiol. Biochem.* **2010**, *48*, 443. [CrossRef] [PubMed]

24. Beuve, N.; Rispail, N.; Laine, P.; Cliquet, J.B.; Ourry, A.; Le Deuneff, E. Putative role of gamma-aminobutyric acid (GABA) as a long-distance signal in up-regulation of nitrate uptake in *Brassica napus* L. *Plant Cell Environ.* **2004**, *27*, 1035–1046. [CrossRef]

25. Palanivelu, R.; Brass, L.; Edlund, A.F.; Preuss, D. Pollen tube growth and guidance is regulated by POP2, an Arabidopsis gene that controls GABA levels. *Cell* **2003**, *114*, 47–59. [CrossRef]

26. Deleu, C.; Faes, P.; Niogret, M.F.; Bouchereau, A. Effects of the inhibitor of the g-aminobutyrate-transaminase, vinyl-gaminobutyrate, on development and nitrogen metabolism in Brassica napus seedlings. *Plant Physiol. Biochem.* **2013**, *64*, 60–69. [CrossRef]

27. Niu, L.; Dong, B.; Song, Z.; Meng, D.; Fu, Y. Genome-wide identification and characterization of CIPK family and analysis responses to various stresses in apple (*Malus domestica*). *Int. J. Mol. Sci.* **2018**, *19*, 2131. [CrossRef]

28. Kasuga, M.; Liu, Q.; Miura, S.; Yamaguchi-shinozaki, K.; Shinozaki, K. Improving plant drought, salt, and freezing tolerance by gene transfer of a single stress-inducible transcription factor. *Nat. Biotechnol.* **1999**, *17*, 287–291. [CrossRef]

29. Cheng, B.; Li, Z.; Liang, L.; Cao, Y.; Zeng, W.; Zhang, X.; Ma, X.; Huang, L.; Nie, G.; Liu, W.; et al. The γ-Aminobutyric Acid (GABA) alleviates salt stress damage during seeds germination of white clover associated with Na$^+$/K$^+$ transportation, dehydrins accumulation, and stress-related genes expression in white clover. *In. J. Mol. Sci.* **2018**, *19*, 2520–2528. [CrossRef]

30. Alqarawi, A.A.; Hashem, A.; AbdAllah, E.F.; Al-Huqail, A.A.; Alshahrani, T.S.; Alshalawi, S.A.R.; Egamberdieva, D. Protective role of gamma amminobutyric acid on *Cassia italica* Mill under salt stress. *Legume Res.* **2016**, *39*, 396–404. [CrossRef]

31. He, F.; Shen, H.; Lin, C.; Fu, H.; Sheteiwy, M.S.; Guan, Y.; Huang, Y.; Hu, J. Transcriptome analysis of chilling-imbibed embryo revealed membrane recovery related genes in maize. *Front. Plant Sci.* **2017**, *7*, 1978. [CrossRef]

32. Grattan, S.; Zeng, L.; Shannon, M.; Robert, S.R. Rice is more sensitive to salinity than previously thought. *Calif. Agric.* **2002**, *56*, 189–195. [CrossRef]

33. Centritto, M.; Lauteri, M.; Monteverdi, M.C.; Serraj, R. Leaf gas exchange, carbon isotope discrimination, and grain yield in contrasting rice genotypes subjected to water deficits during the reproductive stage. *J. Exp. Bot.* **2009**, *60*, 2325–2339. [CrossRef] [PubMed]

34. Ji, K.X.; Wang, Y.Y.; Sun, W.N.; Lou, Q.J.; Mei, H.W.; Shen, S.H.; Chen, H. Drought-responsive mechanisms in rice genotypes with contrasting drought tolerance during reproductive stage. *J. Plant Physiol.* **2012**, *169*, 336–344. [CrossRef] [PubMed]

35. Wang, Y.; Gu, W.; Yao, M.; Xie, T.; Li, L.; Jing, L.; Shi, W. γ-Aminobutyric acid imparts partial protection from salt stress injury to maize seedlings by improving photosynthesis and upregulating osmoprotectants and antioxidants. *Sci. Rep.* **2017**, *7*, 43609. [CrossRef] [PubMed]

36. Li, Z.; Peng, Y.; Zhang, X.Q.; Ma, X.; Hang, L.K.; Yan, Y.H. Exogenous spermidine improves seed germination of white clover under water stress via involvement in starch metabolism, antioxidant defenses and relevant gene expression. *Molecules* **2014**, *19*, 18003–18024. [CrossRef] [PubMed]

37. Fu, Y.; Gu, Q.; Dong, Q.; Zhang, Z.; Lin, C.; Hu, W.; Pan, R.; Guan, Y.; Hu, J. Spermidine enhances heat tolerance of rice seeds by modulating endogenous starch and polyamine metabolism. *Molecules* **2019**, *24*, 1395. [CrossRef]

38. Ghoulam, C.; Foursy, A.; Fares, K. Effects of salt stress on growth, inorganic ions and proline accumulation in relation to osmotic adjustment in five sugar beet cultivars. *Environ. Exp. Bot.* **2002**, *47*, 39–50. [CrossRef]

39. Hamoud, Y.A.; Wang, Z.; Guo, X.; Shaghaleh, H.; Sheteiwy, M.; Chen, S.; Qiu, R.; Elbashier, M. Effect of Irrigation regimes and soil texture on the potassium utilization efficiency of rice. *Agronomy* **2019**, *9*, 100. [CrossRef]

40. Hamoud, Y.A.; Shaghaleh, H.; Sheteiwy, M.; Guo, X.; Elshaikh, N.A.; Khan, N.; Oumarou, A.; Rahim, S.F. Impact of alternative wetting and soil drying and soil clay content on the morphological and physiological traits of rice roots and their relationships to yield and nutrient use-efficiency. *Agric. Water Manag.* **2019**, *223*, 105706. [CrossRef]

41. Trovato, M.; Mattioli, R.; Costantino, P. Multiple roles of proline in plant stress tolerance and development. *Rend. Lincei* **2008**, *19*, 325–346. [CrossRef]

42. Singh, N.K.; Bracker, C.A.; Hasegawa, P.M.; Handa, A.K.; Buckel, S.; Hermodson, M.A. Characterization of osmotin: A thaumatin-like protein associated with osmotic adaptation in plant cells. *Plant Physiol.* **1987**, *85*, 529. [CrossRef] [PubMed]

43. Ashraf, M.; Harris, P. Potential biochemical indicators of salinity tolerance in plants. *Plant Sci.* **2004**, *166*, 3–16. [CrossRef]

44. Wang, C.; Fan, L.; Gao, H.; Wu, X.; Li, J.; Lv, G.; Gong, B. Polyamine biosynthesis and degradation are modulated by exogenous gamma-aminobutyric acid in root-zone hypoxia-stressed melon roots. *Plant Physiol. Biochem.* **2014**, *82*, 17–26. [CrossRef] [PubMed]

45. Ahmed, I.M.; Nadira, U.A.; Bibi, N.; Cao, F.; He, X.; Zhang, G.; Wu, F. Secondary metabolism and antioxidants are involved in the tolerance to drought and salinity, separately and combined, in Tibetan wild barley. *Environ. Exp. Bot.* **2015**, *111*, 1–12. [CrossRef]

46. Hsu, S.Y.; Kao, C.H. Differential effect of sorbitol and polyethylene glycol on antioxidant enzymes in rice leaves. *Plant Growth Regul.* **2003**, *39*, 83–90. [CrossRef]

47. Shivakrishna, M.P.; Reddy, K.A.; Rao, D.M. Effect of PEG-6000 imposed drought stress on RNA content, relative water content (RWC), and chlorophyll content in peanut leaves and roots. *Saudi J. Biol. Sci.* **2018**, *25*, 285–289. [CrossRef]

48. Zgallaï, H.; Steppe, K.; Lemeur, R. Photosynthetic, physiological and biochemical responses of tomato plants to polyethylene glycol-induced water deficit. *J. Integr. Plant Biol.* **2005**, *47*, 1470–1478. [CrossRef]

49. Seif, S.N.; Tafazzoli, E.; Talaii, A.; Aboutalebi, A.; Abdosi, V. Evaluation of two grape cultivars (*Vitis vinifera* L.) against salinity stress and surveying the effect of methyl jasmonate and epibrassinolide on alleviation the salinity stress. *In. J. Biosci.* **2014**, *5*, 116–125.

50. Heber, U.; Tyankova, L.; Santarius, K.A. Stabilization and inactivation of biological membranes during freezing in the presence of amino acids. *BBA Biomembr.* **1971**, *241*, 578–592. [CrossRef]

51. Das, I.; Krzyzosiak, A.; Schneider, K.; Wrabetz, L.; Antonio, M.; Barry, N.; igurdardottir, A.; Bertolotti, A. Preventing proteostasis diseases by selective inhibition of a phosphatase regulatory subunit. *Res. Rep.* **2015**, *348*, 1–5. [CrossRef]

52. Juan, M.; Rivero, R.M.; Romero, L.; Ruiz, J.M. Evaluation of some nutritional and biochemical indicators in selecting salt-resistant tomato cultivars. *Environ. Exp. Bot.* **2005**, *54*, 193–201. [CrossRef]

53. Tejera, N.A.; Soussi, M.; Lluch, C. Physiological and nutritional indicators of tolerance to salinity in chickpea plants growing under symbiotic conditions. *Environ. Exp. Bot.* **2006**, *58*, 17–24. [CrossRef]

54. Turkyilmaz, B.; Aktas, Y.; Guven, A. Salinity induced differences in growth and nutrient accumulation in five barley cultivars. *Turk. J. Field Crops* **2011**, *16*, 84–92.

55. Assaha, D.V.M.; Uedam, A.; Saneoka, H.; Al-Yahyai, R.; Yaish, M.W. The role of Na^+ and K^+ transporters in salt stress adaptation in glycophytes. *Front. Physiol.* **2017**, *8*, 509. [CrossRef]

56. Qui, L.; Wu, D.Z.; Ali, S.; Cai, S.; Dai, F.; Jin, X.; Wu, F.B.; Zhang, G.P. Evaluation of salinity tolerance and analysis of allelic function of *HvHKT2* in Tibetan wild barley. *Theor. Appl. Genet.* **2011**, *122*, 695–703.

57. Yeo, A.R. Molecular biology of salt tolerance in the context of whole-plant physiology. *J. Exp. Bot.* **1998**, *49*, 915–929. [CrossRef]

58. Davenport, R.J.; Munoz-Mayor, A.; Jha, D.; Essah, P.A.; Rus, A.; Tester, M. The Na^+ transporter AtHKT1; 1 controls retrieval of Na^+ from the xylem in Arabidopsis. *Plant Cell Environ.* **2007**, *30*, 497–507. [CrossRef]

59. Berthomieu, P.; Conejero, G.; Nublat, A.; Brackenbury, W.J.; Lambert, C.; Savio, C.; Uozumi, N.; Oiki, S.; Yamada, K.; Cellier, F. Functional analysis of *AtHKT1* in Arabidopsis shows that Na 1 recirculation by the phloem is crucial for salt tolerance. *EMBO J.* **2003**, *22*, 2004–2014. [CrossRef]

60. Senadheera, P.; Singh, R.; Maathuis, F.J. Differentially expressed membrane transporters in rice roots may contribute to cultivar dependent salt tolerance. *J. Exp. Bot.* **2009**, *60*, 2553–2563. [CrossRef]

61. Nayyar, H.; Kaur, R.; Kaur, S.; Singh, R. γ-Aminobutyric acid (GABA) imparts partial protection from heat stress injury to rice seedlings by improving leaf turgor and upregulating osmoprotectants and antioxidants. *J. Plant Growth Regul.* **2014**, *33*, 408–419. [CrossRef]

62. Vijayakumari, K.; Puthur, J.T. γ-Aminobutyric acid (GABA) priming enhances the osmotic stress tolerance in *Piper nigrum* linn. plants subjected to PEG-induced stress. *Plant Growth Regul.* **2015**, *78*, 1–11. [CrossRef]

63. Krishnan, S.; Laskowski, K.; Shukla, V.; Merewitz, E.B. Mitigation of drought stress damage by exogenous application of a non-protein amino acid gamma aminobutyric acid on perennial ryegrass. *J. Am. Soc. Hortic. Sci.* **2013**, *138*, 358–366. [CrossRef]

64. Mittler, R. Oxidative stress, antioxidants and stress tolerance. *Trends Plant Sci.* **2002**, *7*, 405–410. [CrossRef]

65. Sharma, P.; Jha, A.B.; Dubey, R.S.; Pessarakli, M. Reactive oxygen species, oxidative damage, and antioxidative defense mechanism in plants under stressful conditions. *J. Bot.* **2012**, *26*, 217037. [CrossRef]

66. Swigonska, S.; Amarowicz, R.; Król, A.; Mostek, A.; Badowiec, A.; Weidner, S. Influence of abiotic stress during soybean germination followed by recovery on the phenolic compounds of radicles and their antioxidant capacity. *Acta Soc. Bot. Pol.* **2014**, *83*, 209–218. [CrossRef]

67. Ma, Y.; Wang, P.; Chen, Z.; Gu, Z.; Yang, R. GABA enhances physio-biochemical metabolism and antioxidant capacity of germinated hulless barley under NaCl stress. *J. Plant Physiol.* **2018**, *231*, 192–201. [CrossRef]

68. Kovacik, J.; Klejdus, B.; Hedbavny, J.; Backor, M. Salicylic acid alleviates NaCl-induced changes in the metabolism of Matricaria chamomilla plants. *Ecotoxicology* **2009**, *18*, 544–554. [CrossRef]

69. Kim, S.K.; Son, T.K.; Park, S.Y.; Lee, I.J.; Lee, B.H.; Kim, H.Y.; Lee, S.C. Influences of gibberellin and auxin on endogenous plant hormone and starch mobilization during rice seed germination under salt stress. *J. Environ. Biol.* **2007**, *27*, 181–186.

70. Dai, H.X.; Cao, F.B.; Chen, X.; Zhang, M.; Ahmed, I.M.; Chen, Z.H.; Li, C.; Zhang, G.P.; Wu, F.B. Comparative proteomic analysis of aluminum tolerance in Tibetan wild and cultivated barleys. *PLoS ONE* **2013**, *8*, e63428. [CrossRef]

71. Ahammed, G.J.; Choudhary, S.P.; Chen, S.; Xia, X.; Shi, K.; Zhou, Y. (2013) Role of brassinosterods in alleviation of phenanthrene-cadmium co-contamination-induced photosynthetic inhibition and oxidative stress in tomato. *J. Exp. Bot.* **2013**, *64*, 199–213. [CrossRef]

72. Thipyapong, P.; Hunt, M.D.; Steffens, J.C. Antisense down regulation of polyphenol oxidase results in enhanced disease susceptibility. *Planta* **2004**, *220*, 105–117. [CrossRef] [PubMed]

73. Hong, J.K.; Hwang, B.K. Induction by pathogen, salt and drought of a basic class II chitinase mRNA and its in situ localization in pepper (*Capsicum annuum*). *Physiol. Plant.* **2002**, *114*, 549–558. [CrossRef] [PubMed]

74. Yeh, S.; Moffatt, B.; Griffith, A.; Xiong, M.; Yang, F.; Wiseman, D.S.; Sarhan, S.B.; Danyluk, F.J.; Xue, Y.Q.; Hew, C.L. Chitinase genes responsive to cold Encode antifreeze proteins in winter cereals. *Plant Physiol.* **2000**, *124*, 1251–1264. [CrossRef] [PubMed]

75. Xiang, Y.; Huang, Y.; Xiong, L. Characterization of stress-responsive CIPK genes in rice for stress tolerance improvement. *Plant Physiol.* **2007**, *144*, 1416–1428. [CrossRef]

76. Seki, M.; Narusaka, M.; Ishida, J.; Nanjo, T.; Fujita, M.; Oono, Y.; Kamiya, A.; Nakajima, M.; Enju, A.; Sakurai, T.; et al. Monitoring the expression profiles of 7000 Arabidopsis genes under drought, cold and high-salinity stresses using a full-length cDNA microarray. *Plant J.* **2002**, *31*, 279–292. [CrossRef]

77. Xiong, L.; Yang, Y. Disease resistance and abiotic stress tolerance in rice are inversely modulated by an abscisic acid-inducible mitogen-activated protein kinase. *Plant Cell* **2003**, *15*, 745–759. [CrossRef]

78. Anderson, J.P.; Badruzsaufari, E.; Schank, P.M.; Manners, J.M.; Desmond, O.J.; Ehlert, C.; Maclean, D.J.; Ebert, P.R.; Kazan, K. Antagonistic interaction between abscisic acid and jasmonate-ethylene signaling pathways modulates defense gene expression and disease resistance in Arabidopsis. *Plant Cell* **2004**, *16*, 3460–3479. [CrossRef]

79. Bowler, C.; Fluhr, R. The role of calcium and activated oxygen as signals for controlling cross-tolerance. *Trends Plant Sci.* **2000**, *5*, 241–246. [CrossRef]

80. West, G.; Inze, D.; Beemster, G.T. Cell cycle modulation in the response of the primary root of Arabidopsis to salt stress. *Plant Physiol.* **2004**, *135*, 1050–1058. [CrossRef]

81. Zhao, L.; Wang, P.; Hou, H.; Zhang, H.; Wang, Y.; Yan, S.; Huang, Y.; Li, H.; Tan, J.; Hu, A.; et al. Transcriptional regulation of cell cycle genes in response to abiotic stresses correlates with dynamic changes in histone modifications in maize. *PLoS ONE* **2014**, *9*, e106070. [CrossRef]

82. Sheteiwy, M.S.; An, J.; Yin, M.; Jia, X.; Guan, Y.; He, F.; Hu, J. Cold plasma treatment and exogenous salicylic acid priming enhances salinity tolerance of *Oryza sativa* seedlings. *Protoplasma* **2018**, *256*, 79–99. [CrossRef] [PubMed]

83. Sheteiwy, M.S.; Shen, H.; Xu, J.; Guan, Y.; Song, W.; Hu, J. Seed polyamines metabolism induced by seed priming with Spermidine and 5-aminolevulinic acid for chilling tolerance improvement in rice (*Oryza sativa* L.) seedlings. *Environ. Exp Bot.* **2017**, *137*, 58–72. [CrossRef]

84. Li, Z.; Xu, J.; Gao, Y.; Wang, C.; Guo, G.; Luo, Y.; Huang, Y.; Hu, W.; Sheteiwy, M.S.; Guan, Y.; et al. The Synergistic priming effect of exogenous salicylic acid and H2O2 on chilling tolerance enhancement during maize (Zea mays L.) seed germination. *Front. Plant Sci.* **2017**, *8*, 1153. [CrossRef] [PubMed]

85. Hu, Q.; Fu, Y.; Guan, Y.; Lin, C.; Cao, D.; Hu, W.; Sheteiwy, M.S.; Hu, J. Inhibitory effect of chemical combinations on seed germination and pre-harvest sprouting in hybrid rice. *Plant Growth Regul.* **2016**, *80*, 281–289. [CrossRef]

86. Hu, Q.; Lin, C.; Guan, Y.; Sheteiwy, M.S.; Hu, W.; Hu, J. Inhibitory effect of eugenol on seed germination and pre-harvest sprouting of hybrid rice (*Oryza sativa* L.). *Sci. Rep.* **2017**, *7*, 5295. [CrossRef]

87. Sheteiwy, M.S.; Gong, D.; Gao, Y.; Pan, R.; Hu, J.; Guan, Y. Priming with methyl jasmonate alleviates polyethylene glycol-induced osmotic stress in rice seeds by regulating the seed metabolic profile. *Environ. Exp. Bot.* **2018**, *153*, 236–248. [CrossRef]

88. Sheteiwy, M.S.; Guan, Y.; Cao, D.; Li, J.; Nawaz, A.; Hu, Q.; Hu, W.; Ning, M.; Hu, J. Seed priming with polyethylene glycol regulating the physiological an molecular mechanism in rice (*Oryza sativa* L.) under nano-ZnO stress. *Sci. Rep.* **2015**, *5*, 14278.

89. Sheteiwy, M.S.; Fu, Y.; Hu, Q.; Nawaz, A.; Guan, Y.; Zhan, L.; Huang, Y.; Hu, J. Seed priming with polyethylene glycol induces antioxidative defense and metabolic performance of rice under nano-ZnO stress. *Environ. Sci. Pollut. Res.* **2016**, *23*, 19989–20002. [CrossRef]

90. Sheteiwy, M.S.; Dong, Q.; An, J.; Song, W.; Guan, Y.; He, F.; Huang, Y.; Hu, J. Regulation of ZnO nanoparticles-induced physiological and molecular changes by seed priming with humic acid in *Oryza sativa* seedlings. *Plant Growth Regul.* **2017**, *83*, 27–41. [CrossRef]

91. Zhou, W.J.; Leul, M. Uniconazole-induced tolerance of rape plants to heat stress in relation to changes in hormonal levels, enzyme activities and lipid peroxidation. *Plant Growth Regul.* **1999**, *27*, 99–104. [CrossRef]

92. Li, H.S. *Principle and Technology of Plant Physiological and Biochemical Experiments*; Higher Education Press: Beijing, China, 2000; pp. 169–172.

93. Velikova, V.; Yordanov, I.; Edreva, A. Oxidative stress and some antioxidant systems in acid rain treated bean plants. *Plant Sci.* **2000**, *151*, 59–66. [CrossRef]

94. Jiang, M.; Zhang, J. Effect of abscisic acid on reactive oxygen species, antioxidative defence system and oxidative damage in leaves of maize seedlings. *Plant Cell Physiol.* **2001**, *42*, 1265–1273. [CrossRef] [PubMed]

95. Zheng, H.Z.; Cui, C.L.; Zhang, Y.T.; Wang, D.; Jing, Y.K.; Kim, Y. Active changes of lignification-related enzymes in pepper response to Glomus intraradices and/or Phytophthora capsici. *J. Zhejiang Univ. Sci. B* **2005**, *6*, 778–786. [CrossRef] [PubMed]

96. Wyrambik, D.; Grisebach, H. Purification and properties of isoenzymes of cinnamyl-alcohol dehydrogenase from soybean cell-suspension cultures. *Eur. J. Biochem.* **1975**, *59*, 9–15. [CrossRef] [PubMed]

97. Zhang, C.H.; Ying, G.E. Response of Glutathione and Glutathione S-transferase in Rice Seedlings Exposed to Cadmium Stress. *Rice Sci.* **2008**, *15*, 73–76.

98. Zhao, X.; Wang, W.; Zhang, F.; Deng, J.; Li, Z.; Fu, B. Comparative metabolite profiling of two rice genotypes with contrasting salt stress tolerance at the seedling stage. *PLoS ONE* **2014**, *9*, e108020. [CrossRef]

99. Hu, Y.; Zhang, L.; Zhao, L.; Li, J.; He, S.; Zhou, K.; Yang, F.; Huang, M.; Jiang, L.; Li, L. Trichostatin A selectively suppresses the cold-induced transcription of the *ZmDREB1* gene in maize. *PLoS ONE* **2011**, *6*, e22132. [CrossRef]

International Journal of **Molecular Sciences**

Article

Dynamic Transcriptome Analysis of Anther Response to Heat Stress during Anthesis in Thermotolerant Rice (*Oryza sativa* L.)

Gang Liu [†], Zhongping Zha [†], Haiya Cai, Dandan Qin, Haitao Jia, Changyan Liu, Dongfeng Qiu, Zaijun Zhang, Zhenghuang Wan, Yuanyuan Yang, Bingliang Wan, Aiqing You and Chunhai Jiao *

Hubei Key Laboratory of Food Crop Germplasm and Genetic Improvement, Food Crops Institute, Hubei Academy of Agricultural Sciences, Wuhan 430064, China; liug1112@163.com (G.L.); zhongpingzha@163.com (Z.Z.); ytchy@126.com (H.C.); hnqdd@163.com (D.Q.); jiahaitao.1986@163.com (H.J.); Liucy0602@163.com (C.L.); qdflcp@163.com (D.Q.); zjzhang0459@aliyun.com (Z.Z.); zhwan168@163.com (Z.W.); YYY681205@126.com (Y.Y.); ricewanbl@126.com (B.W.); aq_you@163.com (A.Y.)
* Correspondence: jiaoch@hotmail.com
† These authors contributed equally to this work.

Received: 13 January 2020; Accepted: 7 February 2020; Published: 10 February 2020

Abstract: High temperature at anthesis is one of the most serious stress factors for rice (*Oryza sativa* L.) production, causing irreversible yield losses and reduces grain quality. Illustration of thermotolerance mechanism is of great importance to accelerate rice breeding aimed at thermotolerance improvement. Here, we identified a new thermotolerant germplasm, SDWG005. Microscopical analysis found that stable anther structure of SDWG005 under stress may contribute to its thermotolerance. Dynamic transcriptomic analysis totally identified 3559 differentially expressed genes (DEGs) in SDWG005 anthers at anthesis under heat treatments, including 477, 869, 2335, and 2210 for 1, 2, 6, and 12 h, respectively; however, only 131 were regulated across all four-time-points. The DEGs were divided into nine clusters according to their expressions in these heat treatments. Further analysis indicated that some main gene categories involved in heat-response of SDWG005 anthers, such as transcription factors, nucleic acid and protein metabolisms related genes, etc. Comparison with previous studies indicates that a core gene-set may exist for thermotolerance mechanism. Expression and polymorphic analysis of agmatine-coumarin-acyltransferase gene *OsACT* in different accessions suggested that it may involve in SDWG005 thermotolerance. This study improves our understanding of thermotolerance mechanisms in rice anthers during anthesis, and also lays foundation for breeding thermotolerant varieties via molecular breeding.

Keywords: rice; heat stress; transcriptome; anther; anthesis

---

1. Introduction

Rice (*Oryza sativa* L.) is one of the most important and widely cultivated crops for global food security. However, rice farming is constantly subjected to various abiotic and biotic stresses; heat stress is a major abiotic stress that significantly affects rice growth and development [1,2]. It is estimated that rice grain yields decline by 10% for each 1 °C increase in minimum temperature during the growing season [3]. A report from the Intergovernmental Panel on Climatic Change (IPCC) predicted that, by the end of this century, average surface temperatures would increase by approximately 2.0–4.5 °C [4] (p. 151). Therefore, there is an urgent need to develop rice varieties with thermotolerance to cope with global climate change.

Rice production is particularly susceptible to high temperature, especially during the flowering and grain-filling stages, which directly affects grain yields and quality [2,5]. Even a short period of high temperature during these stages, such as >35 °C for 5 d at anthesis, could cause sterility [6,7]. Anther dehiscence is one of the most sensitive physiological processes affected by high temperature during anthesis; an increase in the basal pore length in a dehisced anther is critical for successful pollination [8,9]. Furthermore, differences in pollen numbers and germinating pollen and spikelet fertility between different rice genotypes have been associated with different levels of thermotolerance [8,10], leading to differences in yield under high-temperature stress.

Therefore, a comprehensive understanding of the mechanisms of thermotolerance at the reproductive stage is crucial for developing heat-tolerant varieties that are adapted to global warming. Many physiological studies have contributed to the understanding of heat responses during anthesis—one of the most heat sensitive stage—but molecular data are lacking.

Transcriptomics have been used to study the molecular mechanisms of thermotolerance in wheat [11], tomato [12], potato [13], and carnation [14]; consequently, multiple genes and pathways have been identified as heat-responsive. This information has helped us to understand how plants sense and respond to heat stress. In rice, some transcriptomic analyses have been conducted to investigate heat responses at the flowering stage [1,15–19]. However, most of these analyses were performed in tissues, such as spikelets or flag leaves, with only a few conducted on anthers or pistils [16,17]. To further clarify heat responses at the molecular level, additional studies are needed.

SDWG005 is a landrace from Africa that was identified as thermotolerant in previous study by our group [20]; its relative seed setting rate in the heat treatment (38 °C) was 98.5% of that in the control (28 °C). SDWG005 performs much better than N22, a well-known heat-tolerant rice germplasm, which relative seed setting rate was 64%–86% at 38 °C [5]. Rice variety 9311 is heat-sensitive based on its relatively lower seed setting rate (31.2%). By observing the morphology and microstructure of anthers in SDWG005 and 9311 before and after heat stress at anthesis, the anthers of SDWG005 were more tolerant to heat stress than those of 9311. To illustrate the key molecular mechanism underlying the thermotolerance of SDWG005, we conducted a transcriptional profile analysis under different time courses of heat treatment on the anthers of SDWG005 at anthesis based on RNA-seq. The findings reported here not only provide additional information for understanding the mechanisms of thermotolerance in rice at the reproductive stage but also lay the foundation for breeding heat-tolerant rice varieties. By using this germplasm, rice varieties with better thermotolerance could be developed through modern molecular breeding strategies.

2. Results

2.1. Thermotolerance Assay of SDWG005 and 9311 in Growth Chamber

Our previous field work has identified SDWG005 as thermotolerant but 9311 as thermosensitive. In this study, a growth chamber was used to mimic high temperature treatment on SDWG005 and 9311. The results showed that 9311 had more than 60% sterile spikelets after 5 d of heat treatment (relative seed setting rate 43.8%). In contrast, the spikelets of SDWG005 did not significantly differ between the control and heat treatment (relative seed setting rate 96.4%) (Figure 1). These findings suggested that SDWG005 was heat tolerant and 9311 was heat susceptible.

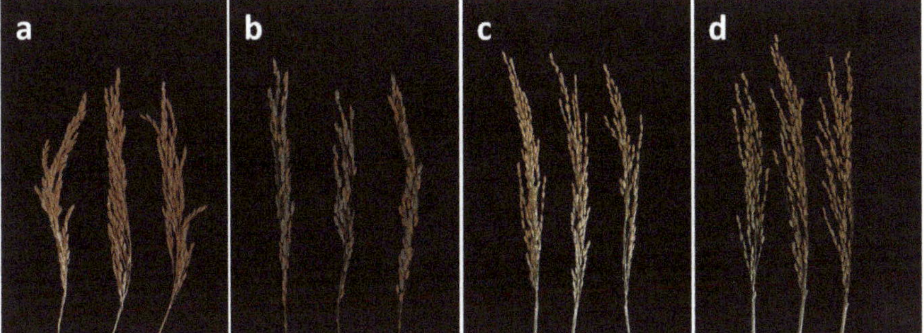

Figure 1. Mature spikes of 9311 and SDWG005 in the control (28 °C for 12 h per day) and heat treatment (38 °C for 12 h per day, for 5 days) conditions. (**a,b**) Panicles of 9311 plants under normal and heat stress conditions, (**c,d**) Panicles of SDWG005 plants under control and heat stress conditions.

2.2. Microstructure of Anthers in Thermotolerant SDWG005 under Heat Stress

To examine the status of rice anthers under heat stress, morphology and microstructure of SDWG005 and 9311 anthers under normal conditions and after 12 h of heat treatment were observed under a microscope (Figure 2). As shown in Figure 2, there were no open anthers in either genotype under normal conditions (Figure 2a,c). After 12 h of heat treatment, the anthers of SDWG005 opened slightly (Figure 2d). In contrast to SDWG005, 9311, as a heat-sensitive variety at anthesis, showed widely opened anthers after the same treatment, but there was no pollen released from these opened anthers (Figure 2b).

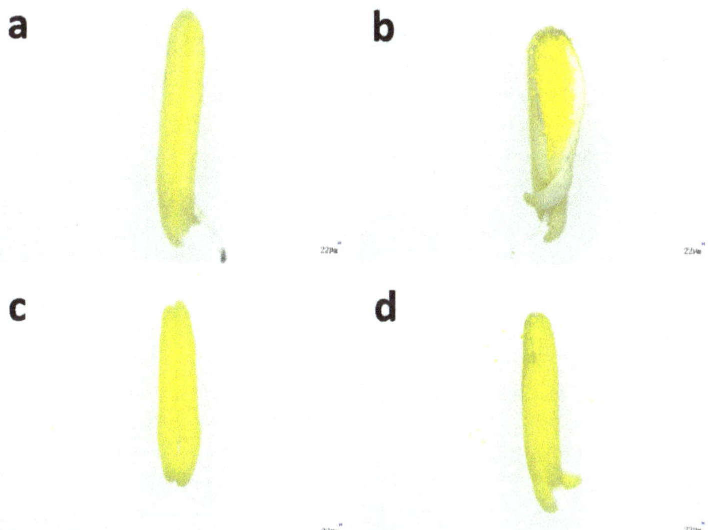

Figure 2. Anther morphology of SDWG005 and 9311 in the control and heat (38 °C for 12 h) treatments. Anthers of 9311 plants in the (**a**) control and (**b**) heat treatments, and anthers of SDWG005 plants in the (**c**) control and (**d**) heat treatments. Scale bar = 22 μm.

In addition, the walls of the anther sacs of 9311 had wilted significantly after 12 h of heat treatment, which damaged the anther structure due to water loss (Figure 3a,b). In contrast, the anthers of the tolerant genotype SDWG005 remained intact under heat stress (Figure 3c,d) and did not differ from those in the control.

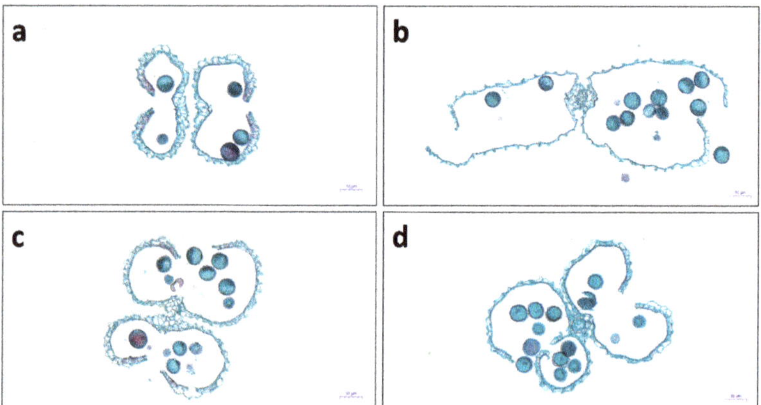

Figure 3. Microstructure of anthers of SDWG005 and 9311 in the control and after 12 h of heat treatment. Anthers of 9311 plants in the (**a**) control and (**b**) heat treatments, and anthers of SDWG005 plants in the (**c**) control and (**d**) heat treatments. Scale bar = 50 μm.

2.3. Quality of RNA-Seq Data

To investigate the transcriptional profiles of heat stressed and unstressed anthers of SDWG005, 15 RNA samples representing four different time points (1, 2, 6, and 12 h) under heat treatment (38 °C) and one control (28 °C) were sequenced individually with Illumina technology; the generated reads were used to assemble the transcriptome for each sample. Approximately 7.17 Gb of data (at least 20 million clean reads, Q30 > 93.2%) were obtained for each sample. The clean reads were used to assemble the transcriptome for each sample by mapping the reads to rice reference genome (MSU7.0). More than 90.0% of the reads could be mapped to the rice reference genome, and approximately 92% and 2% of the mapped reads were mapped to exon and intron regions, respectively. The mapped reads from all samples were then remapped to the reference genome; 1629 new genes/transcripts were discovered.

2.4. Identification and Validation of DEGs

To adapt to stresses such as drought and high temperature, plants usually actively regulate the expression of endogenous genes. To investigate the dynamic transcriptional profiles in response to heat in anthers of SDWG005, changes in the expression of gene transcripts during heat treatment (including WD_1, WD_2, WD_6, WD_12, and WD_0 as a control) were profiled at the genome-wide level by RNA-seq. The expression level of each transcript/gene under each treatment (e.g., WD_1) was identified by the mean value of three biological replicates (e.g., WD_1_1, WD_1_2 and WD_1_3). All of the correlation coefficients between the three biological replicates for each treatment based on all of the transcripts were more than 0.9, indicating that the expression data were highly reproducible (Table S1).

Genes that showed at least a two-fold change in expression in the heat-treated samples compared to the control sample were referred to as DEGs in this study. In total, 3559 genes were identified as DEGs at a minimum of one time-point. Based on the 3559 DEGs, three biological replicates for each sample could be clustered together by hierarchical cluster analysis (Figure S1), which indicated that the DEGs identified were reliable. The number of upregulated and downregulated DEGs varied considerably in different treatments. An increase in the number of DEGs was clearly observed as the heat treatment period lengthened, with 477, 869, 2335, and 2210 DEGs identified after 1 (WD_1), 2 (WD_2), 6 (WD_6), and 12 (WD_12) h of heat treatment, respectively (Figure 4a). All four heat treatments produced more downregulated genes than upregulated genes, especially in the short-term heat treatments (Figure 4b).

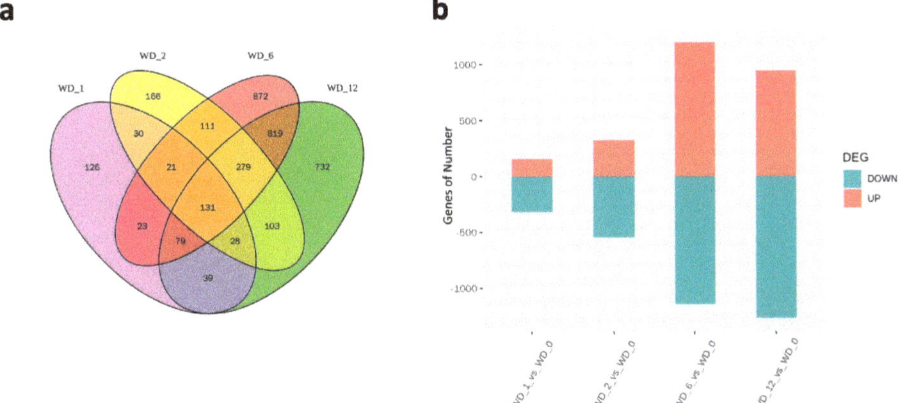

Figure 4. Analysis of differentially expressed genes in response to heat stress. (**a**) Venn diagram analysis of significantly differentially expressed genes at the four heat stress time points. (**b**) Numbers of significantly regulated genes at the four heat stress time points. Red and green columns represent the number of up- and down-regulated genes at different heat stress time points, respectively.

To further validate the RNA-seq results of heat stress responsive genes in this study, qRT-PCR was conducted on anthers of SDWG005 under the same heat treatments as for RNA-seq to examine expressions of ten randomly selected DEGs. The expression patterns of all ten genes examined by qRT-PCR coincided with those determined by RNA-seq analysis (Figure 5), also indicating that the RNA-seq analysis was reliable. For example, RNA-seq analysis revealed 2–4.5 times higher expression levels of LOC_Os04g01740 (*OsHSP1*) in the heat shock (HS) samples than in the control, and the qRT-PCR analysis revealed 1.5–4 times higher expression levels of this gene relative to the control.

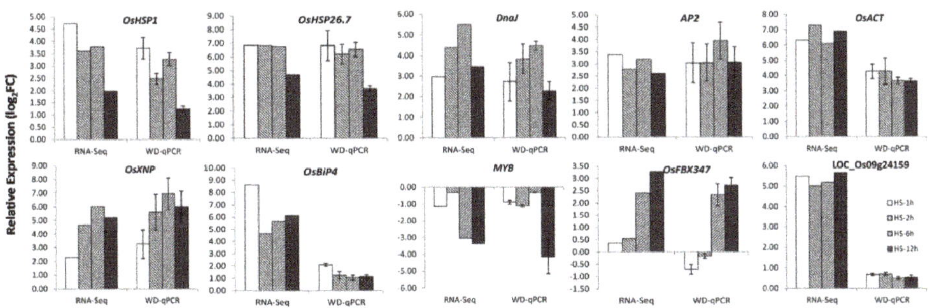

Figure 5. Expression levels of ten randomly selected DEGs by RNA-seq and qRT-PCR. RNA-seq: expression change of these genes in anthers of SDWG005 under heat treatments as compared to control by RNA-seq. WD-qPCR: expression change of these genes in anthers of SDWG005 under heat treatments as compared to control by real-time PCR.

2.5. DEGs in Response to Different Heat Treatment Exposure

The 3559 DEGs were further compared in all four heat treatments; 210, 542, and 1308 common DEGs were found between WD_1 and WD_2, WD_2 and WD_6, and WD_6 and WD_12, respectively (Figure 4a).

To further investigate the dynamic expression of the 3559 DEGs at different time points of heat treatment, heatmap was constructed using *R* package firstly (Figure S2), and then genes with similar expression pattern were plotted using *R* language (Figure 6). As shown in Figure 6, the 3559 DEGs were mainly classified into nine clusters according to their expression level changes in the four heat treatments as compared to the control. Cluster 1 contained 129 DEGs that were shared across all treatments, accounting for less than 4% of the total DEGs. Further analysis showed that 33 and 96 of these genes were consistently up- or down-regulated, respectively, with different fold changes from 1 to 12 h of heat treatment. Cluster 2 contained 20 genes that were regulated by short and intermediate heat treatments but not by long-term treatments, while Cluster 3 contained 279 genes that were not regulated by 1 h heat shock (WD_1), but by longer time of heat treatments after that (WD_2, WD_6, and WD_12). Cluster 4 contained 30 genes that were regulated by the two short-term heat treatments, but not by the intermediate and long-term heat treatment. Cluster 5 contained 819 genes that were regulated by the intermediate and long-term heat treatments, but not by the short-term heat treatments. Clusters 6 to 9 contained 126, 166, 872, and 732 DEGs were identified to be specifically regulated by 1, 2, 6, and 12 h of heat treatment, respectively (Figure 6, Table S2). Expressions of the remaining 386 DEGs were fluctuant with the time of treatments increase (other cluster), e.g., LOC_Os04g59440 was induced by WD_1 and WD_6, but not by WD_2 and WD_12.

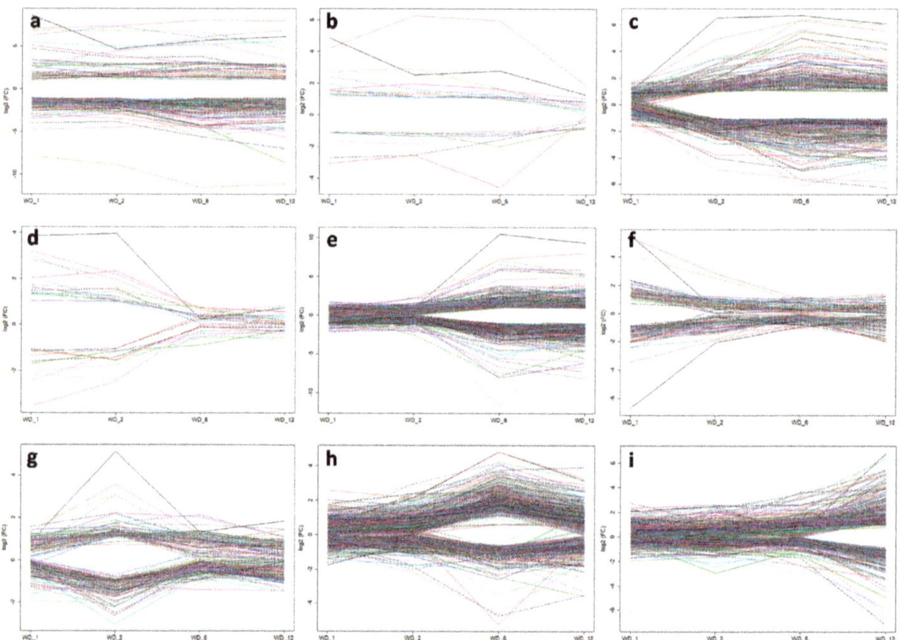

Figure 6. Expression of genes in different clusters. (**a**) cluster 1, genes regulated by all treatments; (**b**) cluster 2, genes regulated by short and intermediate heat treatments but not by long-term treatments; (**c**) cluster 3, genes not regulated by 1 hour heat shock, but by longer time of heat treatments after that; (**d**) cluster 4, genes regulated by the two short-term heat treatments, but not by the intermediate and long-term heat treatment; (**e**) cluster 5, genes regulated by intermediate and long-term heat treatments, but not by short-term heat treatments; (**f–i**) clusters 6–9, genes specifically regulated by 1, 2, 6, and 12 h of heat treatment, respectively. Different color lines indicated different genes.

2.6. Genes Involved in Multiple Biological Processes Are Responsive to Heat Stress

To investigate which categories of genes may be involved in the thermotolerance of SDWG005, all detected DEGs were subjected to functional annotation and GO enrichment analysis. The results showed that the two most dominant GO terms in the molecular function category—catalytic activity and binding, and multiple biological processes, including metabolic process, cellular process, response to stimulus and biological regulation categories were significantly regulated by heat treatment (Figure 7).

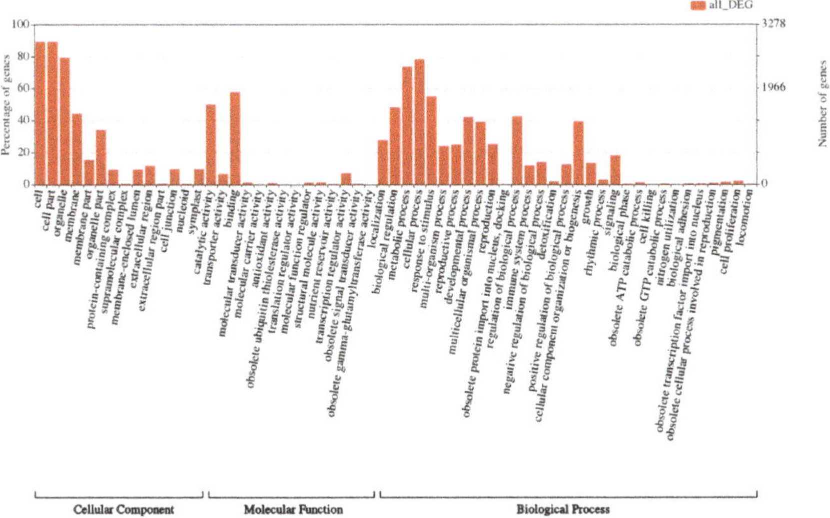

Figure 7. Enrichment of GO classification of 3559 differentially expressed genes.

2.6.1. Transcription Factors

Among the annotated DEGs, nine genes annotated as heat shock factors were upregulated by heat treatment, especially by the short-term treatment (WD_1). In addition to HSFs, a multiprotein bridging factor (LOC_Os06g39240) was significantly induced by all heat treatments. Moreover, more than 270 transcription factors responded to heat stress, predominantly 109 zinc finger-containing, 42 MYB, 41 F-box, 37 AP2/ERF superfamily, 22 bZIP, 14 helix-loop-helix DNA binding domain-containing, and 12 WRKY genes. However, the expression of most of these genes was repressed by heat stress.

2.6.2. Nucleic Acid Synthesis and Modification

About 229 genes involved in nucleic acid synthesis and modification were identified as DEGs in this study. Of these genes, 139 were annotated as putative proteins containing pentatricopeptide repeats, which were mainly upregulated in the intermediate (WD_6) and long-term heat treatments (WD_12) (Table S2). It is notable that 73 genes were also induced specifically by these two heat treatments that participated in RNA biosynthesis and metabolism, including reverse transcriptase, RNA helicase, RNA polymerase, RNA recognition motif, RNA methylase, and tRNA synthetase genes. 17 DEGs encoding DNA polymerases, DNA mismatch repair proteins, DNA ligases, DNA-directed RNA polymerases, and exonucleases were also detected. In addition to the above genes, the expression of other genes essential for DNA and RNA synthesis and repair, such as ABC transporter and ATPase genes, were altered by the heat treatments in our study.

2.6.3. Protein Synthesis and Posttranslational Modification

The expression of about 220 genes that participate in translation, including ribosome biogenesis, translation initiation and elongation, was altered by heat stress. For instance, 35 genes related to ribosomal proteins and ribosome biogenesis were upregulated specifically by WD_6 and WD_12, but not by the short-term heat treatments, WD_1 and WD_2. Furthermore, 28 ubiquitin metabolism-related genes were inducible under heat treatment, such as U-box domain-containing protein, ubiquitin carboxyl-terminal hydrolase, ubiquitin-conjugating enzyme, and ubiquitin-protein ligase genes. Amino acid permeases, amino acid transporters, and proton-dependent oligopeptide transporter (POT) family members involved in the transport of amino acids into cells were also identified as DEGs. It is telling that not only was protein synthesis affected by heat stress, but regulation of protein degradation also occurred under heat stress.

Among the DEGs observed in this study, 144 and 44 were annotated as protein kinases and phosphatases, respectively, according to the Nr database. Interestingly, 113 of the 144 protein kinases were tyrosine kinases. Moreover, these heat-responsive phosphatases mainly included protein phosphatase 2C proteins. In addition to phosphorylation, some genes participating in acylation and methylation were also heat-responsive in SDWG005 anthers, including histone deacetylase, acyltransferase, acyl carrier protein and methyltransferase genes.

As expected, the expression of 23 heat shock protein genes was similar to that of the HSFs, being upregulated by at least one of the heat treatments; more than half were small heat shock proteins, which were distinctly induced by 1 h of heat stress. In addition to the heat shock proteins, other chaperone proteins were heat-responsive in our study, including *ClpB1*, *DnaJ* domain-containing genes, and peptidyl-prolyl cis-trans isomerase.

2.6.4. Physiological Processes Involved in the Heat Stress Response in Rice Anthers

Heat stress damages cell membranes in plants, and causes ion leakage from cells. Fatty acids are important components of the cell membrane. Hence, the relative electrical conductivity and concentration of fatty acids are commonly used as indicators of thermotolerance. In this study, we noted that six genes involved in fatty acid metabolism, including fatty acid desaturase and hydroxylase genes, were responsive to heat stress. In addition, 39 genes associated with ion transfer were regulated, such as heavy-metal-associated domain-containing, sodium/hydrogen exchanger, iron transporter, potassium channel, K^+ potassium transporter, magnesium transporter, and sodium/hydrogen exchanger genes.

Phytohormones play a central role not only in plant development but also in biotic and abiotic stress responses. 26 hormone synthesis and signal transduction-related genes were identified as heat stress responsive, encoding proteins responsible for auxin synthesis signal transduction (inositol-3-phosphate synthase, auxin response factor, auxin-responsive protein, AUX/IAA family gene), gibberellin-regulated proteins, and proteins related to ethylene insensitivity (Table S2).

Our previous physiological study suggested that the excellent tolerance of SDWG005 to heat stress at the seedling stage be attributed to its strong ability to degrade reactive oxygen species (ROS) under heat treatment. Consistent with this notion, about 126 genes responsible for the clearance of ROS or involved in antioxidant defenses showed altered expression levels under heat stress, such as glutathione S-transferase, oxidoreductases, P450, peroxidase, thioredoxin, cysteine protease, and dehydrogenase genes.

As one of the most heat-sensitive physiological processes in the plant kingdom, the expression of photosynthesis-related genes was inhibited by heat stress in this study, especially by the prolonged heat treatments. For example, 19 chlorophyll A-B-binding proteins were induced by 1 h of heat stress but repressed by 6 and 12 h of heat stress. We observed a similar expression pattern for other 22 genes involved in photosynthesis, such as photosynthetic reaction center protein, oxygen-evolving enhancer protein, and photosystem I and II reaction center subunit genes.

Carbohydrates play important roles in anther and pollen development during heat stress. In this study, 67 glycosyl hydrolase genes, 24 UDP-glucosyl transferase genes and 18 glycosyltransferase genes were regulated by heat stress. Additionally, our study showed that 42 genes participating in sugar or glucose metabolism and other types of carbohydrate metabolism were regulated by heat stress, such as genes annotated as sugar transporters, trehalose-phosphatases, glucose/sorbosone dehydrogenases, alpha/beta hydrolases, and galactosyl transferases. It is worth mentioning that most of these genes were down-regulated.

2.7. Characterization of Anther Specific Gene OsACT

To further find some candidate genes that may be involved in rice thermotolerance, real time PCR was also conducted on expression of some DEGs in anthers of 9311 with the same heat treatments as SDWG005. It was revealed that *OsACT* (LOC_Os03g47860) showed quite different expression pattern between SDWG005 and 9311 under heat treatments (Figure 8a). Then, further analysis was conducted to understand the role of this gene in heat stress response. In silico study showed that this *OsACT* gene expressed specifically in anther and 5 DAPs' seeds (http://rice.plantbiology.msu.edu), and our analysis suggested that there was no detectable expression of *OsACT* in leaves of SDWG005 and 9311 at seedling stage either in control or high temperature condition (Data not shown). Coding region of *OsACT* was 1389 bp in length without intron, and the putative protein sequence contained 462 amino acids. The encoded sequence contained the HXXXD domain and DFGGG domain for BAHD family at C-terminal and N-terminal, respectively.

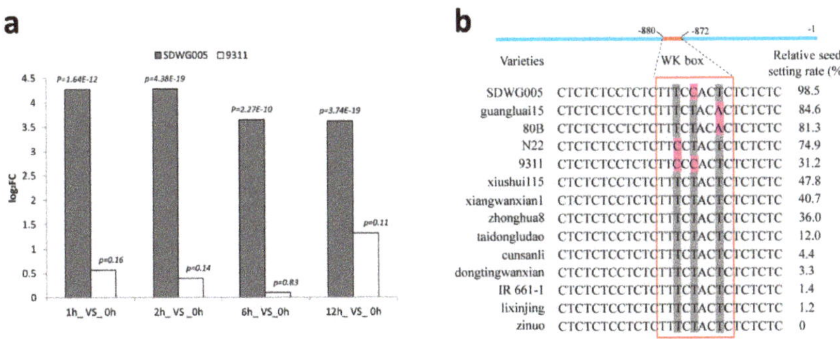

Figure 8. Expression and polymorphic analysis of *OsACT*. (**a**) Expression levels of *OsACT* in anthers of SDWG005 and 9311 under heat treatments by qRT-PCR. (**b**) Polymorphic analysis of WK-box in SDWG005, 9311, N22, and rice core collections with different thermotolerance. Pink indicates minor allele, gray indicates major allele; Relative seed setting rate data was from Zha et al. [20].

OsACT was further cloned from SDWG005 and 9311, and it was found that there were one and three SNPs in 5′ up-stream and coding region of *OsACT* between them, respectively. What is more, all the three SNPs in coding region lead to change of amino acid, but neither in HXXXD domain nor in DFGGG domain. However, the SNP (T/C) in 5′ up-stream region (−878 bp) located in the cis-element called WK-box (TTTTCCAC) [21]. Then, *OsACT* was cloned from 12 other lines with different thermotolerance ability, including N22 and 11 lines from rice core collections. As shown in Figure 8b, totally five haplotypes were detected for WK-box, including TCCACT, TCTACA, and CCTACT for thermotolerant lines, while CCCACT for 9311 and TCTACT for the other nine thermo-sensitive lines.

3. Discussion

Extremely high temperatures are challenging for rice production, as heat stress usually reduces yields, along with the increasing global mean temperature and shortage of varieties with thermotolerance. For example, rice variety 9311 with its excellent agronomic trait performance has been widely planted in China. However, this variety set fewer seeds when subjected to heat stress at anthesis in this study, with a seed setting rate of only 31.2% [20]. Thermotolerant varieties are becoming a necessity for rice production.

Rice germplasm SDWG005, belonging to *Oryza sativa* L. spp. *xian*, is more tolerant to high temperature at the flowering stage than the well-known thermotolerant germplasm N22. In addition, SDWG005 was more thermotolerant at the seedling stage than 9311, as indicated by physiological and growth factors related to thermotolerance, such as photosynthesis and shoot and root fresh and dry weights. Therefore, SDWG005 has promise for breeding thermotolerant rice varieties, and N22 has been successfully used as a source of thermotolerance and drought tolerance in rice breeding [22,23].

It has been well documented that anthesis is the most temperature-sensitive stage in rice [10]. Many physiological processes during this stage are negatively influenced by heat stress, such as anther dehiscence, pollination, pollen germination on the stigma, and pollen tube growth to reach the ovule [24], resulting in reduced fertility. Thus, understanding the thermotolerance of anthers at anthesis is essential to elucidate thermotolerance in rice, which is directly associated with production under high temperature. Our morphological analysis showed that the anthers of SDWG005 and 9311 differ in their response to heat stress. Heat stress not only led to severe wilting of the anther sac wall in 9311 but also caused anther dehiscence without mature pollen release (Figure 2). However, SDWG005 anthers only slightly differed in the control and heat treatments (Figure 2). This finding suggests that heat stress damages rice anthers, particularly in the temperature-sensitive genotype 9311, because pollen development is the most sensitive process to heat stress. In rice, spikelet sterility occurs when temperatures exceed 35 °C for just 1 h [15]. Therefore, differences in the structure and response of SDWG005 and 9311 anthers under heat stress may explain their differences in the timing of seed set.

Transcriptional profiling analysis of stressed and unstressed plants would help to identify genes involved in acclimation and protection against heat stress. Since the relative seed setting rates of SDWG005 and 9311 differed, theoretical transcriptome analyses need to be performed for rice anthers. Although the current study focused on the dynamic gene expression profiles of anthers of the thermotolerant SDWG005 cultivar under heat treatment, we also analyzed the expression of some of these DEGs in 9311 under the same heat treatments using qRT-PCR. We found that these genes showed different expression patterns in SDWG005 and 9311 (Figure S3). This suggests that genotypes with different thermotolerances may respond differently at the molecular level. It was reported previously that induction levels of some heat-responsive genes in anthers correlated well with heat tolerance in rice [16]. Since both of the two studies focused on the genotypes with different genetic background, it was essential to conduct analysis on near-isogenic lines with different heat tolerances to illustrate the key factors responsible for thermotolerance of SDWG005 in the future.

In this study, 3559 genes were modulated at the transcription level under high-temperature stress; these genes were involved in various processes, including transcription regulation, nucleic acid synthesis and metabolism, protein synthesis and modification, hormone signal transduction, reactive oxygen species (ROS) elimination, and photosynthesis. All of these enriched functional categories have been associated with heat responses or thermotolerance in many species, including rice, wheat, maize, tomato, barley, and brassica [11,12,25–28], suggesting a relatively conserved mechanism in response to this kind of stress. For example, heat shock proteins (HSPs) are the most commonly detected molecules and act as an intermediate in protein folding or determination of the protein conformation during stress [29,30]. Moreover, the expression of HSPs and various other heat-responsive genes is controlled by heat shock transcription factors (HSFs) [31]. In the current study, most of the genes annotated as HSP (e.g., LOC_Os04g01740, LOC_Os03g14180, LOC_Os01g42190 and LOC_Os05g35400) and heat shock factors (e.g., LOC_Os05g49310, LOC_Os10g03730, LOC_Os02g34260) were also significantly induced by heat stress.

Since other studies have focused on heat-responsive gene profiles in rice reproductive tissues, including anthers [17] and pollinated pistil [16], common or specific responsive genes were identified by comparing our results to those of previous studies. We detected 48 heat-responsive genes in rice anthers, as done by Li et al. [17] and in our study, and 168 common DEGs in rice anthers (this study) and pollinated pistils under heat treatment [16] (Table S3). However, only 20 DEGs overlapped in the three studies mentioned above, 13 of which were HSPs, as expected (Table 1). These 20 genes are probably a core gene set that responds to heat stress in the floral organs of rice. Moreover, the few overlapping DEGs suggested that the response of rice to heat treatments varies greatly and could be dependent on the stress duration, tissue, or genotype.

Recent years have witnessed a breakthrough in the analysis of the molecular mechanisms of thermotolerance in plants. Increases or decreases in some genes could enhance the thermotolerance of rice or other plant species, which could be used to improve thermotolerance using modern breeding methods (reviewed by Grover et al. [32]). As mentioned above, LOC_Os03g47860 (*OsACT*) was significantly upregulated in thermotolerant SDWG005 at all time points examined under high-temperature stress but not in 9311 (Figure 8a). The *OsACT* gene may specifically express at the reproductive stage according to in silico data and real-time PCR data. Actually, in order to illustrate whether these DEGs were anther specific or not, we also analyzed expressions of the ten genes mentioned above in heat treated leaves in SDWG005 and 9311 at seedling stage. It was very interesting that only four of the ten genes were identified to be responsive to heat stress at seedling stage both in SDWG005 and 9311, however, the other six genes could not be detected either under the control or high temperature condition in both SDWG005 and 9311 at the seedling stage. It was worth mentioned that three of the four genes were heat shock proteins genes, which indicated that this kind of genes may be important to heat response and tolerance in the whole life of rice. While the other six genes could not be detected in leaves at seedling stage, suggesting that there may be unique mechanisms related to heat tolerance at certain development stages.

On the other hand, agmatine coumarin acyltransferase (ACT) is a member of the BAHD acyltransferase family that is unique to plants and participates in the acylation of phenolamides [33]. Phenolamines are not only involved in pollen development but also in resistance to abiotic stresses such as extreme temperature, drought, high salinity, and UV [34–36]. Phenolamine, as a substrate of peroxidase, also participates in the scavenging of hydrogen peroxide and strengthening of the extraplastid cell wall, so it functions as an antioxidant and free radical scavenger, which can improve the ability of plants to resist abiotic stress [37]. The most important is that spermidine, a precursor in phenolamine synthesis, can enhance thermotolerance in rice seeds by modulating endogenous starch and polyamine metabolism [38]. Therefore, to characterize the possible role of the *OsACT* in thermotolerance of rice anther at anthesis, this gene was cloned and sequenced from SDWG005, 9311 and N22. Sequences of *OsACT* were compared between SDWG005, 9311, N22, and other 11 rice core collections with different thermotolerance. The results showed that there were five haplotypes in WK-box in promoter region of *OsACT* (Figure 8b), which was located in the −872 bp to −880 bp upstream region. It was interesting that the core sequence in 9 out of the 10 thermo-sensitive lines was TCTACT, except for 9311, nevertheless, a considerable variation in WK-box was observed in these thermotolerant lines (Figure 8b). WK-box has been reported to be WRKY12 binding site in tobacco [21]. Variation of WK-box in these accessions indicated that *OsACT* may play an important role in anther thermotolerance during rice anthesis. Functional analysis of *OsACT* is still essential to investigate its importance in thermotolerance of rice.

Overall, a dynamic heat-responsive transcriptome analysis of the anthers of the thermotolerant rice cultivar SDWG005 provided the basis for illustrating the molecular mechanism of its thermotolerance. On the other hand, the DEGs identified in this study pave the way for further research regarding the use of heat-tolerant genes in rice breeding, aimed at improving the performance of rice varieties under challenging environments.

Table 1. Common DEGs compared with the transcriptional profiles of the pollinated pistil [16] and anther [17].

Gene ID	WD_1		WD_2		WD_6		WD_12		Functional Annotation
	$\log_2$FC	*p* Value	$\log_2$FC	*p* Value	$\log_2$FC	*p* Value	$\log_2$FC	*p* Value	
LOC_Os11g13980	7.07	8.01×10^{-5}	7.39	4.48×10^{-18}	8.09	2.01×10^{-31}	8.05	1.87×10^{-29}	hsp20
LOC_Os03g14180	6.86	2.02×10^{-17}	6.83	1.65×10^{-134}	6.77	2.44×10^{-99}	4.69	9.23×10^{-32}	hsp20
LOC_Os04g01740	4.71	2.27×10^{-28}	3.60	1.00×10^{-38}	3.77	4.50×10^{-43}	1.97	9.91×10^{-9}	hsp90
LOC_Os04g36750	3.57	3.35×10^{-22}	2.60	3.67×10^{-18}	1.64	2.63×10^{-5}	2.13	6.51×10^{-5}	hsp20
LOC_Os02g52150	3.40	5.69×10^{-32}	3.00	4.65×10^{-40}	2.56	3.06×10^{-24}	3.08	2.04×10^{-5}	hsp20
LOC_Os04g45480	3.08	9.62×10^{-21}	2.58	5.52×10^{-54}	2.10	5.23×10^{-20}	1.42	5.63×10^{-6}	uncharacterized protein
LOC_Os03g16920	2.05	5.14×10^{-7}	1.93	1.36×10^{-18}	2.35	2.28×10^{-20}	2.63	1.54×10^{-12}	hsp70
LOC_Os04g28420	2.41	1.36×10^{-41}	1.63	5.46×10^{-21}	1.22	4.19×10^{-8}	–	–	70 kDa peptidyl-prolyl isomerase
LOC_Os02g15930	1.94	1.90×10^{-15}	1.07	2.10×10^{-8}	1.42	1.23×10^{-16}	–	–	uncharacterized protein
LOC_Os01g08860	3.05	1.20×10^{-28}	1.88	5.74×10^{-24}	–	–	-1.71	2.36×10^{-9}	hsp20
LOC_Os11g05170	2.76	1.00×10^{-11}	–	–	-2.15	5.63×10^{-5}	-2.14	1.20×10^{-3}	uncharacterized protein
LOC_Os11g32890	2.56	6.08×10^{-20}	–	–	-1.68	7.10×10^{-7}	-1.87	1.65×10^{-7}	uncharacterized protein
LOC_Os02g54140	1.95	2.28×10^{-7}	–	–	-1.80	8.55×10^{-6}	-3.68	1.23×10^{-17}	hsp20
LOC_Os01g55270	1.37	3.70×10^{-12}	–	–	-1.31	1.22×10^{-14}	-1.26	1.24×10^{-13}	calcyclin-binding protein
LOC_Os01g04370	2.85	2.73×10^{-17}	1.98	7.64×10^{-19}	–	–	–	–	hsp20
LOC_Os03g16020	2.09	2.88×10^{-7}	1.02	3.61×10^{-5}	–	–	–	–	hsp20
LOC_Os05g46480	–	–	–	–	-2.58	1.42×10^{-16}	-3.56	1.35×10^{-27}	late embryogenesis abundant protein
LOC_Os03g16030	3.98	1.97×10^{-3}	–	–	–	–	–	–	hsp20
LOC_Os03g16040	2.20	5.80×10^{-12}	–	–	–	–	–	–	hsp20
LOC_Os01g18080	–	–	-1.11	1.33×10^{-5}	–	–	–	–	uncharacterized protein

4. Materials and Methods

4.1. Plant Material and Experimental Treatments

Two rice (*Oryza sativa* L. spp. *xian*) genotypes, SDWG005 and 9311, were used in this study. SDWG005 is an African landrace and 9311 is a widely used restorer line in China. The seeds were soaked in water at 28 °C for 24 h and subsequently incubated at 28 °C for 48 h for germination. The germinated seeds were transferred to 96-well plates filled with vermiculite with one seedling per well. The seedlings were irrigated with distilled water until the two-leaf stage and Yoshida solution thereafter [39] (pp. 53–57). The 30-day-old seedlings were transplanted into plastic barrels (30 cm diameter) filled with paddy soil. Three plants were planted in each barrel, with five tillers retained for each plant. The seedlings and plants were grown in a growth chamber with a relative humidity of 75% and a natural photoperiod.

There were two temperature treatments: (1) control, 28 °C during the day (6:00 am to 6:00 pm) and 22 °C at night (6:00 pm to 6:00 am); and (2) heat stress (HS), 38 °C during the day (6:00 am to 6:00 pm) and 28 °C at night (6:00 pm to 6:00 am).

4.2. Microscopic Observations of Rice Anthers

Florets of each genotype in the control and heat stress treatment were subjected to microscopic observation. Anthers which ascending to the top of glume from the middle part of the first branch in every genotype under different treatments were carefully collected using tweezers. Some of these anthers were morphologically observed under a VHX-2000 digital microscope (Keyence, Osaka, Japan), with the remainder fixed in a 50% FAA solution (40% formalin:glacial acetic acid:50% ethanol = 1:1:18) and then dehydrated with 100% ethanol, followed by washing with xylene. The anthers were embedded in paraffin before being sectioned (thickness, 10 microns) and stained with safranine O-Fast Green to produce permanent slices to examine the microstructure of anther sacs under a NIKON ECLIPSE E100 positive feedback microscope (Nikon, Tokyo, Japan).

4.3. Seed Setting Rate

After maturation, the marked spikes from ten plants were harvested separately for each genotype within each treatment. The number of empty and filled grains was recorded, with the seed set rate represented as the percentage of filled grains in all grains. The effect of heat stress on seed set was indicated by the relative seed setting rate, being seed setting rate at high temperature/seed setting rate at normal temperature × 100%.

4.4. Sample Collection for RNA-Seq and Real-Time PCR

A total of 75 plants per genotype were used for sample collection. When three or more panicles extended from flag leaves by approximately 2 cm, the panicles were labeled, and the plants transferred to the growth chamber for the HS treatment. The labeled spikes were harvested before HS treatment (0 h) and 1, 2, 6, and 12 h after the initiation of the HS treatment, respectively, and placed on ice before anther extraction. The treatments were designated WD_0 and 9311_0, WD_1 and 9311_1, WD_2 and 9311_2, WD_6 and 9311_6, and WD_12 and 9311_12 for SDWG005 and 9311, respectively. One hour and 2 h were considered as short-term, 6 and 12 h were considered as intermediate and long-term heat treatments, respectively. For each treatment, 15 independent plants were sampled, and mature anthers which ascending to the top of glume from five independent plants pooled as one biological replicate, such that there were three biological replicates per treatment. The labeled panicles from each time point in each treatment were collected on ice, with the mature anthers from the panicles isolated, immediately suspended in liquid nitrogen, and stored at −80 °C until RNA extraction.

4.5. Total RNA Extraction and RNA Sequencing

Total RNA was extracted from each sample using an RNAprep Pure Plant Kit (Tiangen Biotech, China) following the manufacturer's instructions. The quality and quantity of RNA samples were assessed by gel electrophoresis and a Nanodrop (Thermal Fisher, Waltham, MA, USA). The resultant 15 RNA samples for SDWG005 were used for RNA-seq analysis. Upon treatment with DNase I, 1 μg of RNA from each sample was used for sequencing library preparation with a NEB Next Ultra TM RNA Library Prep Kit for Illumina (NEB, Ipswich, MA, USA). Library preparations were sequenced from paired ends to obtain 150 bp reads (PE150) on an Illumina platform.

4.6. Quality Control and Comparative Analysis of RNA-seq Data

Raw data were processed through in-house Perl scripts to remove low-quality reads (more than 20% of bases presenting a Q value $\leq$ 20 or ambiguous sequence content ("N") exceeding 5%). Clean reads were then obtained by removing reads containing adapters and poly-N sequences. Clean reads were mapped to the rice reference genome sequence (MSU7.0, http://rice.plantbiology.msu.edu/) with Hisat2 tools [40]. Only reads with a perfect match or one mismatch were further analyzed.

4.7. Differential Expression Analysis

Gene expression levels were quantified as fragments per kilobase of transcript per million fragments mapped (FPKM) values. Pearson's correlation coefficient [41] was used to calculate the correlation coefficients of FPKM of all transcripts among three replications for each sample. Differential expression analysis of the two conditions was performed using DEseq2 [42]. The resulting p values were adjusted using the Benjamini and Hochberg approach [43] to control the false discovery rate. Genes with an adjusted p < 0.05 and |$\log_2$FC| > 1 identified by DEseq2 were assigned as differentially expressed. Hierarchical clustering analysis of DEGs was conducted using cluster *R* package based on lgFPKM of them among three replications for each sample (https://cran.r-project.org/web/packages/cluster). Heatmap was plotted by pheatmap *R* package (https://cran.r-project.org/web/packages/pheatmap), Venn was plotted by Venndiagram [44] *R* package. Expression pattern of genes in different clusters were plotted using our in house scripts based on *R* language (File S1).

4.8. Quantitative Real-Time PCR

Quantitative real-time PCR (qRT-PCR) was applied to quantify the expression levels of 10 selected DEGs in SDWG005 and 9311. RNA for all the samples was reverse-transcribed using a RevertAid First Strand cDNA Synthesis Kit (Thermo Scientific) following the manufacturer's protocol. qRT-PCR was performed using ABI StepOnePlus™ Real-Time System. Relative gene expression levels were calculated using the $2^{-\Delta\Delta Ct}$ method, as described by Min et al. [45]. The expression of each replicate was normalized according to the reference gene *OsActin* (LOC_Os03g50885). For randomly selected genes, specific primers (listed in Table S4) were designed and tested by qRT-PCR. The mean of three replicates represents the relative expression level.

4.9. Gene Functional Annotation and Enrichment Analysis

Gene functions were annotated based on the following databases: Nr (non-redundant protein sequence database), Pfam (database of homologous protein family), KOG/COG (Clusters of Orthologous Groups of proteins), Swiss-Prot (manually annotated, non-redundant protein sequence database), KO (KEGG Ortholog) and GO (Gene Ontology database). GO enrichment analysis of the DEGs was implemented with clusterProfiler *R* packages [46]. We defined significant enrichment based on an FDR < 0.05.

4.10. Cloning and Sequence Analysis of OsACT Gene in Rice

Coding region as well as upstream region of *OsACT* was extracted from Rice Genome Annotation Project (http://rice.plantbiology.msu.edu/) using LOC_Os03g47860 as the query. Six overlapped primer pairs (Table S4) were designed and used to amplify *OsACT* from genomic DNA of SDWG005, 9311 and N22. DNA was extracted using Plant DNA extraction kit according to the manufacture's instruction (Tiangen, Beijing, China). Specific PCR product with expected size was subjected to Sanger sequencing (Tianyi, Wuhan, China). Sequences of *OsACT* in other 11 rice core collections were downloaded from http://www.rmbreeding.cn. Alignment of gene sequence was conducted using DNAMAN software (version 8.0, Lynnon Biosoft, San Ramon, CA, USA).

Supplementary Materials: Supplementary materials can be found at http://www.mdpi.com/1422-0067/21/3/1155/s1. Figure S1. Cluster analysis of all samples based on lgFPKM of 3559 DEGs; Figure S2. Heatmap clustering of the global patterns of DEGs expressed at one or more time points in the heat treatment. The color scale represents the lgFPKM (Mean values of FKPM of three replicates); Figure S3. Expression levels of ten randomly selected DEGs by RNA-seq and qRT-PCR. RNA-seq: expression change of these genes in anthers of SDWG005 under heat treatments as compared to control by RNA-seq. WD-qPCR: expression change of these genes in anthers of SDWG005 under heat treatments as compared to control by real-time PCR. 9311-qPCR: expression change of these genes in anthers of 9311 under heat treatments as compared to control by real-time PCR; Table S1. Correlation of all the samples based on FKAM of all transcripts; Table S2. Differentially expressed genes (DEGs) at a minimum of time points during the heat treatment; Table S3. Differentially expressed genes (DEGs) compared with the transcriptional profiles of the pistils and anthers. Table S4. Primers for qRT-PCR and cloning of *OsACT*; File S1. R script for expression pattern of genes in different clusters.

Author Contributions: Conceptualization, C.J. and G.L.; validation, D.Q. (Dandan Qin), H.J., and C.L.; formal analysis, G.L., Z.Z. (Zhongping Zha); investigation, G.L., Z.Z. (Zhongping Zha), H.C., D.Q. (Dandan Qin), H.J., C.L., D.Q. (Dongfeng Qiu), Z.W., and Y.Y.; resources, C.J., Z.Z. (Zaijun Zhang); data curation, H.C., B.W., A.Y.; writing—original draft preparation, G.L., Z.Z. (Zhongping Zha); writing—review and editing, C.J., Z.Z. (Zaijun Zhang); visualization, G.L., Z.Z. (Zhongping Zha); supervision, C.J.; project administration, C.J.; funding acquisition, C.J. All authors have read and agreed to the published version of the manuscript.

Funding: This research was supported by grants from China Postdoctoral Science Foundation (Grant No. 2019M652607), Natural Science Foundation of Hubei Province of China (Grant No. 2019CFB579).

Acknowledgments: The authors would like to thank Zhongfu Ni and Huiru Peng in China Agricultural University for their valuable advice on this study and the manuscript. The authors would like to thank Jihua Cheng at Yuan Longping Hi-Tech Agriculture Co., LTD. for reading the manuscript and providing suggestions for its improvement.

Conflicts of Interest: The authors declare no conflict of interest. The funders had no role in the design of the study; in the collection, analyses, or interpretation of data; in the writing of the manuscript, or in the decision to publish the results.

References

1. Mittal, D.; Madhyastha, D.A.; Grover, A. Genome-wide transcriptional profiles during temperature and oxidative stress reveal coordinated expression patterns and overlapping regulons in rice. *PLoS ONE* **2012**, *7*, e40899. [CrossRef]

2. Shi, W.; Lawas, L.M.F.; Raju, B.R.; Jagadish, S.V.K. Acquired thermo-tolerance and trans-generational heat stress response at flowering in rice. *J. Agron. Crop Sci.* **2016**, *202*, 309–319. [CrossRef]

3. Peng, S.; Huang, J.; Sheehy, J.E.; Laza, R.C.; Visperas, R.M.; Zhong, X.; Centeno, G.S.; Khush, G.S.; Cassman, K.G. Rice yields decline with higher night temperature from global warming. *Proc. Natl. Acad. Sci. USA* **2004**, *101*, 9971–9975. [CrossRef]

4. Pachauri, R.K.; Allen, M.R.; Barros, V.R.; BBroome, J.; Cramer, W.; Christ, R.; Church, J.A.; Clarke, L.; Dahe, Q.; Dasgupta, P.; et al. *Climate Change 2014: Synthesis Report. Contribution of Working Groups I, II and III to the Fifth Assessment Report of the Intergovernmental Panel on Climate Change*; Pachauri, R.K., Meyer, L.A., Eds.; Intergovernmental Panel on Climate Change: Geneva, Switzerland, 2014; p. 151.

5. Jagadish, S.V.K.; Craufurd, P.Q.; Wheeler, T.R. Phenotyping parents of mapping populations of rice for heat tolerance during anthesis. *Crop Sci.* **2008**, *48*, 1140–1146. [CrossRef]

6. Coast, O.; Murdoch, A.J.; Ellis, R.H.; Hay, F.R.; Jagadish, K.S.V. Resilience of rice (*Oryza* spp.) pollen germination and tube growth to temperature stress. *Plant Cell Environ.* **2016**, *39*, 26–37. [CrossRef]

7. Satake, T.; Yoshida, S. High temperature-induced sterility in indica rices at flowering. *Jpn. J. Crop Sci.* **1978**, *47*, 6–17. [CrossRef]

8. Matsui, T.; Omasa, K. Rice (*Oryza sativa* L.) Cultivars tolerant to high temperature at flowering: Anther characteristics. *Ann. Bot.* **2002**, *89*, 683–687. [CrossRef] [PubMed]

9. Matsui, T.; Kagata, H. Characteristics of floral organs related to reliable self-pollination in rice (*Oryza sativa* L.). *Ann. Bot.* **2003**, *91*, 473. [CrossRef] [PubMed]

10. Prasad, P.V.V.; Boote, K.J.; Allen, L.H.; Sheehy, J.E.; Thomas, J.M.G. Species, ecotype and cultivar differences in spikelet fertility and harvest index of rice in response to high temperature stress. *Field Crops Res.* **2006**, *95*, 398–411. [CrossRef]

11. Qin, D.; Wu, H.; Peng, H.; Yao, Y.; Ni, Z.; Li, Z.; Zhou, C.; Sun, Q. Heat stress-responsive transcriptome analysis in heat susceptible and tolerant wheat (*Triticum aestivum* L.) by using Wheat Genome Array. *BMC Genom.* **2008**, *9*, 432. [CrossRef] [PubMed]

12. Bita, C.E.; Zenoni, S.; Vriezen, W.H.; Mariani, C.; Pezzotti, M.; Gerats, T. Temperature stress differentially modulates transcription in meiotic anthers of heat-tolerant and heat-sensitive tomato plants. *BMC Genom.* **2011**, *12*, 384. [CrossRef] [PubMed]

13. Ginzberg, I.; Barel, G.; Ophir, R.; Tzin, E.; Tanami, Z.; Muddarangappa, T.; De Jong, W.; Fogelman, E. Transcriptomic profiling of heat-stress response in potato periderm. *J. Exp. Bot.* **2009**, *60*, 4411. [CrossRef] [PubMed]

14. Wan, X.L.; Zhou, Q.; Wang, Y.Y.; Wang, W.E.; Bao, M.Z.; Zhang, J.W. Identification of heat-responsive genes in carnation (*Dianthus caryophyllus* L.) by RNA-seq. *Front. Plant Sci.* **2015**, *6*, 519. [CrossRef] [PubMed]

15. Endo, M.; Tsuchiya, T.; Hamada, K.; Kawamura, S.; Yano, K.; Ohshima, M.; Higashitani, A.; Watanabe, M.; Kawagishi-Kobayashi, M. High temperatures cause male sterility in rice plants with transcriptional alterations during pollen development. *Plant Cell Physiol.* **2009**, *50*, 1911–1922. [CrossRef]

16. González-Schain, N.; Dreni, L.; Lawas, L.M.F.; Galbiati, M.; Colombo, L.; Heuer, S.; Jagadish, K.S.V.; Kater, M.M. Genome-wide transcriptome analysis during anthesis reveals new insights into the molecular basis of heat stress responses in tolerant and sensitive rice varieties. *Plant Cell Physiol.* **2016**, *57*, 57–68. [CrossRef]

17. Li, X.; Lawas, L.M.F.; Malo, R.; Glaubitz, U.; Erban, A.; Mauleon, R.; Heuer, S.; Zuther, E.; Kopka, J.; Hincha, D.K.; et al. Metabolic and transcriptomic signatures of rice floral organs reveal sugar starvation as a factor in reproductive failure under heat and drought stress. *Plant Cell Environ.* **2015**, *38*, 2171–2192. [CrossRef]

18. Zhang, X.; Li, J.; Liu, A.; Zou, J.; Zhou, X.; Xiang, J.; Rerksiri, W.; Peng, Y.; Xiong, X.; Chen, X. Expression profile in rice panicle: Insights into heat response mechanism at reproductive stage. *PLoS ONE* **2012**, *7*, e49652. [CrossRef]

19. Zhang, X.; Rerksiri, W.; Liu, A.; Zhou, X.; Xiong, H.; Xiang, J.; Chen, X.; Xiong, X. Transcriptome profile reveals heat response mechanism at molecular and metabolic levels in rice flag leaf. *Gene* **2013**, *530*, 185–192. [CrossRef]

20. Zha, Z.; Yin, D.; Wan, B.; Jiao, C. An analysis on the heat resistance of rice germplasm resources during flowering period. *Asian Agric. Res.* **2016**, *55*, 17–23. (In Chinese)

21. Van Verk, M.C.; Neeleman, L.; Bol, J.F.; Linthorst, H.J.M. Tobacco transcription factor *NtWRKY12* interacts with *TGA2.2* in vitro and in vivo. *Front. Plant Sci.* **2011**, *2*, 32. [CrossRef]

22. Vikram, P.; Swamy, B.P.M.; Dixit, S.; Ahmed, H.U.; Teresa Sta Cruz, M.; Singh, A.K.; Kumar, A. *qDTY 1.1*, a major QTL for rice grain yield under reproductive-stage drought stress with a consistent effect in multiple elite genetic backgrounds. *BMC Genet.* **2011**, *12*, 89. [CrossRef]

23. Ye, C.; Tenorio, F.A.; Redoña, E.D.; Morales-Cortezano, P.S.; Cabrega, G.A.; Jagadish, K.S.V.; Gregorio, G.B. Fine-mapping and validating *qHTSF4.1* to increase spikelet fertility under heat stress at flowering in rice. *Theor. Appl. Genet.* **2015**, *128*, 1507–1517. [CrossRef]

24. Jagadish, S.V.K.; Raveendran, M.; Oane, R.; Wheeler, T.R.; Heuer, S.; Bennett, J.; Craufurd, P.Q. Physiological and proteomic approaches to address heat tolerance during anthesis in rice (*Oryza sativa* L.). *J. Exp. Bot.* **2010**, *61*, 143–156. [CrossRef]

25. Dong, X.; Yi, H.; Lee, J.; Nou, I.S.; Han, C.T.; Hur, Y. Global gene-expression analysis to identify differentially expressed genes critical for the heat stress response in *brassica rapa*. *PLoS ONE* **2015**, *10*, e0130451. [CrossRef]

26. Frey, F.P.; Urbany, C.; Hüttel, B.; Reinhardt, R.; Stich, B. Genome-wide expression profiling and phenotypic evaluation of European maize inbreds at seedling stage in response to heat stress. *BMC Genom.* **2015**, *16*, 123. [CrossRef]

27. Mangelsen, E.; Kilian, J.; Harter, K.; Jansson, C.; Wanke, D.; Sundberg, E. Transcriptome analysis of high-temperature stress in developing barley caryopses: Early stress responses and effects on storage compound biosynthesis. *Mol. Plant* **2011**, *4*, 97–115. [CrossRef]

28. Sarkar, N.K.; Kim, Y.K.; Grover, A. Coexpression network analysis associated with call of rice seedlings for encountering heat stress. *Plant Mol. Biol.* **2014**, *84*, 125–143. [CrossRef]

29. Boston, R.S.; Viitanen, P.V.; Vierling, E. Molecular chaperones and protein folding in plants. *Plant Mol. Biol.* **1996**, *32*, 191–222. [CrossRef]

30. Morimoto, R.I. Regulation of the heat shock transcriptional response: Cross talk between a family of heat shock factors, molecular chaperones, and negative regulators. *Genes Dev.* **1998**, *12*, 3788–3796. [CrossRef]

31. Kotak, S.; Larkindale, J.; Lee, U.; von Koskull-Döring, P.; Vierling, E.; Scharf, K.D. Complexity of the heat stress response in plants. *Curr. Opin. Plant Biol.* **2007**, *10*, 310–316. [CrossRef]

32. Grover, A.; Mittal, D.; Negi, M.; Lavania, D. Generating high temperature tolerant transgenic plants: Achievements and challenges. *Plant Sci.* **2013**, *205*, 38–47. [CrossRef] [PubMed]

33. D'Auria, J.C. Acyltransferases in plants: A good time to be BAHD. *Curr. Opin. Plant Biol.* **2006**, *9*, 331–340. [CrossRef] [PubMed]

34. Bassard, J.E.; Ullmann, P.; Bernier, F.; Werck-Reichhart, D. Phenolamides: Bridging polyamines to the phenolic metabolism. *Phytochemistry* **2010**, *71*, 1808–1824. [CrossRef] [PubMed]

35. Grienenberger, E.; Besseau, S.; Geoffroy, P.; Debayle, D.; Heintz, D.; Lapierre, C.; Pollet, B.; Heitz, T.; Legrand, M. A BAHD acyltransferase is expressed in the tapetum of Arabidopsis anthers and is involved in the synthesis of hydroxycinnamoyl spermidines. *Plant J.* **2009**, *58*, 246–259. [CrossRef] [PubMed]

36. Onkokesung, N.; Gaquerel, E.; Kotkar, H.; Kaur, H.; Baldwin, I.T.; Galis, I. MYB8 controls inducible phenolamide levels by activating three novel hydroxycinnamoyl-coenzyme A: Polyamine transferases in *Nicotiana attenuata*. *Plant Physiol.* **2012**, *158*, 389–407. [CrossRef] [PubMed]

37. Edreva, A.M.; Velikova, V.B.; Tsonev, T.D. Phenylamides in plants. *Russ. J. Plant Physiol.* **2007**, *54*, 287–301. [CrossRef]

38. Fu, Y.; Gu, Q.; Dong, Q.; Zhang, Z.; Lin, C.; Hu, W.; Pan, R.; Guan, Y.; Hu, J. Spermidine enhances heat tolerance of rice seeds by modulating endogenous starch and polyamine metabolism. *Molecules* **2019**, *24*, 1395. [CrossRef]

39. Yoshida, S.; Forno, D.A.; Cock, J.H.; Gomez, K.A. Routine procedure for growing rice plants in culture solution. In *Laboratory Manual for Physiological Studies of Rice*; The International Rice Research Institute: Los Banos, Philippines, 1976; pp. 53–57.

40. Kim, D.; Langmead, B.; Salzberg, S.L. HISAT: A fast spliced aligner with low memory requirements. *Nat. Methods* **2015**, *12*, 357–360. [CrossRef]

41. Schulze, S.K.; Kanwar, R.; Gölzenleuchter, M.; Therneau, T.M.; Beutler, A.S. SERE: Single-parameter quality control and sample comparison for RNA-Seq. *BMC Genom.* **2012**, *13*, 524. [CrossRef]

42. Love, M.I.; Huber, W.; Anders, S. Moderated estimation of fold change and dispersion for RNA-seq data with DESeq2. *Genome Biol.* **2014**, *15*, 550. [CrossRef]

43. Benjamini, Y.; Hochberg, Y. Controlling the false discovery rate: A practical and powerful approach to multiple testing. *J. R. Stat. Soc. Ser. B* **1995**, *57*, 289–300. [CrossRef]

44. Chen, H.; Boutros, P.C. VennDiagram: A package for the generation of highly-customizable Venn and Euler diagrams in R. *BMC Bioinform.* **2011**, *12*, 35. [CrossRef] [PubMed]

45. Min, L.; Li, Y.; Hu, Q.; Zhu, L.; Gao, W.; Wu, Y.; Ding, Y.; Liu, S.; Yang, X.; Zhang, X. Sugar and auxin signaling pathways respond to high-temperature stress during anther development as revealed by transcript profiling analysis in cotton. *Plant Physiol.* **2014**, *164*, 1293. [CrossRef] [PubMed]

46. Yu, G.; Wang, L.G.; Han, Y.; He, Q.Y. ClusterProfiler: An *R* package for comparing biological themes among gene clusters. *OMICS* **2012**, *16*, 284–287. [CrossRef] [PubMed]

International Journal of
Molecular Sciences

Article

Season Affects Yield and Metabolic Profiles of Rice (*Oryza sativa*) under High Night Temperature Stress in the Field

Stephanie Schaarschmidt [1], Lovely Mae F. Lawas [1,2,†], Ulrike Glaubitz [1], Xia Li [1,3], Alexander Erban [1], Joachim Kopka [1], S. V. Krishna Jagadish [2,4], Dirk K. Hincha [1] and Ellen Zuther [1,*]

[1] Max-Planck-Institute of Molecular Plant Physiology, 14476 Potsdam, Germany;
schaarschmidt@mpimp-golm.mpg.de (S.S.); lfl0008@auburn.edu (L.M.F.L.);
glaubitz@mpimp-golm.mpg.de (U.G.); rainbowleelx@hotmail.com (X.L.);
Erban@mpimp-golm.mpg.de (A.E.); Kopka@mpimp-golm.mpg.de (J.K.);
Hincha@mpimp-golm.mpg.de (D.K.H.)
[2] International Rice Research Institute, Metro Manila 1301, Philippines; kjagadish@ksu.edu
[3] Institute of Crop Science, Chinese Academy of Agricultural Science, Beijing 100081, China
[4] Department of Agronomy, Kansas State University, Manhattan, KS 66506, USA
* Correspondence: zuther@mpimp-golm.mpg.de
† Present address: Department of Biological Sciences, Auburn University, Auburn, AL 36849, USA.

Received: 6 March 2020; Accepted: 29 April 2020; Published: 30 April 2020

Abstract: Rice (*Oryza sativa*) is the main food source for more than 3.5 billion people in the world. Global climate change is having a strong negative effect on rice production. One of the climatic factors impacting rice yield is asymmetric warming, i.e., the stronger increase in nighttime as compared to daytime temperatures. Little is known of the metabolic responses of rice to high night temperature (HNT) in the field. Eight rice cultivars with contrasting HNT sensitivity were grown in the field during the wet (WS) and dry season (DS) in the Philippines. Plant height, 1000-grain weight and harvest index were influenced by HNT in both seasons, while total grain yield was only consistently reduced in the WS. Metabolite composition was analysed by gas chromatography-mass spectrometry (GC-MS). HNT effects were more pronounced in panicles than in flag leaves. A decreased abundance of sugar phosphates and sucrose, and a higher abundance of monosaccharides in panicles indicated impaired glycolysis and higher respiration-driven carbon losses in response to HNT in the WS. Higher amounts of alanine and cyano-alanine in panicles grown in the DS compared to in those grown in the WS point to an improved N-assimilation and more effective detoxification of cyanide, contributing to the smaller impact of HNT on grain yield in the DS.

Keywords: high night temperature; rice; grain yield; wet season; dry season; metabolomics

---

1. Introduction

Rice is a staple food for more than half of the world's population and the demand is steadily increasing with the growing human population [1]. Climate change is a significant limiting factor for enhancing food production, because increasing abiotic and biotic stresses negatively affect the yield of all major crops [2–4]. During the past century, the global surface temperature has increased by an average of 0.85 °C, and a further increase of up to 3.7 °C has been predicted by 2100 [3]. This temperature increase develops asymmetrically, with a faster rise in daily minimum compared to daily maximum temperatures [5–9], leading to "high night temperature" (HNT) conditions. Asymmetric warming causes a reduction in the temperature difference between daily maximum and minimum temperatures, i.e., the diurnal temperature range (DTR), with a negative influence on both wild and crop plant species [10]. In particular, the main rice-growing countries in Asia, including China [11], the Philippines [12,13] and India [14,15], are affected.

Several studies have reported a strong decrease in yield and grain quality, such as increased chalk formation, and altered grain growth dynamics in rice under HNT [16–21]. HNT can have a stronger impact on grain weight than high day temperatures in rice and wheat [22–24]. Field studies at the International Rice Research Institute (IRRI) in the Philippines showed that rice grain yield was reduced by 10% per 1 °C increase in night temperatures during the dry season (DS), whereas the effect of increasing day temperatures was not significant within the investigated time period [12].

Differences in HNT sensitivity among various rice cultivars based on grain yield [25–27], yield-related parameters, or phenotypes in the vegetative stage [28] have been reported, indicating natural variation in HNT tolerance. In addition, HNT reduces the starch content in panicles and negatively affects grain yield and quality (chalk and amylose content) in the sensitive cultivars Gharib and IR64, but not in the tolerant cultivar N22 [29].

Different factors may cause HNT sensitivity. Physiological effects reported under HNT include higher rates of respiration in leaves [28,30,31] and panicles [29], whereas photosynthesis is not affected [28] or may be decreased as well [32]. A reduction in nitrogen and carbohydrate translocation after flowering as a possible cause of yield reduction in HNT sensitive cultivars was also discussed [25]. Reduced grain weight and quality may be caused by lower sink strength due to lower cell wall invertase and sucrose synthase activity in sensitive cultivars, accompanied by higher sugar accumulation in the rachis [29].

Despite the increasing knowledge of the physiological responses to HNT, only little is known about the metabolomic responses of rice under these conditions. The metabolic status is important for growth, development and stress tolerance, and additionally influences important traits such as flavor, biomass, yield and nutritional quality [33–35]. Therefore, the assessment of the metabolomic status of wild and crop species can help to evaluate natural variation [33]. Additionally, the metabolome integrates molecular and environmental effects as endpoints of biological processes [36]. Moreover, metabolites constitute potential markers for the selection of tolerant crop genotypes in breeding programs. Several studies investigated metabolic changes in rice in response to abiotic stress conditions, such as salinity [37–41], osmotic stress [42], drought [43–47], heat [44,48], and combined drought and heat stress conditions [49,50].

In a corresponding study on rice under HNT conditions, sucrose and pyruvate/oxaloacetate-derived amino acids were shown to accumulate while sugar phosphates and organic acids involved in glycolysis/gluconeogenesis and the tricarboxylic acid (TCA) cycle decreased in developing caryopses [48]. A dysregulation of central metabolism and an increase in polyamine biosynthesis was described for sensitive cultivars, whereas existing metabolic pre-adaptation under control conditions was found for tolerant cultivars [51,52]. Furthermore, in sensitive cultivars, 4-amino butanoic acid (GABA) signaling—and in tolerant cultivars, the jasmonate precursor *myo*-inositol—were linked to the HNT responses [52]. A metabolomics study investigating early seed development and the early grain-filling stage in six rice cultivars reported a sugar accumulation peak seven days after flowering and 19 significantly different metabolites under HNT compared to under control conditions, with a special focus on the generally higher abundance of sugars and sugar alcohols under HNT [53].

The goal of this study was to investigate the seasonal effects of HNT responses by assessing the metabolic responses to HNT stress in flag leaves and panicles during the DS and wet season (WS) in contrasting rice cultivars under field conditions. Previous studies of the comparison of HNT's effects during the WS and DS were limited to agronomic traits [13,14,20,26,54], while the influence of HNT on the rice metabolome has not been reported yet. The present study sheds new light on the responses of rice to an important climatic stress factor that may severely limit grain yield and quality, and therefore the global food supply.

2. Results

Two field experiments were performed at the IRRI in the Philippines during the WS and DS with eight rice cultivars (Table 1). These cultivars comprised the *indica* and *japonica* subspecies and included HNT tolerant, intermediate and sensitive cultivars, as determined during the vegetative growth stage from a study under controlled environmental conditions [28].

Table 1. Experimental set-up for high night temperature (HNT) field experiments for eight contrasting *Oryza sativa* cultivars. Mean temperatures and relative humidity (RH) are given from the beginning of HNT treatment till the sampling time, when panicles reached 50% flowering.

	Experiment 1		Experiment 2	
Season	Wet		Dry	
Conditions	Control	HNT	Control	HNT
Cultivars	CT9993-5-10-1M IR123 IR62266-42-6-2 IR64 IR72 M202 Moroberekan Taipei309			
T_{day} (°C)	27.7		26.1	
T_{night} (°C)	22.2	27.6	22.2	27.8
RH (%)	98.3	89.4	96.4	80.9
Sampling time	Panicle at 50% flowering			
Samples	Flag leaves, panicles			

The WS experiment was performed for 84 to 104 days from transplanting till maturity, and the DS experiment, for 87 to 118 days, depending on the staggered sowing (Figure A1). Samples for metabolite analysis were taken at 59 to 78 days after transplanting in the WS, and in the DS, between 58 and 88 days. During the day (6 a.m.–6 p.m.), plants were exposed to ambient conditions, with an average temperature of 27.7 °C during the WS and 26.1 °C during the DS. The mean daytime temperature ranged from 25.4 to 29.7 °C during the WS, and from 21.8 to 30.9 °C during the DS, with maximum daily temperatures from 26.5 °C to 34.7 °C during the WS and from 24.3 °C to 36.1 °C during the DS (Figure A2e).

During the night, plots were covered by tents, and the temperature was kept constant by air conditioners, set to 22 °C for the control and 28 °C for HNT conditions. The average temperatures measured in the tents were 27.64 °C (± 0.77 °C) and 27.82 °C (± 1.07 °C) under HNT conditions and 22.24 °C (± 0.99 °C) and 22.25 °C (± 0.46 °C) under control conditions during the WS and DS, respectively (Figure 1A,B). The corresponding ambient night temperatures outside the tents are shown in Figure A2f. As the average day temperatures for both seasons were very similar, the night temperature difference of around 5 °C is the main temperature factor driving the physiological and metabolic changes in all cultivars.

Average radiation was about 22% lower in the WS than in the DS and sunshine duration in the WS reached only 45% of the values measured in the DS (Figure A2A,B). Daily rainfall in the WS was recorded between 85 and 0 mm, while it was approximately zero in the DS (Figure A2C. Accordingly, average relative air humidity was lower in the DS with values between 73% and 95%, compared to those between 76% and 98% in the WS (Figure A2D).

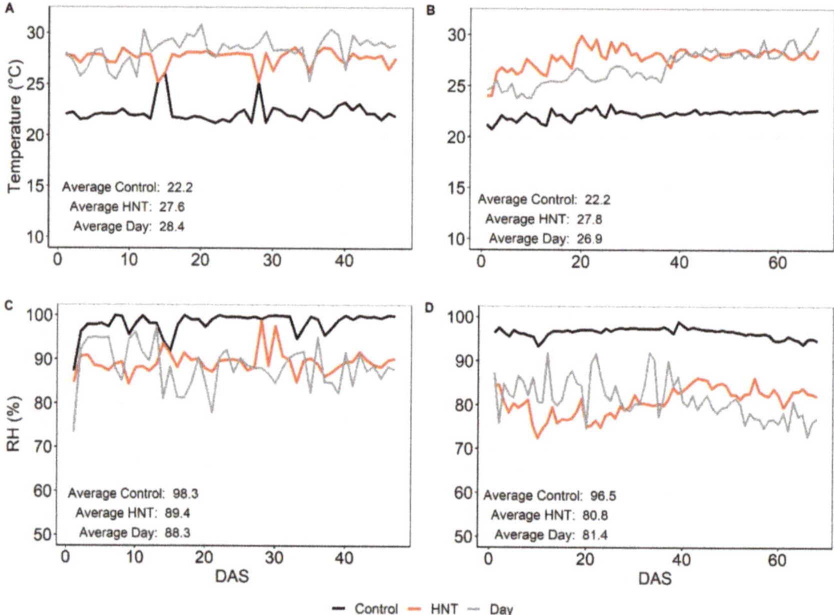

Figure 1. Average temperature (**A**,**B**) and relative humidity (RH) (**C**,**D**) during the night (6 p.m.–6 a.m.) in the wet season (WS) (**A**,**C**) and dry season (DS) (**B**,**D**) under control and HNT conditions, measured till the end of sampling at 50% flowering. For comparison, day temperature and humidity are included (grey lines). Measurements, which were done every 30 min, were averaged. DAS—Days after stress; WS—wet season; DS—dry season.

2.1. Influence of HNT on Agronomic Parameters

For all agronomic parameters, a significant genotype effect was found in both seasons when comparing samples from plants grown under HNT with control conditions (Table 2). Furthermore, a significant seasonal effect was recorded for almost all agronomic parameters. The influence of HNT conditions on the growth response was recorded as differences in plant height. No significant treatment effect but a significant Genotype x Treatment (GxT) effect of HNT on plant height was found over all cultivars for both seasons (Table 2). On average, plant height was slightly lower in the DS compared to in the WS, but cultivar-specific patterns were conserved (Figure 2). In both seasons, plant height was significantly ($p < 0.05$) increased under HNT in three cultivars (IR123, IR64 and IR72), while it was decreased in Moroberekan. IR62266-42-6-2 and M202 showed reduced plant height only in the WS, and Taipei309, only in the DS.

Total grain yield under control conditions was significantly lower in the WS, with a maximum yield among all cultivars of about 617 g·m^{-2} compared to that in the DS of 762 g·m^{-2} (Figure 3A,B). A significant effect of HNT treatment on the grain yield of all eight cultivars compared to control was only detectable in the WS (Table 2), where yield reduction varied between 23% in M202 and 4% in IR123 (Figure A3A). In the DS, yield was only reduced between 8% and 3% in four cultivars, while it was slightly increased (1%–5%) in the other four (Figure A3B). No correlation was found between the yield reduction in our experiments in the WS or DS and the HNT sensitivity rank of the same cultivars in the vegetative stage under controlled environmental conditions determined for the same cultivars in a previous study [28] (not shown).

Table 2. Analysis of variance (ANOVA) on selected agronomic parameters. Sixty plants for the DS and 24 plants for the WS were considered for plant height, tiller number and panicle number. For the remaining parameters, two replicates pooled from twelve plants each were considered for the WS and five replicates pooled from twelve plants each were considered for the DS. Spikelets/panicle represents the number of spikelets per panicle. The seed set was calculated as follows: seed set (%) = filled grains/(filled+half-filled+unfilled grains) × 100. Harvest index was calculated as percentage of dry weight of filled grains relative to total above-ground biomass. The significance of the influence of genotype (G), HNT-treatment (T), season (S) or the interaction between two influences (GxT or TxS) on differences between HNT and control conditions across all eight cultivars is indicated by asterisks: 0.001 < ***; 0.001 > ** < 0.01; 0.01 > * < 0.05. ns—not significant. Original data for plant height, tiller number, panicle number and all yield components for the WS (2011) and the DS (2014) are available in Table S1.

Parameter	WS	WS	WS	DS	DS	DS	Both Seasons		
	G	T	GxT	G	T	GxT	T	S	TxS
Plant Height (cm)	***	ns	**	***	ns.	*	ns	ns	ns
Biomass g/m^2	***	*	ns	***	ns	ns	ns	*	ns
Straw (g)	***	ns	*	***	ns	ns	ns	ns	ns
Rachis (g)	***	ns	ns	***	ns	ns	ns	*	ns
Tiller No	***	ns	ns	***	ns	ns	ns	**	ns
Panicle No	***	ns	ns	***	ns	ns	ns	**	ns
Panicle/m^2	***	ns	ns	***	ns	ns	ns	**	ns
Spikelet/m^2	***	*	ns	***	ns	ns	ns	***	ns
Spikelets/Panicle	***	**	ns	***	ns	ns	ns	***	ns
Seed set (%)	***	ns	ns	***	ns	ns	**	*	ns
Grain yield (g/m^2)	***	***	ns	***	ns	ns	ns	**	ns
1000 grain weight (g)	***	***	***	***	***	ns	ns	***	ns
Harvest Index	***	***	ns	***	**	ns	ns	***	ns

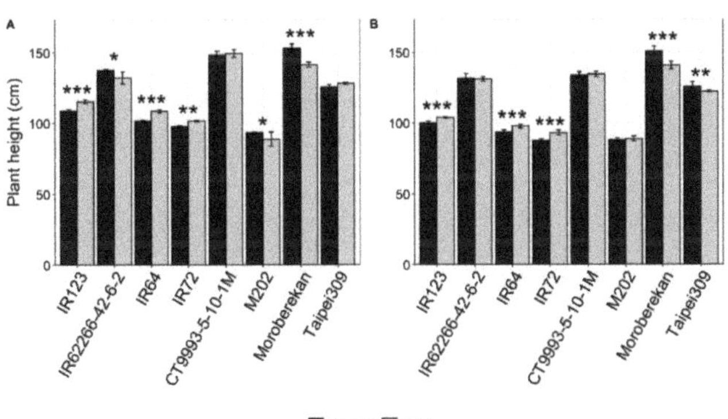

Figure 2. Plant height of the investigated rice cultivars under control and HNT conditions in the WS (**A**) and DS (**B**). Bars for the WS represent means ± SEM of 24 plants per condition, and bars for the DS, those of 60 plants per condition. Cultivars are sorted alphabetically within the respective *O. sativa* subspecies *indica* (1–4) and *japonica* (5–8). Significance levels are indicated by asterisks: 0.001 < ***; 0.001 > ** < 0.01; 0.01 > * < 0.05.

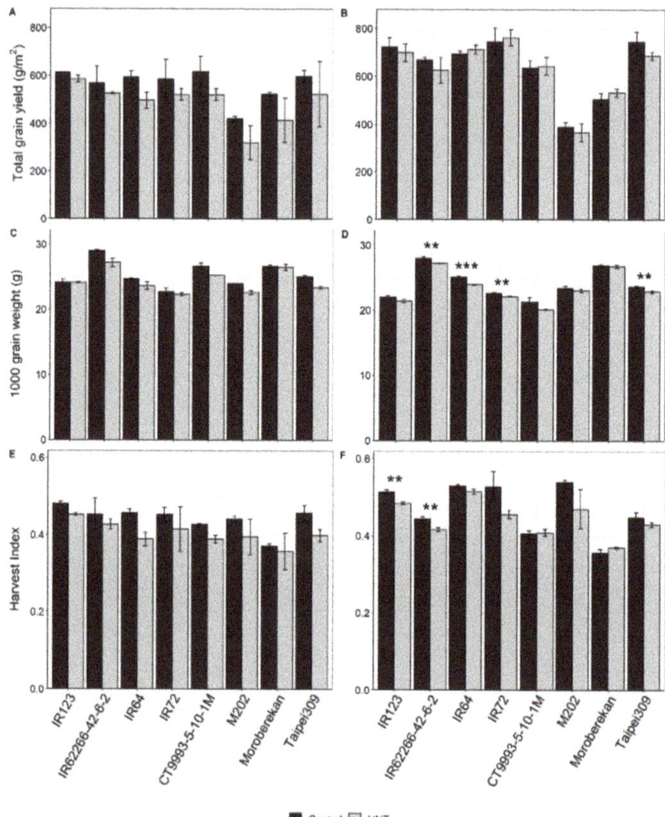

Figure 3. Grain yield (**A,B**), 1000-grain weight (**C,D**) and harvest Index (**E,F**) of eight rice cultivars under control and HNT conditions in the WS (**A,C,E**) and DS (**B,D,F**). For the WS, bars represent the means and error bars, the range of two replicates generated from 12 plants, each. For the DS, the bars represent the means ± SEM of five replicates generated from 12 plants, each. Cultivars are sorted alphabetically within the respective *O. sativa* subspecies *indica* (1–4) and *japonica* (5–8). Significance levels were only calculated for the DS due to an insufficient replicate number in the WS and are indicated by asterisks: $0.001 < ***; 0.001 > ** < 0.01; 0.01 > * < 0.05$.

A significant negative HNT treatment effect was also found for the 1000-grain weight in both growth seasons (Table 2), with the highest reductions in the WS of about 1.7 and 1.8 g for Taipei309 and IR62266-42-6-2, respectively (Figure 3C,D, Table 2).

The harvest index was significantly affected by HNT across all cultivars in both seasons (Table 2) and showed an overall reduction, except for Moroberekan in the WS and DS and CT9993-5-10-1M only in the DS (Figure 3E,F, Table 2). Furthermore, a significant treatment effect was determined for biomass, spikelets per m^2 and spikelets per panicle only in the WS, but not in the DS (Table 2). For cultivar-specific changes in these parameters, see Figure A4.

2.2. HNT's Effects on the Metabolome Are More Pronounced in Panicles Than in Flag Leaves

Profiling of hydrophilic small metabolites was performed by gas chromatography-mass spectrometry (GC-MS) on flag leaves and panicles of all eight cultivars grown in both seasons. Since it has been shown previously that the metabolite profiles of rice flag leaves and panicles differ widely, making meaningful direct comparisons impossible [49], we treated the data from the two organs separately. After the pre-processing of both data sets, a total of 76 metabolites for flag leaves and 69 for panicles were determined that were detected in both seasons. Principal Component Analysis (PCA) indicated that metabolite profiles of flag leaves were not strongly affected by HNT conditions in either the WS or DS (Figure 4A,B).

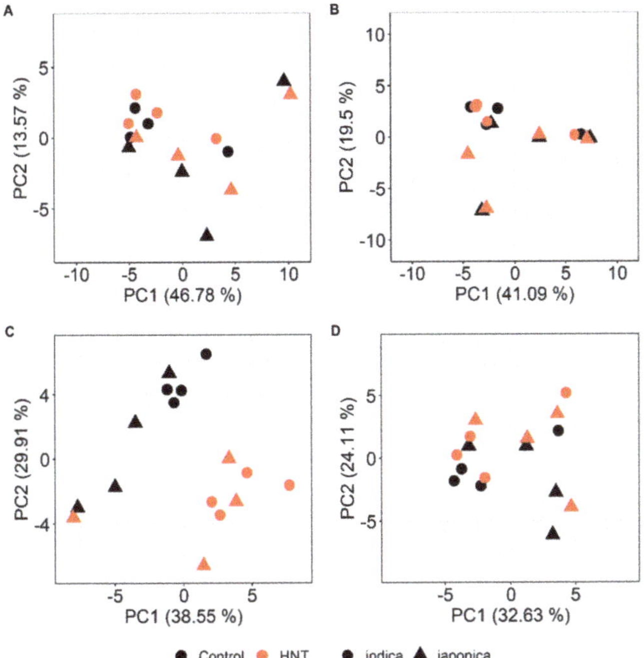

Figure 4. Score plots of the first two Principal Components (PC1 and PC2) from the Principal Component Analysis (PCA) of the metabolite profiles of rice flag leaves (**A**, **B**) and panicles (**C**, **D**) of the eight investigated rice cultivars under control and HNT conditions in the WS (**A**, **C**) and DS (**B**, **D**). For flag leaves, means of the median-normalized and log$_2$-transformed mass spectral intensities of 76 metabolites, and for panicles, those of 69 metabolites, were used. Numbers in parentheses indicate the fractions of the total variance explained by the respective PCs.

Instead, a separation between cultivars belonging to the subspecies *indica* and *japonica* was visible for both seasons and treatments. By contrast, a clear separation along PC1, explaining 38.55% of the total variance in the data set, between samples from plants grown under control or HNT conditions was observed for panicles collected in the WS (Figure 4C). The single outlier represents the *japonica* cultivar Moroberekan under HNT conditions. For the DS experiment, samples from panicles under different night temperature conditions were separated by PC2, explaining 24.11% of the variance, while PC1 separated the subspecies, explaining 32.63% of the total variance (Figure 4D).

The metabolite composition already varied under control conditions between the two growth seasons in both flag leaves and panicles (Figure 5). Of the 76 metabolites identified in flag leaves, 48 (63%) showed a significantly different content in at least three cultivars in this analysis, while of the 69 metabolites in panicles, 28 (41%) differed between seasons. Only eight of these metabolites (malic acid, A159003, A221004, cis-4-hydroxycinnamic acid, trans-4-hydroxycinnamic acid, fructose-6-phosphate, glyceric acid 3-phosphate and raffinose) were identical in both organs, indicating highly organ-specific metabolic reactions to seasonal variations in rice. In addition, there was variation among the cultivars, which was, however, largely independent of the subspecies that the cultivars belong to.

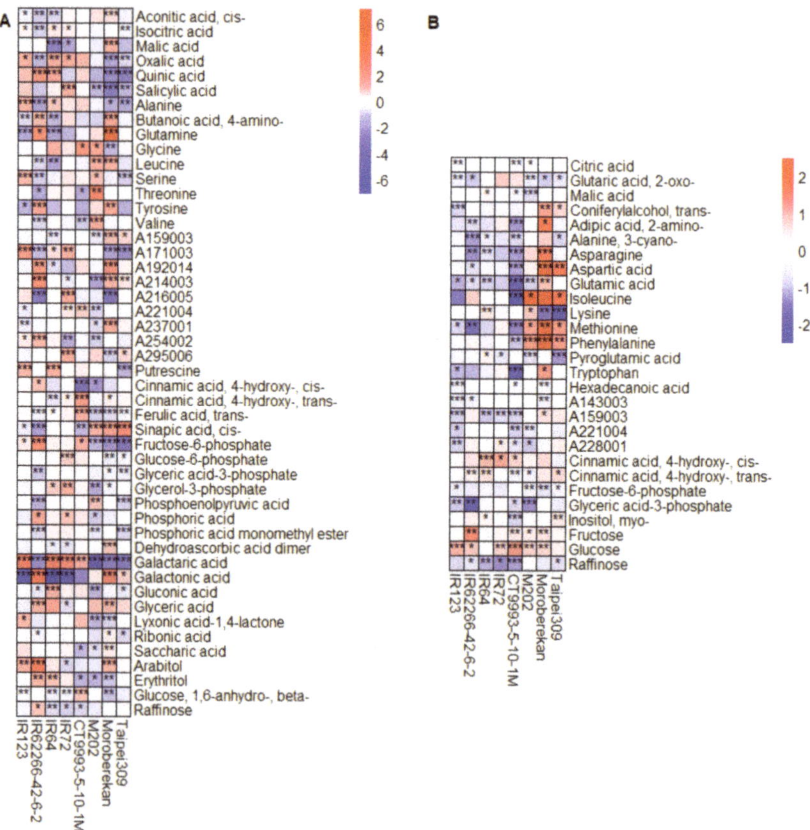

Figure 5. Heat maps showing the $\log_2$ fold changes in metabolite levels under control conditions in the DS compared to the WS for flag leaves (**A**) and panicles (**B**). Only metabolites with a significant change in at least three out of the eight cultivars are displayed. The level of significance is indicated by asterisks (* $p < 0.05$; ** $p < 0.01$; *** $p < 0.001$) and the $\log_2$ fold change is represented by the indicated color code. Blue indicates a lower metabolite level in the DS compared to the WS, and red, a higher level. Metabolites are listed alphabetically within the metabolite classes (compare Supplementary Table S2). Cultivars are sorted alphabetically within the respective *O. sativa* subspecies *indica* (1–4) and *japonica* (5–8).

Under HNT conditions, only three metabolites in flag leaves were significantly changed relative to control values in at least three cultivars in the WS, compared to 17 metabolites that were so in the DS (Figure 6). Only erythritol was significantly affected by HNT in both growth seasons. However, while it was increased or unchanged in the DS, it showed a cultivar-specific increase (strongest in Taipei309) or decrease (strongest in IR72) in the WS. In the DS, all metabolites were either reduced/unchanged or increased/unchanged across all cultivars, except for fructose, which was significantly increased in Taipei309 and CT9993-5-10-1M, and significantly decreased in IR64. In addition, while most metabolites showed significant changes in only three or four cultivars, glucose-6-phosphate was significantly reduced under HNT conditions in seven out of the eight cultivars (Figure 6B).

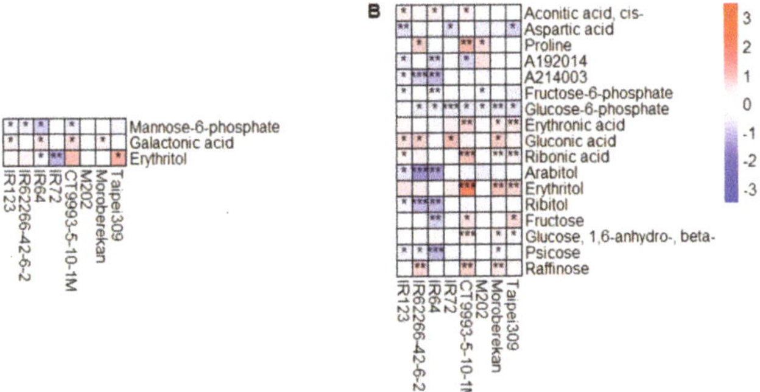

Figure 6. Heat maps showing the $\log_2$ fold changes in metabolite pool sizes in flag leaves under HNT compared to control conditions for the WS (**A**) and DS (**B**). Only metabolites with a significant change in at least three out of the eight cultivars are displayed. The level of significance is indicated by asterisks (* $p < 0.05$; ** $p < 0.01$; *** $p < 0.001$), and the $\log_2$ fold change is represented by the indicated color code. Blue indicates a lower metabolite level under HNT compared to under control conditions, and red, a higher level. Cultivars were sorted alphabetically within the respective *O. sativa* subspecies *indica* (1–4) and *japonica* (5–8).

In panicles, metabolite changes caused by HNT conditions were more pronounced than in leaves, with higher $\log_2$ fold changes and a larger number of significantly changed metabolites—25 during the WS and 12 during the DS. In addition to the larger number of metabolites that were significantly affected by HNT in the WS than in the DS, opposite to what we observed in flag leaves (Figure 6), changes were generally also larger in the WS than in the DS in panicles (Figure 7).

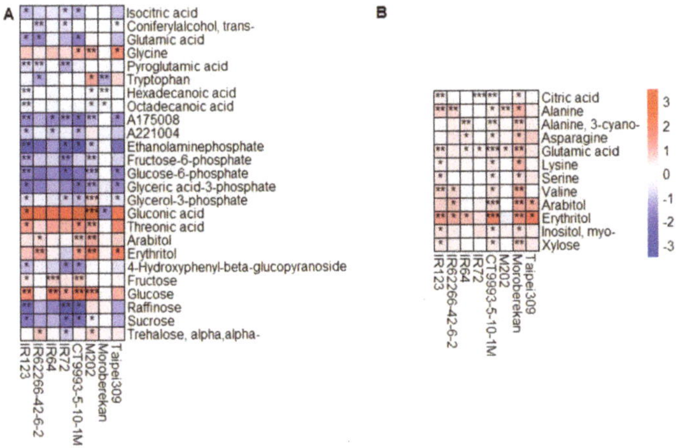

Figure 7. Heat maps showing the $\log_2$ fold changes in metabolite pool sizes in panicles under HNT compared to control conditions for the WS (**A**) and DS (**B**). Only metabolites with a significant change in at least three out of the eight cultivars are displayed. The level of significance is indicated by asterisks (* $p < 0.05$; ** $p < 0.01$; *** $p < 0.001$), and the $\log_2$ fold change is represented by the indicated color code. Blue indicates a lower metabolite level under HNT compared to control conditions, and red, a higher level. Cultivars were sorted alphabetically within the respective *O. sativa* subspecies *indica* (1–4) and *japonica* (5–8).

A comparison of the significantly changed metabolites in at least three of the eight cultivars in the DS with those in the WS revealed an overlap of glutamic acid, arabitol and erythritol (Figure 7,

Figure A5). Glutamic acid content was mainly reduced in the WS but increased in the DS, while the polyols arabitol and erythritol were mainly increased by HNT in both seasons. There was very little overlap in the metabolites significantly affected by HNT between flag leaves and panicles, with only erythritol affected in the WS and arabitol and erythritol, in the DS. Interestingly, arabitol showed an opposite behavior in response to HNT in the two organs, with decreased levels in flag leaves and increased levels in panicles.

In the WS, the levels of organic acids; amino acids (except glycine); the phosphorylated intermediates fructose-6 phosphate, glucose-6-phosphate, glyceric acid-3-phosphate and glycerol-3-phosphate; and the sugars raffinose and sucrose were in general reduced during HNT in panicles compared to under control conditions. On the other hand, glycine, gluconic acid, threonic acid, arabitol, erythritol, and fructose and glucose were increased (Figure 7A). In a direct comparison of these significantly changed metabolites in the WS with the metabolite levels in the DS, no reduction of any of these metabolites could be observed in the DS (Figure A6). In the DS, all 12 of the significantly influenced metabolites (mainly amino acids, arabitol, erythritol, citric acid, glutamic acid and xylose) were increased under HNT conditions (Figure 7B).

Alanine and 3-cyano alanine were among the metabolites that were significantly changed under HNT conditions in panicles in the DS, but not in the WS. Alanine is a major storage amino acid under stress conditions [55], and the activity of the alanine biosynthetic enzyme alanine aminotransferase (AlaAT) can influence rice yield [56]. In the WS, the activity of AlaAT was generally reduced under HNT to values of 62% to 96% (except for Moroberekan) compared to under control conditions, which was significant at $p < 0.05$ for IR123 and IR72 (Figure 8A). By contrast, AlaAT activity in the DS reached values of 77% to 137% higher under HNT in comparison to under control conditions and was increased in five out of the eight cultivars, although none of the differences were statistically significant (Figure 8B).

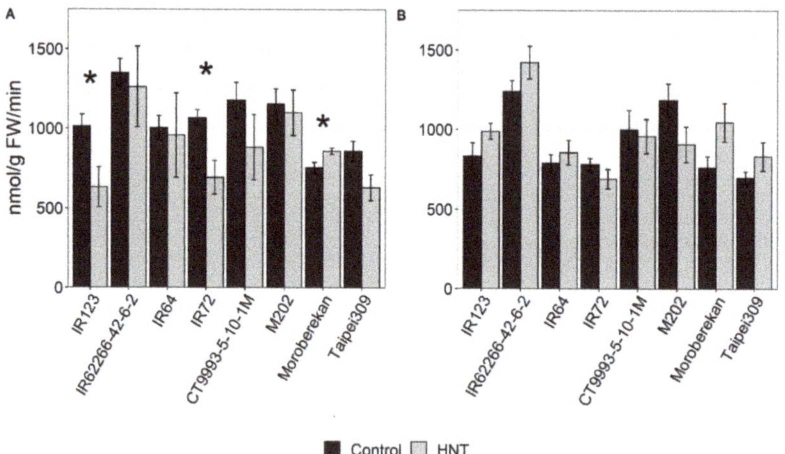

Figure 8. Activity of the enzyme alanine aminotransferase (AlaAT) in panicles of the indicated rice cultivars under control and HNT conditions for the WS (**A**) and DS (**B**). Values are averages of three replicates per cultivar and condition, with four exceptions with two replicates. The level of significance is indicated by asterisks (* $p < 0.05$; ** $p < 0.01$; *** $p < 0.001$). Cultivars were sorted alphabetically within the respective *O. sativa* subspecies *indica* (1–4) and *japonica* (5–8).

To obtain insight into the potential function of particular metabolites in HNT tolerance in the field and to identify possible candidate marker metabolites for HNT tolerance, we performed correlation analyses between the grain yield reduction in eight cultivars under HNT compared to under control conditions and the change in relative metabolite pool sizes ($\log_2$ fold change) under HNT in the WS, wherein HNT significantly affected grain yield. While we only identified one significant correlation for metabolites detected in flag leaves (ribitol), we found seven such correlations among panicle metabolites (Figure 9). In addition to one yet unidentified compound, the others comprised four amino acids (including 3-cyano alanine), pyroglutamic acid (representing the sum of pyroglutamate, glutamine and glutamate pools) and fructose-6-phosphate. All eight metabolites showed positive correlations, i.e., a larger change in metabolite pool size indicates a smaller yield loss.

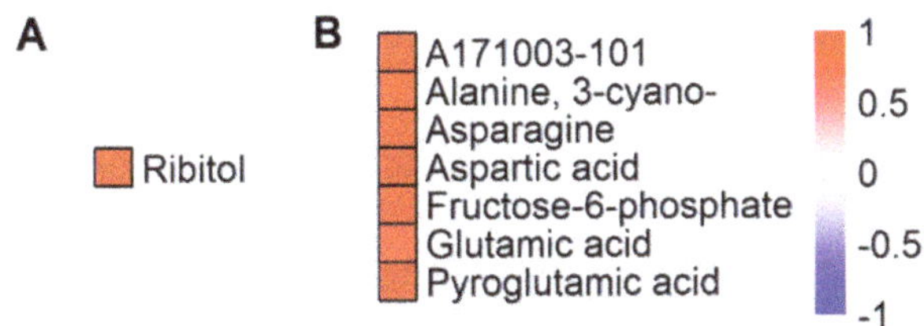

Figure 9. Metabolites with significant correlations (Spearman's rank correlation, $p < 0.05$) between total grain yield reduction under HNT in the WS and the corresponding changes in metabolite contents ($\log_2$ fold change) in flag leaves (**A**) and in panicles (**B**). Red color indicates positive correlations. Metabolites are sorted alphabetically.

3. Discussion

The response of agronomic parameters and metabolic patterns to HNT have been analyzed for eight rice cultivars with different HNT tolerance under field conditions at the IRRI in two different seasons. A comparison of the weather data for both seasons and the respective agronomic parameters identified a slightly longer time to maturity in the DS than in the WS as an important difference. During the DS, plants were exposed to higher radiation intensity and sunshine duration, but lower rainfall and relative humidity compared to in the WS. Similar differences for radiation and sunshine have been reported for a comparison of the DS and the WS from 2005 to 2009 at the IRRI [13]. Furthermore, temperature data for the two growth seasons largely agree between our study and two earlier reports for the same location [13,20], indicating that the plants in our study were exposed to normal climatic conditions without any extreme weather events.

Under control conditions, total grain yield was higher for most cultivars in the DS than in the WS, in agreement with published data [13]. Under HNT conditions, no clear changes in grain yield were observed during the DS, while it was reduced to different degrees in all cultivars in the WS. Under controlled environmental conditions, a yield reduction caused by HNT was previously observed for the cultivars IR62266-42-6-2 and CT9993-5-10-1M, while IR123 showed no change, and IR72 even showed an increased grain yield [28]. During the WS, under our field conditions, IR62266-42-6-2 and CT9993-5-10-1M also showed clear yield reductions of about 22% and 12%, respectively. However, IR123 and IR72 behaved differently under field than under climate chamber conditions, with yield reductions of 16% and 11%, respectively, emphasizing the need for field experiments to determine the effects of stress treatments on rice yield.

A similar influence of the growing season on yield reduction under HNT was previously reported for the *indica* cultivar Gharib and six tropical hybrid cultivars [20]. Additionally, for the tolerant *aus* cultivar N22, a significantly lower yield under HNT was only recorded in the WS. This yield variation was mainly attributed to a reduced grain weight and number of spikelets per m^2, parameters also with significant negative treatment effects during the WS in the present study. Grain yield was reduced in both seasons by around 11% under HNT except for tolerant cultivars in four consecutive years, again partially attributable to a decrease in grain weight [26]. Other authors also highlighted the combination of decreased grain weight, spikelet number per panicle, and biomass production together with a decreased seed set as important for the decline in grain yield under HNT [57]. In general, a reduction in grain weight under HNT conditions was demonstrated for field-grown rice when exposed to HNT stress from panicle initiation to maturity [18,25–27,58]. In agreement with this, we also found a negative HNT effect on the 1000-grain weight in both seasons under similar stress conditions as used in the previous studies.

Grain yield is influenced by carbon and nitrogen supply to the grain, which are affected by HNT [59]. Temperature-sensitive respiration, known to be increased under HNT (e.g., [28]), might have resulted in increased respiratory carbon loss, previously described to be important during the ripening period [60]. Dark respiration was also considered by other reports to be the main factor affecting biomass and yield under HNT conditions [12,20,25,61] and might be responsible for a decline in assimilation supply to developing grains [57].

This hypothesis is in agreement with the metabolite data obtained during the WS. We found a lower abundance of sucrose and the intermediates of glycolysis, such as glucose-6-phosphate, fructose-6-phosphate, glyceric acid-3-phosphate and glycerol-3-phosphate, whereas the monosaccharides glucose and fructose were increased in panicles. A similar decrease in sugar phosphates, but not in sucrose, was also reported for developing rice caryopses exposed to HNT during the milky stage [48]. Likewise, we also found a significant correlation between the magnitude of the changes in the fructose-6-phosphate content of the panicles under HNT conditions and the yield reduction in the WS. This emphasizes the importance of glycolysis for HNT tolerance in rice.

Glycolysis generates biosynthetic intermediates for respiration. Therefore, a high turnover of glycolysis, as indicated by reduced levels of intermediates, could be expected as respiration is highly increased under HNT. In addition, the products of glycolysis also feed into the TCA cycle, which was shown to be dysregulated in leaves under HNT conditions in climate chamber experiments [51,62]. On the other hand, no significant differences in the metabolites associated with the TCA cycle were found in the developing seeds of different rice cultivars under HNT [53]. Likewise, our data did not provide evidence for an altered TCA cycle under HNT conditions in either panicles or flag leaves.

Interestingly, the effects on glycolysis that we found in panicles in the WS were not observed in either flag leaves in our present study or previously in leaves of the vegetative stage [52]. Apparently, photosynthesis, which is unimpaired under HNT conditions, results in largely unaltered carbohydrate pools in leaves [28,52]. It is therefore reasonable to assume that the carbohydrate supply to the panicles is limiting for grain yield under HNT conditions. Lower sink capacity [26], possibly related to a reduction in the activity of enzymes involved in starch synthesis, has been discussed as a reason for the reduction in grain weight under HNT [63], which we have also observed. A further possibility is an impaired import of sucrose into the panicles under HNT conditions, as has been shown in rice under heat stress [44]. Further experiments will be necessary to test these hypotheses.

The larger reduction in grain yield in the WS compared to in the DS may nevertheless, at least in part, be related to carbohydrate availability. One factor may be faster development during a slightly shorter growing period in the WS, caused by higher daytime T_{min}, preventing the accumulation of sufficient biomass, as shown previously in simulation models [54]. In addition, irradiance levels in the WS were much lower than in the DS, resulting in lower photosynthesis rates [64]. This may have led to a lower overall carbon supply for grain filling [20], leading to lower yield in the WS than the DS under control conditions and a more pronounced effect of HNT on yield in the WS [25] that was mitigated by the higher carbohydrate supply in the DS.

The amino acid alanine was among the significantly increased metabolites in panicles under HNT in the DS but not in the WS. Similarly, alanine was also increased under HNT during early seed development and in the early grain-filling stage in six rice cultivars [53] and in wheat spikes [65]. Alanine is synthesized by the enzyme AlaAT, which catalyzes the reversible synthesis of alanine and 2-oxoglutarate from pyruvate and glutamic acid [66]. It is therefore considered an intercellular nitrogen and carbon shuttle involved in both carbon fixation and nitrogen metabolism [67]. AlaAT is localized in various plant organs and is active in developing rice seeds [68]. The activity of AlaAT is increased in developing rice seeds under heat stress [48], and we observed a moderate increase under HNT conditions in the DS and a moderate decrease in the WS. While the overexpression of *AlaAT* from barley in rice or canola results in increased nitrogen uptake efficiency and a higher biomass and seed yield compared to in wild type plants [56,66,69–71], a rice mutant of *AlaAT1* exhibits decreased kernel weight [69]. The higher AlaAT activity in the DS may have led to increased nitrogen uptake and assimilation, as described for plants overexpressing *AlaAT* [56], while reduced activity in the WS may have had the opposite effect.

Another metabolite that was significantly increased in response to HNT in the DS, but not in the WS, specifically in panicles, was 3-cyano alanine. This compound is generated by the enzyme 3-cyano alanine synthase (EC 4.4.1.9) during the detoxification of cyanide, which is generated as a by-product of ethylene biosynthesis [72], when the precursor 1-aminocyclopropane-1-carboxylic acid (ACC) is converted into ethylene and hydrogen cyanide (HCN) by the activity of the enzyme ACC synthase [73]. The resulting 3-cyano alanine is then enzymatically converted to asparagine [74], which was also increased under HNT in the DS, indicating a functional detoxification process. Ethylene is a volatile plant hormone that is important for plant growth and development, and various biotic and abiotic stress responses [75]. HCN, on the other hand, is toxic to cells and therefore needs to be efficiently removed [74]. The lower amounts of 3-cyano alanine and asparagine in the panicles collected in the WS might point to a less efficient detoxification of HCN. This is in agreement with the finding that the magnitude of the reduction of both 3-cyano alanine and asparagine in panicles in the WS is significantly correlated with the reduction in grain yield in the WS observed across the eight cultivars. This may indicate that HCN toxicity plays an important role in the HNT sensitivity of panicles. Additionally, however, HCN may play a direct regulatory role in gene expression in low, non-toxic concentrations [76]. Whether this has any impact on HNT tolerance is currently not known.

Two polyols, arabitol and erythritol, were significantly increased in the flag leaves and panicles of almost all cultivars under HNT in both seasons. Both metabolites were also increased under HNT in the vegetative leaves of 12 rice cultivars, including the eight in the present study, in climate chamber experiments [51]. Polyols generally function as compatible solutes and antioxidants under abiotic and biotic stress conditions [77]. Furthermore, arabitol accumulates in flowering spikelets and developing seeds under combined drought and heat stress in the tolerant *aus* cultivar N22 and has a higher content in N22 compared to in sensitive cultivars in flag leaves in the field under control conditions [49]. Similarly, erythritol is accumulated in flowering spikelets and flag leaves under the same conditions, while it is decreased in developing seeds under combined drought and heat stress. Increased levels of erythritol were also found under drought conditions in Arabidopsis [78,79] and in flag leaves of 292 rice accessions [80]. In fact, arabitol and erythritol were both identified as potential metabolic markers for combined drought and heat tolerance [49], and erythritol content under control conditions was the best predictor of drought-induced yield loss in rice [80]. In the present study, however, no correlation between changes in arabitol or erythritol levels and grain yield under HNT was found. The accumulation of these sugar alcohols may therefore be an unspecific response to HNT stress.

4. Materials and Methods

4.1. Plant Material, Cultivation and HNT Stress Treatment

Eight *Oryza sativa* ssp. *indica* (IR123, IR62266-42-6-2, IR64 and IR72) and *japonica* (CT9993-5-10-1M, M202, Moroberekan and Taipei309) cultivars with different HNT tolerance in the vegetative stage under controlled environmental conditions [28] were used (Table 1). IR72, Taipei309 and Moroberekan were characterized as HNT tolerant; IR64, IR123 and CT9993-5-10-1M showed intermediate tolerance; and M202 and IR62266-42-6-2 were sensitive to HNT under these conditions [28]. The seeds for all cultivars were produced at the IRRI. The experiments were carried out during the WS and DS at the IRRI (14°11′N, 121°15′E, 21 MASL) in the Philippines. The seeds were pre-germinated in water after incubation at 50 °C for 3 d to break dormancy and were then sown in seeding trays. Fourteen-day old seedlings were transplanted to the field to a spacing of 0.2 × 0.2 m. The WS experiment was started in June 2011, with four seedlings per hill and each cultivar (42–48 hills) randomly assigned to two replicate plots per treatment. Phosphorus (15 kg·ha^{-1} p as single superphosphate), potassium (20 kg·ha^{-1} K as KCl), and zinc (2.5 kg·ha^{-1} Zn as zinc sulfate heptahydrate) were applied to all plots as a basal fertilizer a day before transplanting. Nitrogen (N as urea) was incorporated in four splits (30 kg·ha^{-1} as basal, 20 kg·ha^{-1} at mid-tillering, 30 kg·ha^{-1} at panicle initiation (PI), and 20 kg·ha^{-1} just before heading). For the DS experiment, seedlings were transplanted in a staggered approach with one batch in December 2013 and two batches in January 2014. The stagger sowing was based on the phenology data from the first experiment. Each cultivar was randomly assigned to five replicate plots per treatment with one seedling per hill and a total of 28–40 hills per plot. Basal fertilizer (30 kg·ha^{-1} P as single superphosphate, 40 kg·ha^{-1} K as KCl, and 5 kg·ha^{-1} Zn as zinc sulfate heptahydrate) was applied one day before transplanting. N fertilizer as urea was applied in four splits (45 kg·ha^{-1} as basal, 30 kg·ha^{-1} at mid-tillering, 45 kg·ha^{-1} at PI, and 30 kg·ha^{-1} just before heading).

During the day (6 a.m–6 p.m.), plants were exposed to ambient conditions (compare Figure A2 and Table 1). Overnight (6 p.m.–6 a.m.), plants were exposed to the temperature treatments in manually-covered tents with temperature-control devices as described previously [25]. Air conditioners were programmed to maintain the temperature setting at control (22 °C) or HNT (28 °C). Temperature and relative humidity were monitored by sensors connected to data loggers (HOBO, Onset Computer Corporation, Bourne, MA, USA). Temperature treatments started at the panicle initiation stage and lasted until physiological maturity (Figure A1). During the flowering stage, panicles that had flowered for at least 50% were identified and tagged. These were then collected, together with the corresponding flag leaves, the next morning just before the tents were opened (~4 a.m.–6 a.m.). All samples were collected in liquid nitrogen and stored at −80 °C until use.

4.2. Weather Data

Weather data (radiation, sunshine duration, rainfall, relative humidity, maximum temperature (T_{max}) and minimum temperature (T_{min})) recorded by the IRRI wetland agrometeorological station were obtained from the IRRI Climate Unit.

4.3. Growth Analysis, Grain Yield and Yield Components

Twelve hills from each replicate plot were harvested at physiological maturity for the determination of plant height, tiller number, panicle number, and straw and rachis weight and processed for the analysis of yield components [81]. Sixty plants for the DS and 24 plants for the WS were considered for plant height, tiller number and panicle number. For the remaining parameters, two replicates pooled from twelve plants each were considered for the WS, and five replicates pooled from twelve plants each were considered for the DS. The number of panicles per hill was counted for the calculation of panicles per m^2. Afterwards, plants were separated into straw and panicles and panicles were manually threshed. Filled and unfilled grains were submerged in water and separated with a seed blower. Filled, half-filled and empty grains were counted to obtain spikelets per m^2, spikelets per panicle, seed set and 1000-grain weight. Total above ground biomass was determined from the dry weight of straw; rachis; and filled, half-filled and empty grains after drying at 70 °C until constant weight. The harvest index was calculated as the percentage of the dry weight of filled grains relative to the total above ground biomass. Plants from central areas of two m^2 from each plot (two for the WS and five for the DS, per condition and cultivar) were also harvested for the determination of grain yield. Grain weight data were adjusted to a standard moisture content of 0.14 g H_2O g^{-1}.

4.4. Metabolite Profiling and Data Processing

A fraction enriched in small polar metabolites was prepared from 120 mg of fresh weight of snap-frozen and ground flag leaves or panicles from five biological replicates per cultivar and condition and analyzed by gas chromatography coupled to electron impact ionization-time of flight-mass spectrometry (GC/EI-TOF-MS) as described in [82]. Chromatograms were acquired and baseline corrected by the ChromaTOF software (LECO Instrumente GmbH, Mönchengladbach,

Germany). TagFinder [83], the NIST08 software, (http://chemdata.nist.gov/dokuwiki/doku.php?id=start) (U.S. Department of Commerce, Gaithersburg, USA, MD) and the mass spectral and retention time index reference collection of the Golm Metabolome Database [84,85] were used for the manually supervised annotation of metabolites. Mass spectral intensities were normalized to fresh weight and $^{13}C_6$-sorbitol (Sigma-Aldrich, Taufkirchen, Germany) as internal standard. The normalized data are available in Table S2.

Data pre-processing was done separately for both organs and included the omission of metabolites with more than 75% missing values and a missing value imputation for the remaining metabolites with half the minimum amount of the respective mass spectral intensity. Furthermore, contaminations were identified using hierarchical clustering and correlation matrices with a set of known contaminating compounds and removed. A batch effect correction of different measurements of the whole data set was performed using an ANOVA tool [86]. The intensities of each metabolite were divided by the median intensity across all measurements and $\log_2$-transformed to approximate a normal distribution. All presented metabolite data thus represent relative metabolite abundance measures. Outliers were detected with the function *grubbs.test* included in the R-package *outliers* [87] using a threshold of $p < 0.0001$. Finally, 132 metabolite intensities were detected for panicles and 161 metabolite intensities were detected for flag leaves for the DS, and 195 metabolites were detected for both tissues for the WS. For further analysis, the overlap of metabolites per tissue was determined, showing 69 metabolites for panicles and 76 metabolites for flag leaves.

To enable direct comparison, overlapping metabolites for each tissue between both experiments were determined, resulting in 69 metabolites for panicles and 76 for flag leaves.

4.5. Enzyme Activity

The activity of alanine aminotransferase (AlaAT, E.C.2.6.1.2) was measured according to a published method [88]. Ground panicle material (20 mg) was used from three biological replicates per cultivar and condition. In four cases (IR72, IR62266-42-6-2—C, HNT, Moroberekan—C, Moroberekan—HNT), only two replicates were available.

4.6. Statistical Analysis

PCA was perfomed with the R-package *pcaMethods* [89]. For the data processing and visualization, *R v3.4.2* [90] and *R-Studio v1.1.383* [91] were used including the following packages: *ggplot2* [92], *grid* [93], *gridExtra* [94] and *reshape2* [95].

Changes in metabolite content were investigated by calculating the $\log_2$ fold change between the averages of metabolite levels under control conditions in the DS compared to in the WS, or under HNT compared to under control conditions. Unpaired, two-sided t-tests were performed over all replicates, comparing control and HNT conditions to determine the statistical significance of the observed changes. For agronomic data, t-tests were applied for the DS. For the WS, only two replicates were available for most parameters and t-tests were only applied for plant height, tiller number and panicle number. To test the significance of the influence of genotype (G), treatment (T) and GxT interactions across all cultivars, a 2-way ANOVA design was used.

The statistical significance of differences in enzyme activity between control and HNT treatments were evaluated by an unpaired two-sided *t*-test, performed in RStudio [91].

Correlations between total grain yield reduction under HNT in the WS and the corresponding changes in metabolite content ($\log_2$ fold change) were done in R with the package *cor.test* using Spearman Rank Correlation with $p < 0.05$.

Supplementary Materials: The following are available online at http://www.mdpi.com/1422-0067/21/9/3187/s1.

Author Contributions: Conceptualization, S.V.K.J., D.K.H., E.Z.; Methodology, L.M.F.L., X.L., A.E., J.K.; Formal analysis, S.S., U.G., X.L., A.E.; Data curation, S.S., A.E., J.K.; Writing, S.S.; D.K.H., E.Z.; Review and editing, all authors, Funding acquisition, S.V.K.J., E.Z., D.K.H. All authors have read and agreed to the published version of the manuscript.

Funding: This research was funded by the German Federal Ministry for Economic Cooperation and Development through Contracts No. 81141844 and 81206686.

Acknowledgments: We thank Ines Fehrle for her excellent technical assistance with the GC-MS measurements and Jessica Alpers for her excellent support with the enzyme activity measurements.

Conflicts of Interest: The authors declare no conflict of interest. The funders had no role in the design of the study; in the collection, analyses, or interpretation of data; in the writing of the manuscript, or in the decision to publish the results.

Abbreviations

ACC	1-aminocyclopropane-1-carboxylic acid
AlaAT	Alanine aminotransferase
DAS	Days after stress
DAT	Days after transplanting
DS	Dry season
DTR	Diurnal temperature range
FC	Fold change
G	Genotype
GABA	4-amino butanoic acid
GC-MS	Gas chromatography – Mass spectrometry
HCN	Hydrogen cyanide
HNT	High night temperature
IRRI	International Rice Research Institute
PC	Principal Component
PCA	Principal Component Analysis
RH	Relative humidity
S	Season
T	Treatment
TCA	Tricarboxylic acid
WS	Wet season

Appendix A

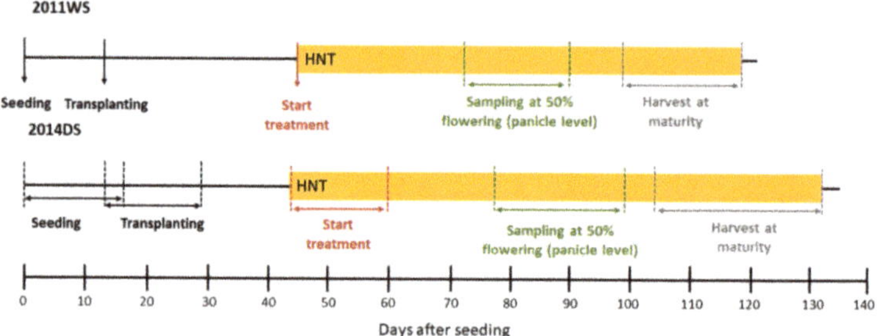

Figure A1. Experimental set-up for the DS and WS experiment. WS—wet season, DS—dry season.

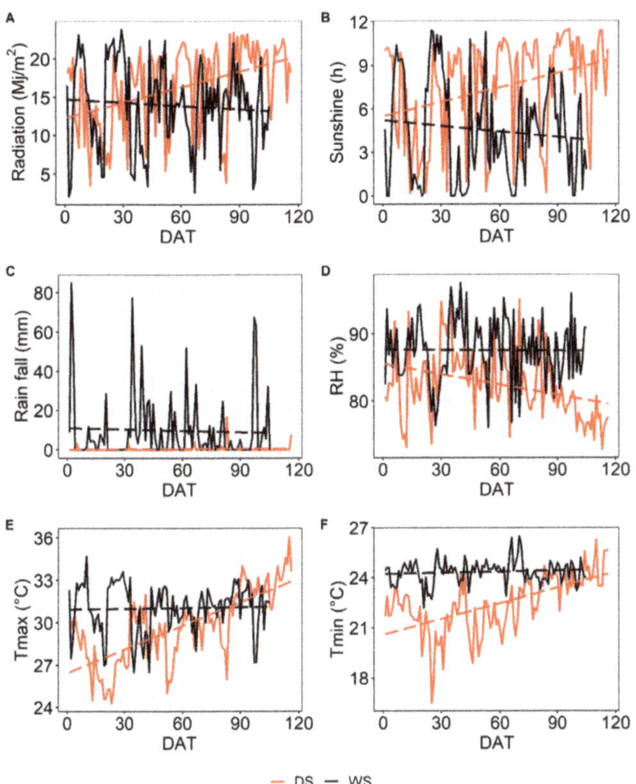

Figure A2. Weather data for the DS and WS experiment measured at the IRRI weather station as average values per day. Radiation (**A**), sunshine duration (**B**), rainfall (**C**), relative humidity (**D**), maximal temperature T_{max} (**E**), minimal temperature T_{min} (**F**). Broken lines represent a trend line for the respective data set. DAT—days after transplanting. Average values for all weather parameters were significantly different ($p < 0.05$) between WS and DS.

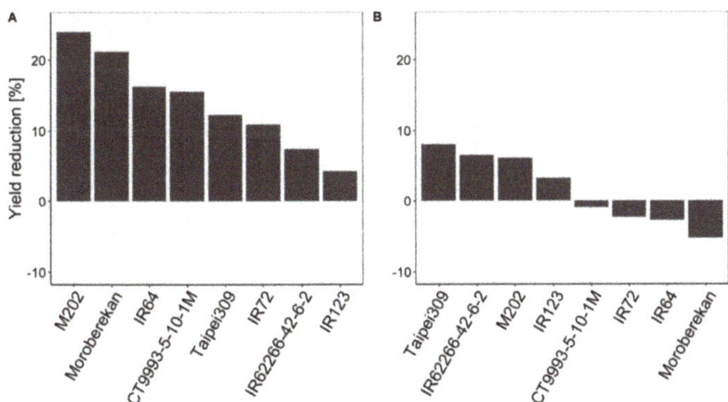

Figure A3. Yield reduction under HNT in the WS (**A**) and DS (**B**). Cultivars are sorted from highest to lowest yield reduction.

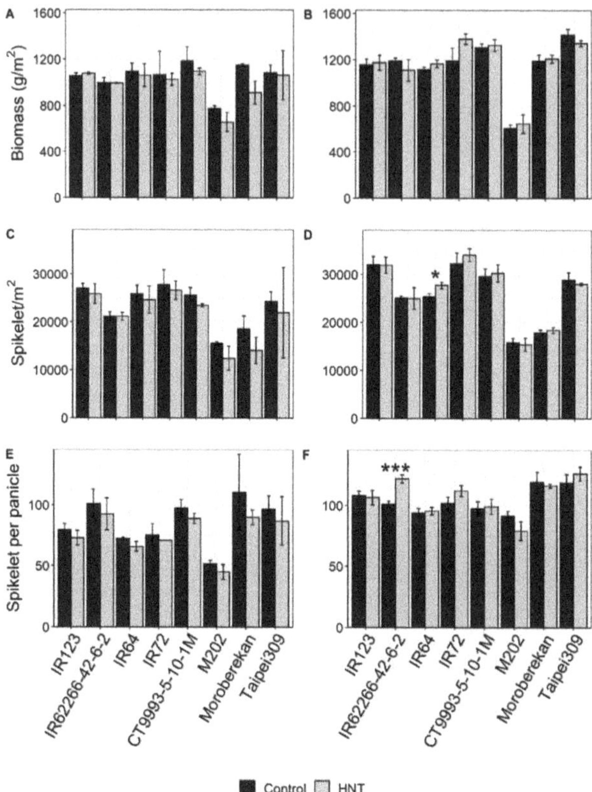

Figure A4. Biomass (**A**, **B**), spikelets per m^2 (**C**, **D**), and spikelets per panicle (**E**, **F**) of eight rice cultivars in response to HNT stress for the WS (**A**, **C**, **E**) and DS (**B**, **D**, **F**). For the WS, variance is displayed as range between means of two replicates with 12 plants each; for the DS, the standard error of the mean of five replicates with 12 plants each is shown. Cultivars are sorted alphabetically within the respective *O. sativa* subspecies *indica* (1-4) and *japonica* (5–8). Significance levels were only calculated for the DS due to the insufficient replicate number in the WS and are indicated by asterisks: 0.001 < ***; 0.001 > ** < 0.01; 0.01 > * < 0.05.

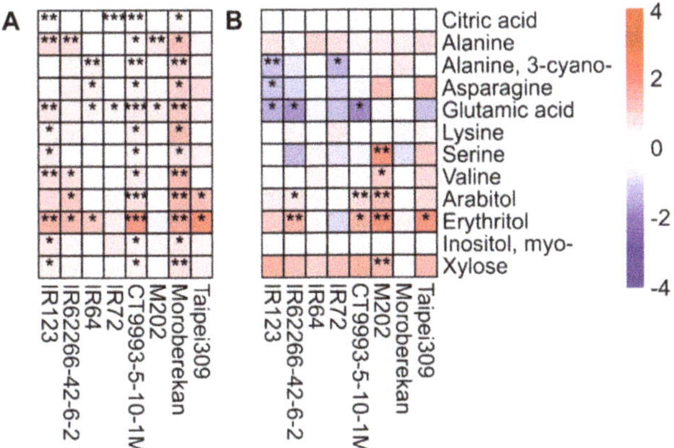

Figure A5. Log_2 fold changes in significantly changed metabolite pools under HNT compared to under control conditions in panicles for the DS (**A**). For comparison, the same metabolites are shown for the WS (**B**) independent of a significant change. For the DS, only metabolites that showed a significant change in at least three out of eight cultivars are displayed in (**A**). The level of significance is indicated by asterisks (* $p < 0.05$; ** $p < 0.01$; *** $p < 0.001$), and the log_2 fold difference is indicated by the color code. Blue indicates a lower metabolite level compared to under the control condition, and red, a higher level. Cultivars were sorted alphabetically within the respective *O. sativa* subspecies *indica* (1–4) and *japonica* (5–8).

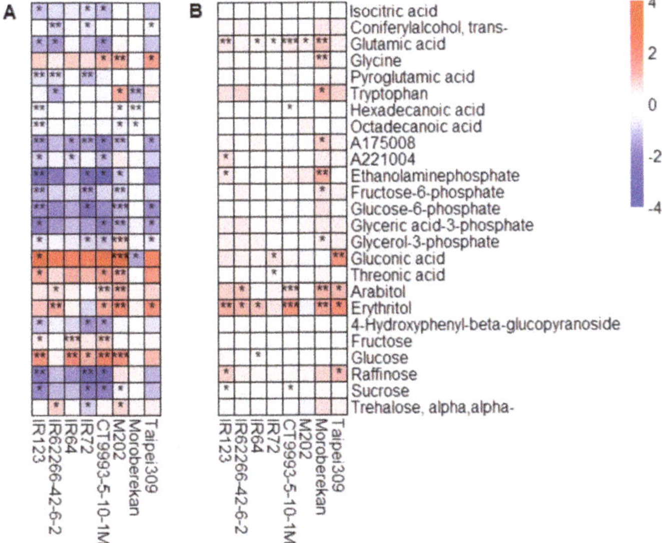

Figure A6. Log_2 fold changes in significantly changed metabolite pools under HNT compared to under control conditions in panicles for the WS (**A**). For comparison, the same metabolites are shown for the DS (**B**) independent of a significant change. For the WS (**A**), only metabolites that showed a significant change in at least three out of eight cultivars are displayed. The level of significance is indicated by asterisks (* $p < 0.05$; ** $p < 0.01$; *** $p < 0.001$), and the log_2 fold difference is indicated by the color code. Blue indicates a lower metabolite level compared to under the control condition, and red, a higher level. Cultivars were sorted alphabetically within the respective *O. sativa* subspecies *indica* (1–4) and *japonica* (5–8).

References

1. GRISP. *Global Rice Science Partnership*; International Rice Research Institute: Los Baños, Philippines, 2013.
2. FAO. FAOSTAT Database 2009. Available online: http://faostat.fao.org/ (accessed on 23 October 2019).
3. IPCC. *AR5 Climate Change 2014: Impacts, Adaptation, and Vulnerability*; Cambridge Univ. Press: Cambridge, UK, 2014.
4. FAO. FAOSTAT Database 2014. Available online: http://faostat.fao.org/ (accessed on 2 November 2019).
5. Easterling, D.R.; Horton, B.; Jones, P.D.; Peterson, T.C.; Karl, T.R.; Parker, D.E.; Salinger, M.J.; Razuvayev, V.; Plummer, N.; Jamason, P.; et al. Maximum and minimum temperature trends for the globe. *Science* **1997**, *277*, 364–367. [CrossRef]
6. Donat, M.G.; Alexander, L.V. The shifting probability distribution of global daytime and night-time temperatures. *Geophys. Res. Lett.* **2012**, *39*, L14707. [CrossRef]
7. Vose, R.S.; Easterling, D.R.; Gleason, B. Maximum and minimum temperature trends for the globe: An update through 2004. *Geophys. Res. Lett.* **2005**, *32*, L23822. [CrossRef]
8. Sillmann, J.; Kharin, V.V.; Zhang, X.; Zwiers, F.W.; Bronaugh, D. Climate extremes indices in the CMIP5 multimodel ensemble: Part 1. Model evaluation in the present climate. *J. Geophys. Res. Atmos.* **2013**, *118*, 1716–1733. [CrossRef]
9. Davy, R.; Esau, I.; Chernokulsky, A.; Outten, S.; Zilitinkevich, S. Diurnal asymmetry to the observed global warming. *Int. J. Climatol.* **2017**, *37*, 79–93. [CrossRef]
10. Peng, S.; Piao, S.; Ciais, P.; Myneni, R.B.; Chen, A.; Chevallier, F.; Dolman, A.J.; Janssens, I.A.; Peñuelas, J.; Zhang, G.; et al. Asymmetric effects of daytime and night-time warming on Northern Hemisphere vegetation. *Nature* **2013**, *501*, 88–92. [CrossRef]
11. Zhou, Y.; Ren, G. Change in extreme temperature event frequency over mainland China, 1961–2008. *Clim. Res.* **2011**, *50*, 125–139. [CrossRef]
12. Peng, S.; Huang, J.; Sheehy, J.E.; Laza, R.C.; Visperas, R.M.; Zhong, X.; Centeno, G.S.; Khush, G.S.; Cassman, K.G. Rice yields decline with higher night temperature from global warming. *Proc. Natl. Acad. Sci. USA* **2004**, *101*, 9971–9975. [CrossRef]
13. Zhao, X.; Fitzgerald, M. Climate change: Implications for the yield of edible rice. *PLoS ONE* **2013**, *8*, e66218. [CrossRef]
14. Rao, B.B.; Chowdary, S.P.; Sandeep, V.M.; Rao, V.U.M.; Venkateswarlu, B. Rising minimum temperature trends over India in recent decades: Implications for agricultural production. *Glob. Planet. Chang.* **2014**, *117*, 1–8. [CrossRef]
15. Padma Kumari, B.; Londhe, A.L.; Daniel, S.; Jadhav, D.B. Observational evidence of solar dimming: Offsetting surface warming over India. *Geophys. Res. Lett.* **2007**, *34*, L21810. [CrossRef]
16. Ambardekar, A.A.; Siebenmorgen, T.J.; Counce, P.A.; Lanning, S.B.; Mauromoustakos, A. Impact of field-scale nighttime air temperatures during kernel development on rice milling quality. *Field Crops Res.* **2011**, *122*, 179–185. [CrossRef]
17. Mohammed, A.R.; Tarpley, L. Effects of night temperature, spikelet position and salicylic acid on yield and yield-related parameters of rice (*Oryza sativa* L.) plants. *J. Agron. Crop Sci.* **2011**, *197*, 40–49. [CrossRef]
18. Nagarajan, S.; Jagadish, S.V.K.; Prasad, A.S.H.; Thomar, A.K.; Anand, A.; Pal, M.; Agarwal, P.K. Local climate affects growth, yield and grain quality of aromatic and non-aromatic rice in northwestern India. *Agric. Ecosyst. Environ.* **2010**, *138*, 274–281. [CrossRef]
19. Shi, W.; Yin, X.; Struik, P.C.; Solis, C.; Xie, F.; Schmidt, R.C.; Huang, M.; Zou, Y.; Ye, C.; Jagadish, S.V.K. High day- and night-time temperatures affect grain growth dynamics in contrasting rice genotypes. *J. Exp. Bot.* **2017**, *68*, 5233–5245. [CrossRef] [PubMed]
20. Shi, W.; Yin, X.; Struik, P.C.; Xie, F.; Schmidt, R.C.; Jagadish, K.S.V. Grain yield and quality responses of tropical hybrid rice to high night-time temperature. *Field Crops Res.* **2016**, *190*, 18–25. [CrossRef]
21. Lanning, S.B.; Siebenmorgen, T.J.; Counce, P.A.; Ambardekar, A.A.; Mauromoustakos, A. Extreme nighttime air temperatures in 2010 impact rice chalkiness and milling quality. *Field Crops Res.* **2011**, *124*, 132–136. [CrossRef]
22. Morita, S.; Yonemaru, J.; Takanashi, J. Grain growth and endosperm cell size under high night temperatures in rice (*Oryza sativa* L.). *Ann. Bot.* **2005**, *95*, 695–701. [CrossRef]

23. Rehmani, M.I.A.; Wei, G.; Hussain, N.; Ding, C.; Li, G.; Liu, Z.; Wang, S.; Ding, Y. Yield and quality responses of two *indica* rice hybrids to post-anthesis asymmetric day and night open-field warming in lower reaches of Yangtze River delta. *Field Crops Res.* **2014**, *156*, 231–241. [CrossRef]

24. Lobell, D.B.; Ortiz-Monasterio, J.I.; Asner, G.P.; Matson, P.A.; Naylor, R.L.; Falcon, W.P. Analysis of wheat yield and climatic trends in Mexico. *Field Crops Res.* **2005**, *94*, 250–256. [CrossRef]

25. Shi, W.; Muthurajan, R.; Rahman, H.; Selvam, J.; Peng, S.; Zou, Y.; Jagadish, K.S. Source-sink dynamics and proteomic reprogramming under elevated night temperature and their impact on rice yield and grain quality. *New Phytol.* **2013**, *197*, 825–837. [CrossRef] [PubMed]

26. Zhang, Y.; Tang, Q.; Peng, S.; Zou, Y.; Chen, S.; Shi, W.; Qin, J.; Laza, M.R.C. Effects of high night temperature on yield and agronomic traits of irrigated rice under field chamber system condition. *Aust. J. Crop Sci.* **2013**, *7*, 7–13.

27. Yang, Z.; Zhang, Z.; Zhang, T.; Fahad, S.; Cui, K.; Nie, L.; Peng, S.; Huang, J. The effect of season-long temperature increases on rice cultivars grown in the central and Southern regions of China. *Front. Plant Sci.* **2017**, *8*, 1908. [CrossRef] [PubMed]

28. Glaubitz, U.; Li, X.; Köhl, K.I.; van Dongen, J.T.; Hincha, D.K.; Zuther, E. Differential physiological responses of different rice (*Oryza sativa*) cultivars to elevated night temperature during vegetative growth. *Funct. Plant Biol.* **2014**, *41*, 437–448. [CrossRef]

29. Bahuguna, R.N.; Solis, C.A.; Shi, W.; Jagadish, K.S.V. Post-flowering night respiration and altered sink activity account for high night temperature-induced grain yield and quality loss in rice (*Oryza sativa* L.). *Physiol. Plant.* **2017**, *159*, 59–73. [CrossRef] [PubMed]

30. Liang, J.; Xia, J.; Liu, L.; Wan, S. Global patterns of the responses of leaf-level photosynthesis and respiration in terrestrial plants to experimental warming. *J. Plant Ecol.* **2013**, *6*, 437–447. [CrossRef]

31. Mohammed, R.; Cothren, J.T.; Tarpley, L. High night temperature and abscisic acid affect rice productivity through altered photosynthesis, respiration and spikelet fertility. *Crop Sci.* **2013**, *53*, 2603–2612. [CrossRef]

32. Dong, W.; Chen, J.; Wang, L.; Tian, Y.; Zhang, B.; Lai, Y.; Meng, Y.; Qian, C.; Guo, J. Impacts of nighttime post-anthesis warming on rice productivity and grain quality in East China. *Crop J.* **2014**, *2*, 63–69. [CrossRef]

33. Fernie, A.R.; Schauer, N. Metabolomics-assisted breeding: A viable option for crop improvement? *Trends Genet.* **2009**, *25*, 39–48. [CrossRef]

34. Oikawa, A.; Matsuda, F.; Kusano, M.; Okazaki, Y.; Saito, K. Rice Metabolomics. *Rice* **2008**, *1*, 63–71. [CrossRef]

35. Nadella, K.D.; Marla, S.S.; Kumar, P.A. Metabolomics in agriculture. *Omics A J. Integr. Biol.* **2012**, *16*, 149–159. [CrossRef] [PubMed]

36. Krumsiek, J.; Bartel, J.; Theis, F.J. Computational approaches for systems metabolomics. *Curr. Opin. Biotechnol.* **2016**, *39*, 198–206. [CrossRef] [PubMed]

37. Zuther, E.; Koehl, K.; Kopka, J. Comparative metabolome analysis of the salt response in breeding cultivars of rice. In *Advances in Molecular Breeding toward Drought and Salt Tolerant Crops*; Jenks, M.A., Hasegawa, P.M., Jain, S.M., Eds.; Springer: Dordrecht, The Netherlands, 2007; pp. 285–315. [CrossRef]

38. Liu, D.; Ford, K.L.; Roessner, U.; Natera, S.; Cassin, A.M.; Patterson, J.H.; Bacic, A. Rice suspension cultured cells are evaluated as a model system to study salt responsive networks in plants using a combined proteomic and metabolomic profiling approach. *Proteomics* **2013**, *13*, 2046–2062. [CrossRef] [PubMed]

39. Zhao, X.; Wang, W.; Zhang, F.; Deng, J.; Li, Z.; Fu, B. Comparative metabolite profiling of two rice genotypes with contrasting salt stress tolerance at the seedling stage. *PLoS ONE* **2014**, *9*, e108020. [CrossRef] [PubMed]

40. Nam, M.H.; Bang, E.; Kwon, T.Y.; Kim, Y.; Kim, E.H.; Cho, K.; Park, W.J.; Kim, B.G.; Yoon, I.S. Metabolite profiling of diverse rice germplasm and identification of conserved metabolic markers of rice roots in response to long-term mild salinity stress. *Int. J. Mol. Sci.* **2015**, *16*, 21959–21974. [CrossRef]

41. Ma, N.L.; Che Lah, W.A.; Abd Kadir, N.; Mustaqim, M.; Rahmat, Z.; Ahmad, A.; Lam, S.D.; Ismail, M.R. Susceptibility and tolerance of rice crop to salt threat: Physiological and metabolic inspections. *PLoS ONE* **2018**, *13*, e0192732. [CrossRef]

42. Baldoni, E.; Mattana, M.; Locatelli, F.; Consonni, R.; Cagliani, L.R.; Picchi, V.; Abbruscato, P.; Genga, A. Analysis of transcript and metabolite levels in Italian rice (*Oryza sativa* L.) cultivars subjected to osmotic stress or benzothiadiazole treatment. *Plant Physiol. Biochem.* **2013**, *70*, 492–503. [CrossRef]

43. Degenkolbe, T.; Do, P.T.; Kopka, J.; Zuther, E.; Hincha, D.K.; Köhl, K.I. Identification of drought tolerance markers in a diverse population of rice cultivars by expression and metabolite profiling. *PLoS ONE* **2013**, *8*, e63637. [CrossRef]

44. Li, X.; Lawas, L.M.F.; Malo, R.; Glaubitz, U.; Erban, A.; Mauleon, R.; Heuer, S.; Zuther, E.; Kopka, J.; Hincha, D.K.; et al. Metabolic and transcriptomic signatures of rice floral organs reveal sugar starvation as a factor in reproductive failure under heat and drought stress. *Plant Cell Environ.* **2015**, *38*, 2171–2192. [CrossRef]

45. Shu, L.; Lou, Q.; Ma, C.; Ding, W.; Zhou, J.; Wu, J.; Feng, F.; Lu, X.; Luo, L.; Xu, G. Genetic, proteomic and metabolic analysis of the regulation of energy storage in rice seedlings in response to drought. *Proteomics* **2011**, *11*, 4122–4138. [CrossRef]

46. Nam, K.H.; Shin, H.J.; Pack, I.S.; Park, J.H.; Kim, H.B.; Kim, C.G. Metabolomic changes in grains of well-watered and drought-stressed transgenic rice. *J. Sci. Food Agric.* **2016**, *96*, 807–814. [CrossRef] [PubMed]

47. Barnaby, J.Y.; Rohila, J.S.; Henry, C.G.; Sicher, R.C.; Reddy, V.R.; McClung, A.M. Physiological and metabolic responses of rice to reduced soil moisture: Relationship of water stress tolerance and grain production. *Int. J. Mol. Sci.* **2019**, *20*, 1846. [CrossRef] [PubMed]

48. Yamakawa, H.; Hakata, M. Atlas of rice grain filling-related metabolism under high temperature: Joint analysis of metabolome and transcriptome demonstrated inhibition of starch accumulation and induction of amino acid accumulation. *Plant Cell Physiol.* **2010**, *51*, 795–809. [CrossRef] [PubMed]

49. Lawas, L.M.F.; Li, X.; Erban, A.; Kopka, J.; Jagadish, S.V.K.; Zuther, E.; Hincha, D.K. Metabolic responses of rice cultivars with different tolerance to combined drought and heat stress under field conditions. *Gigascience* **2019**, *8*, giz050. [CrossRef]

50. Lawas, L.M.F.; Erban, A.; Kopka, J.; Jagadish, S.V.K.; Zuther, E.; Hincha, D.K. Metabolic responses of rice source and sink organs during recovery from combined drought and heat stress in the field. *Gigascience* **2019**, *8*, giz102. [CrossRef]

51. Glaubitz, U.; Erban, A.; Kopka, J.; Hincha, D.K.; Zuther, E. High night temperature strongly impacts TCA cycle, amino acid and polyamine biosynthetic pathways in rice in a sensitivity-dependent manner. *J. Exp. Bot.* **2015**, *66*, 6385–6397. [CrossRef]

52. Glaubitz, U.; Li, X.; Schaedel, S.; Erban, A.; Sulpice, R.; Kopka, J.; Hincha, D.K.; Zuther, E. Integrated analysis of rice transcriptomic and metabolomic responses to elevated night temperatures identifies sensitivity- and tolerance-related profiles. *Plant Cell Environ.* **2017**, *40*, 121–137. [CrossRef]

53. Dhatt, B.K.; Abshire, N.; Paul, P.; Hasanthika, K.; Sandhu, J.; Zhang, Q.; Obata, T.; Walia, H. Metabolic dynamics of developing rice seeds under high night-time temperature stress. *Front. Plant Sci.* **2019**, *10*, 1443. [CrossRef]

54. Van Oort, P.A.J.; Zwart, S.J. Impacts of climate change on rice production in Africa and causes of simulated yield changes. *Glob. Chang. Biol.* **2018**, *24*, 1029–1045. [CrossRef]

55. Good, A.G.; Muench, D.G. Long-term anaerobic metabolism in root tissue (Metabolic products of pyruvate metabolism). *Plant Physiol.* **1993**, *101*, 1163–1168. [CrossRef]

56. Shrawat, A.K.; Carroll, R.T.; DePauw, M.; Taylor, G.J.; Good, A.G. Genetic engineering of improved nitrogen use efficiency in rice by the tissue-specific expression of alanine aminotransferase. *Plant Biotechnol. J.* **2008**, *6*, 722–732. [CrossRef] [PubMed]

57. Xiong, D.; Ling, X.; Huang, J.; Peng, S. Meta-analysis and dose-response analysis of high temperature effects on rice yield and quality. *Environ. Exp. Bot.* **2017**, *141*, 1–9. [CrossRef]

58. Jagadish, S.V.; Murty, M.V.; Quick, W.P. Rice responses to rising temperatures—challenges, perspectives and future directions. *Plant Cell Environ.* **2015**, *38*, 1686–1698. [CrossRef]

59. Mohammed, A.R.; Tarpley, L. Effects of high night temperature and spikelet position on yield-related parameters of rice (*Oryza sativa* L.) plants. *Eur. J. Agron.* **2010**, *33*, 117–123. [CrossRef]

60. Kanno, K.; Makino, A. Increased grain yield and biomass allocation in rice under cool night temperature. *Soil Sci. Plant Nutr.* **2010**, *56*, 412–417. [CrossRef]

61. Coast, O.; Ellis, R.H.; Murdoch, A.J.; Quiñones, C.; Jagadish, K.S.V. High night temperature induces contrasting responses for spikelet fertility, spikelet tissue temperature, flowering characteristics and grain quality in rice. *Funct. Plant Biol.* **2015**, *42*, 149–161. [CrossRef]

62. Liao, J.-L.; Zhou, H.-W.; Peng, Q.; Zhong, P.-A.; Zhang, H.-Y.; He, C.; Huang, Y.-J. Transcriptome changes in rice (*Oryza sativa* L.) in response to high night temperature stress at the early milky stage. *Bmc Genom.* **2015**, *16*, 18. [CrossRef]

63. Dong, W.; Tian, Y.; Zhang, B.; Chen, J.; Zhang, W. Effects of asymmetric warming on grain quality and related key enzymes activities for *japonica* rice (Nanjing 44) under FATI facility. *Acta Agron. Sin.* **2011**, *37*, 832–841. [CrossRef]

64. Chen, Y.; Murchie, E.H.; Hubbart, S.; Horton, P.; Peng, S. Effects of season-dependent irradiance levels and nitrogen-deficiency on photosynthesis and photoinhibition in field-grown rice (*Oryza sativa*). *Physiol. Plant.* **2003**, *117*, 343–351. [CrossRef]

65. Impa, S.M.; Vennapusa, A.R.; Bheemanahalli, R.; Sabela, D.; Boyle, D.; Walia, H.; Jagadish, S.V.K. High night temperature induced changes in grain starch metabolism alters starch, protein, and lipid accumulation in winter wheat. *Plant Cell Environ.* **2020**, *43*, 431–447. [CrossRef]

66. Good, A.G.; Johnson, S.J.; De Pauw, M.; Carroll, R.T.; Savidov, N.; Vidmar, J.; Lu, Z.; Taylor, G.; Stroeher, V. Engineering nitrogen use efficiency with alanine aminotransferase. *Can. J. Bot.* **2007**, *85*, 252–262. [CrossRef]

67. Beatty, P.H.; Shrawat, A.K.; Carroll, R.T.; Zhu, T.; Good, A.G. Transcriptome analysis of nitrogen-efficient rice over-expressing alanine aminotransferase. *Plant Biotechnol. J.* **2009**, *7*, 562–576. [CrossRef] [PubMed]

68. Kikuchi, H.; Hirose, S.; Toki, S.; Akama, K.; Takaiwa, F. Molecular characterization of a gene for alanine aminotransferase from rice (*Oryza sativa*). *Plant Mol. Biol.* **1999**, *39*, 149–159. [CrossRef] [PubMed]

69. Zhong, M.; Liu, X.; Liu, F.; Ren, Y.; Wang, Y.; Zhu, J.; Teng, X.; Duan, E.; Wang, F.; Zhang, H.; et al. FLOURY ENDOSPERM12 encoding alanine aminotransferase 1 regulates carbon and nitrogen metabolism in rice. *J. Plant Biol.* **2019**, *62*, 61–73. [CrossRef]

70. Beatty, P.H.; Carroll, R.T.; Shrawat, A.K.; Guevara, D.; Good, A.G. Physiological analysis of nitrogen-efficient rice overexpressing alanine aminotransferase under different N regimes. *Botany* **2013**, *91*, 866–883. [CrossRef]

71. Selvaraj, M.G.; Valencia, M.O.; Ogawa, S.; Lu, Y.; Wu, L.; Downs, C.; Skinner, W.; Lu, Z.; Kridl, J.C.; Ishitani, M.; et al. Development and field performance of nitrogen use efficient rice lines for Africa. *Plant Biotechnol. J.* **2017**, *15*, 775–787. [CrossRef]

72. Lai, K.W.; Yau, C.P.; Tse, Y.C.; Jiang, L.; Yip, W.K. Heterologous expression analyses of rice OsCAS in Arabidopsis and in yeast provide evidence for its roles in cyanide detoxification rather than in cysteine synthesis in vivo. *J. Exp. Bot.* **2009**, *60*, 993–1008. [CrossRef]

73. Lim, P.O.; Kim, H.J.; Nam, H.G. Leaf senescence. *Annu. Rev. Plant Biol.* **2007**, *58*, 115–136. [CrossRef]

74. Siegień, I.; Bogatek, R. Cyanide action in plants—from toxic to regulatory. *Acta Physiol. Plant.* **2006**, *28*, 483–497. [CrossRef]

75. Helliwell, E.E.; Wang, Q.; Yang, Y. Ethylene biosynthesis and signaling is required for rice immune response and basal resistance against *Magnaporthe oryzae* infection. *Mol. Plant Microbe Interact.* **2016**, *29*, 831–843. [CrossRef]

76. Yu, L.; Liu, Y.; Xu, F. Comparative transcriptome analysis reveals significant differences in the regulation of gene expression between hydrogen cyanide- and ethylene-treated *Arabidopsis thaliana*. *BMC Plant Biol.* **2019**, *19*, 92. [CrossRef] [PubMed]

77. Tian, L.; Liu, L.; Yin, Y.; Huang, M.; Chen, Y.; Xu, X.; Wu, P.; Li, M.; Wu, G.; Jiang, H.; et al. Heterogeneity in the expression and subcellular localization of POLYOL/MONOSACCHARIDE TRANSPORTER genes in *Lotus japonicus*. *PLoS ONE* **2017**, *12*, e0185269. [CrossRef] [PubMed]

78. Pires, M.V.; Pereira Junior, A.A.; Medeiros, D.B.; Daloso, D.M.; Pham, P.A.; Barros, K.A.; Engqvist, M.K.; Florian, A.; Krahnert, I.; Maurino, V.G.; et al. The influence of alternative pathways of respiration that utilize branched-chain amino acids following water shortage in Arabidopsis. *Plant Cell Environ.* **2016**, *39*, 1304–1319. [CrossRef] [PubMed]

79. Fabregas, N.; Fernie, A.R. The metabolic response to drought. *J. Exp. Bot.* **2019**, *70*, 1077–1085. [CrossRef] [PubMed]

80. Melandri, G.; AbdElgawad, H.; Riewe, D.; Hageman, J.A.; Asard, H.; Beemster, G.T.S.; Kadam, N.; Jagadish, K.; Altmann, T.; Ruyter-Spira, C.; et al. Biomarkers for grain yield stability in rice under drought stress. *J. Exp. Bot.* **2020**, *71*, 669–683. [CrossRef]

81. Yoshida, S.; Forn, D.A.; Cock, J.H.; Gomez, K.A. *Laboratory Manual for Physiological Studies of Rice*; International Rice Research Institute: Los Banos, Philippines, 1976.

82. Dethloff, F.; Erban, A.; Orf, I.; Alpers, J.; Fehrle, I.; Beine-Golovchuk, O.; Schmidt, S.; Schwachtje, J.; Kopka, J. Profiling methods to identify cold-regulated primary metabolites using gas chromatography coupled to mass spectrometry. *Methods Mol. Biol.* **2014**, *1166*, 171–197. [CrossRef] [PubMed]

83. Luedemann, A.; Strassburg, K.; Erban, A.; Kopka, J. TagFinder for the quantitative analysis of gas chromatography—mass spectrometry (GC-MS)-based metabolite profiling experiments. *Bioinformatics* **2008**, *24*, 732–737. [CrossRef]

84. Kopka, J.; Schauer, N.; Krueger, S.; Birkemeyer, C.; Usadel, B.; Bergmuller, E.; Dormann, P.; Weckwerth, W.; Gibon, Y.; Stitt, M.; et al. GMD@CSB.DB: The Golm Metabolome Database. *Bioinformatics* **2005**, *21*, 1635–1638. [CrossRef]

85. Hummel, J.; Strehmel, N.; Selbig, J.; Walther, D.; Kopka, J. Decision tree supported substructure prediction of metabolites from GC-MS profiles. *Metabolomics* **2010**, *6*, 322–333. [CrossRef]

86. Lisec, J.; Romisch-Margl, L.; Nikoloski, Z.; Piepho, H.P.; Giavalisco, P.; Selbig, J.; Gierl, A.; Willmitzer, L. Corn hybrids display lower metabolite variability and complex metabolite inheritance patterns. *Plant J.* **2011**, *68*, 326–336. [CrossRef]

87. Komsta, L. Outliers: Tests for Outliers. R Package Version 0.14. 2011. Available online: https://cran.r-project.org/web/packages/outliers/outliers.pdf (accessed on 16 August 2019).

88. Gibon, Y.; Blaesing, O.E.; Hannemann, J.; Carillo, P.; Höhne, M.; Hendriks, J.H.M.; Palacios, N.; Cross, J.; Selbig, J.; Stitt, M. A robot-based platform to measure multiple enzyme activities in Arabidopsis using a set of cycling assays: Comparison of changes of enzyme activities and transcript levels during diurnal cycles and in prolonged darkness. *Plant Cell* **2004**, *16*, 3304–3325. [CrossRef] [PubMed]

89. Stacklies, W.; Redestig, H.; Scholz, M.; Walther, D.; Selbig, J. pcaMethods—a Bioconductor package providing PCA methods for incomplete data. *Bioinformatics* **2007**, *23*, 1164–1167. [CrossRef] [PubMed]

90. RCore, T. *R: A Language and Environment for Statistical Computing, 3.4.2*; R foundation for statistical computing: Vienna, Austria, 2017.

91. RStudio, T. *RStudio: Integrated Development for R*; RStudio: Boston, MA, USA, 2016.

92. Wickham, H. *Ggplot2: Elegant Graphics for Data Analysis*; Springer: New York, NY, USA, 2009.

93. Murrell, P. *R Graphics*; Chapman and Hall/CRC: Boca Raton, FL, USA, 2005.

94. Auguie, B. Gridextra: Miscellaneous Functions for "Grid" Graphics. R Package Version 2.3. Available online: http://CRAN.R-project.org/package=gridExtra (accessed on 16 August 2019).

95. Wickham, H. Reshaping Data with the reshape Package *J. Stat. Softw.* **2007**, *21*, 1–20.

MDPI
St. Alban-Anlage 66
4052 Basel
Switzerland
Tel. +41 61 683 77 34
Fax +41 61 302 89 18
www.mdpi.com

International Journal of Molecular Sciences Editorial Office
E-mail: ijms@mdpi.com
www.mdpi.com/journal/ijms